ADVENTURE SPORT PHYSIOLOGY

ADVENTURE SPORT PHYSIOLOGY

Nick Draper

University of Canterbury, NZ

Chris Hodgson

University of Chichester, UK

A John Wiley & Sons Ltd., Publication

This edition first published 2008

© 2008 by John Wiley & Sons, Ltd

Wiley-Blackwell is an imprint of John Wiley & Sons, formed by the merger of Wiley's global Scientific, Technical and Medical business with Blackwell Publishing.

Registered office: John Wiley & Sons Ltd, The Atrium, Southern Gate, Chichester, West Sussex, PO19 8SQ, UK

Other Editorial Offices:
9600 Garsington Road, Oxford, OX4 2DQ, UK
111 River Street, Hoboken, NJ 07030-5774, USA

For details of our global editorial offices, for customer services and for information about how to apply for permission to reuse the copyright material in this book please see our website at www.wiley.com/wiley-blackwell

Library of Congress Cataloging-in-Publication Data:

Draper, Nick (Nicholas)
   Adventure sport physiology / Nick Draper, Chris Hodgson.
       p. cm.
   Includes bibliographical references and index.
   ISBN 978-0-470-01510-0 (cloth) – ISBN 978-0-470-01511-7 (pbk.)
1. Sports – Physiological aspects. 2. Human physiology. 3.
Body, Human.
I. Hodgson, Chris. II. Title.
   RC1235.D73 2008
   613.7′1 – dc22

                                         2008021443

ISBN 978-0470-01510-0 (HB)   978-0470-01511-7 (PB)

A catalogue record for this book is available from the British Library.

Typeset in 10/12pt Times by Laserwords Private Limited, Chennai, India
Printed and bound in Singapore by Fabulous Printers Pte Ltd
First impression   2008

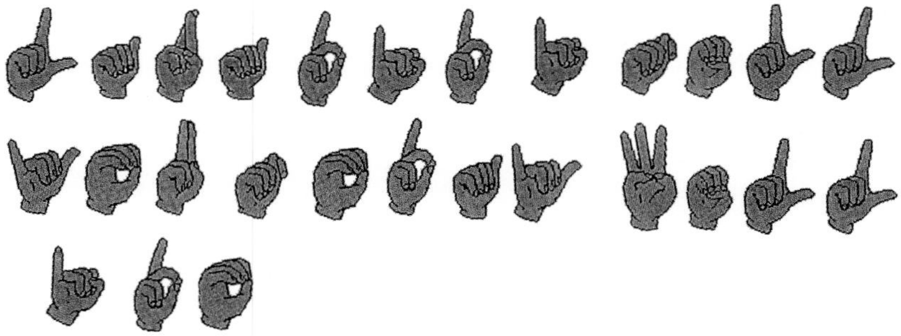

Contents

Acknowledgements

Thank you:

To Professor Chris Laws and Tony Weaden for their support and encouragement during the writing of this book.

To all the performers who gave up their time to write pieces for the book and the photographers who took the excellent shots that accompany each of the contributions. The adventure sports performers who have written pieces for the book are: Ben Ainslie, Matt Berry, Tim Brabants, Naomi Buys, Neil Gresham, Simon Hammond, Anna Hemmings, Fiona Jarvie, Dave MacLeod, Des Marshall, Katherine Schirrmacher, Bob Sharp, Jan Sleigh, Nick Smith and Ian Tordoff – thank you very much.

To Vicky Cummings and Nik Pitharides for their excellent help with the flexibility exercises and Darren Sherwood for his help with the strength training section. Also to Pete Cunningham (sailing) and Chris Bussell (canoe polo) for help with completing the water-based sports sections.

To Emma Sandham, Amy Hart, John Handyside, Alan Edge, Gordon Burton, Chris Wilkinson, Glenys Jones, Carolina Mischiati, Dilys Balaam, Chester Hill, Nicky Norris and Roz Hall for their invaluable help with a variety of technical aspects.

Preface

'It is highly dishonourable for a reasonable soul to live in so divinely built a mansion as the body she resides in, altogether unacquainted with the exquisite structure of it' (Robert Boyle, 1664)

Physiology concerns how a living organism functions, *Human Physiology* how the human body functions and *Exercise Physiology* how the body responds to exercise. *Environmental Physiology* focuses upon how our surroundings affect the functioning of our bodies and *Adventure Sport Physiology*, the theme for this textbook, concerns how the body functions in response to the combined exercise and environmental demands of adventure sports.

The adventure sports that are the focus of this book place unique physiological demands on an individual exercising in extreme environmental conditions. During the ascent of a significant peak, a mountaineer has to maintain their exercise through increasingly difficult environmental conditions. As the ascent is made the climber will experience an increase in energy demand due to the increased altitude and decreasing temperature. The mountaineer will have to overcome the physical (exercise) and environmental demands to reach the summit – the halfway point of their journey. The alpine kayaker may have to contend with sustained heat during a day's paddle yet capsize and have to deal with a sudden cold water immersion. Cavers have to avoid the effects of dehydration through sweat loss while working underground sometimes waist deep in water. The mountaineer, kayaker and caver often find themselves in extreme environmental conditions and through their sports activity place sustained exercise demands upon their bodies.

The purpose of *Adventure Sport Physiology* is to enable you to develop a clear understanding of the exercise and environmental issues that relate to the adventure sports in which you take part. The intention is that, using this textbook as a starting point, you should be able to: break down the fundamental exercise demands of any adventure activity, design assessment tools to measure physiological performance and create training programmes to improve fitness to participate. In addition, knowledge developed regarding the environmental conditions in which adventure activities take place and the physiological demands they can place on the body will enable individuals to make more informed choices about performance and safety during any activity.

Adventure Sport Physiology is aimed at undergraduate students studying for degrees in adventure and outdoor education and all individuals who instruct, coach or take part in outdoor and adventurous activities. By their very nature, there is a physiological component to all adventure activities. With more knowledge about the physiological demands of any sport it is possible for the surfer to better understand the environment in which they are performing, the coach to develop a more appropriate training programme for their climber, the mountain leader to understand the physical and environmental demands for the group he or she is leading on a hill climb and for the canoeing instructor to decide to change the length of a day due to the number of swims a group member had during a river descent.

Physiologists seek to better understand the what, where, when, how and why of the processes that maintain life. In an adventure setting, these

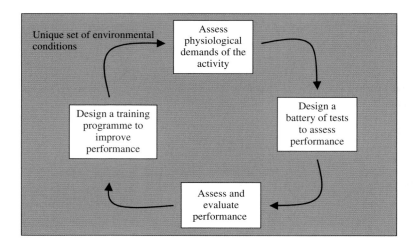

Figure 1   The cyclical role of the adventure sports performer

processes are affected by exercise and environmental demands. The role for the adventure physiologist working with a performer is shown in Figure 1. This role should be viewed as cyclic with the physiologist constantly evaluating the demands of the activity, the battery of tests and the training programme to ensure they are optimum for improving performance.

Adventure Sport Physiology is divided into two parts. The first provides foundations for adventure physiology and the second provides the specific physiological demands of adventure sports. The five chapters in Part I provide a basis for the adventure sport specific chapters in Part II.

Part I covers the history of adventure sport physiology and adventure sports, nutrition for adventure sports, the basic functioning of cells and systems that make up the human body and training fundamentals. Since such a diverse range of activities exists, adventure sports can perhaps be best categorized by the energy demands of the activity. For instance, the energy requirements for a short bouldering problem are very different to those required for a day's sea paddling.

The structure of Part II reflects this with the duration and intensity of exercise being used

to categorize each adventure sport into one of five groups: the explosive activities, anaerobic endurance activities, intermittent activities, high-intensity aerobic endurance activities and lower intensity aerobic endurance activities. Within these chapters there is information on the specifics of energy supply, responses and adaptations to training, fatigue, specific nutritional strategies and where appropriate, specific details of any additional environmental demands for a specific adventure sport. Table 1 provides an overview of the adventure sports covered in this book and the chapter in which the physiology relevant to each is discussed. The placing of an adventure sport in a particular category should be viewed as flexible as the intensity and duration of any adventure sport can be manipulated. Figure 2 provides an illustration of the variation in intensity and duration across the range of adventure sports. The adventure sports shown provide examples of two adventure sports that can be attributed to each of the four quadrants within an intensity-duration continuum diagram.

At the end of each chapter a summary is presented and questions are provided for you to assess your understanding of the material. All

Table 1 Adventure sports categorized by the energy demands placed upon the performer

Ch	Energy system demands	Intensity	Duration	Adventure sports
6	Power and power endurance activities	Very high	Up to 10 s	Dyno moves in climbing, short powerful bouldering problems, individual playboat moves or ski/snowboard moves
7	Anaerobic endurance	High	Up to 90 s	Competitive climbing route, sport climbing route, deepwater solo route, ski run, short wildwater race, short kayak/canoe slalom race or 200 m/500 m sprint kayak/canoe race
8	Intermittent activities	Mixed	Game or session length	Canoe polo game, surfing, surf kayaking, kite surfing, playboat/freestyle session, match racing in sailing (crew) and dinghy sailing (doubles crews, single helms)
9	High intensity aerobic endurance activities	High	90 s +	Mountain marathon, fell running, orienteering, longer wildwater, slalom or sprint kayak races, windsurfing, Olympic racing and wavesailing, marathon paddling, and mountain bike races
10	Lower intensity aerobic endurance activities	Lower	90 s +	Mountaineering, climbing – longer routes, scrambling, hill walking, big wall climbing, caving, mountain biking, inland kayaking, open canoeing, sea kayaking, dinghy sailing (doubles helms), cruising or recreational sailing and diving

references have been included at the end of the book for ease of reading but are presented by chapter to make location of specific references clearer. Each of the five chapters in Part II has summaries of research papers from a specific adventure sport or classic environmental/exercise physiology paper. These examples are provided in **Getting into research** textboxes and are intended

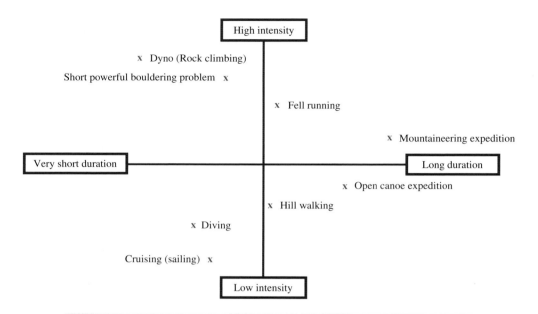

Figure 2 Example adventure sports on the intensity–duration continuum

to assist the reader to begin using journal articles as part of reading for physiology. In addition within Part II, there are *performer contributions* from elite and high-level athletes describing the physiological demands of their adventure sport.

A companion website www.wiley.com/go/ draper contains a large amount of additional material such as an original sample training programme and a set of adventure sport specific exercises to accompany the text in Chapter 5.

Part I
The Foundations of Adventure Sport Physiology

Chapters 1–5 present the foundations for adventure physiology. From the development of the science itself (Chapter 1), we learn how, at its most basic level, the human body comprises elements such as carbon, oxygen and hydrogen. These elements form the basis for human structure and are found within the foods we consume (Chapter 2). The structure of the human body is explored in Chapter 3, from the most basic cell to the complex systems which operate to maintain homeostasis. The fundamental systems for adventure sports performance are presented in Chapter 4 and the basis training programme development is introduced in Chapter 5.

Adventure Sport Physiology   Nick Draper and Chris Hodgson
© 2008 John Wiley & Sons, Ltd

1

A historical perspective: the context of adventure physiology

1.1 The earliest physiologists

Much of the early work carried out to better understand the functioning of the human body related to describing anatomy, what we were made from and to a lesser extent how the parts worked. The Greeks and the development of the Greek empire have been attributed with facilitating the development of some of the earliest physiological knowledge. Greek scholars were very good at assimilating knowledge from other cultures as their empire expanded and they learned from the observations made by scholars in Mesopotamia, Egypt and India. From around 600 BC until the fall of the Greek Empire to Rome in 201 BC, Greek scholars who were often philosophers – such as Empedocles (495–435 BC), Hippocrates (460–377 BC) (Figure 1.1), Plato (427–347 BC) and Aristotle (384–322 BC) – developed existing physiological knowledge, the basis of which remained unchallenged for over 1000 years.

As occurs now in the development of our modern physiological knowledge, each generation built upon the existing accepted theories of the time. In developing his theories of physiological knowledge, Empedocles viewed the world as comprising various elements that had to be held

in balance. He saw humans as comprising four substances: earth, fire, air and water. To maintain health, a balance between the four must be preserved. Ill health was the result of an imbalance between the four substances. The soul, Empedocles argued, was located in the blood.

Hippocrates and his disciples, based on the Island of Cos, wrote many medical papers or treaties on medicine and through this work became known as the founder of western medicine. Doctors today still undertake to follow the Hippocratic Oath. Hippocrates developed Empedocles' theories of the four substances that comprised the human body and described them as operating in four antagonist humours (from Latin, meaning fluids). These humours were hot, cold, moist and dry and to maintain health, these four had to be held in balance. Later, Polybos (Hippocrates' son-in-law) combined the four elements with the four humours to explain human physiology. Figure 1.2 shows these four humours where the black bile was considered cold and dry, yellow bile warm and dry, blood warm and moist and phlegm cold and moist. Maintaining a balance between these components meant good health, whereas an imbalance led to illness. This balance, believed Hippocrates and his followers, was assisted by the heat provided

Adventure Sport Physiology   Nick Draper and Chris Hodgson
© 2008 John Wiley & Sons, Ltd

Figure 1.1   Hippocrates

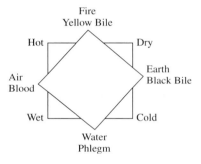

Figure 1.2   The elements, qualities and humours of the human body according to hippocratic medicine

through an internal fire that was maintained in the bloodless left ventricle of the heart.

To maintain the internal fire, life-giving air or 'pneuma' as well as food and drink for nourishment was required. All food and drink consisted of blood, bile (black and yellow) and phlegm. Simple dissections, largely of animals, led Hippocratic physicians to believe the blood ebbed and flowed rather than circulated so that blood only entered the right side of the heart. The left auricle of the heart received pneuma on each inspiration which was supplied to the left ventricle to help nourish the internal fire. The heat created by the internal fire was believed to be responsible for the beating of the heart.

Plato's view of physiology differed little from Hippocrates and Aristotle, although he saw the soul as comprising three parts: the sensual soul located in the liver, the emotional part in the chest and the eternal rational soul contained in the brain. As part of his basic philosophy of humans, Aristotle saw all organisms as comprising two elements: matter and spirit. The combination of matter and spirit were what brought about life. Aristotle believed the soul lay within the heart, which was the central human organ. He believed the brain was an insensitive organ, a fact revealed by the lack of blood found there. Furthermore, he believed the cerebrum was responsible for cooling the heat of the body resulting in the secretion of mucus products that were eliminated from the body through the nose.

Although by modern standards there were many flaws in the thinking of the early physiologists they did identify some of the anatomical structures of the body, such as the ventricles of the heart, the auditory canal, middle and inner ear and various organs such as the liver, brain and blood vessels. Their work paved the way for Galen who was one of the first experimental physiologists.

Galen (130–200 AD)

If Hippocrates can be seen as the founder of western medicine then Galen must be seen as perhaps the most influential physician of all time. He took and developed the Greek theories of physiology and medicine and created a body of medical knowledge that was largely unchallenged for over a thousand years. By the time of Galen's birth, the Greek empire had fallen to Rome where he spent much of his career working on physiological research and as physician to emperors such as Marcus Aurelius.

Born in Pergamos, now part of modern day Turkey, Galen began learning about medicine when he was 16 years old and by the time of his death had produced between 400–500 treaties on medicine, physiology and health. His work has been criticized in the past for the mistakes in some of his findings and beliefs as they predominated for a long period of history. This was perhaps more to do with the control the church had over the development of knowledge until the beginning of the renaissance period as well as the lack of printing presses, rather than the nature of the ideas proposed by Galen.

Galen was responsible for the identification of a wide range of organs and tissues within the body as well as, in some cases, correctly identifying their functioning. He carried out many experiments, occasionally on humans, but mainly on animals such as pigs, cows, goats and horses designed to identify anatomical parts and establish their function. In addition to being physician to emperors he also treated gladiators after they had been in the arena, devising new treatments for injuries and using the observations to further his anatomical knowledge.

In one of his treaties he described the functioning of muscles, correctly identifying the existence and role of muscles, tendons, ligaments, bone and nerve. He explained the functioning of muscles in antagonistic pairs and that the contraction of a muscle was innervated from the brain or spinal cord. He identified previously unidentified muscles such as the popliteus and plantaris of the lower leg. Contrary to previous physicians and philosophers he demonstrated through experiment the existence of blood in the left ventricle, which had previously been thought to contain the heat of life but no blood.

Through ligation of the ureters Galen demonstrated the function of the kidneys excretion of waste products and water. With regard to pulmonary gas exchange, he identified that blood was enriched by a vital element in the air that was subsequently delivered to the tissues around the body. He wrote many essays on health and the recommendations he made then have value today. In his laws of health, he advised people to take regular fresh air, eat proper foods, exercise regularly (he developed training programmes for this purpose) and ensure adequate periods of sleep along with extolling the virtues of a daily bowel movement.

These discoveries represented a major step forward in physiological knowledge. However, there were areas where he mistakenly continued or further developed flawed thinking from previous physiologists. When Galen was unable to demonstrate the existence of a particular finding he would change the language of his papers to include phrases such as 'no sane person can doubt . . .' to implore the reader to accept his assertions. Galen continued to mistakenly refer to the four humours within the body, the existence of the four elements and the necessity to maintain a balance between the elements to maintain health. He believed the life of matter to be brought about by 'spirits' existing within the brain, heart and liver. The blood, he asserted, was formed in the liver and the cardiac septum (the wall between the left and right side of the heart) had pores in it to allow blood to flow from the left to right side of the body.

Despite the errors in some of Galen's findings and reports, he moved physiological understanding forwards and made original discoveries about human and animal physiology. Although it took nearly 1400 years before his work was seriously followed up (and refuted where necessary), he demonstrated, as perhaps the first experimental physiologist, the need to base findings on research observations.

Andreas Vesalius (1514–1564)

When the Belgian physician Andreas Vesalius (Figure 1.3) brought out a new text *De Humani*

Figure 1.3 Vesalius

Corporis Fabrica (the Structure of the Human Body) in 1543, a new epoch in anatomy and physiological discovery began. The origins of the discoveries which Vesalius made can perhaps be traced back to the separation which began between scholastic endeavour and religious control. The 12th and 13th centuries saw some of the great universities make a separation between theological and philosophical faculties for the first time (Bologna in about 1119, Paris in 1200, Padua in 1222, Oxford in 1249 and Cambridge in 1284), allowing for a gradual but sometimes painful separation of science and religion. An example of this was the death of Michael Sevetus (1511–1553), burned at the stake for heresy, after suggesting the true nature of pulmonary circulation which was in disagreement with the Galenist views of medicine as supported by the church.

The development of the printing press by Bi Sheng in China in 1041 and in Europe by Johann Gutenberg in the 1450s brought about a fundamental shift in the transfer of knowledge. One hundred years later, Vesalius greatly benefited from its use for the dissemination of his new textbook of human anatomy and physiology. With the advent of the printing press, Vesalius' work created a watershed in medical thinking.

Andreas Vesalius had a fascination with medicine and anatomy from an early age and was said to have conducted dissections of animals as a boy, a practice that might be frowned upon today and perhaps investigated by psychologists worried about his development. For Vesalius, it led to a career in the medical profession where he started medical school in Paris at the age of 17. He studied under Jacobus Sylvius, a follower of Galen, who continued the practice of having servants carry out the dissections in a rough and untrained manner for students to observe. At one such dissection, Vesalius was annoyed by the poor techniques being used so took over and completed the dissection in a more exact and scientific way. During his time in Paris, he had to acquire bodies from various sources for dissection. Through this, he began to better understand the human anatomy. This led him to question both Sylvius and Galen and the nature of accepted anatomy and physiology. Early in 1537

he moved to Venice to continue his work. Later that same year, at the age of 22, Vesalius was made Doctor of Medicine and put in charge of anatomy and surgery at the University of Padua by the rulers of Venice, who noticed his brilliance. He continued his research in a city and atmosphere where he was allowed access to human bodies to conduct dissections, the results of which were published in *De Humani Corporis Fabrica* in 1543. This was the same year in which Nicolas Copernicus published his work describing the movement of the earth around the sun, rather than the other way round.

Praise for Vesalius' work comes not only for the hundreds of anatomical differences he observed from Galen but from the brilliant illustrations provided for the text by the Belgian artist Jan Stephan (1499–1546). The developments by Vesalius on human anatomy, made available to the medical profession by the invention of the printing press, can be seen in the skeleton depicted in Figure 1.4. This shows the extent of Vesalius' work and knowledge of human anatomy.

There were mistakes in Vesalius' work. For example, he did not refute the pores Galen described

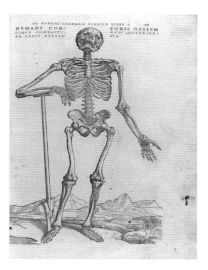

Figure 1.4 Depiction of a skeleton by Jan Stephan, artist for Vesalius' *De Humani Corporis Fabrica*

in the septum wall and he added an extra muscle for controlling the eye, but his work overthrew the dogma of a thousand years and paved the way for later physiologists following his methods of experimentation and observation. From his time, the development of physiological knowledge continued unabated in universities and laboratories in the early centuries across Europe and eventually the world.

William Harvey (1578–1657)

Born in Folkstone, William Harvey studied medicine first at Cambridge and then at the University of Padua (which had continued its support for the development of anatomy and physiology after earlier guidance by Vesalius) before moving to London to begin an medical career.

During his medical work, he was physician to Francis Bacon and Charles I. Harvey had a great fascination with the circulation of the blood, and in 1616 he began lecturing on the subject. It was not until 1628, however, that he set down his ideas in text when he wrote *Exercitatio Anatomica de Motu Cordis et Sanguinis* (an Anatomical Treaties on the Movement of the Heart and Blood). In this work, Harvey explained for the first time how the blood correctly circulated around the body. Through experimentation and observation with animals and humans, Harvey was able to determine that blood was forced into the aorta and pulmonary artery with each contraction of the heart before returning to the heart from the lungs via the pulmonary vein and from the body via all other veins. He concluded that the circulation of the blood was a one-way journey in a closed system. In addition, he denounced the notion of pores in the septum between the left and right side of the heart for the first time, stating:

> 'if the pulmonary vein were destined for the conveyance of air, it has the structure of a blood vessel here... Still less is that opinion to be tolerated which... supposes the blood to ooze through the septum of the heart from the right side to the left ventricle by certain secret pores...

But, in faith, no such pores can be demonstrated, neither in fact, do any such exist' (William Harvey, 1628).

With this statement he correctly rejected over 1500 years of medical thinking. Despite the moves forward in thinking since the times of Vesalius, there were still those who clung to the old beliefs. Jean Riolan, Chair of Anatomy at the University of Paris in the 1640s, claimed that Galen's interpretation of blood circulation was still correct and if Harvey had found differently then his subjects must have been abnormal.

Robert Boyle (1627–1691)

During his career, Irish natural philosopher Robert Boyle devised a number of ingenious experiments based in his laboratories in Oxford and London and was a founding member of the Royal Society. He was the first to carry out a component analysis of blood, describing its properties including taste, colour, weight and temperature. Working with Robert Hooke (1635–1703), he constructed an air pump which they used to examine the nature of air for combustion and living creatures. Using the pump, Boyle and Hooke extracted air from a jar that during different experiments contained mice, birds or candles. With the reduction of air, the animals became unconscious and the flame on the candle died. When they reintroduced air the animals often recovered. From this, the researchers concluded that contained within air was a substance that maintained both life and combustion.

In 1660, Boyle designed and employed a mercury-filled column to determine atmospheric pressure. With this knowledge, he went on in 1662, to examine the effects of pressure on volume of a gas (air) trapped between mercury and the end of a tube, using a J-shaped glass tube that was sealed at one end. Boyle found that the volume of the gas varied proportionally with the pressure of mercury on the gas. It was from this work that *Boyle's Law* was created after a later addition attributed to a French scientist Edme Mariotte (1620–1684) in 1676, who noted that Boyle's finding held where the temperature of the gas remained constant.

Marcello Malpighi (1628–1694) and Anton van Leeuwenhoek (1632–1723)

Further development of physiology was greatly assisted by the development of microscopes and magnifying glasses. Malpighi, born and educated in Bologna, became a celebrated microscopist using two lens. Van Leeuwenhoek, a tradesman born in Delft, was famous for the observation he made using his own hand-crafted (single lens) magnifying glasses.

Following up Harvey's work, Malpighi was able to demonstrate under the microscope the direct link between the arteries and the veins at the lungs of a frog. He also identified for the first time erythrocytes (red blood cells), although at the time he thought they were globules of fat. This was a fact that was later corrected by van Leeuwenhoek who also was the first to notice the striated nature of muscle fibres and the rods and cones in the retina. The advent of microscopes revealed a new world to the scientists of the day.

Joseph Black (1728–1799)

Born in Bordeaux of Scottish descent, Joseph Black studied medicine at Glasgow University before taking a lecturing post at the University of Edinburgh.

Through his research in to gases and their properties, he made a number of discoveries that helped future researchers including Antoine Lavoisier. He proved to be a very popular lecturer with hundreds of students attending his sessions where he often carried out practical experiments for those present. Before Black, air was generally considered to be an element. Through his research, he demonstrated that air was made up of a number of different parts. By burning lime, he found it lost half its weight. He concluded that the lime lost part of its volume to air, which he termed 'fixed air' and we now call carbon dioxide. In doing this he not only discovered carbon dioxide but also demonstrated a gas could combine with a solid, a fact that up until that time had been believed impossible.

Figure 1.5   Joseph Priestley

Joseph Priestly (1733–1804) and Carl Scheele (1742–1786)

Between 1772 and 1774, without knowing the exact nature of it, both Joseph Priestly (Figure 1.5) and Carl Scheele discovered what Scheele termed 'fire air' and we know as oxygen.

Developing older theories, Priestly (a Yorkshireman born near Leeds) and Scheele (from Köping in Sweden) discovered oxygen when they burned red oxide of mercury. In a further experiment, Priestly added plants to a room filled with fixed air (carbon dioxide), in which he demonstrated a lighted flame was extinguished due to the lack of oxygen. He left the room for ten days and then tried to introduce a naked flame again. This time the flame continued to burn from which Priestley concluded plants had the ability to clean air.

William Hewson (1739–1774)

A British physician, William Hewson, continued the work of previous physiologists on the blood and during his research he correctly identified coagulation during blood clotting. As part of this research, he also identified leucocytes (white blood cells) and fibrinogen which he called a coagulable lymph.

Antoine Lavoisier (1743 – 1794)

Working as a tax collector provided Antoine Lavoisier (Figure 1.6), who was born and educated in Paris, with sufficient time to work on the physics and chemistry experiments he loved.

He conducted many experiments into the properties of gases and specifically air. After removing fixed air (carbon dioxide) from air he determined that there remained a gas that was not used as part of respiration or combustion. This gas he termed 'azote' (meaning no life), which we now know as nitrogen. From this work and the work of Priestly and Scheele, who both communicated their results to Lavoisier, he determined that 'fire air' or 'burnable air' was necessary for combustion and respiration. He called this component of air 'oxygéne'. In further research, he recognized that oxygen was converted into carbon dioxide in the body as part of respiration and that heat was generated by this process. He also measured the amount of oxygen used and the carbon dioxide produced during respiration, and determined that the amount

Figure 1.6 Antoine Lavoisier

of carbon dioxide produced was smaller than the oxygen utilized. From this, he deduced that some of the oxygen taken into the body combined with hydrogen within the body to produce water. His work moved respiratory and metabolic physiology forward into a new age. Indeed, as can be seen from the section below, it is also argued he could be the father of exercise physiology. Despite the value of his work, he was guillotined during the French revolution in 1794.

Claude Bernard (1813 – 1878)

Hailed as one of the greatest physiologists of all time, Frenchman Claude Bernard (born in Villefranche near Lyon) often worked in very difficult conditions.

France was not blessed with the number of medical and physiological laboratories that existed in countries such as Britain and Germany at that time, nor did they have the same resources to fund equipment. In fact, the rooms within which he worked at the Collège de France in Paris were cold, damp and dark. Despite these difficulties, and the reluctance of the French to use microscopes, in his 35 year career he produced over 300 scientific writings and was responsible for a large number of significant research findings.

Through experimentation, he revealed that plasma glucose levels varied in healthy individuals. These works led to the identification of the glycogenic (glycogen forming) function of the liver and helped to increase general knowledge of diabetes. He carried out research and identified the nerves responsible for vasodilation and vasoconstriction and identified the deleterious effect of carbon monoxide on arterial and venous oxygenation.

In the 1860s, when he was spending less time in the laboratory due to health reasons, Bernard began to consider the internal environments of living things. He described how cells on the inside of any living matter could not survive if revealed to the outer, external environment. The internal environment was essential for life and components within this environment had to work to maintain these stable conditions. These thoughts were taken forward by Cannon (1871–1945) who, based on Bernard's work, introduced the term *homeostasis*.

Adolf Fick (1829–1901)

Born in Kassell in Germany, Adolf Fick became a renowned physicist, mathematician and physiologist. He graduated from the University of Marburg as a Doctor of Medicine in 1851, and one year later became Chair of Anatomy and Physiology at University of Zurich. He later returned to Germany to a similar position at the University of Würzburg where he remained for the rest of his career.

During a distinguished career, Fick carried out numerous experiments, wrote textbooks on physics and physiology and explained the intensity of diffusion rates in liquids and the distribution patterns of compounds in solvents. Regarding physiology, he is perhaps best known for, in 1870, creating a way of measuring cardiac output, which became known as the Fick Equation. This theory was much ahead of its time, since the required technology did not exist at the time of its development. It was not until 1930 that it was verified by Andre Cournard and Dixon Richards who later received the Nobel prize for their demonstration of the Fick Equation.

Ernest Starling (1866–1927)

Ernest Starling, born in London, was a physiologist who worked within a range of research fields. He is perhaps best known for his development of the Frank-Starling Law, describing the effects of ventricular filling on contraction force, with German physiologist Otto Frank (1865–1944).

In the description of his research it is stated:

> 'Experiments carried out in this laboratory have shown that in an isolated heart... within physiological limits... the larger the diastolic volume... the greater is the energy of the contraction.' (Starling and Visscher, 1926)

Starling was also instrumental in the commencement of research into hormone function, coining the term 'hormone' to describe the chemical messengers at the heart of the endocrine system.

Andrew Huxley (1917–)

Discoverer of two major aspects of physiology, Andrew Huxley was born in London and spent much of his career at Trinity College, Cambridge.

Working with Alan Hodgkin (1914–1998), he identified the roles of potassium and sodium ions in the transmission of nerve impulses, for which they received the Nobel prize for medicine of physiology in 1963. During the 1950s, he and Hugh Huxley (1924–) (no relation) from Birkenhead, Liverpool, and also educated at Cambridge, independently described the nature of muscular contraction in the Sliding Filament Theory.

Frederick Hopkins (1861–1947) and Walter Fletcher (1873–1933)

Frederick Hopkins and Walter Fletcher collaborated in 1906 to isolate lactic acid in the muscle. They published their findings in 1907 in a paper that also detailed the fate of glucose to lactic acid in anaerobic conditions.

Hopkins, born in Eastbourne, studied medicine at Guy's hospital and then worked at Cambridge University. After the 1906 study, he went on to identify the subsequent creation of glycogen from lactic acid.

Otto Meyerhof (1884–1951)

After graduating as a doctor of medicine from the University of Heidelberg, Otto Meyerhof (born in Hannover) initially began a career in psychology before moving on to work in the area of physiological chemistry. With the collaboration of a variety of workers such as Jacob Parnas, Carl Neuberg, Otto Warburg, Carl Cori (1896–1984) and Gerty Cori (1896–1957), Meyerhof was able to demonstrate that suggested in 1933 by Gustav Embden: detailed glycolysis and the complete breakdown of glucose to pyruvate or lactic acid.

Carl and Gerty Cori, born in Prague and graduates from the German University of Prague, later emigrated and continued their careers in the US. In other studies, they went on to identify the pathway for the regeneration of muscle glycogen stores in the liver from lactic acid, in what has become known as the Cori cycle.

Hans Krebs (1900–1981)

Born in Hildesheim, Germany, Hans Krebs spent the early part of his career working in Germany before moving the Britain in 1933. During a distinguished career he worked at Oxford, Cambridge and Sheffield Universities on a variety of research projects based around the area of intermediary metabolism. He is perhaps most renown for his proposal in 1937, in collaboration with William Johnston, of the citric acid cycle or as it is otherwise known the tricarboxylic acid cycle or Krebs cycle.

There were some gaps in the theory at that time; however, further research by Fritz Lipmann (1899–1986) and Nathan Kaplan (1917–1986) published in 1945 identified coenzyme A and completed the cycle as it is known today. Fritz Lipmann was born in Koenigsberg, Germany, and educated there before taking positions in Netherlands, Germany, Denmark and the US. He was also responsible in 1941 for identifying the high-energy bond in adenosine triphosphate (ATP) and the phosphagen system.

1.2 The founders of exercise physiology

The focus for much of the early research in the field of physiology concerned the development of our understanding of human functioning at rest. From this basis, subsequent researchers were able to examine the effects of exercise on human physiology. The earliest studies in this area not only made clear differences in human functioning at rest and during exercise, but led to a fuller understanding of the human physiology and to the development of the specialist field of *Exercise Physiology* which is now taught at universities throughout the world.

Antoine Lavoisier (1743–1794)

Although mentioned as one of the important scientists in the physiologists section above, Lavoisier's fascination with respiratory physiology led him to carry out research in the field of exercise physiology and perhaps makes him the founder of the research area. In order to understand human respiration, Lavoiser measured oxygen consumption at rest, in the cold, after food and during exercise.

Per Henrik Ling (1776–1839) and Hjalmar Ling (1820–1886)

Originally studying divinity, Per Henrik Ling (born in Ljunga, Sweden) became fascinated with exercise as a medium for improving health while teaching fencing at Lund University and taking regular exercise. As a result of this experience, he continued his studies following a medical degree course and then developed his own system of gymnastics. In 1813, with the support of the Swedish Government, he opened the Royal Gymnastics Central Institute in Stockholm and began to teach his method of using exercise to promote health.

Later his son Hjalmar, who studied medicine under Claude Bernard, returned to Stockholm and introduced courses in the biological sciences to complement the teaching of his father's gymnastic methods. This represented perhaps the first exercise physiology programme.

William Prout (1785–1850)

Born in Horton, Gloucestershire, William Prout was not from a rich family and left school aged 13 to work on his father's farm. Realizing he wanted to do more than this he joined a private academy to further his education, working to pay his way and eventually undertook medicine at the University of Edinburgh. During his career, he worked as a urinary specialist at St Thomas and St Guys

Hospital in London but conducted numerous experiments during his non-contact hours.

He was the first person to advocate following a balanced diet including four essential components: 'hydrates of carbon' (carbohydrate), fat, protein and water. In this same paper he advocated milk as the 'great alimentary prototype'. In 1823 he demonstrated that gastric juice was in fact hydrochloric acid, and he also helped to improve the barometer. His inclusion under the founders of exercise physiology is due to his research, on himself, measuring carbon dioxide production during rest, after food and during exercise. With reference to exercise he demonstrated that during walking, carbon dioxide production reached a plateau. This finding formed the basis for the later concept of steady state during exercise at sub-maximal levels.

Heinrich Magnus (1802–1870)

Born in Berlin, Heinrich Magnus studied in Berlin, Stockholm and Paris before returning to teach at the University of Berlin where he was an excellent and popular teacher.

He conducted research within a number of areas, but his most significant finding for the future of exercise physiology was through his demonstration in 1837 that oxygen and carbon dioxide were carried in arterial and venous blood with higher concentrations of oxygen being present in arterial blood and higher concentrations of carbon dioxide being found in venous blood.

Edward Smith (1819–1874)

A physician working in London, Edward Smith began research into the benefits of exercise for prisoners. He introduced a treadmill to Brixton prison in the 1820s, and set exercise protocols for inmates to examine the effects of exercise. Interested by his findings, he designed and built a closed-circuit breathing apparatus to measure carbon dioxide production. Through this and continued work at the prison, which included urine analysis for four inmates, Smith was able to conclude that carbon dioxide production was related to exercise intensity and that nitrogen content of urine was related to the type of food ingested.

August Krogh (1874–1949)

Danish physiologist August Krogh (born in Jutland) studied medicine and then zoology at the University of Copenhagen. After graduation he initially worked as assistant to Christian Bohr (1855–1911) at the laboratory of medical physiology at the University of Copenhagen.

Christian Bohr, born in Copenhagen, became Professor of Physiology when he was just 30 years old; much of his research was in the area of muscular contractions and blood gases. Part of his research legacy was 'the *Bohr effect*' through which the impact of temperature and relative acidity on the oxyhaemoglobin dissociation curve was described (Chapter 10).

Inspired by Bohr's work, August Krogh went on to conduct research in a wide range of areas including disproving the findings of one of his mentor's theories relating to gas exchange into the blood. Bohr and J.S. Haldane (whose contribution to physiology is discussed below) maintained that gases entered the blood via secretion. Through his research, Krogh was able to demonstrate that gas exchange took place via diffusion rather than secretion. In later research he went on to identify the methods by which gases were transported by the blood, and demonstrated that there was an increase in muscle capillary opening in response to exercise. With his colleague Johannes Lindhard (1870–1947) in 1912, he went on to describe the cardiovascular response and relative contribution of fat and carbohydrate to exercise metabolism.

Archibald Vivian Hill (1886–1977)

Educated in Devon and then at Trinity College, Cambridge (where he studied mathematics), Archibald Hill (born in Bristol) made a significant contribution to exercise physiology across a long career at Cambridge and Manchester universities and University College, London. Hill's mentor at Cambridge, Head of Physiology John Langley (1852–1925), drew his attention to the work of Hopkins and Fletcher on lactic acid and the links to oxygen for recovery. This work formed a basis for much of his future study. Working with his colleague Hartley Lupton in 1923 they proposed and demonstrated for the first time a theory of maximum oxygen intake which has now become known as $\dot{V}O_{2\,max}$ or maximal oxygen uptake. The researchers stated 'there is clearly some critical speed for each individual... above which ... the maximum oxygen intake is inadequate'.

Hill and Lupton described for the first time the term *oxygen debt* and the maximal anaerobic level which they termed maximal oxygen debt. They proposed a method for predicting running performance from knowledge of maximal oxygen intake and the running speed at which this was attained. In other research, Hill described the phases in heat production within the muscle that occur in the transition from rest to exercise. He also identified for the first time that heat was produced during the transmission of a nerve impulse.

1.3 The development of environmental physiology

Placing humans in a variety of exercise intensities, durations and modes has led to physiologists having a better overall understanding of human functioning. While some researchers examined responses to exercise to make a comparison with rest, other physiologists began to examine the effects of different environments on physiological function in comparison to normal conditions. The effects of heat, cold, hypobaric and hyperbaric conditions upon human functioning have been examined through the development of the field of environmental physiology.

Blaise Pascal (1623–1662)

The 'Great Experiment', as it was called by Blaise Pascal (Figure 1.7), was perhaps the first in the area of environmental physiology. The experiment

Figure 1.7 Blaise Pascal

was possible as in 1644 Evangelista Torricelli (the unit of pressure, the Torr, being named after him) invented the first mercury barometer which Pascal employed to measure air pressure at altitude. In 1648, using two identical barometers, Pascal left one in the hands of a trusted member of the clergy and persuaded his brother-in-law to take the other to the top of the Puy de Dôme in central France. This experiment revealed that atmospheric pressure fell with increased altitude.

Paul Bert (1833–1886)

A French physiologist born in Auxerre, Paul Bert has been called the father of modern high-altitude physiology and medicine. His research played a significant part in the development of environmental physiology. He completed a law degree in 1857 at the University of Paris and then became interested in zoology. Around 1869 he became friends with a physician who was to fund a great part of Bert's research. Denis Jourdanet (1815–1892), who had a great passion for altitude physiology, wrote papers on the topic and bought barometric chambers and other equipment that was the basis of Bert's research. As part of Bert's research in the area of altitude physiology, he also wrote a history of altitude up to 1878. His main work consisted of studies of human and animal response to changes in barometric pressure. By varying the oxygen levels in hypobaric conditions, he was able to demonstrate that low oxygen partial pressures were responsible for altitude sickness.

Nathan Zuntz (1847–1920)

A German physiologist born in Bonn, Nathan Zuntz was a important scientist in the area of environmental and, specifically, altitude physiology. He studied medicine at the University of Bonn and then worked briefly in Berlin and Bonn before settling for 37 years as the Head of Animal Physiology at the Agricultural University in Berlin. During his career, Zuntz carried out numerous experiments and produced over 430 articles in areas such as blood, gas kinetics, respiration and energy metabolism, as well as the effects of altitude. He led expeditions to Monta Rosa and Tenerife to study altitude physiology. With his colleague August Julius Geppert (1856–1937), he developed an apparatus to analyse expired air gases in the laboratory and even a portable system (carried on the back) which he used to assess energy metabolism. In the laboratory he used his gas analysis equipment in conjunction with several treadmills that were developed specifically for the purpose, including one for cows. As part of his study of expired air, working again with Geppert, he developed the equations necessary to make the Haldane transformation some years before John Haldane (after whom the equations were named) developed his version.

John Scott Haldane (1860–1936)

The Scottish physiologist John Haldane was born in Edinburgh where he went to university to study medicine. Two years after graduating, he moved to the Department of Physiology at Oxford University. After the disappointment of not receiving a promotion, he left the university and opened his own laboratory (the Cherwell Laboratory) in his own back garden.

One of his early research projects, in a career where his main focus was on human respiration and the practical application of physiology, was conducted on the London underground where he set about testing carbon monoxide levels on and around underground trains. The results of his study led directly to the electrification of trains due to the poor air quality. In other work underground, he carried out many studies of mines and was responsible for advocating the use of canaries in mines to detect carbon monoxide, due to their being more susceptible to the effects of the gas and thus presenting an early warning system for miners.

Although he mistakenly continued in his belief about gas secretion in the lungs, he made several very important findings with regard to gas kinetics. In 1905, working with his colleague J.G. Priestley (1880–1941), Haldane identified the role of the partial pressure of carbon dioxide in alveolar gas in the regulation of ventilation. He developed an instrument to measure carbon dioxide and oxygen concentrations in expired gases. He also conducted research in the area of diving and, using goats as subjects (as they were the cheapest and most readily available animal akin to the size of humans), investigated the causes of bends. He subsequently developed tables for decompression that were adopted by the British and American Navies and are still in use today. He also designed decompression chambers for deep-sea divers. He carried out many simulated altitude studies and went on expeditions to examine the effects of altitude on respiration (see the section on adventure physiology below).

His son, J.B.S. Haldane (1892–1964), was a brilliant and colourful British scientist. He received a very early taste of the world of science, being involved in a number of his father's studies including testing an experimental submarine when he was 16 years old, before attending Oxford University. JBS (as he preferred to be called) used his wife to test a variety of pressure studies while researching the effects of nitrogen narcosis. He was very casual about the side-effects of involvement in his research. When commenting upon the incidence of burst eardrums in his decompression studies, he said 'the drum generally heals up; and if a hole remains in it, although one is somewhat deaf, one can blow tobacco smoke out of the ear in question, which is a social accomplishment'.

His achievements included popularizing science through essays and investigations into the effects of breathing pure oxygen when diving (from which guidelines were established). He also studied the effects of elevated carbon dioxide levels on manual dexterity and functioning. As a family, the Haldanes contributed a great deal to the development of environmental physiology.

Sir Joseph Barcroft (1872–1947)

Born in Northern Ireland, Joseph Barcroft moved to Britain to study natural sciences at Cambridge in 1893. After graduating, he continued at Cambridge within the Physiology Department for virtually all his career. He carried out further research into the area of gas diffusion, with his findings later supported by the work of Krogh and the concept of gas diffusion rather than secretion.

He carried out many experiments with himself as the subject, including spending six days in a chamber with low oxygen concentration. He demonstrated the different effects of gases on animals in an experiment, where Barcroft and a dog were exposed to the gas from hydrocyanic acid; although he recovered, the dog was much the worse for the experiment. Barcroft also took part in many high altitude studies including directing a study on Cerro de Pasco in Peru (described in the next section). As part of this research, he successfully examined the effects of decreased oxygen on the dissociation curve for haemoglobin.

1.4 The origins of adventure physiology

There are a number of physiologists interested in the fields of exercise physiology who placed themselves in adventurous environments to further their knowledge. These researchers, by their chosen research methods, were more than physiologists for they were also involved in the advent and development of adventure sports and as such can be rightly thought of as adventure physiologists. Much of the knowledge that has been developed in the area of adventure physiology

has been arrived at through the work of scientists working in expeditionary teams. As a consequence, this section will focus on some of the pioneering teams that have completed research expeditions.

HMS salisbury experiment (1747)

Born and trained in Edinburgh, James Lind (1716–1794) (Figure 1.8) was a naval physician who carried out a number of experiments while working on the many journeys he undertook with the Navy. Lind recognized the effects of cold water immersion and also recognized the need to re-warm victims of body temperature cooling. In 1747, he ran what was probably the first crossover design study to examine the effects of adding citrus to the diet of sailors on board HMS *Salisbury*, where he was serving as physician. Scurvy was responsible for the death of many sailors and Lind found that by adding lemons and oranges to their diet, sailors could avoid this terrible disease. Although his results were ignored for almost 40 years, his ideas were eventually accepted by the Navy and Army.

Figure 1.8   James Lind

First crossing of the english channel by balloon (1786)

The first flight by balloon was made by Pilâtre de Rozier (1756–1785) and the Marquis d'Arlandes (1742–1809) in 1783 in a balloon designed and developed by the Montgolfier brothers Joseph (1740–1810) and Jacques (1745–1799). The period of history in which this flight took place has been referred to as the start of the 'Golden Age of Exploration', a time during which many adventurous journeys took place on land, sea and in the air. The flight took place over Paris, with de Rozier and d'Arlandes reaching a maximum height of 900 m in a flight lasting around 20 min.

This remarkable achievement paved the way for a great number of experimental flights in Europe over the subsequent decades. Just 14 months later, John Jeffries (1745–1819) and François Blanchard (1753–1809) went on to make the first crossing of the English Channel. Jeffries was an American physician who had trained in Britain and Blanchard a French balloonist who for 100 guineas had taken Jeffries on a flight over London. Jeffries was so impressed with the flight he agreed to fund an attempt by the two of them to cross the channel in a balloon. After some back-tracking by Blanchard, who tried initially to make the flight on his own (citing the weight of the pair as a problem on such a long flight over water), the pair set off for France from Dover on 7th January 1785. The flight went very well until they were close to the French coast and the balloon began to descend. In his description of this stage of the flight, Jeffries commented:

> '[We] cast out all the remaining ballast and bags and all; did not rise. Cast out a parcel of Mr. B's books. . . .We cast out all the little things we could find – apples, biscuits. . .but not rising cut away the damask curtains around the car, with the gold cord tassels. . . .Found ourselves still descending, and now approaching the sea, within 120 yards, we proposed and began to strip, Mr. B. first casting away his surtout and coat. I then cast away my coat; then Mr. B. his new coat and long trousers'. (Jeffries, 1786)

After abandoning some of their apparel Blanchard and Jefferies eventually landed in woods 12 miles inland.

Mont Blanc expedition (1787)

Horace-Bénédict de Saussure (1740–1799), a Swiss physicist who had offered a prize for the first to climb Mont Blanc (a feat achieved by Michel-Gabriel Paccard (1757–1827) and Jacques Balmat (1762–1834) in 1786), planned and completed one of the earliest physiological expeditions to altitude one year after the first ascent of Mont Blanc (Figure 1.9). Indeed, during his career he made many studies of the effects of altitude on human functioning and wrote several accounts about the effects of altitude sickness. His comments about his ascent of Mont Blanc reveal much about the practicalities facing those who attempt to collect physiological data in adventurous environments. Having found the last part of the ascent particularly difficult, due to the effects of altitude, on reaching the summit he noted that 'when I had to get to work to set out the instruments and observe them, I was constantly forced to interrupt my work and devote myself entirely to breathing' (de Saussure, 1787).

Figure 1.9 Climbing Mont Blanc in the 18th century

High altitude balloon expedition (1862)

On the 5th September 1862, meteorologist James Glaisher (1809–1903) and full-time balloonist Henry Coxwell (1891–1900) (who had previously been a dentist) took off in a large balloon intent on taking readings at high altitude. The pair reached a height of around 8850 m in less than 1 hr as they took readings from a barometer. Glaisher later described how as they continued to rise after that point, he lost the use of his limbs, went temporarily blind and became unconscious. It was reported by Glaisher that they reached a peak altitude of somewhere around 11 000 m before Coxwell, using his teeth as he had lost the use of his arms, was able to begin their descent by releasing a valve and venting the balloon. The balloonists descended safely and at around 8000 m Glaisher, having regained consciousness, reported rapidly feeling much better – one of the common findings in altitude sickness.

A high altitude flight in 1875 by Gaston Tissandier (1843–1899), Joseph Crocé-Spinelli (1843–1875) and Theodore Sivel (1834–1875) ended in disaster when two of the crew died during the expedition.

Mount Faulhorn expedition (1864)

Physiologists had thought that protein was the major fuel for exercise. To test this finding, two German physiologists, Adolf Fick (whose work was described in more detail in the development of physiology section) and Johannes Wislicenus (1835–1903), set up an experiment using a climb of Mount Faulhorn (1956 m) as the exercise for the study. Having avoided protein from their diet, the two physiologists climbed Mount Faulhorn and on return measured the nitrogen content of their urine and calculated the work done on the hike. After comparison of these results, they were able to conclude that protein could not have provided the energy required for the return journey and so concluded that fats and carbohydrates must have provided the fuel source.

Pike's Peak expedition (1911)

The continued interest of physiologists in the effects of altitude led scientists to build a number of high altitude laboratories. One of the first of these was the Observatoire Vallot on Mont Blanc. The hut was built by over 110 porters and researchers at an altitude of 4350 m and completed in 1890. The hut was used for scientific research by various fields including physiology and as an emergency shelter for anyone attempting to summit on Mont Blanc.

Yandell Henderson (1873–1944) from Yale University knew of a very good four-roomed hut at the summit of Pike's Peak (4300 m) that had the great advantage for researchers of having a cog-railway to the summit. Henderson met C. Gordon Douglas (1882–1963) and John Scott Haldane (1860–1936) at the 1910 Congress of Physiology in Vienna and the conversation held led to the decision to conduct a joint US–British altitude study the following year. The three researchers were joined by Mabel FitzGerald (1872–1973) from Oxford University and Edward Schneider (1874–1954) from Colorado College. The research project was well planned and followed a protocol whereby the team spent 5 days in Colorado Springs before ascending Pike's Peak. The team stayed in the research station for 5 weeks before making a descent and studying the changes following a period at altitude. The research team collected details of altitude sickness (the symptoms of which appeared to last for 2–3 days), data on partial pressures of carbon dioxide and oxygen and measurement of changes in haemoglobin, which rose by 115–154 per cent above normal levels.

One of the interesting points of the research was the role of Mabel FitzGerald, who as a female had taken classes at Oxford but was not allowed to obtain a degree. In fact, she was 100 years old when she finally received an Honorary Masters Degree from Oxford in 1972. She asked to join the research expedition, but never stayed with the group and Haldane devised a study for her working at lower altitudes. She visited various mines and settlements in the Colorado area to collect haemoglobin and carbon dioxide readings along with other data. The results of her study confirmed previous work that those acclimatized to altitude had higher numbers of red blood cells and haemoglobin content in their blood than those living at sea level. Her alveolar carbon dioxide partial pressure (PCO_2) data revealed a linear relationship between PCO_2 and altitude. This data was subsequently used by later researchers to demonstrate that an ascent to the summit of Chomolanga (Everest at 8848 m) without supplementary oxygen would be impossible.

This was later disproved by Reinhold Messner and Peter Habler in 1978 when the reached the summit without supplementary oxygen. Mabel FitzGerald only collected data up to 4270 m and her research showed a linear relationship to this point. The mistake subsequent researchers made was to extrapolate the line to an altitude of 8848 m. The relationship changes after around 5500 m and it is for this reason and the excellent fitness levels of Messner and Habler that they were able to complete the ascent.

The high altitude expedition to Cerro de Pasco, Peru (1921–22)

Led by Joseph Barcroft, a joint US and British expedition team of eight physiologists took the train up to Cerro de Pasco to conduct this high altitude study. The aim of the expedition was to 'investigate the physiological conditions which made considerable muscular and mental effort possible at high altitudes' (Barcroft, 1923). A sub-plot for Barcroft was also to lay to rest Haldane's notion of oxygen secretion by the lungs.

After the study, Barcroft wrote a 129-page report of the research findings, published in 1923. The main results of the research included finding no evidence of oxygen secretion by the lungs, that red blood cell counts were found to be 20–30 per cent higher for acclimatized individuals compared to normal levels and 40–50 per cent higher for natives from the area. Heart rate was found by the researchers to be the same at rest as when at sea

level. However, when exercising at a given work-load, the given heart rate response was higher at altitude than at sea level. The team also found that oxygen saturation levels fell during exercise at altitude, a fact that did not happen at sea level. In addition, and in accordance with the aims of the study, Barcroft and colleagues studied the mental effects of altitude with a number of mental and coordination tests. The research findings indicated that there were no differences in cognitive functioning at altitude, although the subjects felt there were. This was an excellent expedition and with the 1911 Pike's Peak study pointed the way for future adventure physiology studies: physiologists going into the field to examine the exercise and environmental demands in adventure domains.

1.5 The development of adventure sports

While the terms adventure sport and extreme sport are relatively new, many of the activities that have either evolved to become or have inspired adventure sports are actually quite old. Most of the activities we enjoy today have their roots in methods of travel and these can be arbitrarily broken down into three groups, involving travel overland, through water and through the air. Of these three groups, activities involving air travel are clearly the most recent addition but even air travel is a surprisingly old newcomer with ballooning having a history that began in 1783. The history of ballooning has already been covered in the first half of this chapter. The origins of journeying over land and water are lost in prehistory and the technology that prehistoric mankind created to enable travelling to be more effective and more efficient seems remarkable. It is always tempting to speculate that the invention of new technologies also made travelling more fun but since, as a species, we had invented some of the travelling tools that would later become adventure sports equipment thousands of years before we invented writing, this thought can only ever be a romantic notion.

Land-based activities

Skiing

The oldest known ski resides in the Historiska Museet in Sweden and is believed to be between four to five thousand years old. This ski, known as the Hoting ski, was found preserved in a peat bog at Angermanland and although it is the oldest actual ski known, sledge runners have been dated back to 7000 BC suggesting that skiing as a method of cross-country travel may be even older. The first record of skiing is a carving on a Norwegian cave wall created in the region of 2500–5000 BC. Understandably, the ski became a tool used by the military and the legend of the Norwegian army carrying the king's baby to safety during the civil war in 1206 AD has become one of the most famous stories in skiing history.

Sondre Norheim began to experiment with ski design in 1861, developing side-cut, skis that are narrower in the centre and wider at the tip and tail. While this might not seem particularly significant, this was the point at which skis began the evolution from transportation devices to the highly tuned, downhill specific, turning tools we have today. Today's alpine skis are heavily side-cut and this is the feature that allows skiers to perform the dynamic carving turns that define the modern sport. These skis are superb downhill, but unlike the original skis have little uphill capability and most alpine skiers will use powered lifts to climb back to the top of a slope for the next downhill run.

One group of skiers who make a concerted effort to escape from lift-served ski areas are ski mountaineers. This group of adventure sports participants inhabit the mixed ground between alpine skiing and mountaineering. The equipment they use is often a specially designed hybrid of alpine skiing and mountaineering equipment. Ski mountaineers may use climbing skins to ski up snow slopes or might carry their skis strapped to their backpacks while climbing on rock or ice either in bare boot soles or crampons.

While alpine skiing has become the most commonly accepted face of the sport, Nordic or cross-country skiing has remained the closest to the original roots of the activity. Nordic skiing ranges from a relaxed family outing through forested terrain to a snowy version of the marathon. In fact, Nordic ski racers have the highest aerobic capacities of all athletes. Events vary from a straightforward cross-country ski race to biathlon which celebrates the military and hunting lineage of Nordic skiing by combining cross-country skiing with target shooting. This aspect of skiing is covered in Chapter 7.

Snowboarding

Considering skiing has one of the longest histories in adventure sports, snowboarding appears to have one of the shortest. The official start of snowboarding has been attributed to Sherman Poppin's Snow Surfer or Snurfer, which appeared in the 1960s. While sliding sideways down a snow slope is relatively new, it is generally accepted that snowboarding has been heavily influenced by skiing, surfing and skateboarding. Snowboarding has been one of the fastest growing sports of all time and the concept of surfing or skateboarding on snow has captured the imagination of existing snow sports enthusiasts and young people venturing into snow sports for the first time.

The arguments between skiers and snowboarders as to which approach to sliding down snow is 'best' will certainly continue for some time, but what is undeniable is that snowboards are particularly suited to soft snow, off-piste conditions as long as there is little need for traversing across hillsides or over very flat terrain. The legitimacy of snowboarding as a sport in its own right is now clearly accepted and snowboarders competed in the 2006 Olympics in both racing and freestyle formats.

Snowboarding certainly allows the individual performer a lot of choice in their aspirations. Snowboarders range from the backcountry 'powder hound' in pursuit of the perfect soft snow run away from the crowds, through the groomed 'cruiser' to the highly social 'alpine skateboarder' who spends their day in sculpted terrain parks and half pipes perfecting their latest trick to a backdrop of pumping music from a public address system.

Mountaineering

If snowboarding is 'the new kid on the block' then mountaineering is the most accepted and traditional of mountain-based activities. Like skiing, mountaineering has a long heritage due to its emergence from the history of moving through the mountains for utilitarian reasons. Mountaineering as a sport as distinct from a means of travel really began in the Victorian age when affluent gentlemen and ladies began holidaying in the alps and hired local guides to take them up the peaks of a particular area.

Many first assents in the Alps were climbed in this style including the famous ascent of the Matterhorn by Edward Whymper and his party in 1865. Whymper's ascent is well known for two reasons. Firstly, previous attempts on the Matterhorn had been unsuccessful and it was gaining a reputation as a difficult peak. This meant that the first ascent gained importance and Whymper and the Italian Carrel made several attempts on the mountain including some where the two parties were competing to reach the summit first. The other reason that the first ascent of the Matterhorn is famous is more gruesome. On the descent one of the climbers in Whymper's party slipped and fell, pulling four of the climbers from the rock. The rope securing these climbers to the others broke and they were killed.

Mountaineering in the Himalayas became somewhat of a national game around the 1920s when different countries pitted themselves against each other in their hunger to collect first ascents. Everest (Chomolangma) became part of a British agenda and in 1924 Mallory and Irvine made an attempt on the summit which resulted in their deaths. Irvine's ice axe was recovered at 8439 m and despite assertions from some climbing historians that they may actually have made the first ascent, no evidence for this has ever been found. Expeditions continued to attempt to climb the peak, including many further British expeditions, but it was 1953 before Hillary and Tenzing finally reached the summit as part of a British team led by Colonel John Hunt.

In 1978 the mountain was climbed without supplementary oxygen by Reinhold Messner and Peter Habeler. This was a particularly courageous attempt as many believed it impossible to climb this high without oxygen and survive. This ascent was a landmark moment in the development of high-altitude mountaineering as it meant that any point on the earth's surface was now a legitimate target for parties climbing without bottled oxygen supplies.

Messner became one of the key proponents of a new style of mountaineering: *super alpine climbing*, where mountaineers attempt to tackle peaks without fixed camps and porters which had been the standard way to climb mountains in the greater ranges since the turn of the century. This approach has been developed further by climbers like Mark Twight and Steve House who continued to push the ethos of climbing hard routes on high peaks with minimal equipment.

Rock climbing

The exact moment that rock climbing became a discrete activity distinct from mountaineering is unclear. However, most climbers accept that in the early days, rock climbing was thought of as a training activity designed to improve skills and fitness that would be expressed in a mountaineering context. This change meant that the physical location of climbing activity spread to crags, outcrops and sea cliffs. The fact that many of these venues were more accessible and less threatening than full-blown mountain routes meant that the gymnastic and technical elements of rock climbing could be explored.

Stanage Edge in the Peak District is considered to be one of the most popular climbing venues in Britain and typifies the 'cragging' approach. Recorded ascents of rock climbs in the Peak District have been made since 1885, including J.W. Puttrell's Inside Route on the Prow Rock at Wharncliffe. Puttrell is considered a rock climber in the modern sense as he climbed gritstone routes for their own sake which was very different to the accepted training for mountaineering ethos of the time. A year after Putrell's ascent of Prow Rock, Napes Needle in the Lake District had its first ascent by Walter Parry Haskett-Smith and the age of rock climbing had arrived. Rock climbing has always been an activity with a developing and changing value system and the ethics of traditional climbing, sport climbing and big wall climbing are examples of this.

Sport climbing

A key change in the nature of rock climbing was the birth and development of sport climbing. Rock climbers had always placed some fixed gear on routes, such as pegs or pitons hammered into cracks or expansion bolts placed in drilled holes where pegging was not possible. This was generally done 'on the lead' as part of the ascent. In the 1970s, French climbers began placing bolts all the way up very difficult sections of rock to produce technically and physically demanding rock climbs that could be climbed in relative safety since any climber could clip their rope into the fixed gear on the route.

Sport climbing has been controversial, and even today many climbers object to the sport climber's approach of adding fixed gear to change the essential nature of climbing. However, it is probably true to say the growth of sport climbing has had a huge impact on the technical and physical difficulty of today's hardest traditional rock climbing. Many of today's top rock athletes developed their skills and physical attributes on sport routes.

In Britain, climbers generally accept that many areas have very traditional ethics and placing fixed gear for climbing is not the done thing; there are however legitimate areas for the development of bolted sport routes. Generally a good sport route area will have very steep rock that is not possible to protect by traditional methods of leader-placed gear. Some of the more compact limestone cliffs fit these criteria, and areas such as Malham and Raven Tor in the limestone parts of the Peak District and Portland Bill on the south coast have become popular British sport climbing venues. Ben Moon's ascent of Hubble (8c+, Raven Tor) in 1990 put Britain right at the cutting edge of sport climbing standards with a world first. Many climbers will also visit winter destinations for some 'winter sun' rock climbing on Spanish or French limestone bolt routes. Sport climbing has moved forward again and there are now a number of routes around the world with grades starting with a 9, the present limit of sport climbing difficulty.

Bouldering

In Britain and the US, bouldering is often thought of as an activity that evolved relatively recently from a method of training for rock climbing, in the way that rock climbing evolved from training for mountaineering. Interestingly, in France bouldering was already an activity practiced in the 1870s prior to the time the first true rock climbs were being put up in Britain.

Fontainebleau became popular in the 1930s when climbers such as Pierre Allain argued that bouldering was an activity with intrinsic value and not merely a form of training. It seems that the French made the leap to bouldering directly from mountaineering whereas the rest of us needed an intermediate step. In 1947, Fred Bernick realized that scrambling up and down the boulders of Fontainebleau was a good way of replicating the demands of an alpine climb and designed the first 'circuits'. Circuits linked climbs on boulders to create long routes. Climbing bouldering circuits in Fontainebleau is a unique form of the activity still practised today.

Mainstream bouldering is about discrete problems that are often named and have a recorded first ascent. In North America, John Gill was one of the first climbers to be considered a boulderer during the 1950s. Gill had been a gymnast and saw in climbing the opportunity to express his strength and coordination for its own sake. His approach to climbing emphasized these traits and his activities soon evolved into bouldering. He is well known today because of his introduction of the dynamic move or 'dyno' which essentially involves leaping for the next hold.

Big wall climbing

This style of rock climbing takes the opposite approach to bouldering. The objective is to scale a wall of rock of such a size that the logistics of the activity become hugely demanding. The original definition of a big wall was that it would take more than one day to ascend. Early big wall climbs such as Warren Harding's 1958 first ascent of the 3000 ft face of El Cap in Yosemite, which produced the route known as 'The Nose', actually lasted weeks. The average ascent of this 31 pitch route still takes 5 days which means that climbers must transport food, water and sleeping equipment up the wall as they ascend.

The fastest ascent of The Nose is now less than six and a half hours, but it is still classed a big wall since for most parties it is a multi-day affair. The style which climbers adopt on a big wall will vary; some will insist on climbing the entire wall free, but it is more common to use the rope and equipment placed in the rock for aid during some pitches in order to save time or overcome sections that are too difficult for the climber to complete free. This use of aid can vary from resting on protection to the occasional pull on a runner or involve the meticulous placing of a sequence of pieces of equipment, each placement made while hanging from the last.

Ice and mixed climbing

These forms of climbing also evolved from mainstream mountaineering. It is generally accepted that it was in Scotland that ice climbing and the climbing of snowed up rock climbs really took off between 1890 and 1920. Early ice climbers such as Harold Raeburn made first ascents of routes on Ben Nevis, creating classics such as Observatory Ridge that are still enjoyed today. Climbs, at that time, were completed with nailed boots and a single axe used to cut steps and holds into the ice. This was a laborious job that meant routes could take days to complete with the climber occasionally returning to the ground to warm up and recover before climbing back to the high point and adding more steps.

Technology has probably changed the face of ice climbing more than any other mountaineering activity. Crampons with front points were first used on the Eiger in 1932 but it was still standard practice to cut steps on climbs in the 1950s. In the 1960s, the American climber Yvon Chouinard began experimenting with crampon and axe design. Rigid crampons and down-curved picks were born and the basic format of the equipment we see in the sport today became available. The modern climber will rarely cut steps, preferring to rely on a deftly placed axe in each hand and the razor sharp front points of specialist crampons on each foot. Modern climbing crampons have either one or two front points and it is now common among top climbers for these to be permanently bolted to the sole of the boot. Modern ice and mixed climbers use a full repertoire of rock climbing movement and will even hook an axe over their shoulder, or holster it onto the harness in order to free up a bare hand while moving over rock. Jeff Lowe was one of a number of climbers who popularized this approach to ice and mixed climbing in the 1980s and 90s.

Climbers often differentiate between routes that are exclusively snow and ice and routes that involve some climbing on rock. Routes that include any element of rock climbing are designated mixed climbs; this is reflected in the prefix M before the numerical grade. Routes that involve climbing frozen water are designated water ice climbs (Prefix WI) and routes that are compacted snow are referred to as alpine ice (Prefix AI). Mixed climbing has become very popular recently. There seems to be a real opportunity to increase the difficulty of the rock sections of these routes; climbers such as Neil Gresham and Dave Macleod who have contributed to Chapters 9 and 10 have both been involved in this process.

Caving

Underground exploration is a serious and diverse activity. Caves can vary from those where it is quite possible to walk through passageways with little more than a torch, to highly technical caves involving sophisticated rope work or diving. For this reason it is probably best to think of caving as a set of activities that occur in the same kind of arena in the way that mountaineering activities involve hill walking, rock climbing, alpine climbing and Himalayan expeditions. The hardest caves will require rope handling skills that would challenge even the most effective rock climber. Caving ranges from a couple of hours underground to multi-day expeditions with complex logistics involving route finding and the transport of food and equipment. Being underground makes electronic communication systems such as mobile phones relatively useless and so this activity is one where it is possible to be completely out of touch in comparison to mountain-based or water-based activities. Large expeditions will sometimes rig a telephone system as they explore a cave in order to overcome this difficulty.

Geological features such as squeezes can involve wriggling though small gaps and a sump will mean a short section of underwater passageway to negotiate. Cavers will navigate though a system using a survey which is the caving equivalent of a map. The skills required to complete a journey depend very much on the specific nature of the cave involved but ascending and descending ropes is a common activity in advanced caving. Some caves will actually require diving equipment in order to complete sections.

The cutting edge of caving involves journeying into genuinely unknown territory. Krubera cave in Abkhazia Georgia holds the honour of being the deepest cave yet explored; in 2005 a depth record was set at 2080 m. This sort of feat involved four staged camps in order to support the team, camping in tents underground for many days at a time. The Krubera team consisted of 56 cavers from 7 different countries. Unlike mountaineering, it is difficult to know when the true bottom of a cave system has been reached as there may still be unexplored routes which could lead deeper. It seems quite possible that the Krubera cave will be taken deeper in the future by a team who find a new passage.

Orienteering

This adventure sport involves combining the basic mountaineering skills of navigation with cross-country or fell running. Competitors navigate between control points to complete a course against the clock. While there are other forms of orienteering, such as ski orienteering and mountain bike orienteering, the most common is on foot. Generally competitive courses will be set in forests or countryside but it is quite possible to set up courses in urban parks or even playing fields. Orienteering is a worldwide sport but is especially popular in Scandinavian countries.

Similar to orienteering is the mountain marathon, where competitors complete an orienteering-style competition but with full-blown mountainous terrain and distances. A mountain marathon can include more than one day's activity, in which case competitors get an accumulated time.

The popularity of the mountain marathon has led to the introduction of the 'adventure race'. Adventure racing takes the concept one stage further and requires competitors to complete different stages using different modes of travel that can include activities such as running, mountain biking as well as forms of water-based travel. Adventure racing course normally include terrain that is particularly arduous and is often a team event.

Mountain biking

The first mass-produced commercial mountain bike was released in 1983 by Specialized. This is not to say that no mountain biking took place earlier but it was generally by small groups of enthusiasts with hand-made or 'reassigned' cycles such as those who took part in the first ever official mountain bike race, 'The Repack', in 1976. The Repack gained its name from the fact that the 2.1 mile descent cooked the components of the prototype cycles of the day and resulted in competitors having to repack their bearings with grease between runs.

Mountain biking today is an extremely varied activity ranging from downhill competitions on bikes with several inches of front and rear suspension and few gears to cross-country and 'enduro' racing involving lightweight bikes with lots of gears. Technology has come a long way and even the most lightweight cross-country race bike will have at least three inches of front suspension and probably hydraulic disk brakes.

Mountain biking in Britain has now reached a level of maturity where dedicated mountain bike trails are quite common and trail centres such as Coed-y-Brenin in North Wales and the Seven Stanes in Scotland are popular biking venues all year round. Trails can be graded for difficulty, a bit like the piste marking system used in ski resorts. On a worldwide scale, cross-country racing has been an Olympic discipline since 1996 and downhill has its own series of championship races. There are also head-to-head races called 4X (Four Cross) where cyclists compete on very short courses (25–60 s) in knockout rounds.

Away from the competitive scene, the discipline of 'free riding' or specializing in gravity-fed riding over drop-offs, down steep hillsides and jumping huge gaps is very popular and produces an incredible number of videos each year with bigger and bigger stunts. There is also a discipline that seems part mountain biking and part BMX with participants riding specialist jump bikes, spending most of their time in the air above a series of 'dirt jumps'. One thing that seems certain in mountain biking is as technology improves there will always be someone to test it.

Water-based activities

The heritage of water-based activities is probably as long as that of land-based activities. Waterborne transportation offered early man the opportunity to move from place to place and transport loads far in excess of what they could carry over land. The dugout canoe is a popular example of early Stone Age craft, but in fact Stone Age societies were far more inventive and creative and boat-building technologies include rafts constructed from reeds and skin-covered crafts.

Open canoeing

One of the most successful inland craft has been the open canoe. Canoe-like craft are used on rivers and lakes throughout the world. However, the birch bark canoes of North America have become the main blueprint for the open canoe that we use today. Canoes are quite varied, reflecting the diversity of uses by adventure sports performers. Racing disciplines include marathon and sprint as well as the North American 'sit and switch' style events. Open canoes are still used for their more traditional purpose of journeying with loads and many enthusiasts enjoy both the recreation of historic journeys and the exploration of new areas. The capacity of the canoe to swallow equipment means that it is possible to live comfortably for several days or even a week at a time without the need to re-supply. This could probably be pushed to a month if you were prepared to be ruthless. The writers and conservationists Gary and Joanie McGuffin completed a 6000 mile journey and over 7 months of paddling through North America in their tandem canoe. A journey of these parameters will mean a stop for supplies one or twice along the way.

Sea kayaking

The other paddle-based craft that have influenced recreational paddlesport more than any other are the Inuit kayaks of Siberia, North America and Greenland. Kayak means 'hunter's boat', but the modern version is most commonly used for journeying. Traditionally made from seal skins stretched over a wooden frame, the modern sea touring kayak is normally constructed from either expensive composite materials such as glass fibre and kevlar set in resin, or else a more pocket-friendly but slower roto-moulded plastic.

Modern interest in sea kayaking as the activity we know began after Europeans took an interest in the traditional kayak skills of the Greenlanders. Edi Pawlata taught himself to roll after reading accounts of Inuit skills and 'Gino' Watkins learned traditional boat handling skills from the Greenlanders while stationed with the Royal Air Force. These and other examples sparked the imagination of a few adventurers who had until then been using heavy wooden craft on the sea. Victorians such as John 'Rob Roy' MacGregor had completed significant journeys in craft built from wooden planks and weighing over 40 kg in the 1860s and 1870s.

These craft lacked the lightness, agility and sleek lines that we associate with sea kayaks today, whose main influence has undoubtedly been the Greenland kayak.

Today sea kayaking is a popular worldwide activity that includes everything from trips of a few hours duration along sheltered coastlines to multi-day or even multi-week expeditions involving living out of a loaded kayak and travelling great distances with little or no outside support. There are also competitive sea kayaking events such as the Arctic Sea Kayak Race based in Norway, where competitors gather to compete in a race that covers 170 km over 4 days. The choice of boats is wide and ranges from tandem expedition kayaks, capable of transporting heavy loads, to extremely lightweight and unstable racing sea kayaks. Folding sea kayaks, consisting of a soft skin and alloy frame, can be transported easily to remote locations.

Racing

The flat water racing disciplines have been a feature of the Olympics since 1936. Racing includes marathon and sprint categories and the distances of these events range from 200 m to the full 36 km marathon. Flat water events are generally raced using wing paddles with blade profiles that don't just grip the water but are actually drawn forward during the stroke. The events involve K1, in which a single paddler competes in each boat, and K2 and K4 which involve two or four paddlers in each craft.

Whitewater kayaking

Running down rivers with rapids, or whitewater, has become one of the most popular of all paddle sports. Although kayaks now tackle rivers far too rough for canoes, whitewater kayaking is an activity that was inspired and heavily influenced by canoeing. Canoes were often used as transport on rivers that included whitewater, but kayakers started to see running the rapids as an activity itself and by the 1930s kayakers were running rapids in Europe in wooden frame kayaks with canvas skins. Rapids generally needed to be deep and open as contact with rocks would normally mean tearing the canvas skin and the destruction of the craft.

Milo Duffek transformed the sport of whitewater kayaking in the 1950s by re-examining the strokes used in canoeing and transferring these to the kayak. He is most famous for the development of the bow rudder which allowed kayaks to make dynamic and aggressive turns. The use of glass reinforced plastic or GRP in the 1960s and early 1970s allowed the advance of boat design and the repair of damaged boats became more straightforward. In the 1980s, rotomolded plastic began to take over from GRP and composite materials as the main construction for whitewater kayaks. The advantage of plastic was the ability to withstand impact and it was easy to mass produce, allowing plastic-boat whitewater kayaking to become a mass participation sport. Paddlers could hit rocks and continue paddling with generally undamaged boats. This allowed practice and paddling in rocky environments that would have been impractical with GRP or lath and canvas (canvas stretched over a wooden frame) boats.

Whitewater kayaking now includes a range of sub-sports including river running, site-specific freestyle or play boating, expedition paddling, slalom racing and wild water racing. Whitewater paddlers with the latest boats now routinely tackle sections of rapids so severe that they would have been considered 'unrunable' before the development of plastic whitewater boats. Rivers in the Himalayas previously accessible to only a few elite expedition paddlers have become popular adventure tourism destinations for competent recreational paddlers. Slalom and down river racing are still popular competitive whitewater disciplines. Slalom has been an Olympic event since 1992 after a brief appearance in Berlin in 1972.

Freestyle is a highly gymnastic activity where performers are scored for the number and difficulty of tricks they can perform in a whitewater feature. Freestyle competition began as a very informal activity in the 1980s but has now become a major canoe sport disciple with the first official world championship being held in Sydney, Australia, in 2005.

Kayak surfing

The use of kayaks for surfing began as an activity designed to practice the sort of skills required for paddling rough water but soon became an activity in its own right. Surfing was first done in slalom and general purpose kayaks but became heavily influenced by stand up or board surfing. This led to two schools in the paddlesport surf world. There are kayakers who believe that a kayak should look like a slalom boat: long, graceful and without any fins. On the other hand, there are those who believe that what a paddlesport surf craft looks like is not important and that we should adopt the lessons from board surfing, embrace features such as fins and try to emulate the style of surfing that short board surfers have made popular. The latter approach led to the surf 'slipper' kayak; before long the need to sit inside the kayak had been abandoned and the waveski was born. Waveskis were constructed exactly the way surfboards are and rely on fins for lateral grip. The slalom style boats, meanwhile, had kept their length but had developed low sharp rails, flatter hulls and more bow rocker to help takeoffs.

The long boats became the accepted face of the surf kayak and were designated the international class (IC) with surf competitions becoming popular. The ski paddlers held their own separate competitions judged on criteria with different rules, and it looked like the two groups of paddlers would never speak to each other again! Of course this never happened, and many paddlers who surfed but preferred to remain firmly anchored inside their kayaks still hankered after the fins and manoeuvrability of the ski paddlers. The high performance (HP) surf kayak has therefore grown to inhabit the ground between the finless IC boats and the deck-less waveskis. There are now international and HP classes in surf kayak competitions, as well as the separate ski competitions, and it is not uncommon for ski paddlers, IC paddlers and HP paddlers to surf together in the same group.

Surf kayakers will also own IC kayaks, HP kayaks and skis and choose a particular craft because of the conditions of the day. In countries such as the UK where the water temperature is cool and much of the surf arrives in the winter, there are distinct advantages offered by the protection of the kayak and spray deck.

Surfing

Few adventure activities have captured the public imagination like surfing. Arguments may still rage over whether the first surfers used canoes or surfboards but it is certainly board riders who are recognized as the embodiment of surfing. Surfing can be traced to the 1500s and Polynesian kings would surf in order to prove their worthiness. The first contact with western society occurred in 1779 when Captain James Cook and Lieutenant James King observed that Hawaiians participated in the activity of surfing as part of their indigenous culture. The author Jack London popularized surfing around 1907 when he brought a Hawaiian surfer to California. Californian surfing started to take off in the 1950s, beginning the phenomenon of surfing we see today.

Mass production techniques and glass fibre technology have made surfing equipment readily available; high quality effective wetsuits mean that surfing is no longer restricted to warmer waters. Although long board surfing has recently made a comeback, the move has been to shorter more manoeuvrable surfboards and the American Kelly Slater has become the most recognized name in surfing, having been world champion six times. The drive to surfing bigger waves has led to developments in surfing as bigger waves mean higher speeds; it is not possible to paddle a board fast enough to catch some of the waves now being surfed. Instead, surfers such as Laird Hamilton have developed a technique where the surfer is towed in to the wave by a jet ski in a similar fashion to a water skier. This means that much bigger waves can be ridden (up to 70 ft) but the consequences of error are a high risk of serous injury or drowning.

Canoe polo

Within this section on adventure sports there is an activity which stands out as unique. Canoe polo is the only adventure sport played as a team. Played on a 'pitch', two teams of five players attempt to score goals by gaining possession of the ball, passing and dribbling it to move into position and finally shooting in order to place the ball in the opposing team's net. Polo was recognized as an official canoe sport in 1989 and can be played in open water or a swimming pool. Polo boats are light and highly manoeuvrable and top level polo players have astounding boat handling skills, agility and fitness as well as ball handling skills that a basketball player would be proud to display. Polo is a very physical game and it is quite within the rules to push a player over and capsize them if they are in position of the ball. Players wear helmets with face guards to protect them from the dervish of flailing paddles that tend to accompany the game.

Sailing

Under its various guises, sailing is a massively popular activity. While being one of the oldest adventure activities, it also has a vigorously developing crop of newer activities including windsurfing and most recently kite surfing. This variety is one of the strengths of sailing, as participation can be as part of a team on a large cruiser or as an individual who straps a windsurfer to the roof of their car and drives off to the local lake for an afternoon.

Dinghy sailing

One of the most popular forms of sailing is dinghy sailing. Partly this has been because of the accessibility of this form of sailing. The development of plywood meant that strong lightweight dinghies were suitable for small scale and even home construction. Glass fibre has allowed dinghy construction to move forward while maintaining this essential accessibility. It is possible to take part in an activity that can range from the enthusiast pottering around in their local waters to Olympic competition. Dinghies can be sailed either single-handed or in pairs and can provide one of the most athletic forms of the sport. Trapeze boats are a real example of this athletic style of sailing.

Cruisers and keel boats

These types of craft rely on ballast rather than the weight of the crew as the primary source of stability under sail. Dinghies have a light movable centreboard to provide directional stability, whereas the weighted keel of a keel boat is fixed and provides both directional stability and counteracts the tipping forces that wind hitting the rigged sails of a sail boat produces. Keel boats range from lightweight racing boats to boats with cabins suitable for sleeping aboard and cruising from place to place or offshore racing. Offshore racing can take place over courses lasting several days or even weeks, often sailing in arduous conditions.

Multi-hulls

Boats with more than one hull come under the generic term multi-hull. The most common design is the catamaran which has two identical hulls with decks or beams bridging the hulls on which the crew will work. Trimarans have a central hull on which the crew will work and wing hulls called outriggers or floats to provide stability. When under sail trimarans will normally sail on the centre hull and the downwind outrigger, while the upwind outrigger is normally lifted clear of the water. Because the multi-hull can have enormous stability for such a narrow profile through the water, they have significant advantages in terms of speed over a single hull craft. Multi-hulls come in all sizes from dinghy-sized craft to huge cruiser-style craft. The ocean-going multi-hull has been popularized in recent years as sailors such as Ellen MacArthur have used them to great effect in round-the-world endurance-style sailing exploits. In 2005, Ellen completed the fastest solo circumnavigation of the globe in a record 71 days, 14 h, 18 min and 33 s.

Windsurfing

Although the idea of sailing a surfboard was probably conceived earlier, it was in 1968 that Schweitzer and Drake created the windsurfer. Boards and sails have been through many changes since then, but the basic principle of joining the rig to the board through a universal joint has remained the key to the sport of windsurfing and means that the vast variety of equipment we have today is still identifiable as windsurfing equipment. Despite the wide range of designs of windsurfing equipment available, most people today would think of the short, high speed, planning boards as the typical windsurfing set up. This energetic style of windsurfing can take place on relatively flat water or in rough water and surf zones. Wave sailing involves all the high speed skills of short board sailing and also includes jumping and aerial manoeuvres made possible by the moving ramps of water that comprise the surf zone. Windsurfing became an Olympic sport in 1984 at the Los Angeles Games and there are also slalom disciplines, speed sailing, wave sailing competitions and freestyle.

Windsurfing has had its share of eccentric figures and one of the legends of the sport was Baron Arnaud de Rosnay. He successfully completed crossings of the Bering Strait and the Pacific Ocean, but in 1984 set off on a crossing of the Formosa Strait between Taiwan and China and was never seen again.

Kite surfing

The latest evolution in sailing sports is probably kite surfing. The basic concept of kite surfing is similar to windsurfing in that a lightweight board is steered with the lower body while power comes from a rig controlled by the upper body. The essential difference is that in kite surfing, the rig is actually a large power kite that looks far more like a parapont than any conventional sail. The kite is flown on long lines and can provide so much lift that the board has almost no need for buoyancy at all. Kite surfers can launch into the air from the smallest of waves and then hang under the kite as they float back to the surface of the water. The biggest problem for kite surfers is the amount of space that the kite surfer needs and this has led to conflicts between kite surfers and other water users.

SCUBA diving

Human beings have been diving with the assistance of re-breathers and diving bells since the 17th century but it was in 1771 that John Smeaton invented the air pump which allowed divers to be supplied with air from the surface during diving. In 1825, William Jones invented a system which used compressed air to supply the diver and created the first real self-contained air supply for diving. This could easily be seen as the first real SCUBA (self contained underwater breathing apparatus) system. However, it was not until 1943 before Jacques Cousteau and Emile Gagnan designed the 'aqualung' which was capable of providing compressed air to the diver on demand rather than continuously. Jacques Cousteau popularized diving by producing the first commercially available open circuit SCUBA equipment and by producing books and presenting underwater photography to the world.

The history of diving is full of accounts of accidents and in 1960 Connie Limbaugh died in a cave diving incident. This event is significant because diving without training became a thing of the past. Certification of divers became the accepted norm and in 1966 PADI, the Professional Association of Diving Instructors, was formed. Since then, PADI has been responsible for both the training and certification of diving instructors and also ensuring that those who hire diving equipment have been adequately trained.

Freediving

The oldest type of diving is freediving where divers have no artificial oxygen supply and are only able to dive for as long as they can hold their breath. There are several different forms of freediving competition but the most publicized is 'no limits freediving' where the diver rides a weighted sled which runs down a cable. At the bottom of the dive the diver activates an air bag on the sled which inflates and lifts them back towards the surface. The extreme depths and the risks involved have led to an interest in this form of diving which demands high levels of mental control as well as breath-holding ability. Freediving is an activity in which athletes are continually pushing the limits. In 1992, the American Tanya Streeter set a new world record at 160 m which was beyond both women's and men's records at that time. The Austrian Herbert Nitsch has since superseded this with a new men's world record of 172 m. Other free dive events include static apnea which involves holding the breath underwater as long as possible. The record in 2006 for this is 8 min 58 s. A more detailed examination of freediving can be found in Chapter 10.

2

Nutrition for health and performance

2.1 Introduction to nutrition

Humans as omnivores have the capacity to provide the nutrients to sustain life from a wide variety of plant and animal sources. The ability of humans to thrive in their natural environment has largely been due to their adaptability, which includes diet. Throughout history and around the world humans have been able to adapt from vegetarian to extreme meat diets. Inuits have traditionally followed an almost exclusively animal-based diet over the centuries. With this diet and in extreme environmental conditions they have continued to thrive, although they now often have sucrose and lactose intolerances. The Hindu religion, on the other hand, has had a long traditional link with vegetarianism since around the 4th century BC. The *Bhagavada Gita* (Hindu scriptures, literally the 'song of god') first proposed the principle of *ahimsa* which advocates non-violence to all forms of life. Mahatma Gandhi (1869–1948), who led the campaign for independence for India, was one of the most famous vegetarians with the *ahimsa* as a central tenant of his philosophy of life. In the UK, there are currently thought to be over 4 000 000 vegetarians today, compared with 100 000 in 1945. Many cultures now exist on a mixed diet of plant and animal foods; however, it is possible for humans to thrive on a wide variety of diets.

The food we take into our bodies as adventure sports performers is important not only for its impact on our performance for that day or the next, but also for our long-term health. The importance of nutrition for health was recognized over two thousand years ago by some of the earliest researchers interested in health and the human body. Since this early identification, many discoveries about the nature of the foods we take into our bodies have been made. As far as this research has taken us, it still remains apparent that the basis for a sound nutrition lies in the adoption of a *well-balanced diet*.

Traditionally, from a physiological point, six food components are required to support optimal human functioning. Carbohydrates, proteins, fats, vitamins, minerals and water have been identified as essential for healthy living. These macro (carbohydrates, fats and proteins) and micro (vitamins and minerals) nutrients, along with water (the second most important substance needed to maintain life, after oxygen) serve two main functions in humans: to provide the *fuel* for living and to maintain the *structure and functioning* of the body. This chapter examines the nature of these nutrient groups and their role in maintaining health and performance for adventure sports performers.

Adventure Sport Physiology   Nick Draper and Chris Hodgson
© 2008 John Wiley & Sons, Ltd

Table 2.1 The common elements found within the body; some symbols are derived from their Latin, Greek or Arabic names e.g. Potassium (K) from the Latin *kalium*

Chemical element	Percentage	Symbol	Atomic No.	Mass No.
Oxygen	65	O	8	16
Carbon	18	C	6	12
Hydrogen	10	H	1	1
Nitrogen	3	N	7	14
Calcium	1.8	Ca	20	40
Phosphorus	1.0	P	15	31
Potassium	0.4	K	19	39
Sulphur	0.3	S	16	32
Sodium	0.2	Na	11	23
Chlorine	0.2	Cl	17	35
Magnesium	0.1	Mg	12	24
Iron	0.004	Fe	26	56
Zinc	0.002	Zn	30	64
Selenium	0.0003	Se	34	80
Manganese	0.0003	Mn	25	55
Copper	0.0002	Cu	29	63
Molybdenum	Trace	Mo	42	98
Chromium	Trace	Cr	24	52
Fluorine	Trace	F	9	19
Iodine	Trace	I	53	127
Aluminium	Trace	Al	13	27
Boron	Trace	B	5	11
Cobalt	Trace	Co	27	59
Silicon	Trace	Si	14	28
Tin	Trace	Sn	50	119
Vanadium	Trace	V	23	51

Note: Elements in red represent the 'big four' within the body, those in blue the major minerals, green the minor minerals and those in orange the lesser-known minerals (those in green and orange comprise the trace minerals).

The foods that form the basis of our diet comprise chemical elements. Over 112 elements have been identified or produced artificially, the main elements being displayed on the inside cover of this textbook within the periodic table. Ninety-two of these elements are found naturally and of these, 26 are commonly found in humans. Table 2.1 provides details of the elements identified in the body. The elements oxygen, carbon, hydrogen and nitrogen represent 96 per cent of each person's chemical composition and form the basis for the three macronutrients within our diet. Carbohydrates and fats are made up of carbon, hydrogen and oxygen. Proteins also comprise nitrogen. These 26 elements form the structure of our bodies and bring about healthy functioning. They are the building blocks for our cells, tissues and systems and will be the subject of the following two chapters.

Key point

The body requires six main nutrients for healthy living. These nutrients are carbohydrates, fats, proteins, vitamins, minerals and water. In recent years, health experts have also discussed the need to maintain healthy levels of fibre in our diets to facilitate digestion. These nutrients serve to maintain the structure and functioning of the body and provide the energy reactions that take place within the body.

Atomic structure

The chemical elements that are the building blocks for humans and the food we eat are made from atoms which are composed of three types of particles: protons, neutrons and electrons. Each atom has a nucleus made from positively charged protons and neutrons (neutral) with one or more negatively charged electrons orbiting the nucleus (Figure 2.1). Electrons orbit within specific shells or clouds around the nucleus (like the layers of an onion). The first shell can hold a maximum of 2 electrons. The second shell, which has two subshells can hold up to 8 electrons (2 in the first subshell, 6 in the second). The third shell can hold a maximum of 18 electrons (2 in the first subshell, 6 in the second, and up to 10 in the third).

An atom will always remain neutrally charged as there is a balance between the number of protons and electrons. The *atomic number* for an atom refers to the number of protons. For example an atom of oxygen has atomic number 8. This will be matched by the number of electrons so that an oxygen atom is neutrally charged. The number of protons and neutrons in an atom determines its mass and the type of element that it forms. Figure 2.1 depicts the structure of the four most common elements found in our food and bodies.

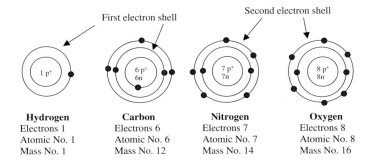

First electron shell		Second electron shell

Hydrogen
Electrons 1
Atomic No. 1
Mass No. 1

Carbon
Electrons 6
Atomic No. 6
Mass No. 12

Nitrogen
Electrons 7
Atomic No. 7
Mass No. 14

Oxygen
Electrons 8
Atomic No. 8
Mass No. 16

Figure 2.1 Atomic structure of the most common elements in humans

Although protons and neutrons are very small, electrons are 1836 times smaller and therefore the mass of an atom is determined by the number of protons and neutrons in the nucleus. The *atomic mass* of an atom refers to the total number of protons and neutrons.

Some elements naturally exist in slightly different versions of the same substance, for example chlorine has two types or *isotopes*. Approximately three-quarters of chlorine atoms exist with 18 neutrons whereas almost a quarter have 20 neutrons. Both isotopes possess 17 protons with the result that the atomic mass is 35 for 75 per cent of chlorine atoms and 37 for the other isotope. The atomic mass for chlorine is calculated as:

Calculation of atomic mass for chlorine with two isotopes ^{35}Cl and ^{37}Cl in nature is 3:1 ratio:

^{35}Cl mass for 3 atoms is $35 \times 3 = 105$
^{37}Cl mass for 1 atom is $1 \times 37 = 37$
Total mass of four atoms $= 142$
Mean mass for chlorine $= 142 \div 4$
$= 35.5$

The atomic mass of chlorine is therefore commonly written as 35.5. In some textbooks and periodic tables chlorine is shown with an atomic mass of 35.4527. This is because the ratio of ^{35}Cl to ^{37}Cl is only approximately 3:1. However, it is important to know that different isotopes of the same element do exist, differentiated by the number of neutrons present in each form.

Chemical bonding

As can be seen from Figure 2.2, atoms which comprise the smallest units of our foods come together in groups to form matter. The force that holds atoms together is called chemical bonding. Through this force, atoms are able to join to form our foods and ultimately our bodies. It is the positive and negative charges of atoms that enable chemical bonding to occur. There are two main forms of chemical bonding, both which involve the use of electrons to form the basis for the bond. *Ionic bonds* involve the donation of one or more electrons from one atom to another and *covalent bonds* involve atoms sharing electrons.

The atoms in Figure 2.1 have one or two shells of electrons. Each of these shells has energy stored within it and the greater the distance electrons are from the nucleus, the more energy they need to have to overcome the forces of attraction to the nucleus. Each shell or subshell of electrons has a maximum number of electrons. Atoms will try to undergo processes to fill their outer shell or subshell. This is due to the fact that when a shell or subshell is filled, the electrons present in that shell share the energy cost of orbiting the nucleus at that position. The number of electrons in the outer shell or subshell of an atom and its need to donate, accept or share electrons with other atoms is known as its *valency*.

An ionic bond is formed between two atoms when one of the atoms donates an electron to another, thus completing the outer shell of electrons for each and creating a stable chemical bond. An example of this is depicted in Figure 2.3 where chlorine accepts one electron from sodium. By

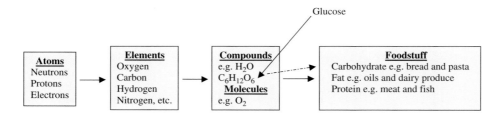

Figure 2.2   From smallest to largest: the components that comprise our food and ultimately our bodies

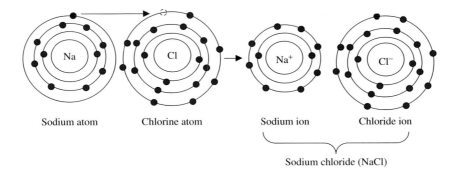

Sodium atom   Chlorine atom   Sodium ion   Chloride ion

Sodium chloride (NaCl)

Figure 2.3 The formation of sodium chloride (table salt) from the donation of electrons by sodium to chlorine is an example of ionic bonding

doing this, chlorine becomes a negatively charged chloride ion (Cl^-) as it has more electrons than protons and sodium becomes Na^+ ion (positively charged) because it has one less electron. Atoms that lose electrons and become positively charged are called *cations* (to remember this think of the t as + and positive) while those that accept electrons and become negatively charged ions are referred to as *anions*. The resultant compound is sodium chloride (NaCl) or table salt.

A covalent bond is formed when two or more atoms share a pair of electrons and form a bonded substance. Figure 2.4 provides examples of covalent bonds where the atoms involved share one pair of electrons (a single covalent bond), two pairs (double covalent bond) or three pairs (a

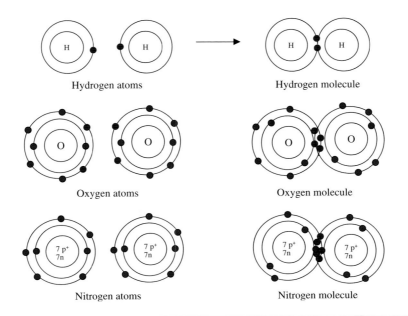

Hydrogen atoms   Hydrogen molecule

Oxygen atoms   Oxygen molecule

Nitrogen atoms   Nitrogen molecule

Figure 2.4 Covalent bonding in single, double and triple bonds to complete outer rings of electrons for hydrogen, oxygen and nitrogen

triple covalent bond). A simple way to remember the specific bonding of the common elements for human nutrition is the acronym HONC: Hydrogen (single bond), Oxygen (double bond), Nitrogen (triple bond) and Carbon (quadruple bond). Each of the gases in Figure 2.4 will exist in this structure when possible rather than as separate atoms. The sharing of electrons enables each atom to complete its outer shell without losing or gaining electrons.

Chemical bonding between atoms of the same type produces *molecules*; bonding between atoms of different types forms a *compound*, e.g. covalent bonding between two hydrogen atoms and an oxygen atom to form water. The formation of water and the covalent bond created is depicted in Figure 2.5.

Key point

Matter (anything that has a mass) is made up of atoms that have joined together through chemical bonding. There are two main types of chemical bonding: ionic and covalent. In ionic bonding atoms donate or accept electrons to complete the outer shell of their electrons, whereas in covalent bonding atoms share electrons to complete the same process. This enables the atoms to reach a chemically more stable (less reactive) state.

Energy from food

The human body is formed by the bonding of atoms to create larger matter, and our food – whether it be from plant or animal sources – is produced in the same way. The food we take into our bodies to maintain health and to perform our adventure sports is broken down and stored in order to provide the energy we need for living. The chemical bonds made between atoms store energy that can be liberated by breaking these bonds. The chemical reactions that take place in the body are of two types: *exergonic* i.e. energy-liberating or *endergonic* i.e. require energy to drive them. The food we eat has energy contained within it and our bodies break the food down and reassemble it in order to harness this energy. The energy stores we create are through endergonic reactions storing energy that we can release later through exergonic reactions.

Energy is the capacity to do work and two types of energy exist: potential (stored) energy or kinetic (motion) energy. There are six forms in which energy can be stored or used (Table 2.2). Energy cannot be destroyed, but can be changed from one form to another but always remaining as either potential or kinetic energy. This represents the *first law of thermodynamics*, i.e. the conservation of energy. In accordance with this principle, the human body does not create or destroy energy, merely transforms it from one form to another. When taking part in an adventure sport, our movements are brought about through chemical

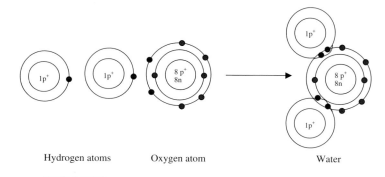

Hydrogen atoms   Oxygen atom   Water

Figure 2.5   Covalent bonding: the formation of water

Table 2.2 The forms of energy

Form of energy	Examples
Chemical	The bonds made in food or by our bodies store and release energy. Fossil fuels also represent a store of chemical energy.
Nuclear	The splitting of atoms through nuclear fusion releases energy that can be harnessed to drive turbines for the creation of electricity, such as that carried out at nuclear power stations.
Mechanical	Physical work or movement in humans and animals. Wind turbines carry out physical work to create electrical energy for our homes and industry.
Heat	The use of electricity and burning of fossil fuels can be used to create heat energy. In our bodies, the reactions that take place to enable us to climb or kayak release heat energy as a by-product and help to keep us warm in cold environments.
Light	The sun during the day, a gas lantern, torch or fire burning at a camp-site in an evening during an expedition or electric power in our homes can provide light energy.
Electric	Created from many sources, energy which can be used to power domestic and industrial machinery.

reactions that allow mechanical work to take place but also release heat energy as a by-product of the process.

It is possible to measure the stored energy in food using a bomb calorimeter; an example of which is shown in Figure 2.6. In a bomb calorimeter, the food to be tested is burned completely within a chamber containing oxygen. The burning of the foodstuff will result in the release of heat energy.

The heat energy released by a particular foodstuff is measured by recording the rise in temperature of the water surrounding the bomb chamber. The rise in water temperature has traditionally been measured in *calories*. A calorie represents the amount of heat required to raise the temperature of 1 g of water by 1 °C. As a gram of water is very small, the term kilocalorie is most commonly used when referring to the energy contained within our foods. A kilocalorie (kcal) represents the amount of heat required to raise the temperature of 1 kg of water by 1 °C. Alternatively, calorie is printed with a lower case c and kilocalorie with an upper case C. When looking at the label on a food product, the energy value of the food is presented in kilocalories. In recent years there has been a move to adopt the joule or kilojoule as the unit of energy value for food. The *système international* (SI) unit is the joule as the measurement of work and food labels therefore also show the energy value of the food in kilojoules (kJ) as well as kcal. To convert from kcal to kJ simply multiply by 4.184, or to convert from kJ to kcal divide by 4.184.

Calculation of kcal and kJ using the 4.184 conversion:

Baked beans contain protein (4.6 g), carbohydrates (12.9 g) and small amounts

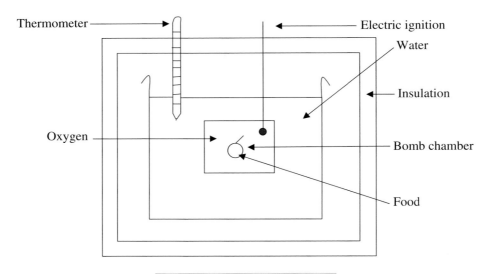

Figure 2.6   Bomb calorimeter

of fat (0.2 g). The energy value listed is 301 kJ or 72 kcal per 100 grams.

To convert from kcal to kJ multiply by 4.184:

$$72 \times 4.184 = 301.248 \text{ kJ}$$

To convert from kJ to kcal divide by 4.184:

$$301 \div 4.184 = 71.94 \text{ kcal}$$

It is possible through the use a bomb calorimeter to identify the energy value of the main macronutrients. When 1 g of an average fat is combusted in a bomb calorimeter, it liberates 9.2–9.5 kcal of energy. Carbohydrates and proteins combusted in the same way yield 3.7–4.2 kcal and 5.65 kcal, respectively.

The human body, however, is not as efficient at digesting and processing the foodstuffs as a bomb calorimeter. As a consequence, research has identified the figures listed in Table 2.3 as appropriate for the energy value of food when combusted within the body. An example of the energy value of the carbohydrates, fats and protein found in baked beans follows.

Table 2.3   Net calorific values of the macronutrients

Macronutrient	Calorific value from 1 g (kcal)
Carbohydrate	4
Fat	9
Protein	4

Calculation of energy value of carbohydrates, fats and protein in baked beans:

Baked beans contain protein (4.6 g), carbohydrates (12.9 g) and small amounts of fat (0.2 g). The energy value listed is 301 kJ or 72 kcal per 100 g.

Energy provided by protein

$$= 4.6 \times 4 \text{ kcal} = 18.4 \text{ kcal}$$

Energy provided by carbohydrates

$$= 12.9 \times 4 \text{ kcal} = 51.6 \text{ kcal}$$

Energy provided by fats

$$= 0.2 \times 9 \text{ kcal} = 1.8 \text{ kcal}$$

Total energy value of baked beans

$$= 71.8 \text{ kcal}$$

As part of this introduction to nutrition, the range of diets that humans can survive and indeed thrive on has been discussed. This was followed by a review of the need for food, what food is made of, how the energy in food is stored and how the energy value of food can be measured. The following sections in this chapter will look at each of the macronutrients and micronutrients necessary for healthy living, how our food is digested and absorbed into our bodies, the necessary components for a healthy diet and general dietary recommendations.

2.2 Carbohydrates

Carbohydrates are compounds formed from carbon, hydrogen and oxygen. The main purpose of carbohydrate in the diet is as an energy source for the chemical reactions that take place in the body. When exercising, particularly during high-intensity short to medium-term exercise, carbohydrates provide the main fuel supply. Red blood cells and the nervous system rely entirely on carbohydrate energy supply. The basic currency for carbohydrate-based energy production in our bodies is the simple sugar glucose. Carbohydrates can be classified as simple sugars or complex carbohydrates. The two forms of simple sugars are *monosaccharides* and *disaccharides*, while complex carbohydrates are formed from chains of simple sugars and are referred to as *polysaccharides*.

Fruits, grains and vegetables all contain carbohydrates. Plants create glucose during photosynthesis, when carbon dioxide (CO_2) and water (H_2O) are combined in the presence of chlorophyll and sunlight. Chlorophyll absorbs sunlight and with the help of a variety of enzymes enables glucose to be produced. A vitally important by-product of this process is the generation of oxygen, which sustains all animal life.

Monosaccharides

As with all carbohydrates, monosaccharides contain carbon, hydrogen and oxygen atoms, hence the short form CHO for the carbohydrate group. There are monosaccharides with three, four and seven carbon atoms and there are over 200 different types. However, the most important for humans are classified into five (pentose) and six (hexose) carbon rings, based on the number of carbon atoms present. The chemical formula for both types of monosaccharide is the same for each group: $C_5H_{12}O_6$ for pentose and $C_6H_{12}O_6$ for hexose. The 'ose' in each of the sugars identifies them as such.

Figure 2.7 provides the chemical linkage for the three nutritionally important hexoses: fructose, glucose and galactose. Often monosaccharides which bond in three-dimensional shapes are shown in a ring notation (Figure 2.8) as this is closer to their structure than as depicted in Figure 2.7. The differences between the simple sugars in this group come from the structural form and the distribution

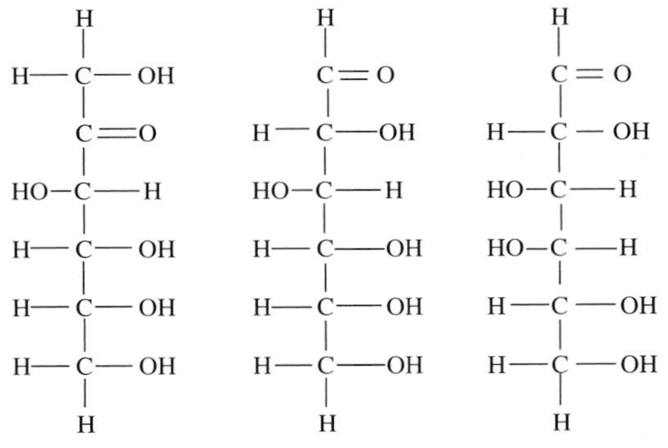

Figure 2.7 The chemical composition of monosaccharides: all three can be represented by the chemical formula $C_6H_{12}O_6$, however each has a different structure

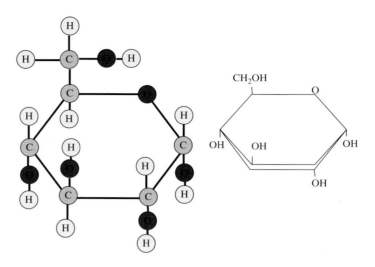

Figure 2.8   Three-dimensional and ring representations of the chemical structure of glucose

of the hydrogen and oxygen atoms. Glucose (sometimes called dextrose) is a building block for more complex sugars such as cellulose (found in plants) and polysaccharides, and is perhaps the most plentiful organic substance on earth. It occurs naturally in plants and animals as part of digestion. Fructose, another common simple sugar is found in fruits and honey. Galactose is most commonly found as the building block for the milk sugar lactose but is also present in peas. Ribose and deoxyribose are important pentoses for humans as they form part of the structure for DNA (deoxyribonucleic acid) and RNA (ribonucleic acid), the blueprint and genetic coding for the development of each cell in our bodies.

Dissaccharides

As the name suggests, disaccharides are formed when two monosaccharides combine. In terms of nutrition, sucrose, maltose and lactose are the most important disaccharides. As can be seen from Figure 2.9, each of these two sugar units is formed when two hexoses are combined. Sucrose is most commonly found in sugar beet or cane, maltose can be found in germinating grain seeds and lactose is produced in the milk of mammals.

Key point

Carbohydrates exist in three main groups based upon the number of single sugar units (saccharides) contained within each molecule. Monosaccharides, the smallest carbohydrate unit, include glucose and fructose. Dissaccharides comprise two sugars and include sucrose (table sugar) and maltose (malt sugar). Polysaccharides occur in plants and animals, where examples of the chain include starch and glycogen, respectively. Polysaccharides are the storage forms of simple sugars and are more complex in their structure due to the number of molecules involved. These carbohydrates provide energy for cells in plants and animals.

Polysaccharides

Polysaccharides are formed from chains of glucose units ranging in size from three to several hundred

Monosaccharide   and   Monosaccharide   =   Dissacharide

$C_6H_{12}O_6$ (Glucose) and $C_6H_{12}O_6$ (Fructose) form $C_{12}H_{22}O_{11}$ **Sucrose** (table sugar) + H_2O

$C_6H_{12}O_6$ (Glucose) and $C_6H_{12}O_6$ (Glucose) form $C_{12}H_{22}O_{11}$ **Maltose** (malt sugar) + H_2O

$C_6H_{12}O_6$ (Glucose) and $C_6H_{12}O_6$ (galactose) form $C_{12}H_{22}O_{11}$ **Lactose** (milk sugar) + H_2O

Figure 2.9 The structure of dissaccharides

thousand. There are two forms of polysaccharide: plant and animal. In animals only a small proportion (approximately 10–20 g for a sedentary adult) of the glucose available to produce energy is stored in the blood. The majority of glucose supplies held within the body are stored in the form of glycogen (around 500 g for a sedentary adult). Glycogen is the storage form of glucose for mammals with anywhere from several hundred to 25–30 000 glucose linked to form chains. The process of converting glucose to glycogen (*glycogenesis*) removes water from the glucose, making it smaller and lighter to store. Glycogen (Figure 2.10) is stored in the liver where it is able to increase levels of blood glucose when required and in the muscles where it is used directly for energy production. Most of the body's glycogen stores are held in the muscles (about 400 g); the remaining 100 g is stored in the liver.

Key point

Each gram of carbohydrate will release approximately 4 kcal of energy, therefore the stores of glycogen within the muscle and liver (around 500 g) can provide 1500–2000 kcal. This supply of energy would be sufficient to climb Ben Nevis at a good speed.

Glycogen is stored in long highly branched chains of glucose molecules, as can be seen in Figure 2.10. The hormones glucagon and insulin control the rate of glycogen formation and recreation of glucose in the liver through the process of *glycogenolysis*. The exact nature of the functioning of these hormones is discussed further in the following chapter.

Plant polysaccharides include starch, fibre and cellulose, all of which can be used to store the glucose created through photosynthesis. Starch is synthesized in two forms, *amylose* and *amylopectin* chains, in order to store the plant's carbohydrate. Food such as pastas, grains, beans and potatoes all contain large amounts of starch. Amylopectin chains are highly branched in a similar way to the mammalian storage of glucose as glycogen, whereas amylose chains are long and non-branched. This branching is linked to the rate at which the starch can be digested when eaten. Amylose starch chains are slower to digest and when present in a foodstuff decrease the glycemic index. Amylopectin chains, on the other hand, digest more readily and are associated with relatively higher glycemic-indexed foods.

Fibre, not traditionally included as one of the six necessary nutrients for healthy living, has been recognized as an important omission from many western diets. There are a variety of different forms of fibre within plants, including cellulose and pectin. Within plants, fibre is used to give the plant its structure and is found within the stems, roots, leaves and fruit. The fibrous, hence the name, nature of this polysaccharide means that it is resistive to the digestive processes within mammals and as such assists with digestive health. It is thought to carry out this function by increasing the bulk of human waste products (fibre contains a relatively high water content) and decreasing

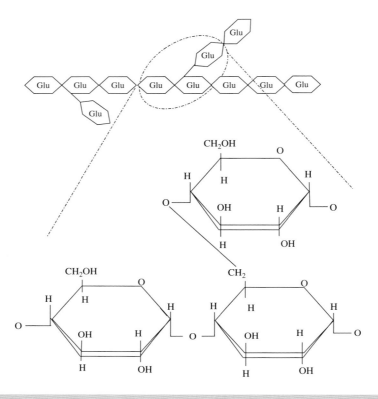

Figure 2.10 The highly branched chains that constitute glycogen, the storage form of glucose

the time taken for food to exit the body. Western diets, often with highly processed foods, have become low in fibre and this is thought to have had a negative impact on digestive tract health with increases in associated cancers and haemorrhoids. Increasing the fibre in diet, by changing to whole wheat grains, pastas, cereals and breads, has been shown to improve digestive function.

Foods containing carbohydrate

There is a wide variety of foods that have a high proportion of carbohydrate within them. Examples of these foods are listed in Table 2.4. When making decisions about which of these foods to include within a healthy diet, it is appropriate to identify the nature of the carbohydrate: how complex it is and its glycemic index (see the following section).

Table 2.4 Foods with a high carbohydrate content

Simple or processed carbohydrates	Complex carbohydrates
Table sugar	Bananas
Honey	Apples
Gatorade	Kidney beans
Powerade	Lentils
Power bars	Chick peas
White bread	Whole wheat or multigrain bread
Quick-cook and white rice	Wholegrain brown rice
Pasta	Whole wheat pasta
Oven-cook chips	Potatoes

Glycemic index

It is recommended for healthy active adults that 55–60 per cent of our food intake should be in the form of carbohydrate. There is another factor that needs to be taken into account when making decisions about which carbohydrates should form the dietary staples, and that is the *type* of carbohydrate. In its simplest context, this relates to the interplay between simple and complex carbohydrates that form the basis of our diet. However, research concerning the glycemic index of foods indicates that this too should be addressed when making decisions about the range of carbohydrates to be eaten. Simple carbohydrates refer to mono and disaccharides and complex carbohydrates to the polysaccharide starch.

Complex carbohydrates generally take longer to be digested and create a lower rise in blood sugar levels; they are therefore beneficial for short-term performance and long-term health. Pasta and potatoes would be examples of complex carbohydrates and sugary drinks are an example of a simple carbohydrate. The division between simple and complex carbohydrates works as a general rule, however, the reality is more complicated. Potatoes, for example, are more quickly absorbed than pasta, and in addition the more processed a starch food item becomes during production the less complex it becomes in its chemical structure. As a consequence, the use of the glycemic index has become popular when describing the complexity of a food item. The glycemic index is based on the rate of digestion and absorption of a food item and the rise in blood sugar level it creates. The glycemic index of a range of foods can be seen in Table 2.5. White bread and glucose achieve a score of 100.

Table 2.5 The glycemic index of a range of foods; glucose and white bread achieve a score of 100

High glycemic index (>80)	Lower glycemic index (<80)
Table sugar	Kidney beans
Honey	Chick peas
Doughnuts	Lentils
Bagels	Baked beans
White rice	Whole grain brown rice
Watermelon	Pasta
Pineapple	Figs
Raisins	Dates
Carrots	Peaches
Potatoes	Sweet potatoes
Rice cakes	Plums
Waffles	Milk
Rice Crispies	Yogurt
Cornflakes	All Bran
Shredded Wheat	Apples
Corn tortilla chips	Peanuts

within the bloodstream should be taken into account. Carbohydrates differ in the rate at which they are broken down and enter the bloodstream. Adventure sports performers need a variety of forms of carbohydrates to provide and restore energy levels.

Key point

When deciding upon the most appropriate carbohydrates to include in your diet, the complexity of the carbohydrate, its glycemic index and the rate at which it is needed

2.3 Fats

Fats or lipids (from *lipos*, the Greek word for fat) are our most dense form of energy store, able to provide on average 9 kcal of energy per gram compared to 4 kcal for carbohydrates and proteins. Similar to carbohydrates, fats comprise carbon, hydrogen and oxygen. However, their structure and the nature of their bonding are different, creating energy-dense molecules that have limited water solubility. As well as an energy source, fats provide protection and support for vital organs, are essential parts of each cell's membrane, help increase the speed of nerve impulse transmission, assist in the preservation of body heat and are the storage and transport system for fat-soluble vitamins. Lipids can be classified as fatty acids, triglycerides (sometimes referred to as triacylglycerols), sterols which include cholesterol, phospholipids and lipoproteins.

Common foods containing fats include meat, diary products and plant-based oils. Fats represent the energy store for the body and excess carbohydrate and protein intake is converted to fat and stored within specialist adipocytes (fat cells) that form the adipose tissue. Fat stores in muscle, blood and adipocytes typically hold 90 000–100 000 kcal of energy. This represents a plentiful energy supply, especially when compared with the 1500–2000 kcal of stored carbohydrate.

Key point

Fats or lipids are the body's main energy store. The stores are held within the adipocytes, specialist fats cells throughout the body, the vast majority (over 90 per cent) held as triglycerides. In comparison to carbohydrates and proteins, fats are energy dense. For every gram of an average fat, 9 kcal of energy can be liberated compared to 4 kcal for every gram of carbohydrate or protein.

Dietary fat is an essential macronutrient. However, excess fat intake has recently been associated with a rise in obesity levels and an increase in the incidence of cardiovascular disease in many western societies. In the UK, a great emphasis has been placed upon the need to decrease dietary fat consumption as a percentage of total food intake. In 1986–1987, fat consumption represented 40 per cent of the average British citizen's food consumption. With an increasing emphasis on decreasing

fat intake this had dropped to 35.3 per cent by 2001 and the Department of Health targeted further decreases to below 35 per cent in their 2004 Choosing Health campaign.

The UK Government also targeted reductions in saturated fat consumption (to decrease from 16–13 per cent) as well as reductions in total fat intake. It is interesting to note that the American Heart Association have further recommended a maximum of 30 per cent total fat in the diet with 6–9 per cent saturated fats. As excess lipid levels have been associated with an increase risk of colon, rectum and prostate cancer, the American Cancer Society have moved further to suggest a total fat intake of 20 per cent to reduce these risks. For an active adventure sports performer, maintaining fat intake at a level below 30 per cent of total calorific intake is believed to be beneficial for health. In addition to this, reducing cholesterol and saturated fat intake as a part of this 30 per cent would appear to be advantageous.

Key point

Fats form an essential part of our diet and perform a large number of roles within the body. However, excessive fat intake and foods high in saturated fats and cholesterol have been shown to be harmful to health. Foods with a higher proportion of monounsaturated or polyunsaturated fats and HDL rather than saturated fats or LDL are healthier choices. As part of a diet for adventure sports performers, a fat intake of up to 30 per cent of total calorific intake is thought to be beneficial for health and performance.

Fatty acids, glycerol and triglycerides

Fat is stored in the adipocytes as triglycerides which are composed of a glycerol molecule (which is not a lipid as it is highly water soluble, but enables fatty acids to link together) and three fatty acids. The process of *lipolysis* occurs when there is a need to access the triglyceride stores such as during low intensity exercise. Through this process, which takes place in the adipocyte, the glycerol molecule is cleaved from the fatty acids allowing them to be used for energy production.

Fatty acids travel in the blood attached to albumin, a blood protein, to where they are needed as *free fatty acids*. Fatty acids exist in saturated and unsaturated forms. The saturation of a fatty acid refers to the number of hydrogen atoms attached to the carbon chain. Figure 2.11 shows the chemical structure of the saturated and unsaturated palmitic and linolenic fatty acids. Carbon makes four covalent bonds to share electrons. In a saturated fat, the carbon atoms are saturated with hydrogens such that each carbon atom is attached to its neighbouring carbon atoms and two hydrogen atoms. It cannot hold any more hydrogen atoms

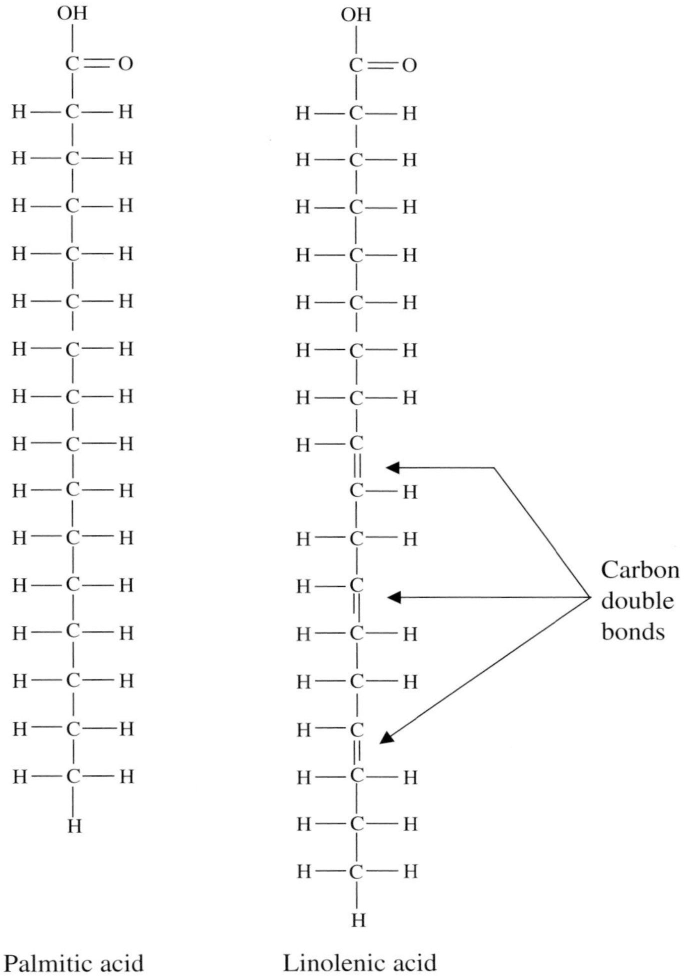

Figure 2.11 Chemical structure of palmitic acid ($C_{16}H_{32}O_2$, a saturated fatty acid) and linolenic acid ($C_{18}H_{30}O_2$, an unsaturated fatty acid)

and remain a part of the fatty acid chain, and is therefore a saturated fatty acid. In an unsaturated fatty acid, one or more of the carbon atoms make a double bond with a neighbouring carbon atom, still creating four covalent bonds. However, it remains unsaturated of hydrogen atoms. In the case of linolenic acid it has three unsaturated (with hydrogen) carbon atoms.

The ratio of hydrogen atoms to oxygen atoms reveals part of the structural differences between carbohydrates and fats. Fats have a higher density of hydrogen atoms than oxygen atoms compared to carbohydrates. For example glucose, with a chemical formula of $C_6H_{12}O_6$ has an H:O ratio of 2:1 whereas the fatty acids in Figure 2.11 palmitic acid ($C_{16}H_{32}O_2$) and linolenic acid ($C_{18}H_{30}O_2$), have ratios of 16:1 and 15:1, respectively.

Fatty acids can be monounsaturated, where they have one carbon double bond, or polyunsaturated, i.e. have more than one carbon double bond. Linolenic acid is a polyunsaturated fatty acid. Sources of saturated fats include cheese, whole milk, beef and pork. Olive oil contains largely monounsaturated fat and sunflower oil and corn oils are predominantly polyunsaturated fats (Figure 2.12). Oils from plants, such as coconut oil, sunflower oil and olive oil contain all three forms of fatty acid but in varying amounts. Figure 2.12 shows the percentages of each fatty acid in these three plant-derived oils.

In terms of health, dieticians recommend that saturated fats are best kept as a small percentage of total food and lipid intake. Research strongly indicates that diets rich in the saturated fats found in items such as red meats and many processed foods are linked with an increased risk of cardiovascular disease. It is interesting to note that in order to create solid fats (at room temperature) for margarine, hydrogen is forced through a vegetable oil to enable it to pick up more hydrogen atoms and solidify. This process increases the hydrogen saturation level of the oil. This is known as hydrogenation, and as a part of this process the carbon chains can move their position and become trans-fatty acids.

The British Heart Foundation (BHF) has identified research to suggest that such trans-fatty acids have a negative effect on cholesterol levels, another factor linked with the development of cardiovascular disease. In 1993, the Food and Drug Administration (FDA) in the US have implemented a law to require food manufacturers to

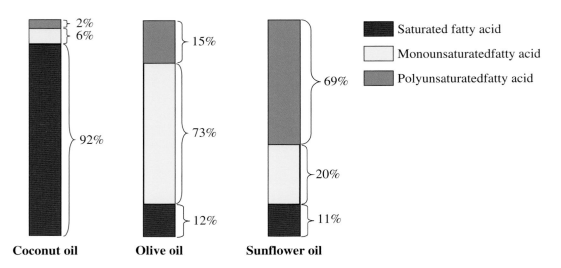

Figure 2.12 The approximate percentage of saturated, monounsaturated and polyunsaturated fats in plant-derived oils

label trans-fats within the contents. However, there is no such law in the UK. The BHF suggest consumers look at the labels of food for hydrogenated fat or hydrogenated oils. The lower they come down the list the smaller percentage of the ingredients they represent. We do however require lipids as part of our diet, but due to the health implications of some fatty acids it is perhaps beneficial to maintain a diet where the majority of fats included are from mono or polyunsaturated sources.

Sterols

Sterols or steroids are lipid derivatives and have a different chemical structure to other lipids, although they still have low solubility in water. Steroids are formed from four ring carbon chains and include cholesterol and the sex hormones oestrogen, testosterone and progesterone. Cholesterol, often seen as only a 'bad' substance, is an important substance found only within animals as it is the precursor to many steroids including the sex hormones. The chemical structure of testosterone and the four carbon rings can be seen in Figure 2.13.

Cholesterol has implications in terms of health and these are discussed in the following section. Cholesterol represents one of the lipoprotein groups and it is in this role that its presence can be harmful to health. The lipoproteins form one group linked with the phospholipids.

Phospholipids

Phospholipids are involved in a variety of functions within the body, including providing the insulating sheath around nerve fibres, assisting in blood clotting and maintenance of cell membranes and structure. Phospholipids, as the name suggests, comprise fatty acids in a similar way to triglycerides with the modification that one of the fatty acids is replaced with a phosphate group and usually a nitrogen group. Figure 2.14 provides an illustration of a phospholipid, with the phosphate head being hydrophillic (water loving) and the fatty acid tail being hydrophobic (water hating). These unique properties enable phospholipids to create a sealed membrane that separates cells from their outside environment. The forces of attraction are strong enough that no covalent bond is needed between adjacent molecules to maintain the structural integrity of the cell membrane.

Lipoproteins

Lipoproteins, usually metabolized in the liver, are formed when phospholipids or triglycerides are joined with a protein and also include cholesterol. The major role of lipoproteins is to enable blood transport of fats. Lipoproteins, unlike lipids, are water soluble and as such can be carried within the plasma of the blood. There are a number of

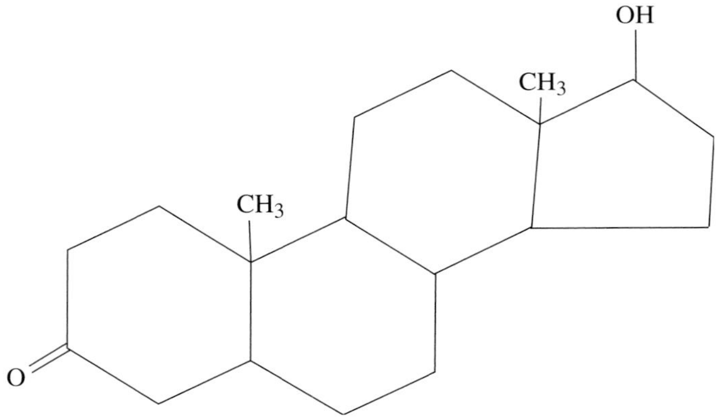

Figure 2.13 The chemical structure of testosterone

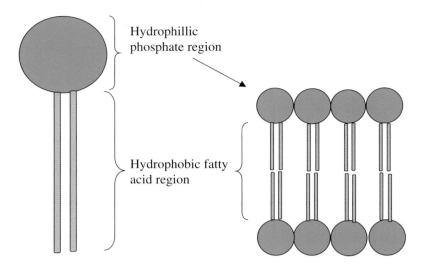

Figure 2.14 The structure of a phospholipid and the close bonding of phospholipids to create the cell membrane for animals

lipoproteins including chylomicrons, low density lipoproteins (LDL) and high density lipoproteins (HDL). Chylomicrons are formed in the small intestine and are the blood lipid carriers for fats that have been digested from our food. Muscle and adipose cells surrounding blood vessels release the enzyme lipoprotein lipase and this enables the chylomicron to release its triglyceride store to those tissues for use as an energy source. The remnants of the chylomicron are then returned via the bloodstream to the liver for reuse and excretion from the body as necessary.

Low density and high density lipoproteins also act as lipid carriers within the blood. However, the proportion of cholesterol they contain differs. Low density lipoproteins are formed in the liver and comprise 45 per cent cholesterol, which is about half the total blood cholesterol, phospholipids and some triglycerides. High density lipoproteins, formed in the liver and small intestine, comprise 55 per cent protein and lesser amounts of cholesterol and phospholipids. When LDL and HDL enter a cell they are broken down into their component parts and used within the cell. Within the blood LDL and HDL have a very different impact on the arterial walls. Seen as the 'bad'

cholesterol, LDL has been found to be strongly attracted to the tissues of the arterial wall where they damage the lining and create plaque (comprising cholesterol-based deposits) that eventually narrow the artery. High density lipoproteins, on the other hand, appear to reverse this process by clearing cholesterol from the arterial walls and blocking the entrance of LDL to the arterial cells. Increasing the mono and polyunsaturated fats in diet as well as not smoking and taking regular exercise appear to decrease LDL and increase HDL levels. Having low total cholesterol levels and high proportions of cholesterol as HDL lowers the risk of cardiovascular disease.

Key point

Cholesterol is carried in the body as part of two main substances: high density and low density lipoproteins. Lipoproteins are formed from the joining of a fat and a protein. Cholesterol within the body is used as a building block for the sex hormones and is a necessary part of our diet. Low density lipoproteins (LDL) carry a higher proportion of cholesterol and appear to attach to the cell walls of the arteries, helping to form plaques that lead to a narrowing of the arteries associated with cardiovascular disease. High density lipoproteins (HDL) appear to be more beneficial to the body, stripping away plaque deposits and blocking LDL attachment to arterial walls. Avoiding smoking, increasing mono and polyunsaturates as a proportion of total fat intake and regular exercise appear to increase the HDL to LDL ratio.

2.4 Proteins

Proteins, from the Greek word *proteios* meaning of first or prime importance, are an essential component of a balanced diet and can be found from plant and animal sources. Within the body, proteins are broken down in the small intestine into amino acids from where they are absorbed to become involved in a myriad of roles throughout the body. Proteins are involved in nearly every reaction that takes place within the body. They play a major role in the structure of all body tissues, being the building blocks for synthesizing hair, skin, bones, tendons and muscles. Proteins form the basis for all enzymes and many hormones, provide the necessary amino acids to enable blood clotting, muscular contraction, growth and repair, regulation of metabolism and immune function. In addition, proteins provide an alternative to carbohydrates and fats for energy production and play a major role in oxygen and carbon dioxide transport in the body.

Key point

Protein plays a crucial role in nearly all physiological processes within the body. Amino acids are the building blocks for the many different proteins synthesized throughout the body. There are 20 amino acids necessary for human life that can be arranged in thousands of different combinations to create a myriad of protein structures. Of the amino acids found in humans, 8 are essential and 12 are non-essential (meaning they can be synthesized within the body from other amino acids) for adults. A varied diet can provide all the necessary amino acids from plant and animal sources.

Amino acids

The building blocks for proteins are *amino acids*. Figure 2.15 reveals the structure of a general amino acid. Amino acids have a similar structure to carbohydrates and fats, in that they comprise atoms of carbon, hydrogen and oxygen. This section of the amino acid (the acid section) is the *carboxyl group* containing one carbon, one hydrogen and two oxygen atoms (COOH). Amino acids differ from carbohydrates and fats in that they contain a nitrogen atom within the *amine group* of the molecule.

Each one also contains an *R group* that is unique to the specific amino acid. For glycine, the simplest amino acid, the R group contains a hydrogen atom covalently bonded to the carbon atom (Figure 2.16). In alanine, a structurally more complex amino acid, the R group includes a further carbon linked with three hydrogen atoms.

Figure 2.17 depicts the structure of the amino acid alanine. The R group differs for all amino acids and can include sulphur, phosphorus, cobalt or iron in its structure. Amino acids, as the building blocks for proteins, are covalently joined by peptide bonds. In Figure 2.17 the formation of a peptide bond between the amino acids glycine and alanine is displayed. The bond is made between the carboxyl group of one amino acid (in this example glycine) and the amine group of the other (alanine). The synthesis of all dipeptide molecules (two bonded amino acids) causes the liberation of one molecule of water.

The amine group which contains nitrogen

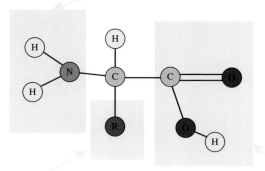

The R group which contains a variety of components for each amino acids; the component differentiates one amino acid from another

The carboxyl group which is the acid part of the amino acid, similar in structure to carbohydrates and fats

Figure 2.15 The general structure of an amino acid

The amine group

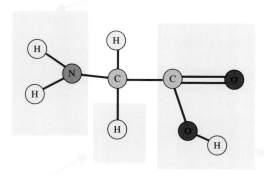

The R group; for glycine the side atom is hydrogen

The carboxyl group

Figure 2.16 The specific structure of glycine, the simplest amino acid

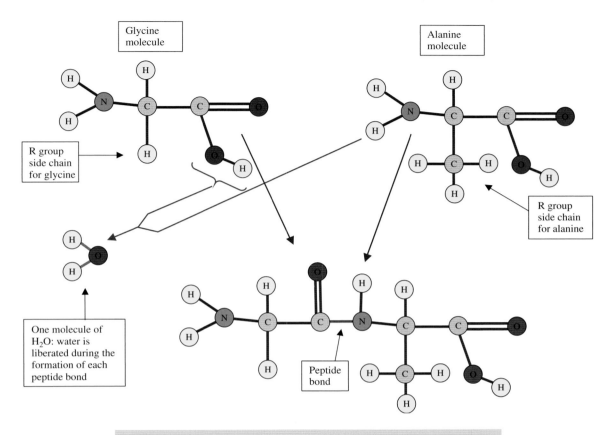

Figure 2.17 The formation of a peptide between glycine and alanine

When three amino acids bond to form a tripeptide molecule and two peptide bonds, two molecules of water are liberated. This continues for each amino acid that joins the chain for polypeptides and eventually proteins. Figure 2.18 displays the formation of protein from amino acids. The named amino acids in the figure are a sample of the 20 amino acids required by the body. The required amino acids are bonded in diverse combinations dictated by the type of protein to be synthesized. There are thousands of different protein structures synthesized from the 20 amino acids to carry out the myriad of roles which proteins play within the body.

In comparison to carbohydrates and fats, proteins are very large molecules. Looking back at the sections on carbohydrates and fats, it is possible to see that a glucose molecule has the formula $C_6H_{12}O_6$ and palmitic acid $C_{16}H_{32}O_2$, respectively. By comparison, the fairly small protein haemoglobin has the chemical formula $C_{2952}H_{4664}O_{832}N_{812}S_8Fe_4$. The structure of a protein has four possible levels in order to create the necessary long chain within a relatively confined space and to allow coupling across amino acids. Figure 2.19 displays the four levels of structure available for amino acids. The example provided is similar to the structure for the protein-based molecule haemoglobin.

The 20 amino acids necessary for human growth and health are shown in Table 2.6. The amino acids are, by convention, usually divided by their essential and non-essential nature. The terms are slightly misleading as all 20 amino acids

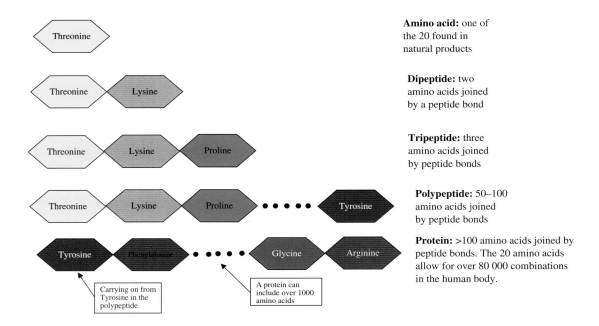

Amino acid: one of the 20 found in natural products

Dipeptide: two amino acids joined by a peptide bond

Tripeptide: three amino acids joined by peptide bonds

Polypeptide: 50–100 amino acids joined by peptide bonds

Protein: >100 amino acids joined by peptide bonds. The 20 amino acids allow for over 80 000 combinations in the human body.

Figure 2.18 The formation of dipeptides, tripeptides, polypeptides and proteins from amino acids

Table 2.6 The essential and non-essential amino acids

Essential amino acids	Non-essential amino acids
Isoleucine	Alanine
Leucine	Arginine
Lysine	Asparagine
Methionine	Aspartic acid
Phenylalanine	Cysteine
Threonine	Glutamic acid
Tryptophan	Glutamine
Valine	Glycine
Histidine*	Proline
	Serine
	Tyrosine

*Histidine is an essential amino acid for children; however, adults have the ability to synthesize it within the body.

are essential for a healthy diet. The terms refer to the fact that our bodies can synthesize the non-essential amino acids and so we do not have to rely on obtaining them from our diet; essential amino acids must be included, however. In terms of a completely balanced diet we should consume foods that contain a full range of amino acids on a regular basis and not wait for our bodies to synthesize them.

Many sports people, particularly those involved in power and strength events, believe mistakenly that protein supplements are necessary for muscle growth and function. Research indicates that this is not the case and we should normally be able to provide all the necessary amino acids within a healthy balanced diet. It is also a misconception that vegetarians cannot obtain all the amino acids from their diet. An increasing number of adventure sports participants have adopted vegetarian or nearly vegetarian diets and it is perfectly possible for them to provide all 20 amino acids within their normal food.

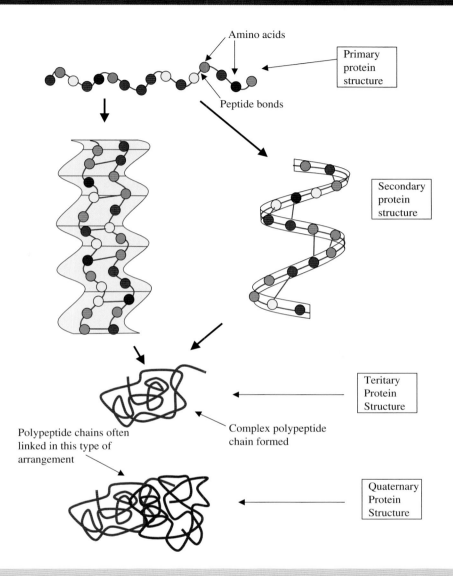

Amino acids

Primary protein structure

Peptide bonds

Secondary protein structure

Teritary Protein Structure

Complex polypeptide chain formed

Polypeptide chains often linked in this type of arrangement

Quaternary Protein Structure

Figure 2.19 The structure of proteins (from Tortora and Derrickson, *Principles of Anatomy and Physiology*, Eleventh Edition, 2006, reproduced by permission of John Wiley & Sons Inc.)

Plant and animal food sources contain protein. The difference between the two sources is that plant sources, with the possible exception of soybean proteins, are less protein dense and often contain an incomplete amino acid range compared to animal sources. Animal-based protein sources such as milk, eggs, chicken and fish are complete proteins as they contain all 20 amino acids. A vegetarian will have to eat a higher volume of food and from a greater range of sources for the same amount of protein compared to someone who consumes animal protein, but they can obtain all the necessary amino acids in a balanced diet. For instance, rice and beans as a meal choice can provide all 20 amino acids and is a common staple in South American countries such as Bolivia.

Key point

There is no evidence to suggest that protein supplements are necessary. A normal healthy diet, providing amino acids through animal, plant or combined sources, is normally sufficient for sedentary people, those who take regular exercise and elite performers.

Protein metabolism

As building blocks, amino acids undergo frequent change and renewal within the body. There are a variety of processes that take place to *anabolize* (create or build) and *catabolize* (dismantle) amino acids, polypeptides and proteins. Dietary protein broken down in the small intestine is absorbed into the bloodstream and from there taken up by cells that require them to anabolize cellular proteins. In the breakdown of a protein it is first catabolized into its amino acid components.

Amino acids can then be further broken down in a process called *deamination* which involves, as the name suggests, the removal of the amine group from the acid. The nitrogen element (the amine) can be removed from the body as urea or in a process called *transamination* be attached to a new *α-keto acid* (the remaining part of an amino acid when the amine has been removed) to create a new necessary amino acid. The remains of the deaminated molecule, the α-keto acid, could then be catabolized to provide energy.

Depending on the duration of the exercise an adventure sports performer is engaged in, and their diet before and during the activity, proteins are thought to provide up to 15 per cent of the necessary energy. Research has indicated that another way the body can do this is through the alanine-glucose cycle. During research into the fate of the amino acid leucine within working muscles, it was determined that during exercise it is oxidized (catabolized). The deamination of leucine allows the amine group to be combined with pyruvate (found within the muscle) to form the amino acid alanine. This enters the bloodstream and is transported to the liver where it is again deaminated, the amine being removed as urea and the pyruvate being converted to glucose for use as an energy source. The remaining α-keto acid within the muscle from leucine is used directly for energy production within the muscle, entering the Krebs cycle which is discussed further in Chapter 4. This process is shown diagrammatically in Figure 2.20.

2.5 Vitamins

Vitamins are a significant group of *micronutrients* and, in small amounts, are vital to optimal physiological functioning. While not formally discovered until the 20th century, it was understood as long ago as during the Greek Empire that there were substances in foods that could prevent diseases.

An example of this was when in 1747 James Lind (see Section 1.4) conducted an excellent experiment on HMS *Salisbury* to test the use of citrus fruit as a protection against the effects of scurvy. Before Lind's experiment and for some time afterwards (until his results were accepted), scurvy was rife in the British Navy. As we now know, scurvy was caused by a lack of Vitamin C in the diet during the long sea voyages. The symptoms of scurvy were bleeding and spongy gums, bleeding from almost every mucous membrane, depression and then physical incapacitation. In his experiment, which included a control group who did not receive the treatment, sailors were given daily rations of citrus fruits to test if it would ward off the effects of scurvy. The study proved successful. Once the results were accepted the British Navy were supplied with rations of citrus fruits, particularly limes, to protect against the disease. It was for this reason that the British acquired the name 'Limeys'. The actual discovery of the chemical structure of Vitamin C was not made until 1928.

The term Vitamin was originally developed by a Polish biochemist, Casimir Funk, who discovered the vitamin Thiamin in 1912. In 1905 Frederick Hopkins (who was also involved in the discovery and isolation of lactic acid; Section 1.2) found that keeping the husks on rice prevented the development of Beriberi. Moving on from this

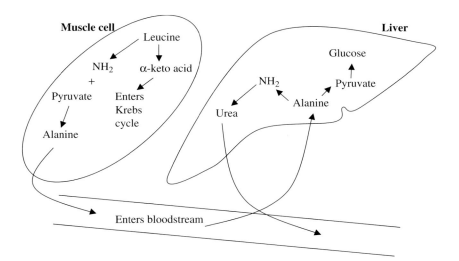

Figure 2.20 The alanine-glucose cycle: leucine is broken down in the muscle to provide an amine group for pyruvate to create the amino acid alanine, which enters the bloodstream and is carried to the liver where it is absorbed

discovery, Funk was able to isolate Thiamin (also called Vitamin B) as the substance responsible for preventing the development of the disease. Funk recognized the importance of this discovery and as a consequence coined the term 'vitamine' from the words *vita* meaning 'life-giving' and amine because the chemical structure included an amine (nitrogen group). This term was later changed to vitamin because, as more were discovered, it was found that some (e.g. Vitamin C) had no amine group as part of their chemical structure.

The dates when the vitamins were discovered are listed in Table 2.7. Vitamins were given alphabetical names according to the date of their discovery and isolation, although for some it is difficult to pinpoint the exact date. Some vitamins continue to be referred to by the alphabetical name, whereas others such as Niacin which was originally named Vitamin B_3 are known by the substance name. There are 13 essential vitamins, although there were more in the past when vitamins B_4, B_7, B_{10} and B_{11} were included. These substances were later found not to be essential to our diet and were consequently removed from the list.

Vitamins are organic compounds that include carbon, hydrogen and oxygen atoms in their structure, in common with carbohydrates, fats and proteins. The 13 essential vitamins are shown in Table 2.7; of these nine are water-soluble and four are fat-soluble vitamins. They were isolated between 1912 and 1941. Vitamins, while not providing an energy source for metabolism, have a wide number of crucial functions within the body. The major roles of vitamins include ensuring appropriate growth and development of cells such as epithelial cells (the tissues associated with the skin) and red blood cells (RBC), the eyes, nerves, muscles, teeth, bones and digestive tract, enabling efficient energy metabolism, assisting in the formation of enzymes and hormones and acting as antioxidants throughout the body. The foods that can provide the essential vitamins, the main function of each and their associated recommended daily allowances (RDA) are also listed in Table 2.7.

Although essential to our diet, they are needed in tiny amounts relative to the volume of food we consume each day. With a few exceptions, such as the need for pregnant women to supplement their diet with folic acid (which has been shown

Table 2.7   Vitamins and their functions

Vitamin	Discovered	Food sources	Function within the body	EU RDA (mg)
Vitamin A (Retinol)	1912	Preformed Vitamin A: milk, butter, egg yolk, liver. Proformed Vitamin A (beta carotene) green leaf vegetables, orange fruit and vegetables e.g. carrots	Repair and growth of body tissues (especially epithelial), maintains night vision, peripheral vision and an antioxidant. **Deficiencies:** Night blindness	0.8
Vitamin B1 (**Thiamin**)	1912	Yeast, wholemeal grains, oranges, whole grain rice, sunflower seeds, nuts, liver and pork	Coenzyme for metabolism, normal function of nervous system, digestion and muscles. **Deficiencies:** Beriberi	1.4
Vitamin B2 (**Riboflavin**)	1933	Milk and dairy products, green leaf vegetables, whole grains and yeast	Co-enzyme for metabolism (FAD), protects vision and skin. **Deficiencies:** cracks on corners of mouth, tongue	1.6
Vitamin B3 (**Niacin**)	1937	Meat, fish, eggs, peanuts, sunflower seeds and whole grains	Co-enzyme for metabolism (NAD), protects nervous function and skin. **Deficiencies:** Skin lesions	18
Vitamin B5 (**Pantothenic acid**)	1933	Fish, liver, chicken, oranges, bananas, nuts, milk, beans, whole grains and whole grain rice	Part of Co-enzyme A for metabolism. **Deficiencies:** Very rare, fatigue and nausea	6
Vitamin B6 (Pyridoxine)	1938	Fish, chicken, bananas, nuts, beans and whole grains	Co-enzyme for metabolism, normal function of nervous system. **Deficiencies:** Irritability, muscle twitching	2

Table 2.7 (*continued*)

Vitamin	Discovered	Food sources	Function within the body	EU RDA (mg)
Vitamin B8 (**Biotin**)	1940	Nuts, peas (black eye), egg yolks and green leaf vegetables	Co-enzyme for metabolism. **Deficiencies:** Very rare, fatigue and nausea	0.15
Vitamin B9 (**Folate, folic acid**)	1941	Name from fact green leaf vegetables (foliage) are rich in folate, also liver, bananas, beans and eggs	Co-enzyme for metabolism and DNA, RBC production, protects baby during pregnancy. **Deficiencies:** Anaemia	0.2
Vitamin B12 (Colbalamin)	1918	Richest sources are from animals, meat, dairy foods, eggs, but also from enriched breakfast cereals	Co-enzyme for DNA, RBC and nerve production. **Deficiencies:** Anaemia and nervous disorders	0.001
Vitamin C (Ascorbic acid)	1912/1928	Best sources are from plants, citrus fruits, broccoli, green leaf vegetables and peppers	Collagen formation, an antioxidant, promotes immune function and cell structure. **Deficiencies:** Scurvy	60
Vitamin D (Calciferol)	1922	Exposure to sunlight 10–20 min, 2–3 per week Fish oils, milk and eggs	Helps with absorption of calcium – promotes bone growth. **Deficiencies:** Rickets and osteomalacia	0.005
Vitamin E (Tochopherol)	1922	Most widely available in food, found in polyunsaturated oils, nuts, eggs and whole grains	Promotes growth and development and works as an antioxidant. **Deficiencies:** Very rare, possibly anaemia	10
Vitamin K (Phylloquinone)	1935	Green leaf vegetables, broccoli, egg yolk and cheese	Promotes correct blood coagulation (clotting). **Deficiencies:** Blood clotting, increased bleeding	–

Note: Water-soluble vitamins are shaded in blue fat-soluble vitamins are shaded in yellow. The name in **bold** is the usual name by which the vitamin is generally referred. Vitamin C was identified as a vitamin in 1912; however, it was not isolated as a substance until 1928.

to decrease the chances of a child developing a neural tube defect), a healthy balanced diet should provide all our vitamin requirements. The vitamin and dietary supplements industry is, however, worth millions of pounds worldwide. We need to be particularly careful about the levels of fat-soluble vitamins (A, D, E and K) which we consume. By taking a vitamin supplement we may consume more water-soluble vitamins than our bodies actually need, but we can excrete these in our urine. This is however not possible with vitamins A, D, E and K and there is a danger of a toxic reaction (vitamin toxicosis) if we include too many of these vitamins in our diet through supplementation.

Key point

There are thirteen essential vitamins, nine of which are water soluble (the B complex and vitamin C) and four fat soluble (Vitamins A, E, D and K). Vitamins are organic compounds that provide a wide number of vital functions to maintain health. Eating a wide range of healthy foods appears to provide sufficient levels of all these vitamins. An excess of the fat-soluble vitamins can cause vitamin toxicosis and should be avoided.

Vitamins as enzymes

Enzymes are chemical substances that act as catalysts during reactions. Catalysts speed up reactions, but remain unchanged at the end, ready to catalyse another reaction. Chemical reactions are necessary to build the tissues that form our bodies and the energy to fuel the work these tissues carry out. The enzymes created within our bodies are designed to speed up the reactions that are taking place to enable us to function more efficiently. The water-soluble vitamins play a key role in the structure of enzymes. Vitamins form coenzymes which combine with protein compounds to form enzymes that are particularly vital in energy

metabolism. Riboflavin forms part of the structure of flavin adenine dinucleotide (FAD), niacin part of nicotinamide adenine dinucleotide, pantothenic acid part of Coenzyme A, Vitamins B6 and B12 are involved in catalysing the breakdown of protein, Thiamin is involved in the breakdown of pyruvate to Acetyl Coenzyme A and Vitamin C forms part of enzymes involved in protein synthesis, all of which we will discuss again in later chapters.

Vitamins as antioxidants

One of the functions performed by vitamins A, C and E is to serve as antioxidants. They perform this antioxidant role against *free radicals* that are developed naturally as a part of metabolism or enter the body through cigarette smoking and air pollutants. In Section 2.1, the nature of electrons and their desire to complete the outer ring of electrons in order to become more stable was discussed. Free radicals act in the same way, seeking out additional electrons. Free radicals are highly reactive molecules (or parts of molecules) that contain missing electrons from their outer rings. Free radicals seek out other compounds with which to bond and can be very destructive to other cellular components as they search for additional electrons. Vitamins A, C and E react with and then remove free radicals from cells. If left unchecked in the body, free radicals can prevent the body removing carcinogens and promote increases in LDL levels, associated with the development of atherosclerosis and cardiovascular disease.

2.6 Minerals

The 26 elements that comprise the human body are listed in Table 2.1. The elements oxygen, carbon, hydrogen and nitrogen represent 96 per cent of the structure of the human body. The remaining 4 per cent is formed from the other 22 elements found within the human body. These 22 elements, which largely comprise metallic substances, are collectively known as the *minerals*. As a group, they are most often subdivided into the *major minerals* and the *minor* or *trace minerals*. The major and minor minerals are listed in Table 2.8 along

Table 2.8 Major, minor and lesser-known minerals and their functions

Mineral	Food sources	Function within the body	EU RDA (mg)
Calcium	Dairy products, green vegetables, beans	Bone and tooth growth, nerve transmission. **Deficiencies:** Osteoporosis	800
Phosphorus	Dairy products, chicken, whole grains	Energy metabolism, Bone and tooth growth. **Deficiencies:** Weakness	800
Potassium	Beans, meat, potatoes, bananas	Nerve conduction, fluid balance. **Deficiencies:** Cardiac arrhythmia	3000
Sulphur	Contained within dietary proteins	Liver function. **Deficiencies:** Very rare	–
Sodium	Contained within salt	Nerve transmission, fluid balance. **Deficiencies:** Muscle cramps	–
Chlorine	Part of salt (NaCl as chloride)	Fluid balance. **Deficiencies**: Very rare	–
Magnesium	Whole grains, green leaf vegetables	Enzyme formation. **Deficiencies:** Growth problems	300
Iron	Red meats, liver, chicken, eggs, whole grains	RBC production, enzyme formation. **Deficiencies:** Anaemia	14
Zinc	Meat, beans	Part of digestive and metabolic enzymes. **Deficiencies:** Growth problems	15
Selenium	Seafood, freshwater fish, sunflower seeds	Antioxidant, links with iodine function. **Deficiencies:** Very rare, anaemia	0.01– 0.075
Manganese	Nuts, whole wheat pasta and rice	Metabolism, skeletal development. **Deficiencies:** Very rare	–
Copper	Eggs, whole wheat grains, beans, liver, nuts	Iron metabolism, enzyme component. **Deficiencies:** Very rare	1.2
Molybdenum	Dairy products, eggs, liver, beans, sunflower seeds	Enzyme function, DNA/RNA synthesis. **Deficiencies:** Very rare	–
Chromium	Beans, liver, yeast, meat, whole grains	Insulin function and glucose metabolism. **Deficiencies:** Very rare	–
Fluorine	Tap water, seafood	Skeletal strength and structure. **Deficiencies:** Very rare	–
Iodine	Seafood, dairy products	Promotes thyroid function, linked with metabolism. **Deficiencies:** Goiter	0.15

Table 2.8   *(continued)*

Mineral	Food sources	Function within the body	EU RDA (mg)
Aluminium	Processed foods, medicines	Not required for health, may be detrimental to bone growth/brain.	–
Boron	Dried fruits, vegetables, fruits, nuts	Assists calcium in bone structure. **Deficiencies:** Excess calcium loss	–
Cobalt	Animal food sources, dairy products, eggs, enriched breakfast cereals	An element within Vitamin B_{12}, RBC production.	–
Silicon	Whole grains, whole grain rice	May help lower Al levels and assist with calcium function. **Deficiencies:** Problems with bone growth	–
Tin	Tinning for foods, low levels in foods	Not known in new born children, may be toxic.	–
Vanadium	Widely distributed in foods	Not known.	–

Note: Major minerals shaded in blue, minor minerals are shaded in green and the lesser-known minerals in orange.

with food sources and functions of most of the elements. While research has indicated the functioning of most minerals (16 of the 22 known to be found within humans) the function of some minerals with the body remains unclear. These minerals – aluminium, boron, cobalt, silicon, tin and vanadium – are listed at the bottom of Table 2.8. The RDA for each mineral (where known) is provided in Table 2.8. Research indicates that as for vitamins there are no apparent health benefits to consuming minerals in excess of these allowances.

Key point

There are 22 major and minor minerals found in the body. These minerals comprise about 4 per cent of the body mass of humans. The minerals found within our bodies originate from within the soils, rivers, lakes and seas around us. Humans obtain these minerals from consuming plants and animals that have previously consumed plants. The exact function of all minerals within the body is not known. However, most are essential for optimum physiological functioning.

Vitamins and minerals can be distinguished by the fact that minerals do not contain carbon; minerals are therefore the inorganic micronutrients and vitamins the organic nutrients. As humans, we obtain the minerals necessary from the soils and water within the earth's crust. By consuming mineralized water and foods from plant or animal sources, humans can obtain all the minerals essential to life.

While not providing a source of calorific energy, minerals carry out three important functions within the body. They provide building blocks for tissues, form components within enzymes and assist in the regulation of processes within the body. The most abundant mineral within the body is calcium and its role includes structural, metabolic, regulatory and enzymatic functions vital to physiological processes. Calcium, along with phosphorus, forms an essential element to healthy bones and teeth and a lack of calcium within the diet has been linked to osteoporosis, a bone mineral deficiency. Research has identified that, particularly for females, weight-bearing exercise and a diet with the RDA of calcium can help to prevent the development of osteoporosis. Calcium, sodium and potassium, as will be discussed in the following chapter, have a particularly important role in nerve impulse transmission and muscle contraction.

Sodium, which often enters our bodies as sodium chloride, has important functions within the body but excess levels have also been associated with the development of hypertension and cardiovascular disease. Sodium occurs widely and naturally within our foods so adding extra salt does not normally provide beneficial increases in sodium levels within the body. About a third of the people in Britain with high blood pressure (hypertension) find that this is as a result of excess salt intake. Sodium within the diet assists with regulation of osmosis (the movement of water between the cells in the body and their outside environment; see following chapter).

Phosphorus assists in lowering levels of acidity in cells (buffering) and with potassium is a major mineral involved in nerve impulse conduction. In this role, the body takes advantage of the potential for sodium and potassium to breakdown in water to the ionic forms and thus create an electrical potential. The third mineral that functions in this role within the body is chlorine (as a chloride ion) and together they form the body's *electrolytes*, substances with electrical charges. Phosphorus, as well as its roles in buffering and bone structure, has a major role in metabolism as a key component of the high energy fuel used by our bodies (adenosine triphosphate or ATP). As is discussed in Chapter 3 and subsequent chapters, ATP is the petrol for our bodies, providing the main fuel for all the reactions that take place in our bodies (that is, metabolism). Carbohydrates, fats and proteins are broken down through digestion and metabolism to create ATP.

An analogy for this would be to think of foods as crude oil and through digestion and metabolism (refining), ATP (petrol) is produced.

Iron, as one of the most understood minor minerals, plays a vital role in the formation of haemoglobin and myoglobin (haemoglobin for the muscles) which facilitate oxygen storage within the blood and muscle, respectively. In addition, iron is involved in some enzyme formation and structural roles within the body. Research indicates, as for calcium, that females in particular need to ensure that their diet includes adequate iron supplies. Where an individual has an iron-deficient diet it can result in anaemia which includes the following symptoms: lack of energy, headaches and loss of appetite. Anaemia is caused by a decrease in the oxygen-carrying potential of the body due to a decrease in the in haemoglobin content of blood.

2.7 Water

Water is not one of the macronutrients or micronutrients yet it is vital to life. Water is second only to oxygen in its importance to the human body. Water balance is therefore a vital aspect when considering nutrition for adventure sports performers. The nature of the activities often involves exercise in extreme environments where issues with water balance regulation are of paramount importance. Water is a major body component, making up between 40–60 per cent of body weight depending on age, gender and body composition. Water balance is regulated by the amount of water consumed each day and the water lost through evaporation and excretion. We obtain most of the water we consume (about 60 per cent) though drinking, a further 30 per cent from our food and the remaining 10 per cent is produced via metabolic pathways such as the water liberated when two amino acids join (see Figure 2.17). Regular exercise in the form of adventure sports creates greater water loss through evaporation than for sedentary individuals. Those involved in adventure sports must replace the additional water loss to maintain their water balance.

Water plays a number of key roles within the body, helping to sustain life. It is a major component of blood plasma and as such is the solution in which red blood cells and the other components of blood are carried. Water loss through evaporation plays a major role in cooling the body during exercise and in hot environments. It is vital that the body can maintain its temperature within a narrow band to maintain optimal functioning and ultimately life itself. Water loss through sweating enables the body to cool its core temperature.

Water, as well as the electrolytes discussed in the previous section (sodium, potassium and chlorine), is vitally important in maintaining the fluid balance and nutrient exchange between the inside and outside of the cell. The fluid inside a cell is the *intracellular* fluid; that outside the cell the *extracellular* fluid. This fluid consists largely of water with about two-thirds of the water in our bodies being contained within cells and the rest surrounding the cells (extracellular), e.g. blood plasma.

In terms of performance during regular exercise, the key concern for the adventure sports performer is water loss and as a consequence of this *dehydration*. Dehydration has a negative effect on performance and as a consequence a status of *euhydration* (normal hydration) should be maintained prior to, during and after exercise. Euhydration involves maintaining a balance between water intake and water loss. Research has indicated that a loss of just 2 per cent in body weight through sweating during exercise will have a negative impact on performance. Loss of fluid from the body results in a decrease in the blood plasma volume which decreases the volume of blood reaching the working muscles. As a consequence, heart rate increases to meet demand. This places a greater strain on the cardiovascular system. Another consequence of the decrease in blood plasma volume is that less blood reaches the skin for cooling purposes, so the body does not lose heat as effectively.

Weight loss during exercise and when in hot environments is a simple way of monitoring fluid loss. Fluid loss must be replaced to maintain performance. It is not advantageous to wait until thirsty before replacing fluid as the thirst mechanism is sluggish and only responds after dehydration and its negative effects on performance

have started. Fluid should therefore be replaced by adventure sports performers *during* exercise to minimize water loss through sweating and its effect on the body. The thirst mechanism is triggered by the accumulation of sodium and chloride ions in the kidneys in response to the hormone aldosterone which is released due to the effects of dehydration. This in turn results in an increase in the sodium and chloride ions in the blood and it is this blood increase that is identified by chemoreceptors, triggering the thirst response.

2.8 Digestion

Once food enters our mouths and we start to chew, the process of digestion – the breakdown of our macronutrients and micronutrients – begins. Digestion represents the initial processing of our food. Once absorbed into our bodies from the digestive tract, the extracted nutrients are either further broken down (carbohydrate, fat and protein in the form of glucose and fatty acids and amino acids can be further broken down to produce ATP) or synthesized to storage forms (glucose to glycogen, fatty acids to triglycerides) or new structures within the body (amino acids to proteins). This section concerns the structures and processes in the initial breakdown of food.

The digestive system comprises the alimentary canal or gastro-intestinal (GI) tract and a number of digestive organs that assist the GI tract with the digestive process. The GI tract consists of a canal or tube which extends from the mouth to the anus and is up to 5–6 m in length. The GI tract begins with the mouth, continues with the oesophagus, stomach, small intestine (duodenum, jejunum, ileum), large intestine (colon), rectum and ends at the anus from which waste products are expelled. The organs that assist in the digestive process are the liver, gall-bladder and pancreas. The structures within the digestive system are depicted in Figure 2.21.

The digestive process involves five actions that help to bring about the initial breakdown of food for absorption into the body and expulsion of waste products: peristalsis, secretion, digestion, absorption and excretion. *Peristalsis*, the movement of

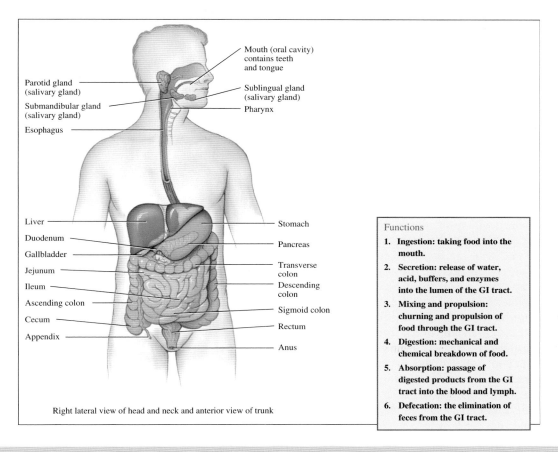

Mouth (oral cavity) contains teeth and tongue

Parotid gland (salivary gland)

Submandibular gland (salivary gland)

Esophagus

Sublingual gland (salivary gland)

Pharynx

Liver

Duodenum

Gallbladder

Jejunum

Ileum

Ascending colon

Cecum

Appendix

Stomach

Pancreas

Transverse colon

Descending colon

Sigmoid colon

Rectum

Anus

Right lateral view of head and neck and anterior view of trunk

Functions

1. **Ingestion: taking food into the mouth.**
2. **Secretion: release of water, acid, buffers, and enzymes into the lumen of the GI tract.**
3. **Mixing and propulsion: churning and propulsion of food through the GI tract.**
4. **Digestion: mechanical and chemical breakdown of food.**
5. **Absorption: passage of digested products from the GI tract into the blood and lymph.**
6. **Defecation: the elimination of feces from the GI tract.**

Figure 2.21 Structures of the digestive system (from Tortora and Derrickson, *Principles of Anatomy and Physiology*, Eleventh Edition, 2006, reproduced by permission of John Wiley & Sons Inc.)

food by muscular contraction of the smooth muscle within the alimentary canal, results in the propulsion of the food through the GI tract and the mixing of the food to enable maximal contact with digestive juices and walls of the tract for absorption. As food is moved through the digestive tract, starting with saliva in the mouth, juices are *secreted* into the canal to enable *digestion* – the breakdown of the macronutrients and micronutrients. Once food has been broken down into its component parts it is made available for *absorption* to the blood and lymph via the walls of the digestive tract. Indigestible food, such as fibre, and other waste products are then *excreted* from the body via the anus.

The organs that assist with digestion – the pancreas, liver and gall-bladder – produce digestive juices which are utilized within the duodenum. The *pancreas* has two main functions: exocrine (the production of pancreatic juice to assist with digestion) and endocrine (the production of the hormones insulin, glucagon and somatostatin which are transported via the blood for use elsewhere in the body). Pancreatic juice neutralizes the acids of digestion in the stomach and further digests carbohydrate, fat and protein ready for absorption. The *liver* produces bile, a yellow-brown or dark green liquid, which as well as containing waste products is also involved in the breakdown of lipids to smaller lipids in the process of readying fats for absorption into the blood. The ducts from the liver that carry bile also link with the *gall-bladder* where bile is stored and concentrated prior to being used in the duodenum. The gall-bladder, a 7–10 cm pear-shaped chamber, is situated just below the liver.

The initial digestion of food begins in the mouth where saliva starts to dissolve food and break down carbohydrate. From the mouth, food is passed as a bolus (lump of food) via the oesophagus into the stomach. The stomach contains hydrochloric acid which kills bacteria and pepsin which begins the breakdown of protein. During this breakdown, the stomach will absorb some nutrients, water ions and fatty acids. Food within the stomach is dissolved and with the digestive enzymes forms the soupy substance chyme. The chyme passes to the small intestine where most of the digestion and absorption of food takes place. To allow time for these processes to take place, the small intestine is up to 3 m in length which increases the absorptive area and the transit time for digestion. Within the small intestine, carbohydrate in the form of the disaccharides sucrose, lactose and maltose are broken down to monossaccharides by the enzymes, sucrase, lactase and maltase ready for absorption. Fat molecules are broken down to smaller units by the action of bile salts (contained within bile) and subsequently acted upon by pancreatic lipase to produce fatty acids and monglycerides for absorption. Protein which began initial breakdown in the stomach is broken down to peptides in the small intestine by trypsin and similar enzymes. Once in the form of peptides, the enzymes aminopeptidase and dipeptidase break protein down to its component amino acids for absorption.

The majority of the absorptive process takes place within the small intestine. The macronutrients are absorbed during the transit of chyme through the small intestine. Ninety per cent of water absorption, as well as the major part of micronutrient uptake, occurs within the small intestine, leaving very few nutrient components in the chyme as it enters the large intestine. When the food enters the large intestine it will have been in the body for between 3–10 hr. During its passage through the small intestine, it will have lost

not only most of its nutrients, but also most of its water and as such is known as the solid substances feces by this time. Very little further digestion and absorption take place within the large intestine. The exceptions are some continued absorption of water (about 10 per cent of total absorption) and further food breakdown to release some additional vitamins for absorption. Once this process is complete, the remaining fecal matter is expelled from the body.

Key point

Digestion begins in the mouth and continues through the stomach and small intestine. The majority of food digestion and absorption takes place in the small intestine. Macronutrients, micronutrients and water are all digested and subsequently absorbed into the blood steam and lymph for use throughout the body during the transit of food through the small intestine. Fibre in our diets assists the passage of the waste products into the large intestine, ready for expulsion from the body.

2.9 Components of a healthy balanced diet

A healthy diet is the basis for sound nutrition and the need for supplements and dietary manipulation is seldom necessary if an adventure sports performer has a sound diet. A healthy balanced diet should include the macronutrients (including fibre as part of carbohydrate intake), micronutrients and water as discussed in this chapter. The RDA for micronutrients – vitamins and minerals – are provided in Tables 2.7 and 2.8, respectively. For a sedentary person, the Department of Health guidelines for the UK recommend an average calorie intake of 1940 kcal for females and 2500 kcal for males per day. For people involved in adventure sports, this amount would obviously need to be increased to account for energy expended during exercise. For those involved in high levels of training, e.g. for an event such as the Devizes to Westminster race, calorie intake requirements could be between 2–3 times the UK guidelines.

Key point

A healthy diet should include carbohydrate (including fibre), fat, protein, vitamins, minerals and water. Carbohydrate should constitute the largest part of our diets at around 55–60 per cent, fats should be maintained at 30 per cent or less in our food and proteins should form around 10–15 per cent of our calorie intake. Total calorie intake for an adventure sports person is dependent upon the level of training and activity completed each week. The more a person exercises the greater the total calorie intake should rise above the 2000 kcal for females and 2500 kcal for males recommended for a sedentary person by the Government. Energy intake should match energy expenditure to maintain weight. The RDA for vitamins and minerals are included in Table 2.7 and 2.8.

Calorie intake to maintain a balance between input and expenditure is highly individual and dependent on factors such as age, gender, frequency, intensity and duration of training and individual metabolic differences. Calorie intake should be matched to energy requirements to maintain a healthy body composition.

Key point

Maintaining fluid balance as well as a nutritional balance is also important to performance. Ideally, water should be taken on board throughout the day (perhaps carried in a water bottle from which sips are taken) rather than only consumed in larger volumes at infrequent times. Water is generally the best form of fluid to maintain levels during the day. Sports drinks with an electrolyte balance may be beneficial during and post training or during an event to increase the speed at which fluid is taken on board. Fluid replacement, however, can be achieved almost as quickly using water and at a fraction of the cost.

In terms of calorie intake from the macronutrients that comprise the bulk of our diet, it is recommended that we consume 55–60 per cent of our calories from carbohydrates, up to 30 per cent, but ideally less than this from fats (<10 per cent from saturated fats) and 10–15 per cent from proteins. The percentage for fats is lower than the Government target of 35 per cent (as is the Government target of <13 per cent from saturated fats). However, research suggests that maintaining a fat intake below 30 per cent is more beneficial to health.

Figure 2.22 below provides an alternative way of viewing these macronutrient recommendations. The food pyramid provides a guide to the distribution of food groups that should form the basis of a healthy diet. Consuming foods from these groups can, without the need for supplements, provide all the nutrients necessary for a healthy athletic life.

Figure 2.22 provides the recommended daily servings for each of the food groups within the pyramid. For the bread group, examples of serving sizes are: 1 slice of bread or half cup (cooked) of pasta or rice. For the fruits and vegetables group, a serving would be an apple, a small banana, a medium tomato or a small stalk of broccoli. For the meat group, example serving sizes would be one egg, one-third of a can (small) of tuna or a 3 oz chicken breast (about the size of a pack of cards).

In Britain people are encouraged to eat 5 portions of vegetables a day. The food pyramid includes both fruit and vegetables; selecting both in your diet helps to achieve a greater variety. It

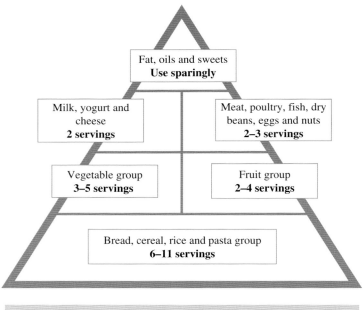

Figure 2.22   Food guide pyramid with suggested daily servings

is better to eat a mixture of fruits and vegetables rather than eating 5 apples a day. Including fibre in your diet will aid digestion and can be found in vegetables, fruits, whole wheat grains, some cereals, beans and peas.

With regard to carbohydrate, it is beneficial to consume a mixture of different forms at different times of the day. During training and immediately after, it is perhaps most useful to consume carbohydrate in the form of simple sugar. Complex carbohydrates should form a basis for recovery meals, main meals during each day and as part of the food during a long day climbing, paddling, skiing or caving. In addition, it is beneficial to pay attention to the glycemic index of the foods to mix with them as well. Lower glycemic index foods will last longer as a fuel source for exercise while out on the hill, mountain, river or sea.

Protein requirements are generally not as high as athletes often presume. A balanced diet should supply ordinary requirements. There normally should be no need for expensive (or inexpensive) protein supplements. For sedentary people, the protein requirement per day is about 0.8 g per kg of bodyweight, e.g. a 70 kg person would require 0.8 g × 70 kg = 56 g of protein a day. An 8 oz (227 g) tin of tuna fish or 7 oz (200 g) chicken breast, for example, contains 67 g and 61 g of protein, respectively. One of these servings would more than satisfy the daily protein requirement. For those involved in adventure sports the requirement rises to 1.0–2.0 g per kg of bodyweight, depending upon the type of activity. However, these requirements can easily be supplied by dietary means.

Regarding lipid consumption, it is essential to maintain some fat within your food to maintain a healthy diet. Fat is a necessary component of our diets. When choosing foods it is beneficial, however, to examine the types of fats they contain. Where possible choose foods with a high proportion of the fat in the form of monounsaturated and polyunsaturated fats. Look on food labels and try to avoid hydrogenated fats – they are very common in processed foods – where possible. It is also useful to check the fat content or percentage relative to carbohydrate and protein. The percentage of fat is presented in a number of different ways, such as by the weight of the food. It is the percentage of total calories that are supplied by fat that is important. A simple way to check the total calorific value of the fat is to use the total calorific value of the food and the grams of fat. Remembering that 1 g of an average fat contains 9 kcal, it is possible to calculate the percentage of calories from fat and, importantly, saturated fat.

Calculation of fat percentage in a breakfast cereal:

One healthy breakfast cereal on the market contains 12 g fat and 4.5 g saturated fat per 100 g. The energy value listed is 1690 kJ or 404 kcal per 100 grams.

Energy provided by fat

$= 12 \times 9$ kcal $= 108$ kcal

Energy provided by saturated fat

$= 4.5 \times 9$ kcal $= 40.5$ kcal

Percentage of energy from fat

$= 108/404 \times 100 =$ **26.7 %**

Percentage of energy from saturated fat

$= 40.5/404 \times 100 =$ **10 %**

This breakfast cereal therefore satisfies the recommended daily fat intake percentages.

Simplified calculation of fat percentage in a breakfast cereal:

To simplify the above calculation, make an allowance of 3 grams of fat per 100 kcal of total energy in any food.

Then per 100 kcal of total energy, 9 kcal $\times 3 = 27$ kcal of energy is from fat i.e. 27 per cent ($<$30 per cent) from fat.

The rule of thumb *allowance* is 3 grams of fat per 100 kcal to stay below 30 per cent fat.

In the breakfast cereal example:

404 kcal $\sim$ 400 kcal per 100 g, of which 12 g is fat

12 g fat per 400 kcal total energy

$\Rightarrow$ **3 g in 100 kcal is fat**

In addition to the basic diet, there are some useful considerations you can make to improve upon these basic guidelines. Variety, moderation and naturalness will add to a healthy diet. There is no one food that can provide all the nutrients we need; having *variety* with meals helps us to gain all the nutrients we need. There is nothing wrong with having occasional treats; these add variety to the diet and increase enjoyment of your food. The secret to this is *moderation* and cutting back in other areas. The more *natural* your food is, that is the less it has been processed by manufacturers before you consume it, the better. Processed food tends to have fewer nutrients than natural foods. Choosing wholemeal bread over white bread, whole grain rice over quick cook rice as well as cooking foods yourself rather than buying processed meals can make a difference. The choice of whole grains and home food preparation add and retain nutrients in the food you consume. It is easy when looking at Tables 2.7 and 2.8 to see how many useful dietary micronutrients are to be found in whole grains.

Chapters 6–10 include further information, where relevant, about specific adaptations that can be made to the diet for certain types of adventure sports activity. For example, there are dietary manipulations that can be made for adventure sports where power or endurance is required.

Key point

A balanced diet including the six food groups becomes a healthy balanced diet when we pay attention to the variety in our diets, moderate our intake of certain food groups and maintain as far as possible the naturalness of our food. What we eat does count with regard to maintaining and improving our performance.

2.10 Summary and study questions

Human beings use food for two distinct purposes: to provide fuel and as building blocks for the maintenance of the body's structure. There are six food groups consisting of:

- carbohydrates;
- proteins;
- fats;
- vitamins;
- minerals; and
- water.

Carbohydrates and fats are made of carbon, hydrogen and oxygen. Proteins include these three elements and nitrogen. Carbon, hydrogen, oxygen and nitrogen make up 96 per cent of our bodies. Elements can bond by receiving or donating electrons (ionic bonding) or by sharing electrons (covalent bonding). These bonds are created and broken in chemical reactions.

The energy values of food are measured in calories or, more commonly, kilocalories. The glucose stores of the body are found mainly in the muscles and liver where glucose is stored as glycogen. One gram of glycogen is capable of providing 4 kcal of energy and the body can store 1500–2000 kcal of energy in the form of glycogen.

Fat is far more effective for energy storage than carbohydrates since each gram of fat can produce 9 kcal of energy rather than the 4 kcal available from carbohydrate. Fat is also easier for the body to store long term as specialist fat cells can be used to store fat for future use. In fact, the body can easily store 90 000–100 000 kcal of energy as fat stores.

While fat is an essential nutrient and building block for the human body, excessive fat consumption has been linked to disease and recommendations are that no more that 30 per cent of the daily calorific intake should come from fat. Fats can be saturated, monounsaturated or polyunsaturated. Saturated fat has been identified as particularly harmful and it seems prudent to limit saturated fat intake to less than 10 per cent of the total calorific intake.

Proteins are much larger and more complex molecules than carbohydrates and fats. While they can be metabolized as an energy source, they have a role in most functions of the body and as building blocks for the body's structure. Essential amino acids are the building blocks of proteins that cannot be synthesized in the body and therefore must be included in food. Non-essential amino acids can be synthesized but a healthy balanced diet should mean that this is not necessary and that all 20 amino acids are part of the regular diet.

Despite strength training mythology, it is easily possible to consume enough protein though the diet without resorting to additional food supplements and it is also possible to include all of the amino acids within a balanced and carefully considered vegetarian diet.

Vitamins are classed as micronutrients which are substances essential for good health but not needed in large volumes such as carbohydrates, fats or proteins. Vitamins are not metabolized for energy but are required to ensure the effective functioning of the body. While vitamin supplementation is popular, it should not really be necessary except in special cases such as pregnancy, where research has shown that supplementation of some vitamins can be beneficial. Vitamins are only needed in tiny amounts and consuming levels that are too high can actually cause problems as some vitamins can become toxic at higher levels.

Minerals are also micronutrients, essential for good health and like vitamins are required in small amounts that should come from a carefully considered and balanced diet. Unlike vitamins, minerals contain no carbon. The 22 major and minor minerals actually make up 4 per cent of the mass of our bodies. Calcium and phosphorus have a particular role in the construction of our bones and teeth. Iron has a role in the oxygen transport function of the blood.

Water is one of the major components of the body making up some 40–60 per cent of our body weight. Water is consumed through drinking and eating as well as being produced during metabolic functions. Exercise will increase the volume of water required by the body and so adventure sports performers will need to actively replace this by drinking more than a sedentary individual. Water has a role in temperature control as sweating and also in blood flow as plasma. Dehydration during activities can therefore present serious problems to the adventure sports athlete.

Digestion is the process of breaking down and absorbing nutrients from food and drink into the body. This essentially involves the movement of food through the digestive tract by the process of peristalsis. The digestive tract is a single continuous tube that processes food in stages on its journey through the body. On this journey, different enzymes are responsible for breaking down food components into forms that can be absorbed at each stage.

Diet is a key component in maintaining the performance and ability to complete training programmes for an adventure sport athlete. A balanced diet should be capable of providing all of the nutritional needs for sporting performance without the need for additional supplementation. Heavy training programmes and physical workloads will mean that the body requires a greater volume of calories and fluids. At times, it may be necessary to consume double or triple the amount of calories compared to a sedentary person. The proportion of carbohydrates, fats and proteins should normally be kept in line with guidelines such as the food pyramid; the composition of carbohydrates should also be considered in terms of complexity and the glycemic index.

Study questions

1. What are the three macronutrients that humans require and what is the basic role of each of these?

2. What elements make up the chemical composition of the macronutrients? How does this relate to the composition of the human body?

3. Explain the difference between an exergonic chemical reaction and an endergonic chemical reaction. Why is it important that both occur within the human body?

4. Give the scientific definition of the calorie. What is a kilocalorie and why is food normally labelled in kilocalories (kcal or Calorie)?

5. What is the glycemic index and why is it important in food choices for adventure sports performers? What would be a good choice of carbohydrates for a mountain biker in a 24-hour endurance race?

6. What chemical differences are there between saturated fats, monounsaturated fats and polyunsaturated fats? What is the advice for long-term health in relation to these?

7. How does a protein molecule such as haemoglobin differ from a carbohydrate molecule like glucose? What do the terms anabolize and catabolize mean in relation to protein structures within the body?

8. What are the two kinds of micronutrient? How is their function in the body different to the macronutrients?

9. Why can dehydration have an impact on cardiovascular performance during adventure activities? What are the three ways our body obtains the water it needs to function effectively?

10. What are the average daily calorific needs of a sedentary individual? How might this be different for an endurance-based adventure sports performer involved in a multi-day sea kayak event?

3

The structure and functioning of the human body

When broken down to into its most basic components, the human body comprises the 26 elements listed in Table 2.1. These elements are combined within the body to form *cells*, the basic unit of living matter. Just as elements form the building blocks for cells, cells form the building blocks for the human body and all the structures it contains. Cells work in cooperative groups to maintain life, forming the basis for *tissues* and *organs* which in turn comprise *systems* that enable the body to function. There are 11 systems commonly identified within the body and these, along with their functions, are listed in Table 3.1.

The primary role of cells and the systems they comprise is to maintain *homeostasis*. This chapter examines the structure and function of cells and the importance of homeostasis in human functioning. Body systems which have an important role in adventure sports performance are briefly described. To better understand the nature of adventure sports performance, the exercise and environmental demands placed on the body, it is essential to have knowledge of the systems that enable human functioning. Human performance in adventure sports relies upon the cells within the body.

3.1 The cellular basis for life

The foundation of all living things is the cell – the smallest unit capable of carrying out all the processes associated with life. Each human being begins as one cell and the process of cell division enables us to grow to adult size. Cells are tiny; five or more cells would fit into the dot above the letter 'i' printed here. In order to study cells, an electron microscope is needed to reveal the structure of these tiny building blocks.

Humans are made up of around 200 different types of cell that work in coordinated groups to accomplish functions within the body. The body's systems – skeletal, muscular, respiratory and nervous – are made up of cells working in cooperative units. All cells have specialized roles that contribute to the functioning of the body and are dependent on the other groups of cells carrying out their work to maintain life. For example, muscle cells use energy to bring about movement, but are dependent on the circulatory system (blood and blood vessels) for supplying oxygen, the food necessary to produce energy and for removing the waste products that result from energy production. Cells within the body have to

Adventure Sport Physiology   Nick Draper and Chris Hodgson
© 2008 John Wiley & Sons, Ltd

Table 3.1   The eleven systems within the human body

System	Function within the body
Cardiovascular system	Includes the heart, blood vessels and blood and is responsible for transport of nutrients and waste products including O_2 and CO_2, hormones and electrolytes. Also involved in temperature regulation and defence against infection.
Digestive system	Responsible for the breakdown and absorption of macronutrients, micronutrients and water into the body, as well as the excretion of waste products.
Endocrine system	The longer-term control system. Responsible for the production of hormones that regulate control of function throughout the body.
Integumentary system	Comprises skin, hair and nails, provides a barrier between the body and outside environment and is involved in temperature regulation.
Immune system	Includes the lymph and white blood cells, is involved in defence against disease and controls the level of interstitial fluid within the body.
Muscular system	Along with the skeletal system, is responsible for bringing about human movement as well as producing heat through the reactions that bring about movement.
Nervous system	Divided into the central nervous system and the peripheral nervous system, it is responsible for receiving, processing and creating responses to sensory input. Along with the endocrine system, it is responsible for control and regulation within the body.
Reproductive system	The reproductive systems are different for males and females and are chiefly responsible for perpetuation of the species and determining individual sex characteristics, but also have endocrine function links.
Respiratory system	Responsible for exchange of O_2 and CO_2 with the outside environment.
Skeletal system	Supports and protects body parts and along with the muscular system is responsible for bringing about movement.
Urinary system	Responsible for the excretion of waste products and maintaining fluid balance within the body.

work cooperatively in this integrated fashion to sustain life.

Although serving different functions, the various cells within our bodies all have the same general structure as depicted in Figure 3.1. There is no one cell that has all these structures or that looks exactly like that in Figure 3.1, but they have many of these aspects in common. The diversity in the functioning of cells can be illustrated through the development of red blood cells (erythrocytes).

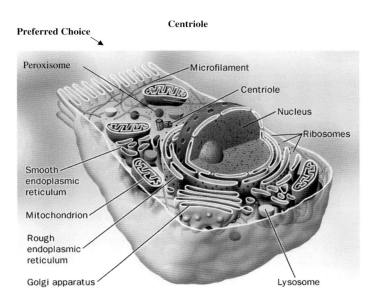

Figure 3.1 Cell structure (from Tortora and Derrickson, *Principles of Anatomy and Physiology*, Eleventh Edition 2006, reproduced by permission of John Wiley & Sons Inc.)

Erythrocytes, which are produced in the bone marrow, have a normal life span of around 120 days before they are replaced by new cells. As new erythrocytes initially grow within the bone marrow, they contain the common organelles shown in Figure 3.1. However, because of their specialist oxygen-carrying role, they expel their nucleus and all other organelles from their structure during their development. In doing this, they greatly increase the space inside their cell for carrying haemoglobin and consequently oxygen. The 200 or so different cells found within the body share the common characteristics of the cell illustrated in Figure 3.1, but all develop uniquely to best serve the function they play within the body.

Figure 3.2 shows the structure of just two cell types within the body, erythrocytes and a leukocyte (white blood cell) which are responsible for transporting oxygen and carbon dioxide and immune function respectively. To get in to perspective the size and number of cells which comprise the human body, there are 5,000,000,000 (five billion) red blood cells in every millilitre of blood. An average human carries about five litres of blood

which means each person has 25,000,000,000,000 (25 trillion) red blood cells within their body and this is just one type of cell.

Cells have three main parts: the cell membrane (outer skin of the cell that separates it from its outside environment), the nucleus and the remaining contents of the cell known as the cytoplasm. The cell membrane is a very thin structure that encloses every cell and keeps it separate from its surrounding environment. The nucleus, normally the largest single unit in the cell, contains the deoxyribonucleic acid (DNA), the genetic blueprint for controlling the operations of the cell and ribonucleic acid (RNA) which controls protein synthesis within the cell. The cytoplasm comprises the inside of the cell except for the nucleus. It comprises a gel-like substance, cytosol, which houses and protects nine main types of small structures called organelles. These organelles serve a variety of roles for the cell. In Figure 3.3 the mitochondria organelle is shown, vital in the process of energy production. Mitochondria are the power plants of the cell and they are responsible for producing about 90 per cent of our energy.

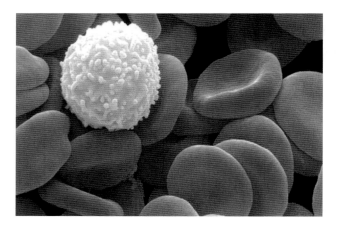

Figure 3.2   Red and white blood cells (from Tortora and Derrickson, *Principles of Anatomy and Physiology*, Eleventh Edition 2006, reproduced by permission of John Wiley & Sons Inc.)

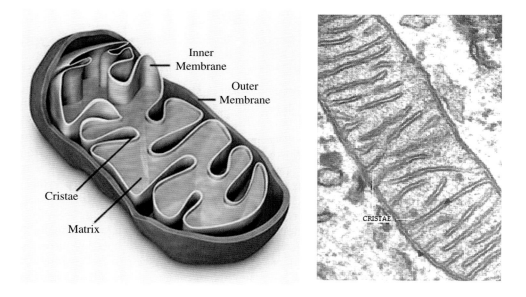

Figure 3.3   Mitochondria structure (from Tortora and Derrickson, *Principles of Anatomy and Physiology*, Eleventh Edition 2006, reproduced by permission of John Wiley & Sons Inc.)

Key point

Cells are the smallest unit within the body capable of sustaining life. Cells comprise a cell membrane that separates the cell from other parts of the body, a nucleus that controls the cell functions and the rest of the cell which is known as the cytoplasm.

Cells work in cooperative groups to build the tissues, organs and systems that make up the human body.

Cell membrane

The cell or plasma membrane maintains a barrier between the internal environment of the cell and the exterior, which consists of interstitial fluid. To carry out this role successfully, cells take advantage of the dually hydrophilic and hydrophobic nature of phospholipids (illustrated in Figure 2.14 and in more detail in Figure 3.4) described in Chapter 2. The benefit of utilizing these two forces of attraction is that no covalent bonds are required to maintain the integrity of the cell membrane and, as a consequence, individual phospholipid molecules are free to move in a fluid fashion on their side of the cell membrane. Phospholipids do not change place on either side of the membrane (inside and outside of cell membrane) but are able to move within their layer. The outside and inside walls of the plasma membrane must thought of as fluid i.e. constantly changing, but always maintaining the integrity of the cell membrane.

The structure of the plasma membrane is, however, more complex than illustrated in Figure 2.14. The nature of the membrane is more accurately depicted in Figure 3.4 which reveals the presence of proteins, cholesterol and carbohydrates, all of which have a role in the functioning of the cell membrane.

As part of or close to the plasma membrane there are peripheral and transmembrane proteins. *Peripheral proteins* have close links with the function of transmembrane proteins. There are four main sorts of *transmembrane protein*; recognition proteins, receptor proteins, channel proteins and transporter proteins. *Recognition proteins* are glycoproteins, meaning that they have carbohydrate (glucose) and protein components. Their role is to set an identity for each cell type so that cells can differentiate between types for attachment purposes. *Receptor proteins* have a binding site, specific to a particular signal molecule, on the outside of the cell membrane which when receiving

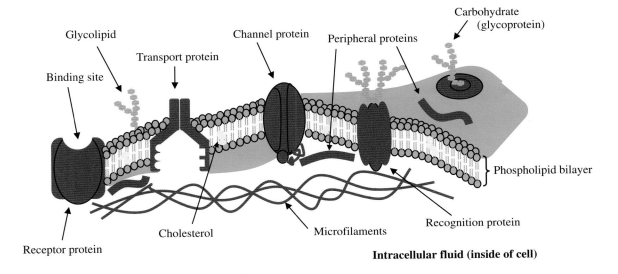

Figure 3.4 The structure of the cell membrane

the appropriate signal will trigger a reaction from the cell. An example for adventure sports relates to muscle cell contraction, which will be covered in more detail in Chapter 5. When the correct chemical signal binds with a muscle cell receptor protein, it will trigger the muscle contraction. *Channel proteins* enable specific ions to diffuse into the cell. Channel proteins are often gated (Figure 3.4) to enable diffusion of a specific ion only when the gate is open. There are usually many channel proteins for potassium and chloride ions, but fewer for calcium and sodium ions. Ions pass through channel proteins into the cell via diffusion (from higher to lower concentration), but sometimes the same ions and larger molecules must be actively transported into the cell. *Transport proteins* make use of adenosine triphosphate (ATP), the energy currency for all cells, to bring ions or molecules into the cytoplasm. Figure 3.5 depicts an example of active transport as necessary for the sodium-potassium pump. ATP is necessary to enable the transport of sodium out of the cell and potassium

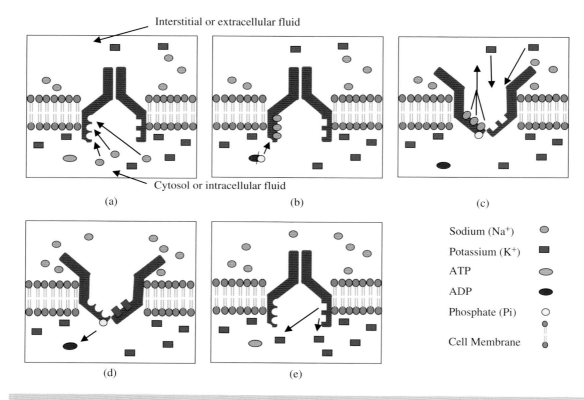

Figure 3.5 The sodium-potassium pump: transport proteins (purple) within the cell membrane are responsible for the movement of molecules into and out the cell. (a) Three Sodium ions bind with the transport protein, which triggers (b) the hydrolysis of ATP to ADP and releases a phosphate to bind to the protein. (c) The binding of the phosphate ion (Pi) alters the shape of the protein, opening it to the extracellular fluid. This change in shape causes a release of the 3 Na^+ as the protein then favours binding with K^+. (d) Two K^+ attach to the protein which triggers the release of the Pi and (e) the protein returns to its original shape and releases the K^+ into the intracellular fluid. The released phosphate is then free to bind with ADP within the intracellular fluid to form ATP once more

into the cell. Cells therefore establish an electrical charge differential between the inside and outside of the cell. This electrical differential is know as a *resting membrane potential* and will be discussed further in the section on the nervous system.

Key point

The cell or plasma membrane is a bilipid layer that encloses the cell. The bilipid layer is composed of phospholipids and cholesterol, with transmembrane proteins contained within. Amongst other things, the transmembrane proteins are responsible for controlling the flow of ions and molecules into and out of the cell.

The carbohydrate components of the cell membrane serve two main purposes: as *glycoproteins*, part of the identity structure of recognition proteins, and as *glycolipids* to create a slippery layer on the outer surface of the cell membrane. Along with the fluidity of the bilipid (two layers of fat molecules) layer, this is an essential component of erythrocytes to enable them to slip easily through the narrow constrictions within capillaries. Cholesterol as a form of lipid is found within the hydrophobic lipid section of the bilipid cell membrane. The presence of cholesterol in the cell membrane is thought to help with both the stability and fluidity of the plasma membrane.

Nucleus

The nucleus, usually found at the core of each cell, is the control centre for the cell. It is encapsulated within a double-layered membrane that separates it from the rest of the cytoplasm and contains nuclear pores enabling substances to pass between the nucleus and the cytoplasm. Most cells have one nucleus, although as already mentioned red blood cells expel their nucleus during development to increase oxygen-carrying capacity and muscle cells and some bone cells are multinucleated. The nucleus represents the single largest component within the cell. As it contains DNA, it governs the cells functions. Deoxyribonucleic acid contains the blueprint or genetic code for the cell and therefore controls replication of the cell and directs protein and enzyme synthesis within the cell. Deoxyribonucleic acid does not, however, leave the nucleus to control the function of the cell RNA, of which there are three types.

The genetic code for a particular protein is passed from DNA to messenger RNA (mRNA). Messenger RNA leaves the nucleus via the nuclear pores, enters the cytoplasm and delivers the genetic transcript to ribosomal RNA (rRNA) where it is interpreted and translated into amino acid sequences for a particular protein synthesis. Transfer RNA (tRNA) is then used to move the necessary amino acids within the cytoplasm for the synthesis of the protein.

Cytoplasm

The cytoplasm comprises cytosol, a cytoskeleton and organelles which carry out a wide range of functions within the cell. The *cytosol* or *intracellular fluid* is a gel-like substance, largely composed of water containing suspended and dissolved particles such as ATP, glucose, lipids, amino acids and a variety of ions. The cytosol is where many metabolic reactions take place, including glycolysis (covered in more detail in the final section of this chapter) which represents one of the systems for energy production. The *cytoskeleton* is literally that, a network of three types of protein filament that among other roles maintain cell structure and hold in place many of the organelles. The *microtubules* are responsible for support and structure within the cell, the *intermediate filaments* give the cell strength and help to maintain the structure, while *microfilaments* (illustrated in Figures 3.1 and 3.4) in their structural role also help to give shape to the cell. The microfilaments (which comprise the protein actin) are also partly responsible for inciting muscular contractions.

There are nine main organelles that provide a variety of functions within the cell. Situated near the nucleus, each cell has a centrosome which consists of two *centrioles* that play an important role in cell division during the creation of new cells. The *rough endoplasmic reticulum* is named as such because its outer membrane is littered with ribosomes. *Ribosomes* are responsible for protein synthesis and are linked with RNA which determines the composition of each protein. Ribosomes are found throughout the cytoplasm. The role of the rough endoplasmic reticulum is linked to the proteins created by the ribosomes on its outer surface. Within the rough endoplasmic reticulum, ribosomal proteins are combined with glucose to form glycoproteins which, among other roles, are a vital component of the cell membrane. The *smooth endoplasmic reticulum*, like its rough counterpart, is a folded sac-like structure similar in appearance to a folded bicycle inner tube. The smooth endoplasmic reticulum can be differentiated from the rough endoplasmic reticulum because it does not have ribosomes attached. As a consequence of this, the smooth endoplasmic reticulum is not involved with protein synthesis but does play an important role in the formation of fatty acids and steroids.

The *Golgi complex* is the packaging and shipping centre for the cell. Proteins formed in the rough endoplasmic reticulum are first collected in the Golgi complex for possible further synthesizing and packaging. Within the Golgi complex, lipids and glucose are added to proteins to form glycoproteins and lipoproteins. These and other proteins are transferred to *vesicles*, plasma membrane-covered round containers used to ship the contents to either the cell membrane or other organelles within the cell, while keeping them separate from the rest of the cytoplasm during transport. *Lysosomes*, a form of vesicle, are responsible for carrying over 50 different enzymes involved in digestion and other processes. *Peroxisomes* are smaller than lysosomes but play a similar enzyme-carrying role. In the case of peroxisomes, the enzymes are linked with the oxidation (removal of hydrogen) of fatty and amino acids. *Mitochondria*, illustrated in Figure 3.3, play a vital role in energy production within the cell. Some cells have several hundred, whereas muscle cells, as a result of the importance of energy production, have thousands of mitochondria. The role of mitochondria will be further discussed later in this chapter and in Chapter 10.

Key point

The cytoplasm comprises all parts of the cell within the cell membrane with the exception of the nucleus. The cytoplasm consists of cytosol, a water-based gel-like mass that surrounds the organelles (tiny cellular organs) and a cytoskeleton that holds the organelles in place. There are nine organelles found commonly within cells. For energy production during adventure sports, one of the most important organelles is the mitochondria.

Homeostasis

French physiologist Claude Bernard (Chapter 1) was one of the first to record the importance of maintaining a stable environment within the body; later the American physiologist Walter Canon coined the term *homeostasis*. Homeostasis is vital to all living organisms, from unicellular amoeba to the most complex multicellular animals including humans. Homeo means 'equal' and stasis 'to stand' or putting the two together 'to remain equal' and refers to the need of an organism to provide constancy to its internal environment. For an amoeba, as a single cell organism, this is a simple process because the cell is in contact with its external environment to exchange nutrients and waste products. For humans, not all cells are immediately in contact with the external environment and as a consequence must work in cooperative units to ensure a stable internal environment. For example, muscle cells do not have contact with the environment outside the body and so must rely on other cells to provide the nutrients and remove the waste products necessary to maintain a stable environment within each muscle cell. Cells within the human body are *mutually dependent* and must work together to sustain life. In order to do this, they must maintain homeostasis.

The co-operative units in which cells function to maintain homeostasis are the tissues and organs that ultimately form the 11 systems shown in Table 3.1. The purpose of each of these systems is to help maintain a stable internal environment and each has a vital role to play. A major role of the systems within the body involves providing a constant *supply* of nutrients and oxygen to each cell and *removal* of waste products such as carbon dioxide produced during the reactions that maintain each cell. As well as a delivery and removal system, the cells must maintain the internal environment by controlling the *pH, electrolyte balance* (of the fluids surrounding each cell) and *temperature* in which the cells function.

The nervous and endocrine systems, which will be described in more detail later in this chapter, play a vital role in the regulation of homeostasis. They are the control mechanisms that regulate the functioning of other systems to maintain the body's internal environment. The *nervous system* is the body's fast response mechanism and the *endocrine system* the slower sustained regulation process. Throughout the body, there is a variety of receptor or sensory mechanisms that pick up and relay homeostatic status reports to the brain. The brain processes the sensory information and provides either a nervous or endocrine response. For example, if receptors detect a lack of glucose in the blood (a necessary energy supply for all cells), the brain interprets this information and stimulates an endocrine response in the form of a release of glucagon (a hormone that stimulates the breakdown of glycogen to release glucose to the circulating blood). Alternatively, when a more immediate response is required, nervous control can be altered. When you start to climb, mountain bike or run there is an immediate increase in the demand for oxygen for the working muscles. Receptors detect this (and the rise in carbon dioxide levels) and send sensory information to the brain where it is processed and a nervous response effected, which would include a increase in respiration (breathing rate) and heart rate.

The most common control mechanism for ceasing a nervous or endocrine response is through a *negative feedback* loop. For example, when climbing, mountain biking or running a mountain marathon, there is a rise in heart rate to respond to increased oxygen demand for the working muscles. The rise in heart rate will continue until the receptors detect that oxygen and carbon dioxide levels have returned to normal limits i.e. supply has met demand. At this stage the brain will stop further nervous stimulation to increase the heart rate and it will be maintained at a level where supply meets demand, a term in exercise physiology called *steady state*.

Homeostatic balance can be endangered by a wide range of *threats* or, as Hans Seyle (discussed further in Chapter 4) refers to them, *stresses*. These stresses can include environmental, exercise, circadian, nutritional threats and illness or disease. These threats can provide significant challenges to human homeostasis, particularly when operating as multiple stressors. Taking part in an adventure sport perhaps provides the greatest challenge to maintaining homeostasis, and it nearly always involves multiple threats to homeostatic balance. Skiers and snowboarders create exercise and environmental (cold and possibly altitude) challenges for the homeostatic mechanisms within the body. Desert marathon runners, such as those taking part in the Marathon des Sables in the Sahara, place exercise, nutritional (water and energy) and environmental (heat) demands upon their bodies. The ultimate challenge to homeostasis perhaps comes from alpine ascents of mountain peaks, where the participant can face exercise, environmental (cold, altitude), circadian (alpine ascents often start at 01:00 or 02:00) and, due to the length of the day, nutritional (water and energy) threats. To survive, the body must maintain homeostasis. As adventure sports performers, it is vital to understand this delicate balance and the threats presented by participating in any activity.

Key point

Homeostasis is vital to sustain life and refers to maintaining constancy within the internal environment of the body. All cells within the body work cooperatively in systems to maintain homeostasis. To maintain homeostasis, systems within the body control the supply of nutrients and removal of waste products as well as controlling the temperature, acidity levels and electrolyte balance of the internal environment. Adventure sports participation can involve the greatest number of challenges to homeostasis e.g. alpine ascents of mountain peaks provide multiple stresses.

3.2   Skeletal system

The skeletal system is more than just bones and includes cartilage, tendons and ligaments. In common with any of the other 10 systems within the body, in its most basic units the skeletal system comprises elements and molecules found within our diet. Along with carbon, hydrogen, oxygen and nitrogen which form the macronutrient elements within skeletal tissue, calcium and phosphorus are vital minerals for the development of bones, comprising two-thirds of their structure.

Nearly all bones begin life as cartilage which, in common with bone, is a form of connective tissue but is more elastic in nature and therefore less rigid. Cartilage is formed by *chondroblasts*, which produce the chondrin and collagen that form cartilage. As cartilage develops, the chondroblasts become trapped in their own network or matrix and become *chondrocytes*. Most of the skeletal cartilage is replaced by bone as the body develops in a process called ossification. Three classes of bone cell involved in the growth and repair of the tissue have been identified. In a similar fashion to cartilage, bone-forming *osteoblasts* which combine calcium and phosphorus to produce

hydroxyapatite crystals – the mineralized form of bone – become *osteocytes* when they become trapped in their own mineral network. *Osteoclasts*, the third form of bone cell, are responsible for the removal and re-absorption of unwanted bone when the minerals are required elsewhere in the body.

The human skeleton comprises 206 bones, most of which can be seen in Figure 3.6. The skeleton is commonly divided into the axial skeleton, which includes the 80 bones that make up the core of the body, the skull, vertebrae and ribs, and the appendicular skeleton or the 126 bones that make up the shoulders, hips and four limbs. The skeletal system serves five main functions within the body:

- to provide *support* and *protection* for the soft tissues and organs of the body (e.g. the rib cage protects the lungs and heart);

- in conjunction with muscles, which gain their leverage from the bones to which they are attached, the skeletal system assists in bringing about human *movement*;

- bones within the skeletal system are a *store* for minerals and triglycerides; and

- marrow at the heart of bones is the *production* centre for red and white blood cells and platelets (that are involved in blood clotting).

Bones are classified into five main types: *long bones* such as the femur; *short bones* like those of the wrist; *flat bones* which include the ribs; *irregular bones* such as the vertebrae; and *sesamoid bones* (literally, shaped like sesame seeds) of which the patella is an example. Figure 3.7 depicts the structure of the femur, a long bone. It is constructed of two basic types of bone: spongy and compact. *Compact bone* has a greater structural density than spongy bone and consequently has few tissue spaces. It provides strength and protection for bones and is a particular feature of long bones. The less dense matrix of *spongy bone* (sponge-like in its appearance) allows bones to be lighter in weight while maintaining rigidity. Short, irregular and flat bones are generally composed of spongy bone. The spongy bone also allows space for the *red bone marrow*, the site of red and white

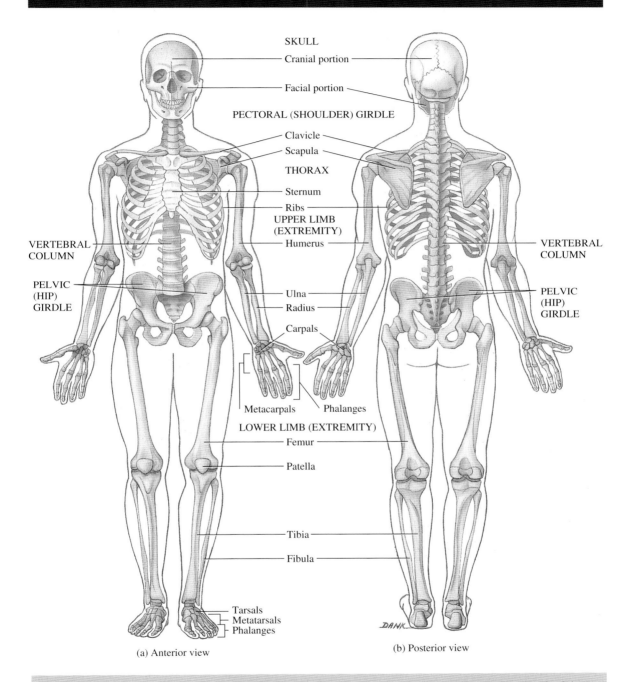

Figure 3.6 The human skeleton (from Tortora and Derrickson, *Principles of Anatomy and Physiology*, Eleventh Edition 2006, reproduced by permission of John Wiley & Sons Inc.)

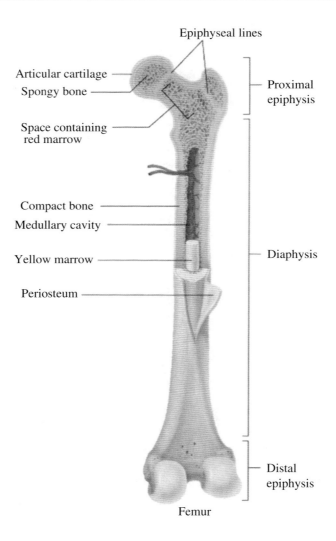

Figure 3.7 The structure of bone (from Tortora and Derrickson, *Principles of Anatomy and Physiology*, Eleventh Edition 2006, reproduced by permission of John Wiley & Sons Inc.)

blood cell and platelet production, and the *yellow marrow* where triglycerides are stored. The yellow marrow is situated in the *medullary cavity*, which is the central cavity within the diaphysis. The medullary cavity is also the central tract for the passage of blood vessels to deliver blood to all part of the bone.

Although much of the cartilage in the body is replaced by bone as the body develops, *articular cartilage* remains and forms a softer tissue layer between the articulating surfaces of a joint. Composed of hyaline cartilage, it is essential to smooth joint movement as it helps to reduce friction and also has a minor role in shock absorption. Each bone is surrounded in a connective tissue sheath, the *periosteum*, which protects the bone, helps with repair and presents the surface for ligament and tendon attachment.

During development, bone growth begins within the *diaphysis* and later, around the time of birth, spreads to the *proximal epiphysis* (nearest to the centre of the body) and *distal epiphysis* (furthest distance from the centre of the body). During growth to adulthood, the *epiphyseal lines* in Figure 3.7 are cartilage-based structures also known as *epiphyseal plates* and are responsible for the growth in length of bones.

Joints

Joints or articulations are formed at the point where two or more bones come together. Fibrous joints such as those found in the scull and cartilaginous joints such as those in the hip do not allow movement. However, synovial articulation is central to the movement that muscles can bring about through contraction. There are a variety of synovial joints throughout the body that enable different degrees of movement. For example, the elbow joint enables less movement than found at the shoulder. Synovial joints comprise a synovial cavity filled with fluid to lubricate the joint as well as articular cartilage, tendons and ligaments to maintain the integrity of the joint. Figure 3.8 provides a diagram of a typical synovial joint: the knee.

Gliding joints, such as between the vertebrae in the spine, are formed when the flat surfaces of bones come together. A *saddle joint* allows more movement than a gliding joint and an example can be found between the trapezium and first metacarpal in the thumb. In this joint, the trapezium forms the saddle and the first metacarpal the 'rider's legs'; the joint allows the thumb to

Anatomy of the Knee

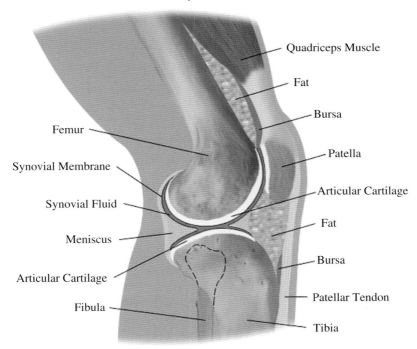

Figure 3.8 The structure of the knee joint (from Tortora and Derrickson, *Principles of Anatomy and Physiology*, Eleventh Edition 2006, reproduced by permission of John Wiley & Sons Inc.)

circumduct (move in a circle) but not rotate. An example of a *condyloid joint* can be found at the wrist. The wrist enables movement in two planes and is formed where a convex surface fits into a concave surface. An example of a *pivot joint* is the atlanto-axial joint in the spine which enables the head to rotate. It is formed between the peg of the axis bone and the ring of the atlas. The elbow and knee are examples of *hinge joints* that enable flexion and extension (bending and straightening) of the arm and leg. The *ball and socket* joints of the shoulder and hip girdles enable the greatest range movement of any synovial joint. These joints enable abduction (moving limb away from midline), adduction (moving limb to midline), flexion, extension, circumduction and rotation.

Key point

The body is composed of 206 bones which serve to support, protect and assist in bringing about human movement. In addition, they are a store of minerals and triglycerides and the production plant for red and white blood cells and platelets. Joints are formed where bones come together and it is through their articulation that movement can occur. The other key component of bringing about movement is the muscular system.

3.3 Nervous system

The endocrine and nervous systems are the control and regulation mechanisms within the body. Their purpose is to alter the functioning of all other systems in the body in order to maintain homeostasis and ultimately life. The endocrine system, described in the next section of this chapter, is the slower, sustained response system. The nervous system, comprising the brain, spinal cord, ganglia, nerves and sensory receptors, represents the body's immediate response mechanism and is able to collect information from the senses, interpret

it and generate a response. It also has an important stimulatory and inhibitory role over parts of the endocrine system. The connection between the nervous and the muscular systems can be found at the neuromuscular junctions described in the next chapter. In a movement response, nerve impulses from the brain or spinal cord link with an effector organ (muscle fibres) at neuromuscular junctions where the nerve stimulation is passed to the muscle fibre to initiate the muscle contraction.

The nervous system, despite its importance in the body, comprises less than 5 per cent of total body weight with a total mass of around 2 kg. It is made up of two main divisions: the *central nervous system* (CNS) which includes the brain and spinal cord and the *peripheral nervous system* (PNS), comprising the 12 pairs of cranial nerves, 31 pairs of spinal nerves, sensory receptors and ganglia, and can be further divided into the afferent and efferent divisions. The *afferent division* comprises sensory or afferent neurones which, as the name suggests, detect internal and external (to the body) signals and relay them to the CNS.

There are five main types of sensory receptors within the body.

1. *Mechanoreceptors* are responsive to stretching and pressure and are associated with the senses of touch, pressure, sound and balance.

2. *Proprioceptors* the awareness of where the body is in space, is a vital sense for many adventure sports brought about through specialised mechanoreceptors referred to as *proprioceptors*.

3. *Photoreceptors* are sensitive to light and are the basis for the sense of vision.

4. *Chemoreceptors* detect chemicals that bind with their cell membranes and are responsible for the senses of smell and taste as well as detecting chemical changes within the body e.g. a CO_2 rise in the blood vessels is a trigger for an increase in heart and breathing rates.

5. *Thermoreceptors* detect both heat and cold and relay this information to the brain and spinal cord.

6. *Nocicereceptors* detect pain and respond to mechanical, thermal and chemical sources of pain which can be from internal or external sources.

The *efferent* or *motor division* consists of nerves that transmit a response (to sensory information that has been processed by the CNS) to be carried out by the relevant tissues, organs or systems in the body. We have conscious control over some motor responses, such as the movement of our limbs (through muscle movements) when rock climbing. However, we have no control over others, such as the change in heart rate that occurs when climbing (or taking part in any adventure sport). The *somatic nervous system* describes the aspects of the nervous system over which we have control: our muscles.

The *autonomic nervous system* relates to those aspects over which the brain and spinal cord have control without us consciously having to think, such as breathing. The final division of the nervous system is of the autonomic nervous system which can be further separated into the sympathetic and parasympathetic divisions. The *sympathetic division* is responsible for speeding up the body systems and is sometimes called the 'fight or flight' response. The *parasympathetic division* is responsible for returning the body to normal functioning and can be thought of as the 'rest and digest' response. The divisions in the CNS are shown diagrammatically in Figure 3.9.

Neurons and neuroglia

At a cellular level, the nervous system is made up of two types of nerve cell: *neurons* and *neuroglia* (neuroglial cells). Neurons and the nerves which they make up, in common with muscle tissue, is *excitable*, meaning that it can conduct an electrical impulse. This nerve impulse or action potential is the signal which passes along a neuron. *Afferent (sensory) neurons* pass information to the brain and spinal cord, and *efferent (motor) neurons* pass these signals from the central nervous system (CNS) to the organ to be effected. Some specialized neurons pass information from one

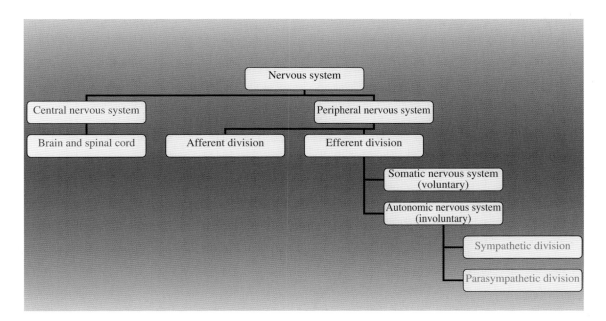

Figure 3.9 Divisions within the nervous system

neuron to another and as a result are called *interneurons* or *association neurons*.

Neurons are normally made up of three main components: dendrites that receive a stimulus which could be sensory or from a neurotransmitter; a cell body; and a single axon. There are three main structural classifications of neurons. Figure 3.10 shows the different structures of neurons. The arrows show the direction in which an impulse will travel. Nerve impulses are received at the dendrites, passed to the cell body and then onwards down the axon to the axon terminal. Neurons are structurally classified by the number of processes extending from the cell body. Although all neurons have only one axon, *multipolar neurons* have many dendrites connecting to the cell body, *bipolar neurons* have separate dendrite processes and axons leaving the cell body, and *unipolar neurons* have only one joint dendrite axon process

connecting with the cell body. Motor neurons are generally multipolar in structure and sensory neurons are mainly unipolar. However, some specialist neurons such as those of the rods and cones of the eye are bipolar.

Neurons and neuroglia are formed during development from neuroblasts, the nervous system equivalent of osteoblasts and myoblasts found in the skeletal and muscle tissue. Neurons vary greatly in both length and diameter, dependent on their function within the body. Some nerve cells, such as those responsible for bringing about muscular contraction of the lower leg and feet (e.g. when pushing on the footrests in a kayak) are over 1 m in length. Figure 3.11 shows a multipolar neuron in more detail. As with other cells in the body, the cell body or *perikaryon* comprises a cell membrane nucleus and cytoplasm. The plasma membrane of the cell body and

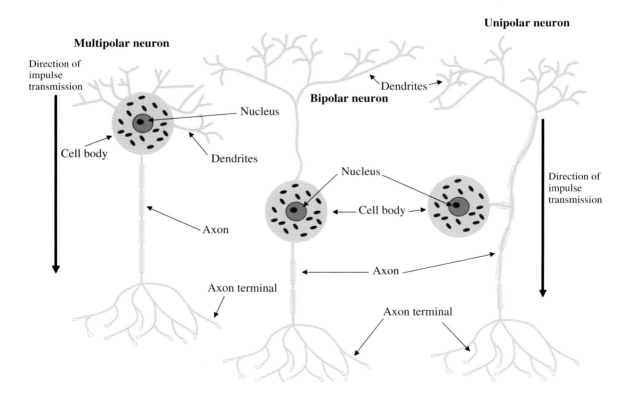

Figure 3.10 Classifications of neurons

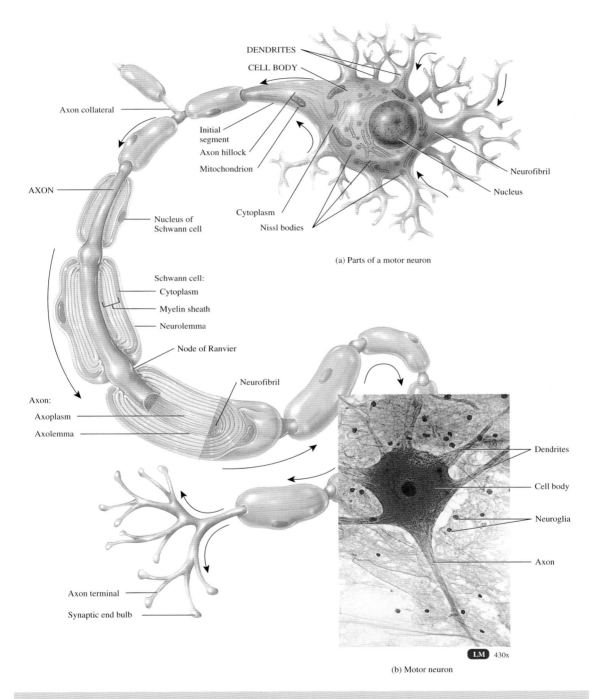

DENDRITES

CELL BODY

Axon collateral

Initial segment

Axon hillock

Mitochondrion

AXON

Nucleus of Schwann cell

Cytoplasm

Nissl bodies

Neurofibril

Nucleus

(a) Parts of a motor neuron

Schwann cell:

Cytoplasm

Myelin sheath

Neurolemma

Node of Ranvier

Neurofibril

Axon:

Axoplasm

Axolemma

Dendrites

Cell body

Neuroglia

Axon

Axon terminal

Synaptic end bulb

LM 430x

(b) Motor neuron

Figure 3.11 The structure of a multipolar neuron (from Tortora and Derrickson, *Principles of Anatomy and Physiology*, Eleventh Edition 2006, reproduced by permission of John Wiley & Sons Inc.)

dendrites is called the *neurolemma*. However, the cell membrane of the axon is usually referred to as the *axolemma* as the function of axons necessitates differences in structure when compared to the rest of the cell. In a similar way, the cytoplasm of the perikaryon is known as axoplasm in the axon of the neuron.

The cytoplasm of neurons contains many organelles common to other cells such as lysosomes, ribosomes, mitochondria and a Golgi complex. As well as these general organelles, they contain some neuron-specific components such as neurofibrils, microtubules and Nissl bodies. *Neurofibrils* comprise protein filaments and are designed to maintain the structure of the cell body. *Microtubules* provide a transport structure through which substances can pass between the cell body and the axon. *Nissl bodies* are small rough endoplasmic reticulum structures that, within the neuron, are responsible for synthesizing proteins for growth and repair.

The axon of each neuron propagates action potentials to the receiving neuron, organ or tissue. Due to their often great length and lack of Nissl bodies (for protein synthesis), axons have two specialist transport systems – slow and fast axonal transport – that enable essential substances to be moved between the axon and the cell body. *Slow axonal transport* (at a speed of about 0.1 mm hr^{-1}) is responsible for the movement of axoplasm and *fast axonal transport* (around 12 mm hr^{-1}) is responsible for moving organelles such as mitochondria and proteins between the cell body and axon terminals. After information from the dendrites is received and processed by the nucleus, the nerve impulse is propagated from between the axon hillock and the initial segment from the area known as the trigger zone. Some axons have branches from them called *collaterals* which lead to another set of *axon terminals*. The axon terminals form the link with other tissues or neurons. When a motor neuron meets a muscle fibre the junction is called a neuromuscular junction, whereas the junction between two neurons is called a *synapse*. Synapses will be described in the following section on nerves.

The axons of any neuron are described by their diameter and whether they are myelinated. The diameter and myelination of the axon also determine the speed of nerve impulse conduction. Type A axons, which comprise all motor neurons and many sensory neurons such as proprioceptors within the somatic nervous system, are the largest diameter fibres (between $4-20 \text{ }\mu\text{m}$) and are myelinated. The larger surface area and myelinated nature of the axons means that nerve impulses travel at the highest speeds in these fibres; as high as 140 m s^{-1} or over 300 mph. Type B fibres are smaller (between $2-4 \text{ }\mu\text{m}$) but still myelinated and as a consequence impulses also travel at relatively quick speeds; up to 18 m s^{-1} or around 36 mph. Type B fibres form the majority of preganglionic autonomic nervous system neurons (described in the section on the PNS). Type C, the smallest axons, are up to $2 \text{ }\mu\text{m}$ in diameter and are unmyelinated. Consequently, nerve impulses travel at slower speeds of $1-2 \text{ m s}^{-1}$ or up to 4 mph. Post ganglionic neurons within the autonomic nervous system comprise predominantly Type C axons.

Neuroglia, although smaller than neurons, are the most common type of nerve cell found in the body and make up over half of the CNS. Neuroglia are found in the CNS and the PNS. There are four types of neuroglia found in the CNS: astrocytes; ependymal cells; microglia; and oligodendrocytes. There are two types found in the PNS: Schwann cells (Figure 3.11) and satellite cells. *Astrocytes* have several functions within the CNS which include, due to their high protein structure, providing support for neurons, maintaining chemical balance and regulating nervous tissue growth. *Ependymal cells* are cube-shaped cells found on the inside edges of the sheaths which enclose the brain and spinal cord and have hair-like structures protruding from them called microvilli and cilia which help with the flow of cerebrospinal fluid (CSF). *Microglia* are involved in the repair of neurological tissue and help by removing debris from the interstitial environment. *Oligodendrocytes* are responsible for creating a myelin sheath around neurons of the CNS. The presence of a myelin sheath helps with the speed of nerve impulse transmission and is discussed in more detail later.

Schwann cells are the PNS equivalent of oligo-dendrocytes and, as can be seen in Figure 3.11, each Schwann cell is responsible for myelination of one section of a neuronal axon. Schwann cells also have a role in providing a coordinated structure and protection for unmyelinated neurons and can enclose up to 10 or 20 of these neurons. *Satellite cells*, the second form of neuroglia found in the PNS, help with the regulation of nutrients within the interstitial fluid around neurons.

Nerves

Neurons are found collected in groups or bundles, unmyelinated neurons forming the grey matter of nervous tissue and myelinated neurons forming the white matter (myelin is formed from lipid and protein). The structure of a myelinated nerve fibre, depicted in Figure 3.12, has some similarities to muscle fibres in that individual fibres or neurons are grouped together into fascicles and are surrounded by three protective sheaths. The protective sheaths for nerve fibres are the epineurium, the perineurium and the endoneurium.

The structure of an unmyelinated nerve also includes Schwann cells that wrap around the fibres but, unlike Figure 3.12, many fibres are encapsulated by one Schwann cell but without myelination. The majority of nerves are myelinated, since as previously mentioned this helps with the speed of nervous transmission. Within the PNS, the presence of Schwann cells and their neurolemma surrounding each axon provides a protective tube in the event of injury (providing the nerve fibres

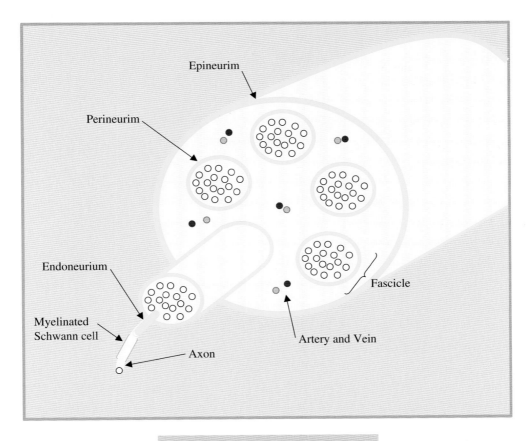

Figure 3.12 The structure of a nerve

remain aligned). In this situation, the protective tube created by the Schwann cell neurolemma enables *neurogenesis* (new tissue growth). However, if the damaged neuron is misaligned during injury the tissue may not be able to repair itself. There is currently little or no chance of nervous tissue regeneration within the CNS, although research into medical stimuli for neurogenisis is progressing.

The brain has 12 pairs of nerves that emerge from and return to the CNS and serve sensory and motor roles in and around the head for somatic (muscular system) and autonomic function. The spinal cord has 31 pairs of nerves that serve somatic and autonomic functional roles for the rest of the body. The cell bodies of the neurons that form these 43 pairs of nerves are normally located within ganglia – small masses of nerve tissue – that exist on the outside of the CNS. It is therefore the axons of each neuron which form the nerves that descend and ascend from the CNS.

A synapse is formed at the junction between two neurons. Figure 3.13 depicts the structure of a synapse. Synapses enable an action potential to pass from one neuron to another. The neuron carrying the action potential (already stimulated

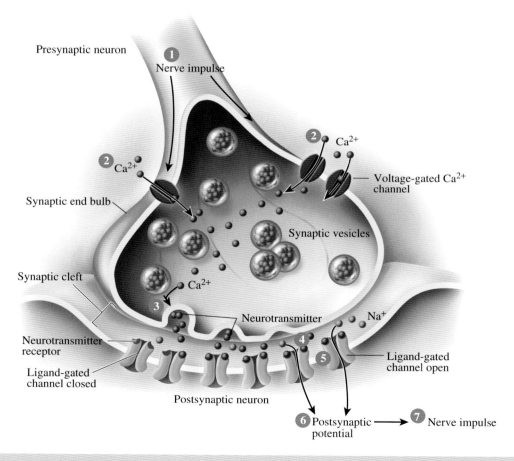

Figure 3.13   The structure of a synapse (from Tortora and Derrickson, *Principles of Anatomy and Physiology*, Eleventh Edition 2006, reproduced by permission of John Wiley & Sons Inc.)

and conducting a nerve impulse) is called the presynaptic neuron; the one to which it will pass is the postsynaptic neurone. The sequence of events for the transfer of the action potential from one neuron to the next is very similar to that at a neuromuscular junction. As can be seen from Figure 3.13, the arrival of the nerve impulse at the axon terminal of the presynaptic neurone causes voltage-gated transmembrane protein channels to open. These channels allow Ca^{2+} to enter the axon terminal. The presence of calcium within the presynaptic neuron causes vesicles containing a chemical message – a neurotransmitter – to fuse with the plasma membrane and open the vesicle to the synaptic cleft.

Key point

The nervous system along with the endocrine system form the control and regulation mechanisms for all other systems in the body. The nervous system comprises the central nervous system (the brain and spinal cord) and the peripheral nervous system (all nervous tissue outside the brain and spinal cord). The nerve cells that form the basis of the nervous system are called neurons and neuroglia. The 43 pairs of nerves that serve the body are formed from bundles of neurons. Each neuron comprises dendrites (for receiving information or stimuli), a cell body and a single axon that propagates nerve impulses. Neuron cell bodies are located in ganglia (knots of nervous tissue situated outside the brain and spinal cord) and it is the axons of each neuron that form the structures for nerves. Neurons are generally sensory (afferent) i.e. relay information to the CNS or motor (efferent) i.e. transmit responses from the brain. The 43 pairs of nerves are composed of sensory and motor axons.

The neurotransmitter is able to cross the interstitial fluid within the synaptic cleft and bind with ligand-gated channels to enable sodium to pass into the postsynaptic neuron and initiate a postsynaptic nerve impulse. Ligand-gated channels are also transmembrane proteins which have special receptor cells on their outside (open to the synaptic cleft) that will only enable a particular neurotransmitter to bind with them and open the channel. The neurotransmitter for neuromuscular junctions and many neurons in the PNS is acetylcholine. As well as acetylcholine for the CNS, there are several proteins that form the basis for synaptic neurotransmitters. Aspartic and glutamic acids and glycine work directly as neurotransmitter and protein tyrosine forms the basis for the synthesis of the catecholamines adrenalin, noradrenalin and dopamine, all of which are neurotransmitters within the brain.

Nerve impulse conduction

The conduction of nerve impulses is essentially linked to the *sodium-potassium pump* described in the section on cell membranes (plasma membranes) above and depicted in Figure 3.5. It is worth reading this section again before reading about the nerve impulse conduction. A nerve impulse or action potential is an *electrical signal* propagated by the *electrolytes* (chiefly sodium and potassium) within the intercellular and extracellular fluids (inside and outside the cell). The sodium-potassium pump creates a differential in the levels of sodium and potassium ions on the inside and outside of the cell membrane. The pump expels sodium ions and imports potassium ions from the cytoplasm. The differential between the ions on the inside and outside of the cell can be shown as an electrical charge and it is this (resting) membrane potential that enables nerve impulse conduction. The selectively permeable nature of the cell's plasma membrane (cell membrane) along with the sodium-potassium pump allow for the *resting membrane potential* to be established.

The sodium-potassium pump expels three sodium ions for every two potassium ions and in addition: (a) the plasma membrane is more 'leaky' to potassium ions (which will therefore move from high concentration in the cell to low concentration outside the cell) than to sodium ions; and (b) there are many large negatively charged protein molecules and phosphate ions within the cell, which result in a greater number of positive ions outside the plasma membrane. The negative ions immediately inside the cell membrane are drawn towards the greater number of positive ions on the outside of the cell, which generates a fine layer of ions across the membrane that have a relatively large difference in charge. This difference in charge between the inside and outside of the cell, the *resting membrane potential*, can be measured by a voltmeter and in most cells is around −70 mV. Cells that maintain a resting membrane potential are said to be *polarized* (there is a difference in the negativity or positivity inside and outside the cell). All cells in the body maintain a resting membrane potential; however, nerve and muscle cells have developed a method to use this electrical differential to propagate an electrical signal.

When a stimulus reaches a muscle fibre or a neuron it changes the immediate electrical charge and the opening of specialist voltage-gated transmembrane protein channel in the cell membrane is triggered. The sodium voltage-gated channels open and enable sodium ions to flood into the cell before they close again. The influx of sodium ions changes the electrical charge from −70 mV to around +35 mV and as a consequence the cell is said to be *depolarized*. This depolarization is the action potential. The action potential does not flow across the cell membrane in one movement, especially as axons and muscle fibres can be quite long, but moves across the cell membrane as a *wave of depolarization*. As the action potential travels along an axon or muscle fibre it depolarizes the next segment in the plasma membrane. As soon as the membrane has depolarized, the sodium voltage-gates are closed and potassium channels open and enable potassium ions to leave the cell. This action coupled with the sodium-potassium pump returns the neuron or muscle fibre to its resting membrane potential i.e. it has been *repolarized*. A repolarized cell is

then ready to conduct a new action potential. The whole process of depolarization/repolarization is very fast, taking a thousandth of a second (1 ms or 0.001 s).

The process of nerve impulse conduction in an unmyelinated neuron is conducted in waves across the axon of a neuron in the way described above. Myelination of an axon is created by a Schwann cell, each of which is about 1 mm in length, and the gaps between the cells and in the myelination are called nodes of Ranvier. Neuron myelination enables the action potential to jump from one node of Ranvier to another. Myelination insulates the myelinated section of fibre and, as a consequence, the depolarization jumps from one node to the next. The jumping of the action potential, called salutatory conduction, results in the speed of travel of the impulse being greatly enhanced. This and the difference in axon diameter is the reason why Type A and B axons conduct nerve impulses much more quickly than Type C axons. Figure 3.14 shows nerve impulse conduction in a myelinated and unmyelinated neuron.

Key point

A nerve impulse or action potential is an electrical message that travels along the axon. Nerve impulses can travel to the brain (sensory) or from the brain (motor). Nerve impulse conduction is enabled through the selectively permeable nature of the cell's membrane, through the transmembrane proteins that enable electrolytes to move between the inside and outside of the cell and the resting membrane potential established by the sodium-potassium pump. An electrical differential known as the resting membrane potential is established between the inside and outside of the cell membrane. During nerve impulse, conduction channels in the cell membrane open and enable sodium ions to flood into

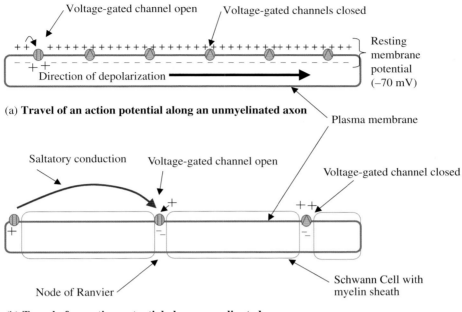

Voltage-gated channel open

Voltage-gated channels closed

Resting membrane potential (–70 mV)

Direction of depolarization

(a) **Travel of an action potential along an unmyelinated axon**

Plasma membrane

Saltatory conduction

Voltage-gated channel open

Voltage-gated channel closed

Node of Ranvier

Schwann Cell with myelin sheath

(b) **Travel of an action potential along a myelinated axon**

Figure 3.14   Nerve impulse conduction in myelinated and unmyelinated neurons

the cell depolarizing it. A nerve impulse travels in a wave along a neuron or muscle fibre by depolarizing the next segment of the plasma membrane.

Central nervous system (CNS)

The CNS is composed of the brain and the spinal column. The brain has a grey outer and white inner and the spinal cord the reverse of this; the grey neurons represent unmyelinated neurons, the white myelinated neurons. The human brain, although making up less than 3 per cent of total body mass, comprises around 100 billion neurons and uses up to 20 per cent of oxygen and glucose consumed at rest. The reason for the relatively high oxygen demand by the brain relates to ATP synthesis by the neurons and neuroglia of the brain, which is almost exclusively through aerobic means.

The cerebrospinal fluid that helps to protect the brain and spinal column is produced from the walls of the ventricles within the brain. It is a clear liquid that as well as providing protective mechanisms also provides the brain with nutrients and electrolytes. The protective structures of the brain very closely match those of the spinal cord. The main differences between the two are that the brain has two dura maters – the outside protective meninge – and no epidural space.

The brain is commonly divided into four regions: the brain stem (composed of the medulla oblongata, pons and midbrain); the cerebellum; the diencephalon (which incorporates the thalamus, hypothalamus and the epithalamus); and fourthly the cerebrum. The *brain stem* is responsible for providing a link between the spinal cord and the brain as well as controlling breathing (medulla oblongata and pons), heart rate, blood vessel diameter and a variety of reflexes such as coughing, vomiting and sneezing. The *cerebellum* is very important to adventure sports performers as it represents the feedback centre of the brain. The cerebellum judges movements made against planned movements and then makes adjustments as necessary. This is very important in the development of fine motor movements. Through the structures that it contains, the *diencephalon* plays an important role in controlling autonomic nervous function, control of circadian rhythms (body clock), regulation of behaviour and emotion (hypothalamus), provides a relay of information to the cerebrum (thalamus) and is involved in endocrine function (hypothalamus and pituitary gland, part of the epithalamus). The *cerebrum*, the largest component of the brain, is responsible for sensory and motor functions including muscular movement as well as higher cognitive functions such as learning, memory, intelligence and personality.

The *spinal cord*, on average 45 cm long, 1 cm wide and comprising around 100 million neurons, is protected by a number of structures including the 33 vertebrae that form the vertebral column, 3 protective sheaths, surrounding fluids, connective tissue and fat. The spinal column is located within the vertebral cavity and is immediately surrounded by three protective sheaths: the dura mater; arachnoid mater; and pia mater which are known collectively as the spinal meninges. Within the epidural space, between the meninges and the vertebrae, are connective tissue and fat as a second level of protection. The dura mater is also constructed of connective tissue and forms the outer layer of the meninges. The subdural space, between the dura mater and the arachnoid mater, is filled with interstitial fluid which serves to maintain electrolyte balance and as a nutrient supply. Between the arachnoid mater and the inner protective layer – the pia mater – is cerebrospinal fluid. The three meninges are continuous with the meninges of the brain and, as such, the pia mater carries blood vessels, nutrients and oxygen to the brain and spinal cord.

Peripheral nervous system (PNS)

The PNS, which structurally includes the ganglia, sensory receptors and nerves, can be further divided into the somatic nervous and autonomic nervous systems. The cranial and spinal nerves provide the afferent and efferent neuron pathways for the somatic and autonomic divisions. *Sensory neurons* for the somatic and autonomic systems have the same structure and function in the same fashion, receiving information from receptors throughout the body and passing it to the CNS. At the CNS, the sensory information is processed and a response generated. The efferent pathways are where structural and functional differences between the autonomic and somatic nervous systems can be identified.

The somatic nervous system motor neurons comprise Type A axons for fast innervation of the relevant skeletal muscle. Within the somatic nervous system, one motor neuron carries the nerve impulse from the CNS to the skeletal muscle fibres. In contrast, for an *autonomic motor response* two synapsed (linked by a synapse) neurons form the efferent pathway. In addition, there are two branches of efferent autonomic neuron: those that have a *sympathetic* effect (speed up the organ e.g. increasing heart rate during exercise) and those that carry a *parasympathetic* response (slowing down the organ e.g. decreasing heart rate). The first of these visceral neurons has its cell body within the brain or spinal cord and conducts a nerve impulse to small clusters of neuron cell bodies called an *autonomic ganglion* (similar to ganglia described above). This type of visceral neuron is called a *preganglionic neuron* because it is the neuron that conducts the autonomic response from the CNS to the autonomic ganglion. As was mentioned in the section on neurons (above), preganglionic neurons normally have Type B axons and, as a consequence, nerve impulse conduction is slower than in somatic (motor) neurons. *Postganglionic autonomic neurons* carry impulses from the ganglion to the tissue or organ over which they have effect, and are most usually Type C fibres. Figure 3.15 provides a diagrammatic representation of differences and similarities between sympathetic visceral, parasympathetic visceral and motor neurons.

Type A motor neurons of the somatic nervous system have their cell bodies within the CNS and their axons form part of one of the 43 paired nerves that exit the CNS. In contrast, sympathetic preganglionic neurons (with Type B axons) have

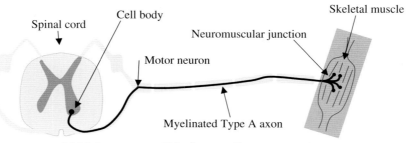

(a) Motor neuron within the somatic nervous system

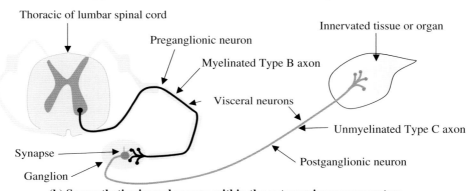

(b) Sympathetic visceral neuron within the autonomic nervous system

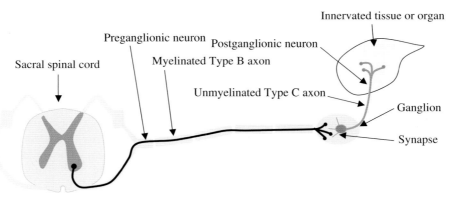

(c) Parasympathetic visceral neuron within the autonomic nervous system

Figure 3.15 Somatic and autonomic efferent neuronal pathways

their cell bodies located only in the thoracic or lumbar regions of the spine and their axons originate from the spinal cord.

There are three *sympathetic autonomic ganglions* which are found either side of the posterior of the vertebrae (two) and anterior of the spinal column (one) from which the Type C postganglionic neurons emerge to complete the nerve impulse journey. *Parasympathetic preganglionic neurons* originate only from cranial nerves or the sacral region of the spinal cord and connect with the parasympathetic autonomic ganglions. The *parasympathetic autonomic ganglions* are located close to or within the walls of the organ or tissue to be innervated, so by the time the nerve impulse reaches the parasympathetic ganglion it has almost completed its journey.

Key point

The nervous system can be divided into the CNS (the brain and spinal cord) and the PNS (the nerves and ganglia outside the CNS). The PNS comprises the somatic nervous system (for control of skeletal muscles) over which we have control (often called voluntary) and the autonomic nervous system (for control of smooth muscle, cardiac muscle and other tissues and organs) over which we do not consciously have control (also named involuntary). The somatic nervous system comprises sensory (afferent) neurons which relay information to the brain and motor neurons that carry signals from the brain to the skeletal muscles. Motor neurons carry an action potential that stimulates the relevant muscle to contract. The autonomic nervous system, in common with the somatic nervous system, has sensory neurons that carry information from receptors to the brain. However, efferent neurons for the autonomic nervous system differ from those of the somatic nervous system. Each autonomic tissue or organ is supplied by two types of visceral (efferent) neurons: those that speed up the organs they affect (sympathetic) and those that slow the organs down (parasympathetic). A further difference between somatic and autonomic efferent neurons is the number of neurons that form the pathway. Motor neurons have one neuron with their cell body in the CNS and axon travelling to the skeletal muscle being innervated. Autonomic visceral neuron pathways comprise two motor neurons: one to carry the nerve impulse from the CNS to the ganglion (a cluster of neuron cell bodies) and a second to carry the signal from the ganglion to the organ or tissue.

3.4 Endocrine system

The endocrine system is the second control system. Working alongside the nervous system, it regulates the function of the nine other systems within the body. The organs and tissues of the main endocrine system are depicted in Figure 3.16. Some have an exclusively endocrine function such as the thymus, whereas other such as the kidneys have an additional role. Endocrine comes from the Greek *endo* meaning inside and *crino* meaning separate, underlying the way it operates. Three elements are found in an endocrine function: a *production tissue* that secretes a *hormone* from within the tissue into the bloodstream, where it is transported to a *target cell* upon which it will have an effect (tissue or organ). A hormone, from the Greek *hormaō* meaning to set in motion, is a chemical message that initiates a change in the function of specific cells. Each hormone is highly specific and will only impact upon the target cell for which it is intended, although some hormones e.g. human growth hormone are intended to impact across a wide number of cells within the body. The total weight of the exclusively endocrine organs is relatively small: around 0.5 kg in an adult. However, their importance to the body as a control system is considerable.

Endocrine organs

The main organs of the endocrine system, the common hormones they produce and the effect of these hormones on the body are shown in Table 3.2. The hypothalamus represents the controller of endocrine function and, along with the pituitary gland, produces 16 hormones (nine from

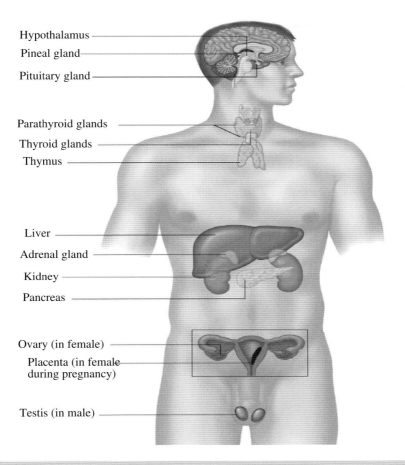

Hypothalamus

Pineal gland

Pituitary gland

Parathyroid glands

Thyroid glands

Thymus

Liver

Adrenal gland

Kidney

Pancreas

Ovary (in female)

Placenta (in female during pregnancy)

Testis (in male)

Figure 3.16 Organs of the endocrine system (from Tortora and Derrickson, *Principles of Anatomy and Physiology*, Eleventh Edition 2006, reproduced by permission of John Wiley & Sons Inc.)

hypothalamus and seven from pituitary gland) which regulate growth, maturation, metabolism and homeostatic balance. The hypothalamus is the interaction point between the endocrine and nervous system, for not only is it the driver of endocrine function, but it also controls autonomic nervous system function. In its role within the nervous system, it is responsible for controlling homeostatic processes such as body temperature, thirst and hunger.

The hypothalamus receives sensory information that it processes to effect endocrine and/or autonomic responses. The sensory input for the hypothalamus originates from a variety of sensory receptors throughout the body, including hormone levels and stresses and emotions affecting the individual.

Function of hormones

Hormones regulate the functioning of the body by impacting upon processes such as growth, metabolism, reproduction and circadian rhythms. There are two main groups of hormone: those that are lipid- (fat) based and those that are protein-based. Table 3.3 lists the most common hormones, their classification and the tissues that produce them. The majority of hormones (all of

Table 3.2 Summary of the major endocrine glands, hormones and their function

Endocrine gland	Hormone	Function	Target
Hypothalamus	Releasing hormones Inhibiting hormones	Increase and decrease the functioning of the pituitary gland	Pituitary gland
Pituitary gland	Growth hormone	Stimulates growth and development of all body tissues	All cells
	Prolactin	Stimulates milk production in breast feeding mothers	Breasts
	Antidiuretic hormone	Plays a major role in reducing water excretion by kidneys	Kidneys
	Endorphins	Involved in providing pain relief for the brain	Brain
	Oxytocin	Assists during childbirth with the regulation of uterus contraction	Uterus
Pineal gland	Melatonin	Regulates sleep cycles by increasing the desire to sleep	Brain
Thyroid	Throxine	Increases metabolism and heart rate	All cells
	Calcitonin	Decreases blood levels of calcium and phosphate	Bones, kidneys
Parathyroid glands	Parathyroid hormone	Increases calcium levels in interstitial fluid and blood	Bones, kidneys small intestine
Thymus	Thymosin	Promotes and maintains immune function	All cells
Adrenal medulla	Adrenalin	Increases breakdown of glycogen and fats, heart rate, O_2 consumption	Most cells

Table 3.2 (*continued*)

Endocrine gland	Hormone	Function	Target
	Noradrenalin	Increases lipid breakdown (lipolysis) and blood pressure	Most cells
Adrenal cortex	Aldosterone	Increases sodium retention and potassium excretion	Kidneys
	Cortisol	Anti-inflammatory and plays a role in control of metabolism	Most cells
Kidneys	Erythropoietin	Stimulates red blood cell (erythrocyte) production	Bone marrow
Pancreas	Insulin	Increases glucose uptake from the bloodstream	All cells
	Glucagon	Promotes release of glucose in to the bloodstream	All cells
	Somatostatin	Decreases glucagons and Insulin production	Pancreas
Gonads			
Testes	Testosterone	Promotes muscle growth and development of male sex characteristics	Sex organs, muscles
Ovaries	Oestrogen Progesterone	Promote development of female sex characteristics, regulates menstrual cycle and adipose tissue growth	Sex organs, adipocytes
Placenta	Gonadotropin	Stimulates testosterone production in the foetus – promoting growth	Foetal testes
Platelets	Serotonin	Binds sites on sodium channels – promotes depolarisation	All cells

Table 3.3 Classification and secretion site of common hormones

Hormone	Hormone classification	Secretion site
Protein-based hormones		
Growth hormone	Protein	Pituitary gland
Insulin	Protein	Pancreas
Prolactin	Protein	Pituitary gland
Antidiuretic hormone	Polypeptide	Pituitary gland
Glucagon	Polypeptide	Pancreas
Somatostatin	Polypeptide	Pancreas
Endorphins	Polypeptide	Pituitary gland
Oxytocin	Ploypeptide	Pituitary gland
Adrenalin	Amino acid derivative	Adrenal medulla
Noradrenalin	Amino acid derivative	Adrenal medulla
Serotonin	Amino acid derivative	Platelets
Melatonin	Amino acid derivative	Pineal gland
Thyroxine	Amino acid derivative	Thyroid
Erythropoietin	Glycoprotein	Kidneys
Parathyroid hormone	Glycoprotein	Parathyroid glands
Thyroid stimulating hormone	Glycoprotein	Pituitary gland
Lipid-based hormones		
Testosterone	Steroid	Testes
Oestrogen	Steroid	Ovaries
Progesterone	Steroid	Ovaries
Aldosterone	Steroid	Adrenal cortex
Cortisol	Steroid	Adrenal cortex
Prostaglandins	Fatty acid derivative	All cells but RBC
Leukotrienes	Fatty acid derivative	All cells but RBC

those in Table 3.3) are produced by a host cell, secreted into the interstitial fluid and from there diffused into the bloodstream where they are carried to their target cell. The hormonal exceptions are: *paracrines* which secrete hormones into the interstitial fluid around their cell, targeting cells very close to the host organ and therefore not entering the bloodstream; and *autocrines* that secrete hormones that have an effect on the cell from which they are secreted. Normally, however, hormones are carried in the bloodstream to the target cells. For example, erythropoietin (erythrocyte – red blood cell – production stimulating hormone) is produced in the kidneys and travels via the bloodstream to the bone marrow.

Lipid-based and protein-based hormones are both carried by the bloodstream to their target cell. However, the methods by which they are transported and how they make contact with target cells are different. Figure 3.17 shows the nature

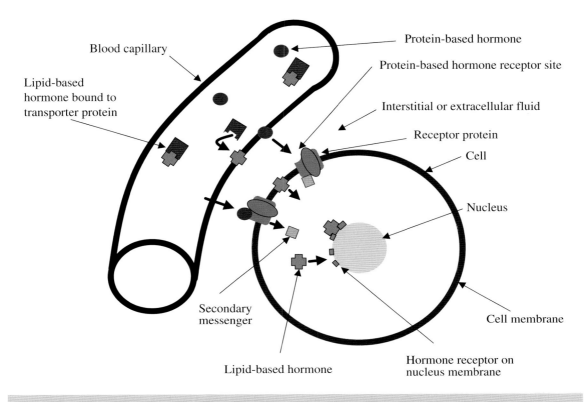

Figure 3.17 Process through which lipid-based and water-soluble hormones pass their message to the nucleus of the receptor cell

of lipid and protein-derived hormone transport and message entry to the target cell. As a result of their insolubility in water, lipid-based hormones have to attach to carrier proteins which enable them to be transported to the target cell. As they reach the target cell they disconnect from the carrier protein, which is left in the bloodstream to bind with other lipid-based hormones and diffuse to the interstitial fluid surrounding the target cell. When they reach the target cell, their lipid nature gives them an advantage over protein-based hormones. Due to the lipid nature of the plasma membrane of each cell, they are able to diffuse into the cell then reach their targets. For lipid-based hormones, the target is normally found on the membrane of the nucleus of each cell (Figure 3.17). Protein-based hormones, such as human growth hormone, travel within the bloodstream as an independent unit. However, they

cannot pass through the cell membrane of the target cell so have to attach themselves to special receptor transmembrane proteins which form part of the plasma membrane. The binding of the protein-based hormone causes the release of a secondary messenger that stimulates the desired response within the cell (Figure 3.17). Cells are able to differentiate between hormones because each cell has specific receptors for each type of hormone. There are normally around 2000–10 000 receptors for each hormone that a cell requires as part of its functioning and regulation.

An example of hormone functioning can be seen through the action of glucagon and insulin, both of which are produced by the pancreas. The degree of glucose within the bloodstream is regulated by the secretion of glucagon and insulin. This control process is carried out by negative feedback. After the food we have eaten has been digested, blood glucose levels rise. If blood glucose levels rise above normal levels we become hyperglycaemic and this stimulates the β (beta) cells of the pancreas to produce insulin. Insulin enters the bloodstream and takes its effect on the many cells throughout the body, especially those of the liver and skeletal muscles, stimulating them to take up glucose from the bloodstream and form glycogen (glycogenesis). Insulin has the effect of reducing the level of glucose in the bloodstream to normal, at which point through negative feedback (being no longer required) the pancreas discontinues insulin secretion.

Glucagon, produced by pancreatic α (alpha) cells, is secreted at times of low blood glucose (hypoglycaemia). The exercise undertaken during a training session or a day's climbing, paddling or caving can result in a decrease in blood glucose levels which stimulate the pancreas to secrete glucagon into the bloodstream where its target cells are the hepatocytes of the liver. The liver and muscles are a major store of glycogen and, when the muscles are already exercising, glucagon stimulate the liver to convert its stores of glycogen back to glucose (glycogenolysis). In addition, the presence of glucagon will stimulate the liver to convert lactic acid and amino acids through gluconeogenesis into glucose. When blood glucose levels reach normal, the production of glucagon is inhibited.

Key point

The endocrine system is the second of the body's control and regulation systems. Each endocrine unit consists of a host gland, a hormone and a target cell. Hormones, the chemical messages endocrine organs produce, target specific cells and change the functioning of those cells. Hormones impact upon metabolism, growth, immune function, the body's internal environment and many other aspects of human functioning.

3.5 Summary and study questions

Cells

Cells are the basic units of living things. Physiologists organise the cells within the body into a hierarchy of tissues, organs and systems. In human beings, around 200 different types of cell make up 11 different systems. Each of these cells share a common general structure but each kind of cell plays a different function and therefore has different characteristics and developed components. Cells are composed of three major parts: a cell membrane that surrounds all of its other components, a nucleus responsible for controlling its functions and cytoplasm that contains organelles that carry out the other work of the cell.

The cell membrane is responsible for maintaining a barrier between the cell's internal environment and the external environment. In addition, the cell membrane also contains transmembrane proteins that allow communication and the transportation of selected ions or molecules into and out of the cell. There are four main kinds of transmembrane proteins: recognition proteins, receptor proteins, channel proteins and transport proteins.

The nucleus is the control centre of the cell. Most cells have one nucleus; however, red blood cells expel their nucleus and bone cells can contain more than one nucleus. The key feature of the nucleus is the DNA (deoxyribonucleic acid) which contains the genetic code, or blueprint, for the cell and directs its protein and enzyme synthesis.

The cytoplasm consists of cytosol (intercellular fluid) and organelles. This intercellular fluid contains dissolved particles essential for the functioning of the cell as well as being the place where many metabolic functions take place. The cytosol also provides the structure and rigidity of the cell through its network of protein filaments, referred to as the cytoskeleton. Other metabolic functions, from cell reproduction to energy production, are carried out by the organelles of which there are nine main types.

Living cells require a relatively narrow band of internal conditions in order to sustain life efficiently. As a result, cells work actively to maintain a constant internal environment and this process is referred to as homeostasis. Cells within the human body work cooperatively in order to achieve homeostasis; in fact, the main function of many cells is supporting the homeostasis of other cells. The human body can be required to work very hard to maintain this homeostasis during adventurous activities where extreme temperatures, altitudes and energy requirements can all threaten the homeostatic balance.

The skeletal system

The skeletal system provides the basic structure and support for the human body. It consists of bones, cartilage, tendons and ligaments. The skeletal system has five main functions: support and protection of the soft tissues, a role in movement, storage of minerals and triglycerides and the production of some blood components.

The nervous system

The nervous system is one of the two control and regulation systems for the human body. The nervous system is responsible for immediate responses and the control and coordination of the muscular system. It also has a key role in the stimulation and inhibition of the body's other control and regulation system (the endocrine system). The nervous system comprises the brain, spinal cord, ganglia, nerves and sensory receptors. These components are often divided into the central nervous system (the brain and spinal cord) and the peripheral nervous system. The peripheral nervous system (PNS) can also be further divided into the afferent division, which carries information to the central nervous system (CNS), and the efferent division, which carries signals away from the CNS.

Another useful distinction is to divide the nervous system into the somatic nervous system, over which we have conscious control, and the autonomic nervous system which functions subconsciously. The autonomic nervous system has sympathetic and parasympathetic divisions, which speed up and slow down the body's autonomic responses, respectively.

There are two types of cell that make up the nervous system: neurons and neuroglia. Neurons are capable of carrying electrical impulses and perform the transmission function of the nervous system. The neuroglia are cells which maintain and support the neurons and allow them to perform their communication function effectively. The junction between two neurons is referred to as a synapse. The electrical impulse in the presynaptic neuron triggers the release of a chemical messenger known as a neurotransmitter, capable of crossing the synaptic cleft, which in turn triggers the electrical impulse in the postsynaptic neuron.

The endocrine system

The endocrine system is the second control and regulation system for the human body. It involves the transmission of chemical messengers called hormones through the bloodstream. Hormones are secreted by production tissues and have an effect on target cells, tissues or organs. Each type of hormone can only impact on specific target cells, although in some cases these can be spread throughout the body.

The hypothalamus provides a junction between the two communication systems of the body controlling autonomic nervous system activity and endocrine functioning. This junction ensures that the two systems respond in harmony. Hormones can be either protein-based or lipid-based messengers and, as a consequence, have different implications for transportation and the effect on the target cell. Protein-based hormones are water soluble and can therefore be dissolved into plasma and carried in the blood directly. Lipid-based hormones need to attach to carrier proteins in order to be transported. Lipid-based hormones can diffuse directly into target cells whereas protein-based hormones attach to receptor proteins on the outside of the cell membrane, causing the release of a new messenger within the cell.

Study questions

1. Describe the relationship between cells, tissues, organs and systems within the human body. How does the differentiation of cells allow the body to function?

2. What are the three main parts of the cell? What are the basic functions of each of these?

3. What is meant by the term *homeostasis*? Why is homeostasis important within the human body?

4. What is the cell membrane composed of? How does the cell membrane allow the functioning of the endocrine system?

5. What role does the nucleus play in cell functioning? How does RNA assist the nucleus in the functioning of the cell?

6. What are the three types of bone cell? What role does each of these play in the maintenance and functioning of the skeletal system?

7. How are messages transmitted by the nervous system? How do electrical and chemical processes contribute to communication?

8. How do myelinated and unmyelinated neurons differ? What advantages do myelinated neurons have?

4

Fundamental systems for adventure sports

In Chapter 3, the links between cells and systems was established. Each of the 11 systems in the body consists of cells which form the tissues and organs for any particular system. Chapter 3 described three important systems maintaining homeostasis and providing the basis for movement; these were the skeletal system and the nervous and endocrine systems. In this chapter, those which represent the fundamental systems for adventure sports performance are described. These include the muscular system, respiratory system and the cardiovascular system. In addition, we introduce the energy systems – the mechanisms through which the body provides the energy for the work that every cell in the body performs. This is a key component for understanding the physiology of adventure sports and provides the basis of the structure of each of the chapters in Part II of the book.

4.1 Muscular system

The connection between the skeletal and muscular systems is through tendons. Muscle attachment to bone is brought about by the tendon intermeshing with the skeletal periosteum sheath. Tendons are formed from three layers of strong connective tissue, composed of the protein collagen, extending the whole length of the muscle. The three connective tissues that form each tendon are the *epimysium* (which surrounds each muscle), *perimysium* (found around each fascicle or bundle of muscle fibres) and the *endomysium* (that encapsulate each muscle fibre or cell). These structures can be seen in Figure 4.1.

There are three types of muscle tissue found within the body: smooth, cardiac and skeletal muscle. *Smooth muscle*, as was described in Chapter 2, is found in the digestive tract and its contractions assist the movement of food along the digestive tract. In a similar fashion, smooth muscle surrounds the blood vessels and airways and again assists with the passage of blood and air through the vascular and respiratory systems. Smooth muscle, in common with cardiac muscle, is enervated by the autonomic nervous system (described in more detail in the previous chapter) and therefore we do not have voluntary control over its contractions. It is not directly possible to voluntarily make the heart beat faster (contract more quickly) or push food through the digestive tract more quickly. *Cardiac muscular tissue* which forms the heart is striated, meaning it has visible light and dark patches within its

Adventure Sport Physiology Nick Draper and Chris Hodgson
© 2008 John Wiley & Sons, Ltd

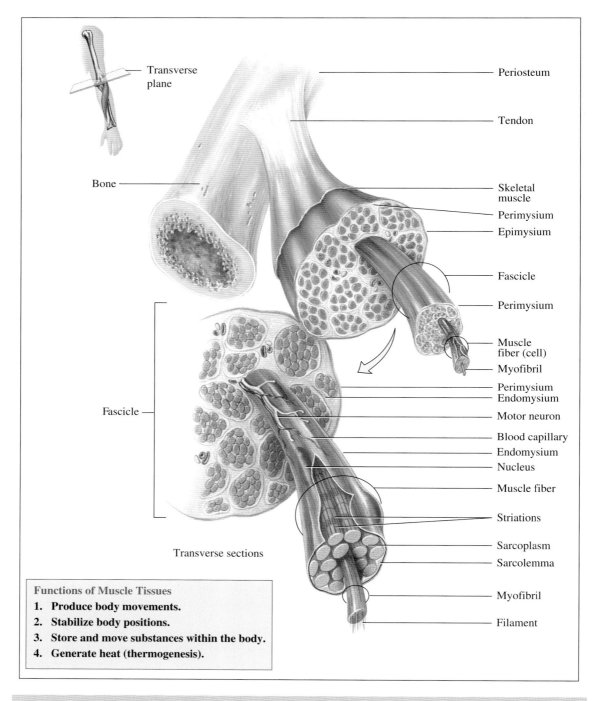

Figure 4.1   Structure of skeletal muscle (from Tortora and Derrickson, *Principles of Anatomy and Physiology*, Eleventh Edition, 2006, reproduced by permission of John Wiley & Sons Inc.)

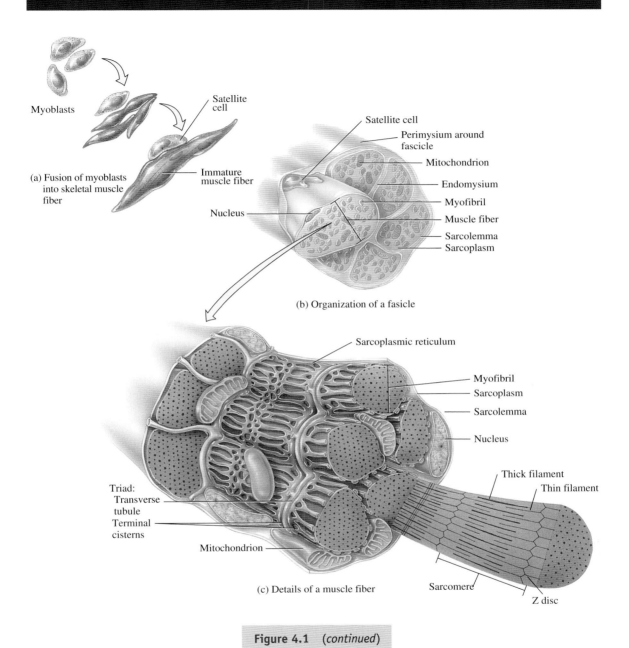

Myoblasts

Satellite cell

(a) Fusion of myoblasts into skeletal muscle fiber

Immature muscle fiber

Nucleus

Satellite cell

Perimysium around fascicle

Mitochondrion

Endomysium

Myofibril

Muscle fiber

Sarcolemma

Sarcoplasm

(b) Organization of a fasicle

Sarcoplasmic reticulum

Myofibril

Sarcoplasm

Sarcolemma

Nucleus

Thick filament

Thin filament

Triad:
Transverse tubule
Terminal cisterns

Mitochondrion

Sarcomere

Z disc

(c) Details of a muscle fiber

Figure 4.1 (*continued*)

structure. These striations, also found in skeletal muscle, are caused by the presence of different proteins that enable muscular contraction. *Skeletal muscular tissue* is the only muscle type that can be voluntarily contracted and is the form of tissue that enables bodily movement by contracting against the skeleton. The human body consists of over 600 muscles which vary in size from 1 mm (such as those found attached to the bones of the ear) to the 30 cm long sartorius muscle in the thigh. The functions of the muscular system are *movement of blood and food* within the body, the ability to *stop the body moving* (to *stabilize* the body when standing or sitting), to *store* oxygen and nutrients such as glycogen for energy production and, through the energy production reactions, to *produce heat* to help maintain body temperature.

The cellular unit of the muscular system is the *muscle fibre*. Muscle fibres are formed from *myoblasts* (similar in function to the osteoblasts of the skeletal system) that fuse together to form multinucleated (with many nuclei) muscle fibres. *Myo* meaning muscle and *sarco* meaning flesh are common prefixes for components of the muscular system and indicate that the cellular mechanism belongs to a muscle. For instance, there is a red-coloured oxygen storing protein, similar in structure and function to haemoglobin found in red blood cells, known within the muscles as *myoglobin*. The myoglobin is stored within the *sarcoplasm* (the cell body for a muscle equivalent to the cytoplasm). Each muscle fibre is enclosed in a muscle cell membrane known as the *sarcolemma* and each muscle fibre contains a *sarcoplasmic reticulum*, the muscular equivalent of the smooth endoplasmic reticulum found in other cells.

The structure of skeletal muscle is, in a physiology textbook on adventure sports, the main focus as it is this tissue that brings about the movements made in any activity (see Figure 4.1). Knowledge of the structure of muscle fibre down to the microscopic, cellular level is essential for understanding of the process of muscular contraction. Muscular contractions are brought about at the cellular level through the interaction of protein filaments within each muscle fibre. Muscular contraction will be discussed in the section after muscle fibre structure.

Each muscle is attached to the skeletal system or other muscles via tendons. Moving from largest to smallest units as can be seen in Figure 4.1, a *muscle* is made up of many hundreds or thousand of *fascicles*. Each fascicle contains from ten to hundreds of individual *muscle fibres* and each muscle fibre is made up of tiny *myofibrils* (little fibres), the contractile units of muscles which are composed of *proteins* and arranged in units called

sarcomeres. As already mentioned, a series of protective sheaths enclose each of these structures and together form the tendon. Each muscle is enclosed within the epimysium, each fascicle within the perimysium and each muscle fibre by an endomysium.

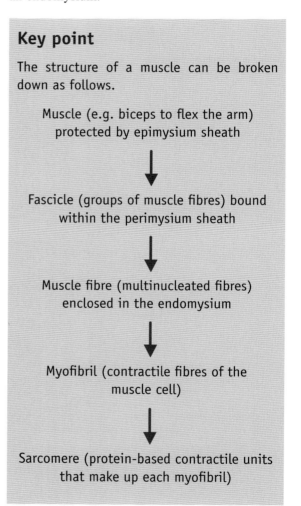

Key point

The structure of a muscle can be broken down as follows.

Muscle (e.g. biceps to flex the arm) protected by epimysium sheath

↓

Fascicle (groups of muscle fibres) bound within the perimysium sheath

↓

Muscle fibre (multinucleated fibres) enclosed in the endomysium

↓

Myofibril (contractile fibres of the muscle cell)

↓

Sarcomere (protein-based contractile units that make up each myofibril)

Muscle fibre structure

As with all other cells, muscle fibres are contained within a cell membrane or sarcolemma. This is situated between the endomysium (which protects the muscle fibre) and the sarcoplasm (the body of the cell). The structure of a muscle fibre can be seen in Figure 4.1. The muscle fibre comprises several hundred myofibrils that run the entire length of the fibre. Each myofibril, the contractile fibre of the muscle cell, consists of thousands of sarcomeres or contractile units. There are, on average, around 4500 sarcomeres to every centimetre of a myofibril. Each myofibril is surrounded by a tube-like structure that, as well as providing structure for the fibre, also contains *calcium ions* which are necessary for muscle contraction and will be described in the next section.

In addition to the sarcoplasmic reticulum, there is a series of tubules that run transversely across each myofibril. These transverse or *T-tubules*, despite their location within the sarcolemma (cell membrane) of the muscle fibre, are continuous with and open to the external environment of the cell. As a result of being open to the exterior

environment of the cell they are filled with interstitial fluid (extra cellular fluid). The nature of the location of T-tubules and the access this provides for interstitial fluid to surround individual myofibrils is essential for coordinated stimulation of muscular contraction. As can be seen from Figure 4.1, a *triad* is formed where the terminal cisternae (sacs at the end of sarcoplasmic reticulum section) form against either side of a T-tubule. Unlike other cells, the nuclei in muscle fibres rest near the edge of the cell on the outside of the myofibrils. Depending upon the type of muscle fibre there can be up to thousands of mitochondria – the powerhouse of the fibre – essential for aerobic energy production.

The structure of a sarcomere can be seen from Figures 4.2a and b. Sarcomeres operate as individual contractile units separated by Z-discs

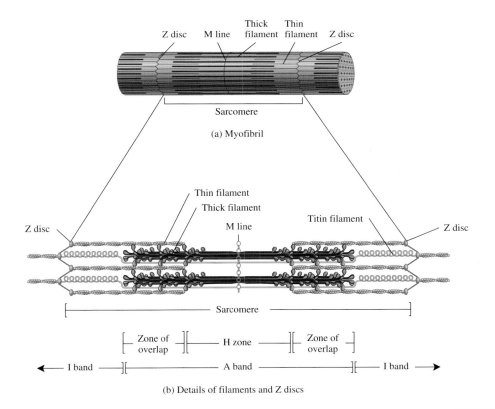

Figure 4.2a Myofibril structure (from Tortora and Derrickson, *Principles of Anatomy and Physiology*, Eleventh Edition, 2006, reproduced by permission of John Wiley & Sons Inc.)

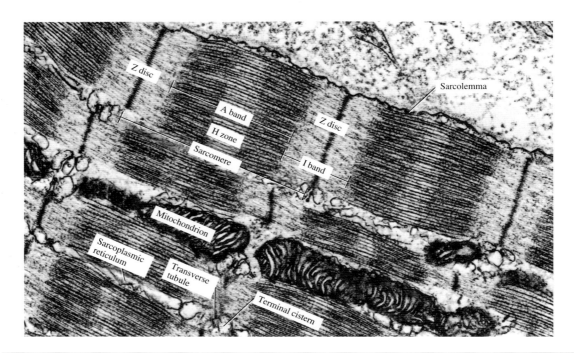

Figure 4.2b Micrograph of myofibril structure (from Tortora and Derrickson, *Principles of Anatomy and Physiology*, Eleventh Edition, 2006, reproduced by permission of John Wiley & Sons Inc.)

or discs and comprise over a dozen different proteins, the most common of which are *actin* and *myosin*. Other proteins found within the sarcomere, important for contraction, include *troponin* and *tropomyosin*. The structure and function of troponin and tropomyosin in relation to actin, and how these proteins interact with myosin to bring about contraction, is described further in the next section. Of the other proteins found within each sarcomere the large elastic protein *titin* (Figure 4.2), which runs the entire length of the sarcomere, is thought to play an important role in the ability of muscle to be stretched without damage. *Desmin* is thought to help in Z-disc alignment.

The striated (striped) nature of skeletal muscle is due to the presence of the thick myosin and the thin actin filaments. The dark stripes appear in the A-band where thicker myosin filaments and actin filaments are found; *the lighter stripes* are the I-bands which contain the thinner actin filaments only. Actin is a structural protein and is the anchored part of the sarcomere, whereas myosin is a motor protein. Each sarcomere is separated from the next by a Z-disc, which comprises connective tissue and from which the actin filaments are anchored. The connective tissues of the Z-discs form a thin dark stripe across the I-bands. The M-line represents the middle of each sarcomere.

As can be seen from Figure 4.2 where a sarcomere is seen at rest, each part of a sarcomere is demarked as a particular zone or band. The I-bands denote that part of each sarcomere where no myosin is present and actin is the main protein filament anchored to the Z-disc. The A-band describes that part of the sarcomere that includes the entire length of the myosin filament. It is at either end of the A-band that an overlap between the myosin and actin filaments is to be found; this partial overlap is essential for contraction. Lastly, the H-zone at the centre of the sarcomere describes that part where only myosin filaments are found. Having described the structure of muscle fibres, myofibrils and sarcomeres, the next section will explain how muscular contraction is thought to take place.

Key point

Muscular contraction is brought about at a cellular level though the contraction of individual sarcomeres, the contractile units of muscle. Actin and myosin are the key proteins involved in the contraction process. Actin filaments provide a structural framework within which the thicker myosin filaments (the motor protein) slide to shorten the sarcomere length.

Muscular contraction

The basis behind muscular contraction was described by Huxley and Huxley in the 1950s (Chapter 1) within the Sliding Filament Theory. To understand this theory it is useful to first understand how a muscle is *stimulated* to contract. The stimulus to contract comes from the nervous system, specifically nerve fibres (cells). When a muscular contraction is required, such as to pick up a paddle or place an ice-axe, a nerve signal is sent to the necessary muscles to initiate the muscle movement. Nerves that send impulses to muscles are called *motor nerves* or *neurons*. The nerve impulse is called an *action potential* (because it has the potential to result in action). Action potentials were described in more detail in the previous chapter on the nervous system. The *sodium-potassium pump* (Chapter 3, Figure 3.5) is a key factor in the movement of an action potential from the brain along a motor neuron to a muscle. During rest, the sodium-potassium pump maintains an *electrical difference* (potential) between the inside and outside of the cell, with potassium ions being sequestered (stored) on the inside of the cell and sodium ions being pumped out from the cell. A nerve impulse travels down its length towards the muscle through the opening of other protein channels, enabling sodium ions to flood into the cytoplasm of the nerve fibre. In this way, the nerve impulse travels the length of the nerve and reaches the muscle fibres it is responsible for stimulating.

Motor neurons divide into finger-like projections that enable one nerve to stimulate more than one muscle fibre at a time. Typically, nerve fibres stimulate 50–60 individual muscle fibres (when fine movements are required) to over a thousand (when gross movements are required e.g. walking). A *motor unit* describes the motor neuron and all the fibres it is responsible for stimulating. The junction between a nerve and the muscle is called the *motor end plate* or *neuromuscular junction*. Neuromuscular junctions are normally located in the middle of a muscle fibre to enable the action potential to travel the length of the fibre in the quickest time possible (enabling sarcomeres to contract at the same time).

Key point

The neuromuscular junction is the point at which a nerve fibre (cell) and a muscle cell meet. The stimulus to contract is passed to the muscle from the brain via the nerve fibre. Nerve cells are not directly connected to the fibres they stimulate but form a junction with a small gap (synaptic cleft) between them and the muscle fibre. To cross this gap, the electrical signal stimulates the release of acetylcholine (a neurotransmitter i.e. chemical nerve message) into the synaptic cleft. Acetylcholine bonds with receptor proteins on the sarcolemma and enables the signal to contract to pass to the muscle fibre.

Unlike the interaction between the skeletal and muscular systems, where the protective muscular sheaths form tendons that intermesh with the periosteum of bone, there is no direct contact between nerve fibres and the muscles they stimulate. The neuromuscular junction is a non-contact link between the nervous and muscular systems. The small gap between the two fibres at a neuromuscular junction is called the *synaptic cleft* (a synapse is a junction between two nerve fibres). Figure 4.3 depicts a neuromuscular junction.

To stimulate the muscle fibres for which the nerve is responsible, the nerve impulse must cross this cleft. The nerve does this by converting the electrical impulse (created by the influx of sodium into the cell) that has travelled the length of the nerve fibre to a chemical message that can cross the interstitial fluid in the synaptic cleft. The arrival of the electrical signal at the *pre-synaptic terminal* (the finger-like end of the nerve fibre before the neuromuscular junction) stimulates the release of the *neurotransmitter* (nerve chemical message)

acetylcholine from the vesicle stores, as shown in Figure 4.4.

Figure 4.4 shows a close-up of one section of the nerve and muscle fibres at a neuromuscular junction. Acetylcholine is stored in vesicles within the pre-synaptic terminal fuse with the cell membrane of the nerve cell, in a process called *excytosis*. This process, bonding with a plasma membrane to release a substance outside the cell, enables the acetylcholine to be released into the synaptic cleft. The acetylcholine can cross the synaptic cleft and, as can be seen in Figure 4.4, bonds with the receptors of special transmembrane channel proteins to open the protein channels and enable the action potential to begin within the muscle fibre.

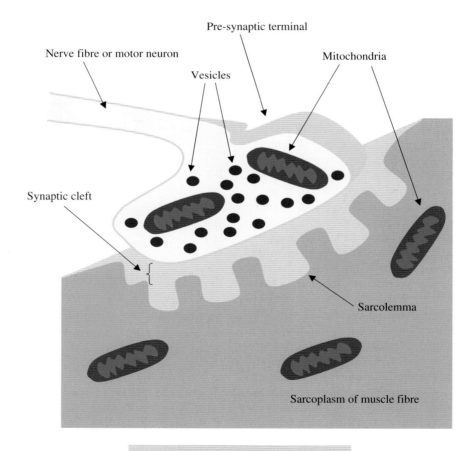

Figure 4.3   A neuromuscular junction

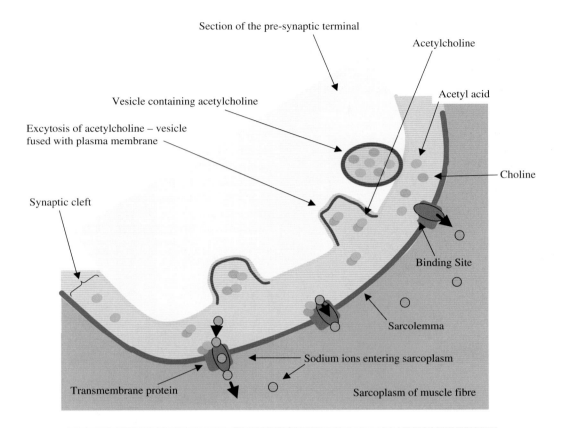

Section of the pre-synaptic terminal

Acetylcholine

Acetyl acid

Vesicle containing acetylcholine

Excytosis of acetylcholine – vesicle fused with plasma membrane

Choline

Synaptic cleft

Binding Site

Sarcolemma

Sodium ions entering sarcoplasm

Transmembrane protein

Sarcoplasm of muscle fibre

Figure 4.4 Mechanism for the release of acetylcholine into the synaptic cleft

Muscle fibres, as with other cells, have sodium-potassium pumps which maintain low levels of sodium ions within the sarcoplasm at rest. When acetylcholine molecules bind with receptors on the channel proteins within the sarcolemma, they open the channels for the influx of sodium ions into the sarcoplasm. Once acetylcholine has completed its role, it is catabolized by the enzyme *acetylcholinesterase* into its component parts. *Acetyl acid* is used by many cells in the body and *choline* is returned via protein channels to vesicles within the motor neuron.

The influx of sodium ions is the signal (action potential) for the muscle to contract. The message travels quickly to the sarcomeres since the neuromuscular junctions are situated in the middle of the muscle fibres and the T-tubules run through the muscle fibres. The passage of the electrical signal to the T-tubules, located between two terminal cisternae (Figure 4.1) of the sarcoplasmic reticulum (forming a triad), is the starting point for muscular contraction. As mentioned above, when a muscle fibre is relaxed the sarcoplasmic reticulum sequesters (stores) calcium ions, essential components for muscular contraction, When the sodium ions – the electrical signal to contract – reach the triads, the T-tubules stimulate the terminal cisternae to release the calcium ions to the sarcoplasm. It is the release of calcium ions that enables muscle contraction. This process of nervous stimulation leading to muscular contraction is sometimes referred to as *excitation coupling*.

Key point

The influx of sodium ions into the sarcoplasm propagates quickly down the length of the muscle fibre and into the T-tubules. The arrival of sodium to the T-tubules which form part of a triad with two terminal cisternae causes the release of calcium from the sarcoplasmic reticulum. The release of calcium enables muscular contraction.

The Sliding Filament Theory of muscular contraction

The structure of the main protein filaments involved with muscular contraction – actin and myosin – are depicted in Figure 4.5. The structural actin filaments comprise two twisted *actin chains* with the protein *tropomyosin* wrapped around the length of the actin chain and the protein *troponin* (which contains forms of troponin) bound to tropomyosin and actin. At rest, tropomyosin (a structural protein that helps to strengthen the actin filament) covers special myosin binding sites and stops myosin from combining with actin. The three

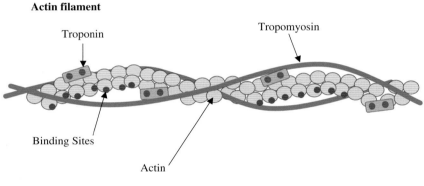

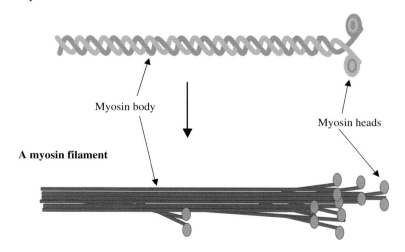

Figure 4.5 The structure of actin and myosin

isomers of troponin are: *troponin T* which binds to tropomyosin; *troponin I* which binds to the actin chain; and *troponin C* which, when available, binds to calcium. (It is important to remember that calcium is released by the sarcoplasmic reticulum and troponin C binds with calcium when available, as this is central to muscular contraction.)

Myosin filaments, thicker than actin filaments, comprise around 200 *myosin molecules* (Figure 4.5). Each of these is formed from intertwined strands of myosin that form a long body or rod and myosin heads which when possible will bind with the myosin binding sites on the actin chains. Individual myosin molecules resemble two karabiners on slings which are twisted together. Extending from the slings (the rod of the myosin molecule) the karabiners (myosin heads) will attempt to clip onto a piece of gear in the cliff face (actin binding site) if available.

The formation of actin and myosin can be seen in Figures 4.2 and 4.6. In two-dimensions, the layout of the filaments appear as in the top of Figure 4.6, however, it is important to understand the three-dimensional nature of sarcomeres and the myofibrils they form. In cross-section, each sarcomere comprises a hexagonal of myosin filaments around each filament and hexagonal actin formations around each myosin filament. The cross-sections are depicted in Figure 4.6. The structure of sarcomeres, with myosin filaments embedded between thinner actin filaments structurally attached to the Z-discs, is what makes muscular contraction through the sliding of filaments possible.

Muscular contraction begins with nervous stimulation. The nerve impulse carried across the synaptic cleft via acetylcholine reaches the muscle fibre and travels along the T-tubules. This causes the release of calcium (stored in the sarcoplasmic reticulum) into the sarcoplasm surrounding each sarcomere. The sequence of muscular contraction from this point is shown in Figure 4.7. The presence of calcium around the sarcomere enables the troponin C component of the troponin to bind with the available calcium. In binding with calcium, tropnin C changes the shape of tropomyosin (which covers the active binding sites on the actin chain) and literally pulls it away from the myosin

binding sites. This makes the binding sites available to bind with the myosin heads of the myosin filament.

As well as calcium to reveal the binding sites, energy in the form of ATP is required for muscular contraction. Myosin heads store energy from the breakdown of ATP and retain ADP and Pi (phosphate) attached to them as they form *crossbridges* i.e. bind with an active site on the actin filament. When the crossbridge is formed, the Pi is released from the myosin head. Myosin is now bound to the actin filament. The energy stored from the breakdown of the ATP molecule is then used by each myosin head to perform a *powerstroke*. The myosin head powerstroke pulls the attached actin filament towards the middle of the sarcomere and shortens its length. In this action the actin slides past the myosin filament, hence *Sliding Filament Theory*. During the powerstroke the ADP molecule is released, leaving the ATP binding site on the myosin head free to bind with another molecule of ATP. The binding of a new ATP molecule to the myosin head and its breakdown to ADP and Pi enables the myosin head to detach from the active binding site on the actin filament and leaves it 'cocked' ready to bind with another actin binding site. The crossbridge formation/powerstroke sequence continues many times during a contraction and as long as there are sufficient calcium ions and ATP molecules available. Figure 4.8 shows the effect of muscular contraction and the sliding filaments on the size of the sarcomeres. During muscular contraction, the size of the I-bands and H-zone is decreased as the muscle shortens.

Key point

Muscular contraction is brought about through the release of calcium from the sarcoplasmic reticulum into the sarcoplasm of the muscle fibre. Calcium binds with troponin C to pull tropomyosin (which covers active binding sites on the actin chain) away from the myosin binding sites,

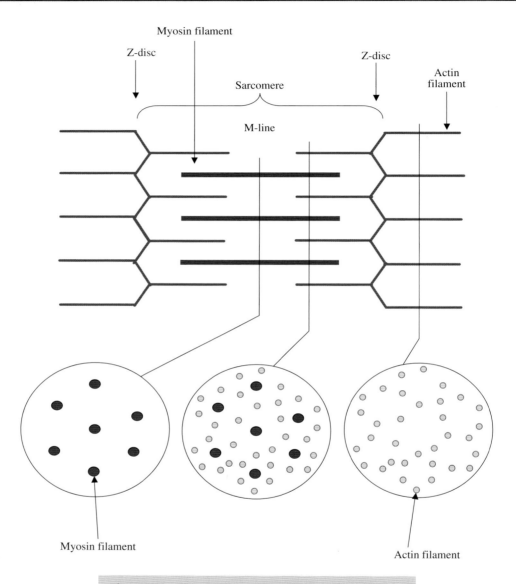

Figure 4.6 The cross-sectional structure of a sarcomere

enabling the myosin to bind with actin. During the sliding process of muscular contraction, myosin heads bind, release and rebind with myosin binding sites to pull the myosin filament along the actin filament.

Muscle fibre types

The strength, power and endurance for adventure sports participation is provided by skeletal muscles. Just as there are different mechanisms for energy production, so there is a variety of muscle cells or fibre types that can be recruited. These muscle fibre types are differentiated by

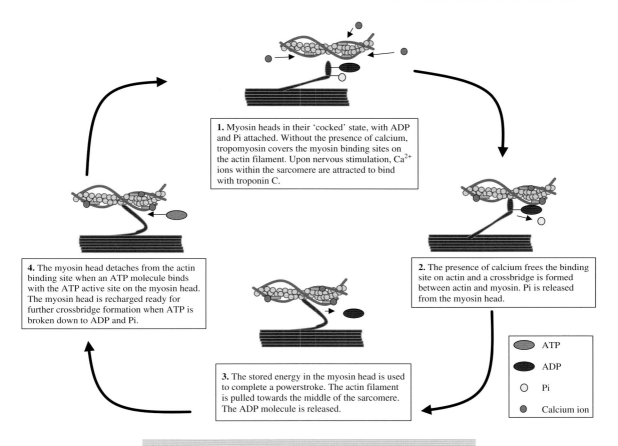

1. Myosin heads in their 'cocked' state, with ADP and Pi attached. Without the presence of calcium, tropomyosin covers the myosin binding sites on the actin filament. Upon nervous stimulation, Ca²⁺ ions within the sarcomere are attracted to bind with troponin C.

4. The myosin head detaches from the actin binding site when an ATP molecule binds with the ATP active site on the myosin head. The myosin head is recharged ready for further crossbridge formation when ATP is broken down to ADP and Pi.

2. The presence of calcium frees the binding site on actin and a crossbridge is formed between actin and myosin. Pi is released from the myosin head.

3. The stored energy in the myosin head is used to complete a powerstroke. The actin filament is pulled towards the middle of the sarcomere. The ADP molecule is released.

ATP
ADP
Pi
Calcium ion

Figure 4.7 Sliding filament crossbridge–powerstroke cycle

the predominant energy system that each favour. Two main distinctions have been made: Type I (slow twitch) and Type II (fast twitch) fibres. Type I fibres have greater aerobic capabilities with a higher density of blood vessels supplying oxygen to the fibres, a larger number of mitochondria and more aerobic enzyme activities (the assisting substances or catalysts which enable reactions to take place) than Type II fibres. Type II fibres have greater anaerobic qualities and are able to generate greater forces than Type I fibres, but fatigue more quickly. Each individual has a different composition of Type I and Type II fibres and this is what makes a performer physiologically more suited to some adventure sports than others. An adventure sports performer with a higher percentage of

Type I fibres would be more suited to endurance activities; one with a predominance of Type II fibres being more suited to power sports. Through training, however, we can affect our fitness and our bodies can make adaptations according to the type of training we undertake. There are further differentiations that can be made between fibre types and these are covered in Chapter 6.

Key point

There are two main muscle fibre types: Type I and Type II. Every adventure sports performer is made up of different proportions

Sarcomere at the start of contraction where crossbridges have just been created

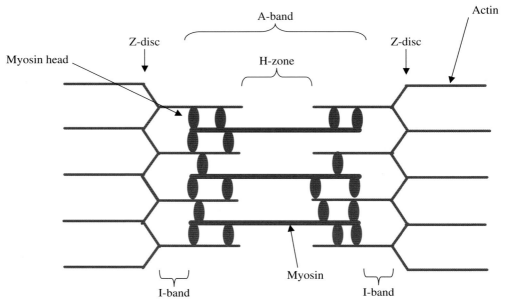

Sarcomere at the end of contraction where mysosin heads have slid along the actin filaments to contract the sarcomere

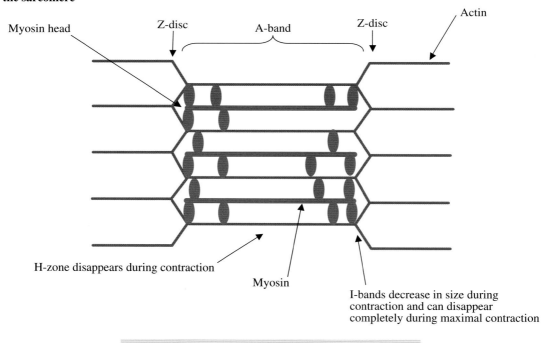

Figure 4.8   Muscle contraction within the sarcomere

of each fibre type. If a performer comprises more Type I fibres they may be more suited to aerobic disciplines such as marathon paddling; if Type II fibres predominate they may be more suited to sprint and power type activities.

Muscles and movement

As already discussed, muscles bring about movement through contracting against their attachments to bones or other muscles. To bring about movement, the muscular contraction and bone operate as levers with a load, effort and balance point or fulcrum. Within the body, it is possible to find examples of first, second and third order levers through the varied attachment of muscles and bones. Figure 4.9 provides an example of each type of lever. *First order levers* have the fulcrum in the middle, *second order levers* have the load in the middle and *third order levers* have the effort in the middle. First order levers can produce a mechanism advantage or disadvantage, depending on whether the fulcrum is closer to the load or effort. The closer the fulcrum is to the load, the greater the mechanical advantage. Second order levers always produce a mechanical advantage and third order levers a mechanical disadvantage.

The points of attachment of a muscle are called the origin and insertion. The *origin* is normally the point of attachment to the bone that does not move and the *insertion* is the attachment to the bone that does move. For example, the origin of the biceps muscle is on the scapula and the attachment is on the radius of the lower arm. Muscles are

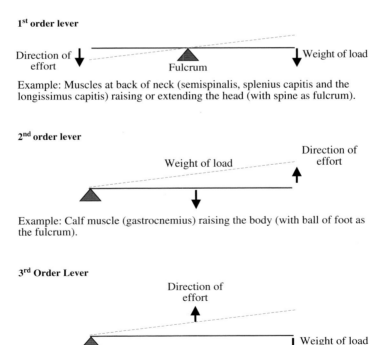

1st order lever

Direction of effort ↓   Fulcrum   ↓ Weight of load

Example: Muscles at back of neck (semispinalis, splenius capitis and the longissimus capitis) raising or extending the head (with spine as fulcrum).

2nd order lever

Weight of load   Direction of effort ↑   ↓

Example: Calf muscle (gastrocnemius) raising the body (with ball of foot as the fulcrum).

3rd Order Lever

Direction of effort ↑   ↓ Weight of load

Example: Biceps muscle raising or flexing the arm with a load in the hand (elbow as fulcrum).

Figure 4.9 Levers created between muscles and bones

normally arranged in antagonistic pairs such that they have the opposite effect of each other. Taking the biceps as an example once more, when bending (flexing) the arm the biceps forms the *prime mover* or *agonist* and the triceps is the *antagonist*. When straightening the arm, the roles are reversed and the biceps becomes the antagonist.

Types of muscular contraction

Knowledge of the following muscular contraction-related terms necessary for understanding muscle function in adventure sports are explained in this section as they will be used throughout the subsequent chapters. When a muscles moves as part of a contraction, the load can be applied when the muscle is shortening (a *concentric contraction*) or when the muscle is lengthening (an *eccentric contraction*). An example of that could be a bench press. The push up of your maximum bench press would be brought about through the concentric contract of the pectoral, front of deltoid and triceps muscles. If, for the next repetition you then added 10 kg in weight you would not be able to lift the weight, but you could using eccentric contraction of the same muscles to lower the weight to your chest, thereby training with a weight higher than your maximum. Eccentric contractions have been used in the past by weightlifters as part of a strategy to increase the overload (supramaximally) to improve maximal lifting. Research has shown that eccentric contractions improve muscle strength during concentric contraction, but they often result in increased post-exercise muscle pain.

In the sport of rock climbing, there are many instances where the climber has to hold *isometric contractions* (*iso* meaning same and *metric* referring to length). Isometric contractions describe muscle contractions *without* muscle movement (shortening or lengthening). Each hold a climber takes with their hand involves an isometric contraction of the forearm muscles. The longer the climber remains holding a specific hold, and the tighter they grip the hold, the greater the isometric contraction and the more quickly the muscle will fatigue.

Isotonic and *isokinetic contractions* are associated with strength training and represent muscular contractions where the muscle moves against a resistance. In a weight training context, *tonic* refers to load and *kinetic* to work, thus they mean same-load and same-work, respectively. An example of these terms can be drawn from a biceps curl. To clarify the example, it is important to know that as the biceps curls (the muscle shortens against the resistance) through the range of motion, the muscle is weakest at the start of the lift when the biceps is at the limit of its range of motion and strongest once it gets past 90°. If an athlete performs a bicep curl with a 20 kg barbell, the load remains the same throughout the lift (isotonic) throughout the lift as the bar moves to the chest. When performing an isokinetic contraction biceps curl, an athlete would make the lift against a machine that can vary the load to ensure the same relative work throughout the range of motion. The use of special cams, the same in concept as the cams used on a camming device in rock climbing (camalot or friend), are used to relatively decrease the load in the early part of the lift (where the biceps muscle is at its weakest) and increase the load as the biceps moves past 90°. This results in the relative work being the same throughout the lift i.e. an isokinetic contraction.

In a *plyometric contraction, plyo* refers to variable or more i.e. the muscle varies in length as it makes its contraction. In the context of weight training, a plyometric contraction is made where the muscle is pre-tensioned prior to the actual lift or movement in order to increase the force generated. The use of a plyometric contraction will improve the power of a movement. A plometric contraction uses an eccentric contraction (pre-tensioning) to bring about a more powerful concentric contraction. The pre-tensioning of the muscle adds an eccentric stretch to the muscle and makes it act like a spring as well as making a muscular contraction. As the muscle stretches, it wants to recoil. If the recoil of the muscle is timed with the start of the concentric contract, the result is increased force generation. An example of this can be seen in a dyno for rock climbing. When a climber rocks up and down on the holds they are starting from, they are pre-tensioning the muscles prior to making the dyno movement. In doing this, the climber will increase the force generated in the dyno and improve their chances of catching the hold for which they are aiming. They are making a plyometric contraction or movement.

Key point

There are over 600 muscles within the body which comprise water, proteins and minerals and contain stores of nutrients for energy production, including oxygen. The oxygen store within the muscles is myoglobin. The proteins within myofibrils bring about muscular contraction. Calcium released from the sarcoplasmic reticulum stimulates muscular contraction after a nerve signal is passed to the muscle via a neuromuscular junction.

4.2   Respiratory system

The primary role of the respiratory system, and the tissues it consists of, is to facilitate the exchange of oxygen and carbon dioxide between the air and the internal cells of the body. Oxygen is essential for the production of energy, a requirement of every cell in the body. Carbon dioxide, a waste product of energy production, must be removed from the body to sustain life. In addition, the tissues of the respiratory system have an important role in speech, coughing, hiccupping, yawning, sneezing and the Valsalva manoeuvre. The respiratory system supplies the oxygen required by cells to perform *cellular respiration* i.e. the use of oxygen by cells to sustain life, resulting in the production of energy, water and carbon dioxide. This makes oxygen the most important substance for sustaining life.

The oxygen necessary for cellular respiration reaches the cells through three stages of *ventilation* or gas exchange. *Pulmonary ventilation* or *breathing* enables oxygen to enter the lungs and carbon dioxide to be expelled. Figure 4.10

is a diagram of the structures of the respiratory tract from the trachea, which leads directly from the mouth and nose, to the terminal structures of the tract i.e. the alveoli. *Alveolar ventilation* enables oxygen brought in to the lungs via pulmonary ventilation to diffuse into the bloodstream and carbon dioxide to be exchanged. Once within the bloodstream, oxygen is transported to cells throughout the body where the final form of gas exchange – *cellular ventilation* – enables it to diffuse across the plasma membrane and enter the cell for use within cellular respiration.

The respiratory tract is most commonly divided into the *conducting zone* (Figure 4.10) and the *respiratory zone* (Figure 4.11). The conducting zones includes the nose, mouth, pharynx (throat), larynx (voice box), trachea (windpipe), bronchi, bronchioles and terminal bronchioles. The respiratory zone comprises the respiratory bronchioles, alveolar ducts, alveolar sacs and the alveoli.

For air to reach the alveoli from the trachea it commonly passes 25 branches, starting with the division of the trachea into the left and right bronchi, before reaching the alveoli. As can be seen from Figure 4.10, after the point where the trachea branches to the left and right primary bronchi, each branch of the respiratory tract is contained within a pleural sac which form the left and right lung. Within the pleural sac created by the parietal pleura (outer membrane) and the visceral pleura (inner membrane) is a pleural fluid. Surface tension, the attraction of surface molecules to other molecules deeper within the fluid, enables the two membranes to stick together but also allows sliding of the membranes across each other during pulmonary ventilation. The further divisions of the respiratory tract create lobes in each lung, each formed of lobules. Each lobe has its own secondary bronchi and lobules containing further branching of the respiratory tract. Due to the presence of the heart on the left side of the body, the left lung has two lobes whereas the right lung comprises three lobes and is therefore slightly bigger (about 10 per cent) than the left.

Pulmonary exchange takes place between the lungs and the pulmonary blood vessels. The pulmonary arteries supply deoxygenated blood

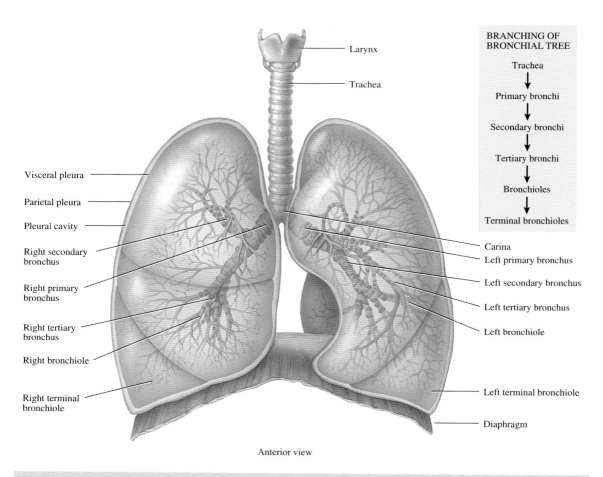

BRANCHING OF
BRONCHIAL TREE

Trachea
↓
Primary bronchi
↓
Secondary bronchi
↓
Tertiary bronchi
↓
Bronchioles
↓
Terminal bronchioles

Larynx

Trachea

Visceral pleura

Parietal pleura

Pleural cavity

Right secondary bronchus

Right primary bronchus

Right tertiary bronchus

Right bronchiole

Right terminal bronchiole

Carina

Left primary bronchus

Left secondary bronchus

Left tertiary bronchus

Left bronchiole

Left terminal bronchiole

Diaphragm

Anterior view

Figure 4.10   Structure of the lungs (from Tortora and Derrickson, *Principles of Anatomy and Physiology*, Eleventh Edition, 2006, reproduced by permission of John Wiley & Sons Inc.)

pumped from the right ventricle of the heart to the lungs. The pulmonary artery that leaves the heart divides into a left and right branch and supplies blood to the left and right lung. The pulmonary arteries then branch into smaller vessels that lead to the tiny capillaries that surround each alveolus and enable gaseous exchange to take place. The capillaries lead to the four pulmonary veins that return the oxygenated blood to the heart (left atrium) for its journey around the rest of the body. The lungs also receive oxygenated blood via the right and left bronchial arteries (a branch of the aorta) as part of the systemic blood flow. This

enables the lungs to exchange oxygen necessary for the cells within the lungs and remove the carbon dioxide produced as a result of work done. The deoxygenated blood from the lungs is returned to the heart via the bronchial veins, which flow into the superior vena cava.

Mechanics of pulmonary ventilation

Pulmonary ventilation comprises inspirations (inhalations) and expirations (exhalations). At rest, inspiration is achieved through *active* mechanisms whereas expiration is *passive* and completed through the relaxation of the muscles involved in inhalation. Each inspiration is brought about through two actions. Firstly, the *diaphragm* (ridge-shaped muscle in Figure 4.10) and the *external intercostals* contract causing the rib cage to rise and the diaphragm to flatten, increasing the volume of the lungs. Secondly, through the decreased air pressure within the lungs (a result of the muscular contraction), air moves from outside the body to the lungs to balance the pressure between the two areas.

Just as electrolytes (ions) in fluids move from areas of high concentration to low to bring balance, air moves from low to high pressure areas to maintain equilibrium. At rest, each inspiration and expiration leads to around 0.5 L of air moving to and from the lungs. During exercise, up to six times this volume i.e. 3.0 L breath^{-1} can be

exchanged. Due to the in and out nature of this flow, the air in each breath is known as *tidal volume*, V_T. The increase in tidal volume during exercise is enabled through an increase in the muscles involved in inspiration and a change to active expiration.

During exercise, the *sternocleidomastoid, scalenes* and *pectoralis minor* muscles add to the force of contraction by the diaphragm and external intercostals, resulting in a further increase in the lung volume with each inspiration. This causes an increased differential between the air pressure in the lungs and that outside the body and, as a consequence, more air enters the lungs with each inhalation. The passive mechanism of exhalation, resulting from the return of the elastic muscle fibres of inspiration to rest, is enhanced by the contraction of the *internal intercostals* and the

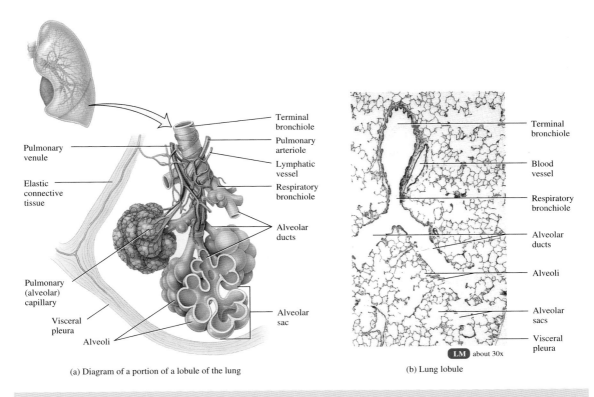

Pulmonary venule

Elastic connective tissue

Pulmonary (alveolar) capillary

Visceral pleura

Alveoli

Terminal bronchiole

Pulmonary arteriole

Lymphatic vessel

Respiratory bronchiole

Alveolar ducts

Alveolar sac

(a) Diagram of a portion of a lobule of the lung

Terminal bronchiole

Blood vessel

Respiratory bronchiole

Alveolar ducts

Alveoli

Alveolar sacs

Visceral pleura

LM about 30x

(b) Lung lobule

Figure 4.11 The structure of aveoli and the facilitation of gaseous exchange (from Tortora and Derrickson, *Principles of Anatomy and Physiology*, Eleventh Edition, 2006, reproduced by permission of John Wiley & Sons Inc.)

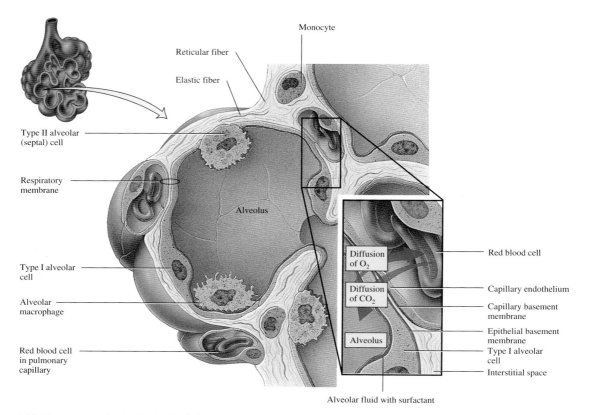

(a) Section through an alveolus showing its cellular components   (b) Details of respiratory membrane

Figure 4.11   (*continued*)

abdominal muscles. In addition to increases in the depth of breathing (tidal volume) exercise also results in an increase in the rate of breathing (*respiratory frequency*).

Each inspiration at rest typically takes 2 s and expiration 3 s, resulting in a respiratory frequency (*f*) of 12 breaths min^{-1} (60 seconds ÷ 5 seconds for each breath). *Minute ventilation* $\dot{V}_E$ (L min^{-1}), the amount of air inspired or expired per minute, is calculated as

$$\dot{V}_E = V_T \times f.$$

where the dot above the V indicates that the *rate* of volume of air being expired or inspired per minute is being calculated. At rest, $\dot{V}_E = 0.5 \times 12 =$ 6 L min^{-1}. During exercise, as well as tidal volume increasing to 3.0 L breath^{-1}, respiratory frequency can rise to 60 breaths min^{-1} for some experienced adventure sport athletes such as cross-country skiers. Minute ventilation values can therefore rise to $\dot{V}_E = 3.0 \times 60 = 180$ L min^{-1}. Changes in the depth and frequency of breathing, which can result in dramatic increases in minute ventilation and indeed in breathing, are controlled by the respiratory centre.

Control of pulmonary ventilation i.e. the rate and depth of breathing, is located in the medulla oblongata and the pons (part of the brain stem). Cranial afferent nerves deliver sensory information from chemoreceptors located at three positions in the body: (i) very close to the medulla oblongata, to detect changes in the partial pressure of carbon dioxide (PCO_2) and hydrogen ion (H^+) levels in the cerebrospinal fluid; (ii) in the aorta (artery); and (iii) common carotid arteries which detect PCO_2, PO_2 and H^+ ions in the blood. Although the sensory neurons pick up information about the PO_2 in the blood, the respiratory driver is more sensitive and responsive to changes in levels of CO_2; PCO_2 is therefore the main stimulant for increasing breathing rate or depth. It is for this reason that freedivers hyperventilate (breath very rapidly) before diving. By hyperventilating prior to diving, they reduce the PCO_2 in the blood which decreases the drive to inspire air and means they can hold their breath for longer. All three chemoreceptor locations respond to H^+ ions as the level of these in the blood and cerebrospinal fluid is indicative of the levels of carbon dioxide in the body. Carbon dioxide, produced in cells, quickly combines with water to form carbonic acid (H_2CO_3) and then dissociates (breaks down) to H^+ ions and HCO_3^- (bicarbonate) ions.

Cranial sensory nerves relay information about levels of oxygen and carbon dioxide to four areas in the brain stem. During breathing at rest the *inspiratory area*, located in the medulla oblongata, controls the rate and depth of breathing for inspiration and expiration. The inspiratory area stimulates the external intercostals and diaphragm to contract for 2 s. When this stimulus is removed (for 3 s) these muscle relax and expiration takes place. When active expiration is required during exercise, the *expiratory area*, also located in the medulla oblongata, sends nerve impulses to stimulate the internal intercostals and the abdominals to contract. The *pneumotaxic* and *apneustic* centres, found within the pons, are responsible for shortening the duration of inspiration during exercise and deepening breathing, respectively.

Key point

Pulmonary ventilation (breathing) differs during rest and exercise. At rest, the external intercostals and diaphragm contract to bring about inspiration. As they relax and fall back to their starting positions, air is expired from the body. During exercise, muscular contraction is required to make the expiration more forceful. During active expiration, the internal intercostals and abdominal muscles contract, forcing air from the lungs. When moving from rest to exercise, both the depth and rate of breathing can be changed to increase the amount of air moving to and from the lungs. The inspiratory area of the brain is responsible for setting the rhythm of breathing during rest and stimulates the inspiratory muscles to contract. Expiration takes place in the gaps between nervous stimulation. During exercise the expiratory area, located near the inspiratory area in the medulla oblongata, stimulates the contraction of the internal intercostals and the abdominals to facilitate active expiration.

Pulmonary gas exchange

Gaseous exchange between the alveoli and the capillaries is enabled through the unique microscopic structure of their cell membranes. The structure of an alveolus and the sites for gaseous exchange are shown in Figure 4.11. Each alveolus is a similar to a rock climbing chalk bag in that it has two layers. The outer layer is a protective membrane and the inner layer, the wall of the alveolar, comprises two types of cell. *Type I* alveolar cells which form the main structure of the wall are interspersed with *Type II* cells which help to maintain the structure of each alveolus by secreting an alveolar fluid, largely composed of water but also containing *surfactant*.

The alveolar fluid, including surfactant which is chiefly composed of lipoproteins and phospholipids, lines the inner surface of the alveolar wall. The lipid composition of the surfactant means that it creates interruptions in the water-predominant alveolar fluid (recall that water and lipids do not mix well). The role of surfactant is vital to pulmonary ventilation as the interruptions it creates help to reduce the surface tension created by the alveolar fluid. As a microscopic structure, the partial deflation of the alveolus achieved through each exhalation could be sufficient to enable the surface tension created by alveolar fluid to collapse the alveolus. The presence of surfactant interruptions in the fluid layer reduces the alveolar fluid surface tension, maintaining the alveolar structure. The third type of cell found within each alveolar structure are *macrophages* (shown in Figure 4.11 with Type I and II cells) which form the clean-up crew, assisting with immune function and maintaining the gaseous exchange surfaces by removing debris from the alveoli.

As can be seen from Figure 4.11, each capillary surrounding the alveoli has a protective membrane and a capillary wall, in common with the alveolar structure. This means that for gaseous exchange to take place, each gas molecule has to cross four layers to move into or away from the lungs. The four layers that form the *gaseous exchange (GE) membrane* are very thin. Although providing protection and integrity for the alveoli and capillaries, gaseous exchanges can easily take place. The unique structure of the lungs, which are thought to comprise over 300 million alveoli, means that there is a vast surface area available for gaseous exchange between the alveoli and the millions of capillaries that surround them. This surface area is about 70 m^2, the equivalent of the area inside about 12–14 open boats.

Gaseous exchange takes place by diffusion, the flow of gases across the GE membrane from an area of high concentration to low concentration. Through this process, carbon dioxide in high concentration in the capillaries diffuses into the alveoli and oxygen molecules in higher concentration in the alveoli make the opposite journey. The concentration of gases is measured through their *partial pressure*, first described by the English scientist John Dalton (subsequently known as Dalton's law). Air is made up of a mixture of three main gases and water vapour. Table 4.1 lists the percentage (concentration) of each of the gases and water vapour in atmospheric air. Dalton's law states that the total pressure of a mixture of gases (such as that found in air) is equivalent to the sum of the partial pressures of each of the gases. The partial pressure of oxygen (PO_2) only needs to be higher in the alveolar air than in the blood for oxygen to diffuse into the bloodstream. Similarly, if the partial pressure of carbon dioxide (PCO_2) is higher in the bloodstream than in the alveoli, carbon dioxide will diffuse into the alveolar space.

The partial pressures of respiratory gases, oxygen and carbon dioxide in the alveolar air differ from those in atmospheric air. The changes in PCO_2 and PO_2, the gases involved in respiration, are shown in Table 4.2. When air enters the mouth it is immediately saturated with water vapour

Table 4.1   The partial pressures of the main gases and water vapour in air

Air constituent	% Concentration	Calculation	Partial pressure (mm Hg)
Nitrogen*	78.63	(78.63 ÷ 100) × 760	597.6
Oxygen	20.93	(20.93 ÷ 100) × 760	159.1
Water Vapour**	0.4	(0.4 ÷ 100) × 760	3
Carbon Dioxide	0.04	(0.04 ÷ 100) × 760	0.3
Total Pressure	100 %		760 mm Hg***

*The percentage concentration for nitrogen includes other trace gases found in air, but not involved in respiration such as argon.
**Water vapour in air is referred to as humidity and varies according to time and location; the figure of 0.4 % is typical.
***760 mm Hg represents mean atmospheric pressure at sea level.

Table 4.2 Changes in PCO_2 and PO_2 between atmospheric and alveolar air (mm Hg)

Air constituent	Atmospheric air	Tracheal air	Alveolar air	Pulmonary*
Nitrogen	597.6	563.49	570.41	N/A
Oxygen	159.1	149.23	103.38	40
Water vapour	3.0	47.0	47.0	N/A
Carbon dioxide	0.3	0.28	39.21	46
Total pressure	760	760	760	N/A

*The partial pressures in the last column are for oxygen and carbon dioxide in pulmonary capillary blood. At sea level, nitrogen is not normally absorbed into the blood and water vapour does not taken part in the respiratory process.

and, as a consequence, the PH_2O rises from 3.0 mm Hg to 47 mm Hg. This leads to a drop in the PO_2 to 149 mm Hg, as when air enters the trachea, oxygen comprises 20.93 per cent of 713 mm Hg (760–47 mm Hg for water vapour). As the percentage of CO_2 in inspired air is so small the effect of water vapour on PCO_2 is negligible (0.3 mm Hg to 0.28 mm Hg). There is a second drop in the PO_2 and an increase in the PCO_2 as air reaches the alveoli. The fall in the PO_2 and rise in PCO_2 in the alveolar air occurs due to the constant diffusion of gases between the blood and alveoli. The continual movement of gases between the bloodstream and alveoli results in the percentage of oxygen being lower and carbon dioxide being higher (around 14.5 and 5.5 per cent, respectively) in the alveoli. Despite the drop in PO_2 as it makes its journey from the atmosphere to the alveoli, oxygen molecules are able to diffuse across the GE membrane because the PO_2 in the bloodstream is lower than that in the alveolar space, at around 40 mm Hg. On the other hand, carbon dioxide diffuses into the alveoli from the bloodstream because the PCO_2 in the bloodstream is around 7 mm Hg higher. Although lower than the differential for oxygen, this is still sufficient for CO_2 to cross the GE membrane.

As can be seen from Table 4.2 there is a reduction from 159 mm Hg to around 100 mm Hg as air moves from the atmosphere to the alveoli. This drop is known as the *oxygen cascade* and

is an important aspect of altitude physiology. The implications of altitude to the oxygen cascade are covered in Chapter 10.

Key point

Gaseous exchange takes place across the GE membrane which is formed between the alveoli walls and the capillaries that surround them. To facilitate diffusion of gases, the GE membrane is very thin and enables oxygen and carbon dioxide to pass easily to and from the lungs. Gases diffuse from high levels of concentration to lower levels. The concentration of the gases in air is measured by their partial pressure. The PO_2 is higher in the alveoli than in the bloodstream and so oxygen diffuses into the blood. Conversely, the PCO_2 is higher in the bloodstream so carbon dioxide diffuses into the alveoli.

Measurement of pulmonary ventilation

For physiologists and medical staff, the movement of air from the lungs and the component gases in expired air can reveal much about human functioning. Resting measurements of lung volumes, along with analysis of the gas composition of expired air, are frequently recorded in research and clinical settings. In addition, gas analysis of expired air samples recorded during exercise is a common place measurement taken in exercise and adventure physiology settings. Figure 4.12 shows the static lung volumes that can be measured with the use of a spirometer where average total lung volume for a male is around 6 L and approximately 5 L for females. The spirometer creates a trace referred to as a spirogram, which records a wide variety of lung volumes including total lung capacity. This is possible via a specialist breathing system that includes the inhalation of a known volume gas (often helium) from which, after recording expired air volumes, the total lung capacity can be calculated.

More readily available in the laboratory, and used more frequently in a research setting, are the peak flow meter and the vitalograph pictured in Figure 4.13. The peak flow meter is used to measure the efficiency of the lungs. The participant exhales as forcefully as possible and the peak flow meter records the maximal force at which air leaves the lungs. The vitalograph can record two lung measurements: the forced vital capacity (FVC), which is also shown in the spirometer trace in Figure 4.12, and the forced expiratory volume in one second (FEV_1). FVC is a measure of the voluntary capacity of the lungs and FEV_1 a measure of the efficiency of the lungs. For an FVC test the participant (normally wearing a nose clip) breathes out as much air as they can force from their lungs after three maximal inhalations. During an FEV_1 test, the participant inhales maximally and then breathes out as much air as quickly as possible. The volume of air expired in the first second is recorded and shown as a percentage of the total FVC. The higher the percentage of the FVC exhaled in one second the more powerful the lungs.

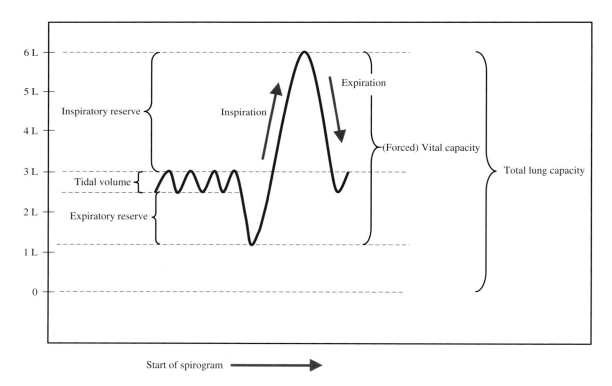

Figure 4.12 Lung volumes

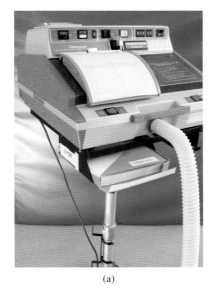

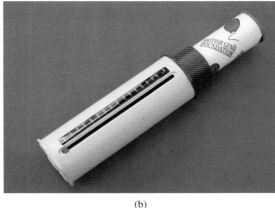

(a) (b)

Figure 4.13   (a) Vitalograph and (b) peak flow meter

The measurement of expired air gases represents a fundamental research tool for exercise and adventure physiologists. The use of Douglas bags and collection tubes to collect expired air samples, a gas analyser to assess the oxygen and carbon dioxide content and a gas meter to record the total air expired in a given time are frequently utilised to examine the demands of an activity.

The tubing and valve boxes are connected to the Douglas bag rack and the participant puts the mouth piece in and the nose clip on. An expired air sample is then collected over a given time by opening and closing the valve to one of the Douglas bags. Figure 4.14 shows the use of Douglas bags during an expired air collection for an athlete using a rowing machine. The collected gas samples are then analysed to record the percentage of oxygen remaining in the expired air sample and the percentage of carbon dioxide present. The percentage of each gas is known as the fraction of expired oxygen F_EO_2 and the fraction of expired carbon dioxide F_ECO_2. Finally, the Douglas bag is emptied using a gas meter to measure the total volume of air expired during the test period. From the results collected it is

Figure 4.14   Douglas bags in use

possible to calculate the minute ventilation $\dot{V}_E$, the volume of O_2 consumed per minute i.e. $\dot{V}O_2$ and the volume of CO_2 produced per minute i.e. $\dot{V}CO_2$ during the test.

The gas collected in any experiment must be standardized before any comparisons can be made between data collected on different days or from a different laboratory. This standardization involves the conversion of gas volume collected at ambient temperature and pressure saturated (ATPS) to standard temperature and pressure dry (STPD). The need to make this conversion arises from the properties of gases. French physiologist Jacques Charles identified that an increase in temperature causes a proportional increase in the volume of a gas (where pressure is held constant). Consequently, collecting a gas samples from the same individual completing the same amount of work but in two different temperature environments would result in different volumes of air expired. This principle of gas properties became know as *Charles' Law*.

Robert Boyle (Chapter 1) identified that at a constant temperature the volume of a gas varies inversely with pressure. This principle of gas behaviour has since been termed *Boyle's law*. The atmospheric pressure on a given day will affect the volume of an expired air sample: the higher the pressure the lower the gas volume. Finally, the humidity in the air and the *water saturation* that occurs while air is in the lungs will increase any gas volume. The conversion to STPD changes a gas volume to that which would have been recorded at $0\,°C$ and at $760\,mm\,Hg$ (sea level). The dry component of STPD takes account of water saturation and provides a dry air conversion which makes an STPD gas collection comparable with data from anywhere in the world, providing that the comparison data is also converted to STPD. The conversion of a gas volume from ATPS to STPD can be calculated by equation, or more simply through the use of a correction factor table (e.g. http://www.sportex.bham.ac. uk/research/hpl/lab8.htm). For the example STPD conversion shown, the correction factor was 0.898 which was found by looking down the barometric pressure column to the row for $760\,mm\,Hg$ and then across to the $22\,°C$ temperature column. By multiplying the gas volume by the correction factor, the expired air gas volume can be found.

Calculation of STPD gas volume from ATPS

In a laboratory, a 60 s expired gas sample was collected from a kayaker during a 1000 m paddle on a kayak ergometer. The laboratory temperature was $22\,°C$ and the barometric pressure was $760\,mm\,Hg$ on the day the data was collected. The gas volume at ATPS was 82.5 L. The correction factor for the temperature and pressure during this experiment (using the table cited above) is 0.898. The conversion of the expired gas volume to STPD is as follows:

Gas volume (STPD) = Gas volume (ATPS) × correction factor

$$= 82.5 \times 0.898\ L$$

$$= 74.085\ L$$

Once the $\dot{V}_E$ STPD has been found, it is possible to calculate $\dot{V}O_2$ and $\dot{V}CO_2$ for an expired air gas sample using F_EO_2 and F_ECO_2. At the most basic level, the rate of oxygen consumed $\dot{V}O_2$ is simply the volume of oxygen in air inspired per minute less the volume of oxygen in air expired per minute. The rate of consumed oxygen from an expired air sample is not known. However, it can be calculated from the F_EO_2 reading from the gas analyser. The calculation of $\dot{V}O_2$ would then be:

$$\dot{V}O_2 = \left(\dot{V}_I \times F_IO_2\right) - \left(\dot{V}_E \times F_EO_2\right)$$

This calculation would be fairly straightforward but for two problems with expired air assessment of oxygen consumption. The first involves the volume of inspired air per minute $\dot{V}_I$ which is not normally recorded during a standard gas analysis procedure, and the second relates to the volume of gas expired. The latter is normally smaller than that consumed, which will lead to errors in the estimation of $\dot{V}O_2$. Early physiologists assessed both $\dot{V}_I$ and $\dot{V}_E$. From this they determined that the number of molecules of oxygen consumed during metabolism is not replaced by the number of molecules of carbon dioxide produced. As a consequence, the gas volume in expired air was smaller than the volume of air inspired.

The development of a clever transformation, named after the Scottish physiologist John Haldane (Chapter 1) but actually developed some 20 years earlier by the German physiologists August Geppert and Nathan Zuntz, has enabled physiologists to calculate $\dot{V}O_2$. This method does not require $\dot{V}_I$ to be measured or the nitrogen content in expired air to be known, and makes an allowance for the change in volume between inspired and expired gas samples. The *Haldane transformation* actually takes advantage of the drop in volume, and uses it as part of the calculation of $\dot{V}O_2$ from an expired air sample. Geppert, Zuntz and Haldane recognized that with the drop in total gas volume, the relative percentage of nitrogen in an expired air sample increased. This was not because nitrogen was consumed during metabolism, but by remaining unchanged during metabolism, the same amount of nitrogen becomes a larger percentage of a smaller total volume of gas. The change in the relative percentage of nitrogen forms the basis of the Haldane transformation. The original Haldane transformation involves a series of steps first to identify $F_E N_2$ and then $\dot{V}_I O_2$ and finally $\dot{V}_E O_2$ or, more simply, $\dot{V}O_2$. The series of equations can be manipulated to create one calculation for the determination of $\dot{V}O_2$ from an expired air sample:

$$\dot{V}O_2 = \dot{V}_E \times \{0.265[1 - (F_E O_2 + F_E CO_2)] - F_E O_2\}.$$

Although computer programmes are usually available in laboratories to apply the Haldane transformation, the following calculation box highlights the steps in the determination of $\dot{V}O_2$. It is easy to enter data into a computer without having an understanding of the computational results, and so the following example (as well as conversion of ATPS data to STPD) should be studied carefully.

Example calculation of $\dot{V}O_2$ from expired gas samples according to the Haldane transformation:

$$\dot{V}O_2 = \dot{V}_E \times \{0.265\,[1 - (F_E O_2 + F_E CO_2)] - F_E O_2\}$$

Using the example from the STPD calculation for the kayaker, $\dot{V}_E = 74.1\ \text{L min}^{-1}$, $F_E O_2 = 17.4\,\%$ (or 0.174) and $F_E CO_2 = 4.1\,\%$ (or 0.041). Based on the Haldane transformation,

$$\dot{V}O_2 = 74.1 \times \{0.265$$
$$\times [1 - (0.174 + 0.041)] - 0.174\}$$
$$= 74.1 \times \{0.265 \times [1 - 0.215]$$
$$- 0.174\}$$
$$= 74.1 \times \{0.208 - 0.174\}$$
$$= 74.1 \times 0.034$$
$$= 2.521$$

i.e. The rate of O_2 consumed is 2.521 L min^{-1}.

The $\dot{V}CO_2$ produced during metabolism, calculated from an expired air sample, is more simply identified than that of $\dot{V}O_2$ due to the very small content of carbon dioxide in inspired air.

The amount of carbon dioxide in the air around us represents around 0.03–0.04 per cent of the total volume. As a result, the $\dot{V}CO_2$ can be calculated as

$$\dot{V}CO_2 = \dot{V}_E \times (F_ECO_2 - F_ICO_2)$$

where $F_ICO_2 = 0.04\% = 0.0004$.

Example calculation of $\dot{V}CO_2$ from expired gas samples according to the Haldane transformation:

$$\dot{V}CO_2 = \dot{V}_E \times (F_ECO_2 - F_ICO_2)$$

where $F_ICO_2 = 0.04\%$ (0.0004) and as in previous worked example, $\dot{V}_E = 74.1$ L min^{-1} and $F_ECO_2 = 4.1\%$ (or 0.041). According to the Haldane transformation,

$$\dot{V}CO_2 = \dot{V}_E \times (F_ECO_2 - F_ICO_2)$$
$$= 74.1 \times (0.041 - 0.0004)$$
$$= 74.1 \times 0.0406$$
$$= 3.008$$

i.e. The rate of CO_2 expired is 3.008 L min^{-1}.

Key point

The capacity and functioning of lungs can be measured through the use of spirometers, vitalographs and peak flow meters. In addition, expired air samples can be collected to provide details about the $\dot{V}O_2$ consumed, the $\dot{V}CO_2$ produced and $\dot{V}_E$ at rest or during exercise. To facilitate comparison with previous data or gas collections from other parts of the world, samples from one

laboratory must be converted from ambient temperature and pressure saturated (ATPS) to standard temperature and pressure dry (STPD). After standardization, the Haldane transformation can be applied to expired air fractions of oxygen and carbon dioxide to calculate $\dot{V}O_2$ and $\dot{V}CO_2$.

Pulmonary ventilation and exercise

The use of gas analysis techniques can be employed to examine the effects of exercise on oxygen consumption, carbon dioxide production and $\dot{V}_E$. From this fundamental data collection method, a number of measures have been developed for assessing exercise performance. Perhaps the most commonly reported of these measures is the *maximal oxygen uptake* $\dot{V}O_{2\,max}$ which has traditionally been the measure of aerobic fitness and is described in more detail in Chapter 5. The *aerobic-anaerobic threshold*, that is, the transition between aerobic and anaerobic exercise (Chapters 5 and 9), can be estimated from $\dot{V}O_2$, $\dot{V}CO_2$ and $\dot{V}_E$ data.

The *respiratory exchange ratio* (RER) provides physiologists with information about the foodstuff that is providing the main energy source for exercise. At low intensities of exercise, lipids (as part of aerobic energy production) form the main fuel source for exercise. As the intensity rises, the body will rely more heavily upon *glycolysis* as an anaerobic method for energy production. This shift in fuels from lipid to glucose can be derived from the ratio of $\dot{V}CO_2$ produced to $\dot{V}O_2$ consumed. The following RER calculations show RER for the saturated lipid palmitic acid and glucose. The RER for fats is 0.7 and the RER for glucose is 1.0. During the transition from low to high-intensity work, the RER increases from 0.7 to 1.0. An RER of 0.85 provides an indicator that mixed fuels are being used: i.e. both fats and carbohydrates are providing the energy for work.

Calculation of RER from $\dot{V}O_2$ and $\dot{V}CO_2$ data:

Lipid oxidation
Palmitic acid $C_{16}H_{32}O_2$
$C_{16}H_{32}O_2 + 23O_2 \rightarrow 16CO_2 + 16H_2O$
   $+129$ ATP
RER $= 16CO_2 : 23O_2 = 16 \div 23 = 0.7$

Glucose oxidation
Glucose $C_6H_{12}O_6$
$C_6H_{12}O_6 + 6O_2 \rightarrow 6CO_2 + 6H_2O + 38$ ATP
RER $= 6CO_2 : 6O_2 = 6 \div 6 = 1.0$

At rest, the RER is sometimes referred to as the *respiratory quotient* (RQ). The RQ was first developed in the early 20th century before the term RER came into use, to identify the predominant fuel for metabolism. Since the early research into RQ it has been discovered that, during exercise, carbon dioxide is produced from additional sources as well as from macronutrient catabolism. As a consequence the term RQ was replaced with RER to acknowledge the possible influence of non-macronutrient produced CO_2 to the expired $\dot{V}CO_2$. In a similar way, further research has indicated that during exercise protein can provide up to 10–15 per cent of the fuel for metabolism. The RER does not distinguish protein catabolism, which has an RQ of around 0.83, from that of fat or carbohydrate. As a result, RER should be thought of as a non-protein RER. The relatively small contribution of protein to metabolism means that it is generally ignored in exercise physiology.

In this section on pulmonary ventilation and exercise, particularly in the case of strength training for adventure sports and in rock climbing, it is useful to consider the breath-holding brought about during the Valsala manoeuvre. The *Valsalva manoeuvre*, named after 17th century physiologist who first described the process, involves initially holding the breath to fixate the chest and abdomen which can enhance strength during heavy lifts or a difficult climbing move. After a full inhalation, the performer closes the vocal cords (the *rima glottidis* space, through which air passes in between the vocal cords of the larynx) but at the same time contracts the expiratory muscles. The net result of this is a fixation of the chest and abdomen cavities and a large increase in pressure which can lead to improved strength in heavy lifts. As part of such a heavy lift, and indeed the Valsalva manoeuvre itself, the athlete performs a forced exhalation to release the air from their lungs and the pressure. This manoeuvre leads to an increase in strength, but also causes an initial drop in blood pressure followed by a rise in pressure once the air is released. This can be a concern for individuals with high blood pressure levels.

4.3   Cardiovascular system

The cardiovascular system, the body's main transport mechanism, comprises three components: the *vascular system*, the continuous circuit of vessels through which blood flows; the *heart*, the system pump; and the 5.0–5.5 L of *blood* which is transported around the body. From very early on in development the cardiovascular system is required for transport; as a consequence, the heart is the first functional organ in the body. The functions of the cardiovascular system are to *transport* nutrients and hormones to cells and waste products from cells for removal from the body, to help *maintain* body temperature and to assist in the *prevention* of infection. The cardiovascular system includes structures that contain two types of muscle, different to skeletal muscle discussed earlier in the chapter. The heart consists of cardiac muscle which enables the heart to beat continuously, without fatigue, around 40 000 000 times per year and therefore between 300–350 billion times in an average lifetime. The blood vessels contain smooth muscle that assists the heart in transporting blood around the body.

Blood

Males and females carry between 5.0 L and 5.5 L of blood throughout their adult lives. The functions of blood are similar to those of the cardiovascular system: to enable *transport* by carrying nutrients and waste products either in solution or combined with cells which form part of the blood; to enable *heat transfer* for cooling or preservation purposes; to carry cells which form a major role in *immune function* and to *minimize blood loss* and *maintain blood pH*.

Blood represents an average of 8 per cent of total body mass and normally comprises 55 per cent plasma and 45 per cent cellular components. The cellular components include erythrocytes (red blood cells), leukocytes (white blood cells) and thrombocytes (platelets) which are all developed in the bone marrow. Blood plasma, which comprises 91 per cent water, holds proteins and other solutes within it and transports the cellular components in suspension. Blood is slightly sticky to touch, tastes slightly metallic and due to its cellular components is denser that water. Using a centrifuge it is possible to separate blood into its component parts. Figure 4.15 provides a diagram of the centrifuged content of blood and provides details of the constituent components of plasma. Centrifuging blood is employed clinically to examine the

haematocrit of blood, defined as the packed cell volume to plasma ratio. This is normally between 40–45 per cent i.e. the cellular components making up 40–45 per cent of the total blood volume. After the sample has settled, the leukocytes and thrombocytes can normally be clearly identified as a buffy coat between the erythrocytes and plasma.

Erythrocytes, mentioned as an example cell in the first section of this chapter, number around 5 billion in every millilitre of blood. Red blood cells have a life cycle of around 120 days. Their lifecycle is limited by the stresses placed on the cell membrane during the journey through the narrow capillaries of the vascular system and, consequently, erythrocytes have to be constantly replaced at rate of 2 million cells every second. As mentioned previously, the specialized oxygen and carbon dioxide carrying function of red blood cells requires a unique structure that is established during their development in the bone marrow.

Erythrocytes remove many of the organelles found within other cells during their development. This occurs to enhance their oxygen-carrying potential through the additional carriage

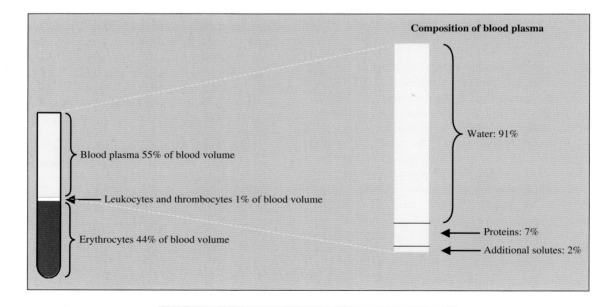

Figure 4.15 Major components of centrifuged blood

of haemoglobin molecules. Each erythrocyte contains in the region of 280 million haemoglobin molecules and every one of those can hold up to 4 oxygen molecules; red blood cells are therefore well equipped for their oxygen-carrying role. To enhance the diffusion rate of oxygen across the cell membrane, erythrocytes are biconcave (see Figure 3.2) which greatly increases the surface area available for diffusion. Each haemoglobin molecule contains four protein-based polypeptides which combine to form the protein globin. Attached to each of the four polypeptides chains is a haeme group which is a ring-like structure containing an iron ion at its centre. Each oxygen molecule binds with iron during its transport within the blood.

White blood cells, with around 7 million per millilitre of blood, are the least numerous of the cellular components. Leukocytes are commonly divided into two types: those having cytoplasmic granules that are visible under a light microscope, and those that do not. There are three types of granular leukocyte: neutrophils, eosinophils and basophils and two forms of agranular white blood cell: lymphocytes and monocytes. The lifespan of a leukocyte is determined by the health of the individual. In a healthy person, they can live months and even years. However, when someone is ill they only survive for a matter of days. The role of leukocytes is one of immune function. White blood cells are responsible for fighting bacterial, viral, and fungal infections as well as chronic disease and are responsible for determining allergic reaction to substances within the body.

Thrombocytes or platelets are formed from megakaryocytes. In cellular terms these are vast cells which, during development in the bone marrow, fragment into around 2500 smaller subcellular units. These units are called thrombocytes and each has its own intact cell membrane within which it stores a wide range of chemicals including ATP, glycogen and serotonin. There are normally around 250 million platelets found in each millilitre of blood. The primary role of thrombocytes is clotting to stop blood loss after trauma to a blood vessel. Thrombocytes arriving at the site of an injury stick to the side of the wound and create a platelet plug which seals the wound. They are assisted in this role by the blood-carried enzyme thrombin which converts the plasma protein fibrinogen to fibrin. Fibrin threads interact with the platelets to create a fibrin mesh between the platelets to seal the wound.

The largest component of the blood is plasma which, as shown in Figure 4.15, comprises over 90 per cent water, 7 per cent plasma proteins and 2 per cent additional solutes such as electrolytes, nutrients, gases, enzymes, hormones and waste products. The electrolytes include sodium, potassium, chloride and bicarbonate. A wide range of macronutrients and micronutrients are carried in solution by the plasma along with the gases oxygen and carbon dioxide. Waste products include those to be excreted through the urinary system and from the digestive tract. The key plasma proteins are synthesized in the liver and include *albumins* which serve as carrier proteins for fat-soluble hormones (such as those shown in Figure 3.17) and fatty acids, *globulins* which are involved in immune function and serve as fat-soluble vitamin carriers and *fibrinogen* which is an essential component of the blood-clotting process.

The link between the respiratory system and the cardiovascular system lies in the transport of oxygen and carbon dioxide. Carbon dioxide, produced by the cells of the body, is transported within the blood for removal from the body via the lungs and at the same time oxygen is taken up by the blood for transport to all cells in the body. Oxygen is either bound with haemoglobin to form oxyhaemoglobin (98.5 per cent) or dissolved in the blood plasma (1.5 per cent). The 1.5 per cent found in the plasma facilitates control of breathing and heart rate by the brain, as it is this store of oxygen that is detected by chemoreceptors in the common carotid artery to relate sensory information about oxygen levels to the brain. Fully-oxygenated blood is able to transport 20 mL oxygen per decilitre of blood, due primarily to the vast number of haemoglobin molecules within each red blood cell.

Carbon dioxide is transported in the blood either combined with haemoglobin to form carbamino-haemoglobin (20 per cent) or dissolved in the plasma (5 per cent). However, the major form of

CO_2 transport (75 per cent) is by chemical reaction with water to produce initially carbonic acid which then dissociates in the plasma to bicarbonate and hydrogen ions. This reversible reaction, which is catalysed by the enzyme carbonic anhydrase, was discussed in **Mechanics of pulmonary ventilation**. Just as dissolved oxygen is detected by sensory chemoreceptors, so the dissolved CO_2 and presence of H^+ (produced through the formation of bicarbonate resulting from the reaction between carbon dioxide and water) are detected by sensory bodies in the aorta and common carotid arteries. The pH of blood (the level of acidity to alkalinity of a solution ranging from 0–14 with 7 being neutral) is around 7.4 so it is very slightly alkaline. However, when exercise intensity is above 50 per cent of maximum, the pH levels begins to decrease. The blood becomes more acidic due to the release of H^+ ions into the blood.

Key point

Blood comprises 55 per cent plasma and 45 per cent cellular components. Plasma, which is largely composed of water, carries plasma proteins and a range of additional solutes essential for life or requiring removal from the body. The cellular components of blood include: red blood cells which, with their high haemoglobin content, are essential for oxygen transport; white blood cells that are vital to immune function; and platelets that are chiefly responsible for blood clotting but also produce serotonin, a hormone of the endocrine system.

The heart

The heart consists of four chambers enclosed by cardiac muscle which is similar to striated skeletal muscle. The *myocardium* (muscle of the heart) differs from skeletal muscle in the way the multinucleated fibres are arranged. In skeletal muscle, the fibres are parallel to each other but in cardiac muscle the fibres are arranged in a network to facilitate coordinated contraction. The heart, which weighs less than half a kilogram and is about the size of a fist, is situated near the midline of the body resting on the diaphragm. It is located such that two-thirds of its volume is on the left side of midline (this is why the left lung is slightly smaller than the right and has only two lobes instead of three) and is angled on a slight diagonal from midline.

Protected by an outer membrane known as the *pericardium*, the heart actually consists of two separate pumps that contract in unison. The right side pumps deoxygenated blood to the lungs (the *pulmonary circulation*) and the left side is responsible for pumping oxygenated blood throughout the body (the *systemic circulation*). Figure 4.16 depicts the heart; note that the heart is always drawn from the perspective of a person looking out from the page such that the left side of the heart appears on the right side of the page as you look into the textbook.

The heart, as mentioned previously, is the first functional organ to commence activity during development and it continues to beat 40 million times per year, without fatigue, throughout our lives. Each day it pumps around 370 litres of blood around the body. Deoxygenated blood from the main veins, the superior and inferior vena cavae and the coronary sinus (which returns blood from the heart's own blood circulation) is returned to the *right atrium* or collecting chamber. At the same time, oxygenated blood from the four pulmonary veins (from the lungs) returns blood to the *left atrium*. Much of the blood returning to the atria flows directly into the ventricles. However, the first phase of contraction of the heart forces blood through the *tricuspid valve* (right side of the heart) and *bicuspid valve* (left side of the heart), together known as the atrioventricular valves, into the ventricles. Due to the nature of the journeys that blood entering each of the ventricles will follow, when seen in cross-section the walls of the left ventricle (the cardiac muscle surrounding the chamber) is much thicker than that of the right ventricle.

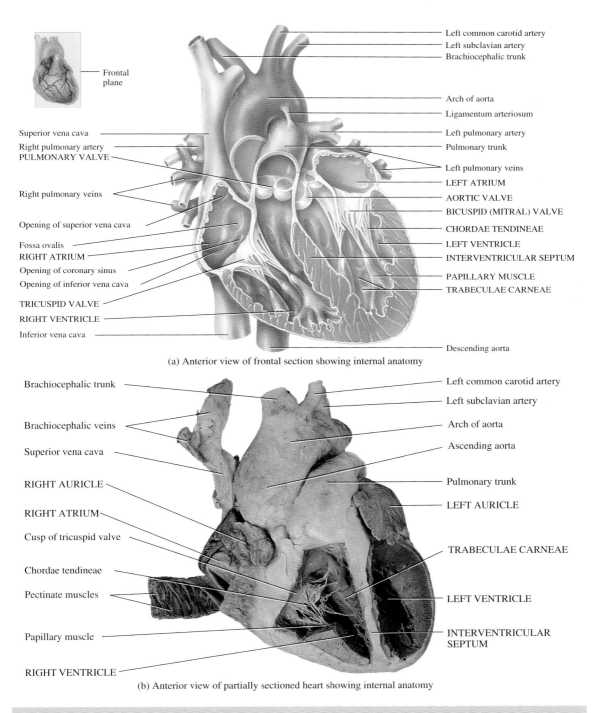

Frontal plane

Left common carotid artery
Left subclavian artery
Brachiocephalic trunk

Arch of aorta
Ligamentum arteriosum

Superior vena cava
Right pulmonary artery
PULMONARY VALVE

Left pulmonary artery
Pulmonary trunk

Left pulmonary veins
LEFT ATRIUM
AORTIC VALVE
BICUSPID (MITRAL) VALVE
CHORDAE TENDINEAE
LEFT VENTRICLE
INTERVENTRICULAR SEPTUM
PAPILLARY MUSCLE
TRABECULAE CARNEAE

Right pulmonary veins

Opening of superior vena cava

Fossa ovalis
RIGHT ATRIUM
Opening of coronary sinus
Opening of inferior vena cava

TRICUSPID VALVE

RIGHT VENTRICLE

Inferior vena cava

Descending aorta

(a) Anterior view of frontal section showing internal anatomy

Brachiocephalic trunk

Brachiocephalic veins

Superior vena cava

RIGHT AURICLE

RIGHT ATRIUM

Cusp of tricuspid valve

Chordae tendineae

Pectinate muscles

Papillary muscle

RIGHT VENTRICLE

Left common carotid artery
Left subclavian artery
Arch of aorta
Ascending aorta

Pulmonary trunk

LEFT AURICLE

TRABECULAE CARNEAE

LEFT VENTRICLE

INTERVENTRICULAR SEPTUM

(b) Anterior view of partially sectioned heart showing internal anatomy

Figure 4.16 The structure of the heart (from Tortora and Derrickson, *Principles of Anatomy and Physiology*, Eleventh Edition, 2006, reproduced by permission of John Wiley & Sons Inc.)

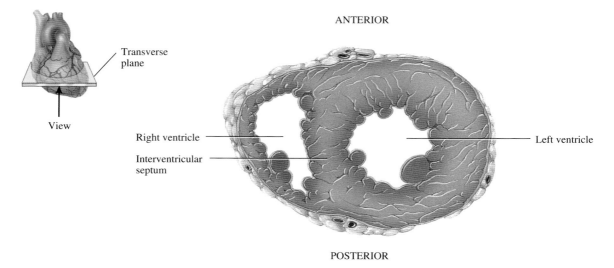

(c) Inferior view of transverse section showing differences in thickness of ventricular walls

Figure 4.16   (*continued*)

Blood ejected from the right ventricle need only travel to the lungs and so the required muscle mass is smaller than that required for the left ventricle which supplies blood throughout the body.

The second phase of cardiac contraction involves the contraction of the ventricles and the ejection of blood from the left ventricle into the aorta and from the right ventricle to the pulmonary trunk (which branches into the left and right pulmonary arteries). As the ventricles contract, the atrioventricular valves (one-way flow regulators) are forced closed to ensure blood is ejected from the ventricles and does not flow back into the atria.

In sequence, a deoxygenated red blood cell returning from the leg would enter the right atrium from the inferior vena cava, pass through the tricuspid valve and into the right ventricle. After contraction of the right ventricle, it would be ejected into the pulmonary trunk and then pass to either the left or right pulmonary artery and onwards to the capillaries of the lungs. The same red blood cell, having been oxygenated at the lungs, would return to the heart via one of the four pulmonary veins and pass into the left atrium. From the left atrium, the erythrocyte would pass into the left ventricle. Upon ventricular contraction, it would be ejected into the aorta and onwards into the systemic blood flow to supply oxygen to another tissue or organ in the body.

As mentioned previously, the heart comprises two pumps. The heart's double pump can be thought of in two ways. Firstly, deoxygenated blood pumped into the right ventricle supplies blood to the lungs and the left side supplies oxygenated blood to systemic circulation. Secondly, the heart contracts in a two-step fashion, which can be inferred from the sound heard when using Doppler ultrasound echocardiography. During echocardiography, the contraction of the heart can be heard as a 'lub-dup' sound. The first sound is created by the closure of the atrioventricular valves at the start of a ventricular contraction (after completion of atrial contraction); the second phase represents the end of the ventricular contraction and the closure of the semi-lunar valves in the pulmonary trunk and aorta which stop blood flowing back into the ventricles (after which the atria will begin to contract again to fill the ventricles).

Control of contraction of the heart is brought about by nervous stimulation and is triggered by two nerve bodies: the *sinoatrial* (*SA*) *node* and the *atrioventricular* (*AV*) *node*. The conduction system for the heart is shown in Figure 4.17. The SA node works as the pacemaker for the heart and the SA and AV together provide the conduction systems for the contraction of the atria and ventricles.

The SA node, situated on the right atrial wall, stimulates autorhythmic contraction of the heart. This means that even without nervous stimulation

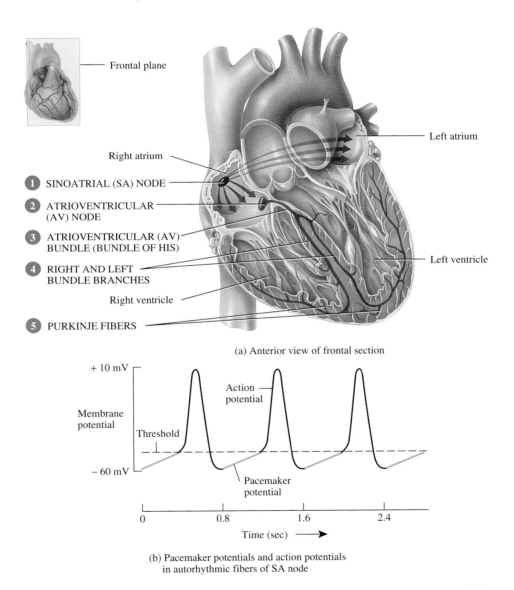

Frontal plane

Right atrium

Left atrium

1 SINOATRIAL (SA) NODE

2 ATRIOVENTRICULAR (AV) NODE

3 ATRIOVENTRICULAR (AV) BUNDLE (BUNDLE OF HIS)

4 RIGHT AND LEFT BUNDLE BRANCHES

Left ventricle

Right ventricle

5 PURKINJE FIBERS

(a) Anterior view of frontal section

+ 10 mV

Membrane potential

Action potential

Threshold

Pacemaker potential

− 60 mV

0   0.8   1.6   2.4

Time (sec) ⟶

(b) Pacemaker potentials and action potentials in autorhythmic fibers of SA node

Figure 4.17 Conduction system of the heart (from Tortora and Derrickson, *Principles of Anatomy and Physiology*, Eleventh Edition, 2006, reproduced by permission of John Wiley & Sons Inc.)

from the autonomic nervous system, the SA node – as the heart's pacemaker – propagates contraction of the heart at rate of around 100 bts min^{-1}. Nervous stimulation of the SA node via the parasympathetic division slows the natural rhythm of the SA node to around 70–75 bts min^{-1}. The effects of the sympathetic nervous system and the release of the hormone adrenalin during exercise can increase SA node rhythm to an individual's heart rate maximum. The SA node triggers an action potential through the muscle fibres of the atria, causing them to contract and eject blood into the ventricles of the heart. The spread of the action potential through the atria reaches the AV node and this triggers the release of an action potential through the *bundle of His* to the Purkinje fibres. The *Purkinje fibres* propagate the action potential to the ventricular cardiac muscles initiating from the apex of the heart. The location of the Purkinje fibres and the nature of cardiac muscle fibres results in a twisting-squeezing contraction by the ventricles ejecting blood from within.

An *electrocardiograph* can be employed to detect the electrical stimulation of the heart and the trace it creates. An *electrocardiogram* (ECG) can be interpreted for medical purposes to examine the functioning of the heart. Figure 4.18 provides an example of an ECG.

The ECG trace is normally described according to the three deflections in the trace: a P wave; the QRS complex; and the T wave. The *P wave* represents the depolarization of the atria, the electrical impulse spread from the SA node that stimulates the atria to contract. The *QRS complex* depicts the depolarization of the ventricles propagated from the AV node via the Purkinje fibres to the ventricular muscle fibres. The *T wave* represents the repolarization of the ventricles ready for the next contraction. The repolarization of the atria, which takes place during the depolarization of the ventricles, is masked by the QRS complex and therefore does not have a separate depiction on an ECG trace. The period between the P and the R wave, called the *P–R interval*, represents the time for the atria to contract and then begin to relax (R is used rather than Q for this interval because the Q wave is often small). The same phase for the ventricles, from the start of contraction to the start

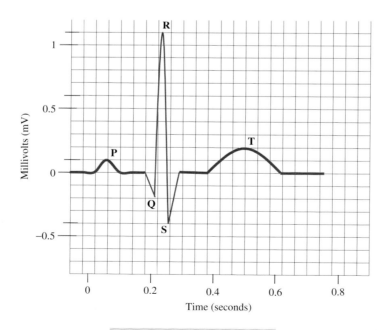

Figure 4.18 A normal ECG

of relaxation, is known as the *Q–T interval*. The contraction phase of the cardiac cycle is known as systole and the relaxation phase as diastole.

In a medical setting, a great deal of information can be gleaned from an ECG. The wave amplitude (size) and wavelength (time between two waves) can be interpreted to identify problems with heart function. As part of the detection of heart dysfunction process, an exercise ECG can be completed to examine the cardiac cycle under physical stress. The stress of exercise can reveal changes in wave traces and arrhythmias that might not be present in a resting ECG. Irrespective of any heart dysfunction, during exercise the cardiac cycle will shorten with corresponding increase in heart rate. As a consequence of this, the gaps between electrical events will decrease. For instance, the T–P interval (the gap between the end of ventricular and atrial depolarization) will shorten as heart rate increases.

At rest, a complete cardiac cycle from P wave to P wave normally takes about 0.8 s for a person with a heart rate (HR) of 75 bts min^{-1} (0.8 × 60 = 75 bts min^{-1}). As well as identification of the phases in a cardiac cycle, an ECG can be used to determine heart rate. An ECG is normally calculated by counting the number of R waves in a given time or by number of squares between R waves. *Tachycardia*, normally defined as an HR above 100 bts min^{-1}, can be cause by a rise in body temperature during illness or other sympathetic response. *Bradycardia*, when resting HR is below 60 bts min^{-1}, is often detected in athletes who have a higher stroke volume (amount of blood ejected in one ventricular contraction) due to the trained increase in ventricular wall thickness, but can also be related to illness.

In a sporting context, HR is normally determined by telemetry through the use of a heart rate monitor. Heart rate monitors still rely on the electrical activity of the heart to determine the rate of contraction, but only display the HR and not any information about the cardiac cycle phases as would be revealed in an ECG. Nevertheless, because of the close relationship between HR and exercise intensity, heart rate monitors provide an excellent method of monitoring effort during training.

Heart rate is a component of a further measure of cardiac function: cardiac output ($\dot{Q}$). Cardiac output represents the total amount of blood ejected from the left ventricle *in one minute* (hence the dot above the Q) and is determined by stroke volume (SV) (amount of blood ejected in one contraction of the left ventricle) and HR. Although HR can be readily determined using a HR monitor, it is more problematic to determine SV and $\dot{Q}$. Stroke volume is theoretically determined by the difference between the volume of blood filling the left ventricle by the end of the diastole (*end-diastolic volume*, EDV) and the amount of blood left in the chamber at the end of the systole (*end-systolic volume*, ESV).

To directly measure SV and $\dot{Q}$, unacceptable invasive techniques would be required. A number of indirect methods have therefore been devised to calculate $\dot{Q}$. The most widely accepted of these was first proposed by Adolf Fick in 1870. Fick noted that by determining the amount of oxygen consumed by an individual in one minute, and the difference between the oxygen content in a known amount of arterial and venous blood (the a–$\bar{v}$ O_2 difference), it is possible to determine the cardiac output and therefore the stroke volume.

Equations

$$\dot{Q} = SV \times HR$$

$$SV = EDV - ESV$$

Fick Equation

$$\dot{Q} = \frac{\dot{V}O_2}{a - \bar{v} O_2 \text{ difference}} \times 100$$

For example, consider an individual with a resting HR of 75 bts min^{-1}, a $\dot{V}O_2$ consumed of 250 mL min^{-1} and a–$\bar{v}$ O_2 of 5 mL min^{-1}. According to the Fick equation,

$$\dot{Q} = \frac{250}{5} \times 100 = 5000$$

and

$$\dot{Q} = SV \times HR \Rightarrow SV = \frac{\dot{Q}}{HR} = \frac{5000}{75} = 66$$

i.e. Stroke volume is 66 mL.

During exercise, both HR and SV rise to supply the increased demands of the activity. Increases in HR, as already mentioned, occur through shortening of such factors as the T–P interval and occur via hormonal and autonomic nervous system stimulation. Stroke volume, the amount of blood ejected with each contraction of the ventricles, is thought to increase due to the effects of two mechanisms. The first, known as the *Frank–Starling Law*, suggests that the increased venous return due to exercise results in an increased filling of the ventricles, placing an additional stretch on the cardiac muscle of the ventricular walls. This preloading creates a stronger contraction and therefore results in an increased stroke volume. The second proposed mechanism suggests that increases in electrical stimulation of the myocardium due to the demands of exercise result in an increase in the ventricular contractions strength and a consequential rise in SV. Both mechanisms are thought to operate during exercise, although research has determined that increases in SV are limited. While SV may increase during exercise up to around 150 per cent of resting level, SV appears to plateau and may even decrease after this point as a result of continually decreasing filling times between contractions.

Key point

The heart is a four chamber pump comprising cardiac tissue, responsible for delivering blood to the lungs and every tissue and organ of the body. The heart is functional from very early on in foetal development and beats continually throughout life. The left side of the heart is responsible for delivering blood to the systemic circulation and the right for the transit of blood to the lungs via pulmonary circulation. Each contraction of the ventricles is known as systole and each relaxation phase as diastole. The phases in heart contraction, the contraction of the atria which force blood into the ventricles and the contraction of the ventricles themselves can be recorded with the assistance of an electrocardiograph. The ejection of blood from the ventricles is known as stroke volume (SV) and the number of ventricle contractions per minute as heart rate (HR). The total volume ejected in one minute is referred to as the cardiac output, the product of SV and HR.

Vascular system

The vascular system comprises the various blood vessels that form a closed system to transport blood to and from the heart. Early physiologists believed the blood ebbed and flowed from the heart and it was not until the work of William Harvey in 1628 (Chapter 1) that the true circulatory nature of blood flow was discovered. The main vessels of the vascular system include: arteries, arterioles, capillaries, venules and veins. Figure 4.19 shows the main circulatory routes, with the aorta being the main artery from which all other arteries flow and the main veins being the superior and inferior vena cavae. Arteries and arterioles carry blood away from the heart to the capillary networks that form the gaseous and nutrient exchange beds with all other cells of the body. Venules and veins return blood back to the right atrium of the heart. With the exception of the pulmonary arteries and veins, arteries carry oxygenated blood (blood that has been supplied with oxygen from the lungs) and veins deoxygenated blood (blood which has had some oxygen removed by the

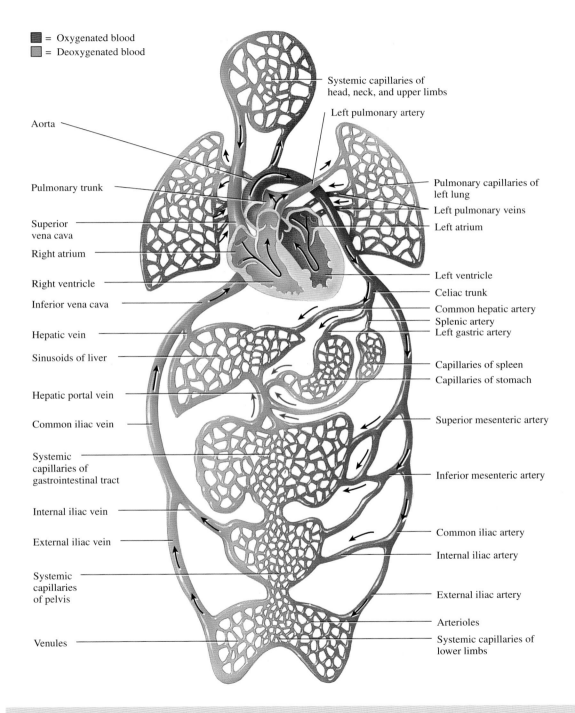

■ = Oxygenated blood
□ = Deoxygenated blood

Systemic capillaries of
head, neck, and upper limbs

Left pulmonary artery

Aorta

Pulmonary trunk

Pulmonary capillaries of
left lung

Left pulmonary veins

Left atrium

Superior
vena cava

Right atrium

Left ventricle

Right ventricle

Celiac trunk

Inferior vena cava

Common hepatic artery

Splenic artery

Left gastric artery

Hepatic vein

Sinusoids of liver

Capillaries of spleen

Capillaries of stomach

Hepatic portal vein

Common iliac vein

Superior mesenteric artery

Systemic
capillaries of
gastrointestinal tract

Inferior mesenteric artery

Internal iliac vein

Common iliac artery

External iliac vein

Internal iliac artery

Systemic
capillaries
of pelvis

External iliac artery

Arterioles

Venules

Systemic capillaries of
lower limbs

Figure 4.19 Main circulatory routes (from Tortora and Derrickson, *Principles of Anatomy and Physiology*, Eleventh Edition, 2006, reproduced by permission of John Wiley & Sons Inc.)

tissues of the body). The pulmonary arteries carry deoxygenated blood from the heart to the lungs for oxygenation and the pulmonary veins return blood from the lungs to the left atrium (as can be seen in Figure 4.19).

Arteries and veins differ in their construction, although both contain smooth muscle which assists with the movement of blood around the body. Veins tend to have thinner walls and larger diameter bores and the smooth muscle fibres present are elastic to enable larger volumes of blood to be carried. Arteries have thicker walls and smaller bores than their venous counterparts, but have elastic smooth muscle fibres to help squeeze blood through the arterial system. The smooth muscle within the arterioles enables the vessels to dramatically alter the bore size (diameter of the vessel) to control blood flow. *Vasodilation*, an increase in the diameter of the arteriole, is triggered by warm temperatures, a decrease in oxygen levels and an increase in CO_2 and H^+ ions. A decrease in vessel diameter, *vasoconstriction*, is promoted in cold temperatures and in response to a rise in oxygen levels or a decrease in CO_2 and H^+ ions. Vasodilation increases blood flow while vasoconstriction has the opposite effect.

Blood leaves the heart under pressure imparted by each contraction of the heart. Arteries carry blood under pressure and blood flows to the capillaries as a result of this pressure. During each heart beat (contraction) at rest, the pressure in the arterial system normally rises to 120 mm Hg. As blood flows further from the heart, the pressure decreases to such an extent that by the time blood enters the veins, the internal pressure is as low as 18 mm Hg.

Without alternative assistive processes, venous return would be seriously compromised. Venous return is assisted by an active pump provided by skeletal muscles which on contraction squeeze blood within the veins. Venous return through muscular squeeze is enhanced by the presence of one-way valves within veins which prevent blood leaking back through a valve. In addition, the pressure created in the abdomen and thoracic regions by respiration places a further squeeze, called the respiratory pump, on veins and promotes improved venous return. Despite these methods

of venous return at rest, over 60 per cent of the total blood volume is contained within the veins at any one time. The veins therefore represent a significant blood store. During rest when the parasympathetic nervous system is controlling nervous response, much of the blood is directed to the liver, kidneys and digestive tract (after a meal) with perhaps only 15 per cent being directed to the active muscles. During exercise, when the sympathetic nervous system triggers response in the body, up to 80 per cent systemic blood flow can be redirected to the skeletal muscles. During exercise, in addition to the redirection of arterial blood flow, the stores of blood in the large veins of the legs are returned to the heart (by skeletal muscle pump) to assist in the delivery of oxygen and nutrients to the working muscles.

The nature of venous pooling at rest makes the employment of a cool-down strategy after exercise all the more important. If, as an adventure sports performer, you finish exercise without a cool-down the metabolites of exercise will pool (gravity led) in the veins of the legs. Completing an active cool-down will more readily enable waste products from metabolism to be removed from the blood for catabolism and re-synthesis or excretion from the body.

As described above, each contraction of the heart ejects blood from the left ventricle along the aorta and into the systemic vascular system. Each surge of blood places pressure on the walls of the arteries and it is this arterial pressure that physiologists and clinicians measure when they take blood pressure. During resting blood pressure measurement, the pressure cuff is placed over the upper arm and normally inflated to around 200 mm Hg which is a sufficient pressure to stop blood flow in the artery. The sounding end of the stethoscope is placed over the brachial artery at the point of the antecubital space (the hollow at the front of the elbow joint) and the pressure slowly released from the cuff. The pressure in the cuff is sufficient to continue to halt the flow of blood until the pressure falls to around 120 mm Hg. At or close to this point a first surge of blood is able to move through the artery and sound like a 'pop' in the stethoscope ear pieces. The reading forms the systolic blood pressure. The popping sound continues in the ear pieces as the pressure in the cuff continues to fall, coinciding with blips at the meniscus of the mercury column on the sphygmomanometer. When the pressure in the cuff falls to around 80 mm Hg, the blood is able to pass through unhindered and the point where the sounds disappear is termed the diastolic pressure. The diastolic pressure reading represents (the lowest level of pressure in the vascular system in between the contractions of the heart) when the walls of the arteries have recoiled after the stretch induced by the contraction. A blood pressure reading is normally given as systolic pressure 'over' diastolic pressure such that a normal blood pressure reading would be spoken as '120 over 80' or written 120/80. A blood pressure reading is used by clinicians as an initial examination of the health of the vascular system.

Hypertension (high blood pressure) normally occurs from a narrowing or hardening (or both) of the arteries which means the heart has to beat harder to pass blood through the vascular system; resting blood pressure readings are therefore higher. This increase in pressure can be seen when you place a thumb over the end of a hosepipe to spray water out. The occlusion of the hosepipe causes an increase in the pressure on the water which then travels further. During exercise, however, systolic blood pressure will rise naturally in response to the heart contracting harder to increase the flow of blood to the active muscles. This is a normal response to exercise and in a healthy individual the diastolic pressure normally remains close to its resting value.

Key point

The vascular system is a closed network of arteries that take blood away from the heart and veins that return blood to the right and left atria. The function of the vascular system is to serve as the transport vehicle for blood, circulating it to every tissue in the body. Blood leaves the heart under the pressure created by each contraction (heart beat) and flows to the capillaries for gaseous exchange and transfer of nutrients and waste products to tissue cells. By the time blood reaches the veins, pressure has reduced so valves (muscular squeeze from active muscles) within the vessels and respiration aid the return of blood to the heart. The movement of blood around the body can be detected by palpating a pulse in arteries that run close to the surface of the skin. The pressure under which blood flows can be determined using a sphygmomanometer.

4.4 Energy systems

This section provides an overview of the energy systems that form the basis for the chapters in Part II. Every cell, and therefore every system within the body, needs energy to perform the functions it carries out to maintain homeostasis. Muscle cells need energy to make the movement necessary for performance in any adventure sport. Although not a system in the same sense as others (e.g. skeletal system) discussed in this and the previous chapter, the energy systems provide mechanisms through which the body can provide the energy needed to sustain life.

In Chapter 2, the role of food as the source for the structure of the human body and energy production was discussed. The initial breakdown of foods by digestive system was also covered in Chapter 2. The current section provides an overview as to how food, once digested and absorbed into the lymph and bloodstream, is synthesized to provide the energy for adventure sports. It is also intended to make clear the distinction between aerobic and anaerobic energy production that is essential knowledge for a performer in making decisions about the components of fitness relevant to their adventure sport. This section is, however, an overview of the energy systems and they will be covered in more detail in subsequent chapters as each becomes relevant in relation to the duration and intensity of the adventure sport.

Energy production

Energy production for use by the body begins with digestion or breakdown of food. The process of digestion can be thought of as drilling for crude oil. The 'crude' forms of fuel produced through digestion are glucose, fatty acids and amino acids. These crude fuels are shipped (via the blood) to cells throughout the body where they are converted into the refined fuel of adenosine triphosphate (ATP). Adenosine triphosphate, shown in Figure 4.20, is the 'petrol' for our bodies – the fuel every cell uses to do work.

Adenosine triphosphate is so named because it has one adenosine molecule bonded to three phosphate molecules. It is these bonds which make ATP the 'petrol' for our bodies since although the bonding of the phosphate molecules to adenosine requires energy, the binding of the final two phosphates also creates a high energy store. When the bond between the adenosine molecule and these final two phosphates is broken, the stored energy is released enabling a cell to do its work e.g. providing energy for muscles to contract and bring about movement. The most common breakdown of ATP is for one phosphate to be broken from the chain to release energy and leave behind the free phosphate and ADP (Adenosine diphosphate). This energy release reaction is shown in Figure 4.21.

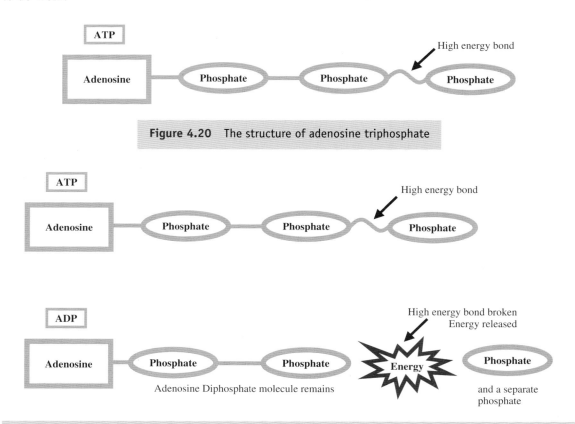

Figure 4.20   The structure of adenosine triphosphate

Figure 4.21   The breakdown of ATP to ADP releases the energy necessary for muscular contraction during exercise

Key point

Adenosine triphosphate is the 'petrol' for our bodies. Ingested carbohydrate, fat and protein are broken down in the body to produce the fuel (ATP) which every cells need to perform its work. Every cell has a mechanism or choice of mechanisms though which it can produce energy for work. There are three mechanisms or systems (two anaerobic and one aerobic) through which cells can produce ATP. Each system acts as a 'gear' for energy production. The first and second gears (the ATP-PCr system and glycolysis) become the dominant mechanisms relatively quickly, but have a limited duration, whereas the overdrive or third gear (aerobic system) is slower to become the predominant system but can enable activity for long periods of time. Each adventure sport will place different demands on the energy systems.

Muscles within the body maintain limited stores of ATP. If they did not, every time we wanted to make a movement there would be a delay until ATP production was enabled. However, these ATP stores are only enough for a few seconds work, so the supply must be replenished quickly. We have three energy systems that can be used to create ATP from the food we eat to produce the energy we need for participation in our adventure sport. There are two anaerobic energy systems and one aerobic system. Each of these systems requires one or more chemical reactions to create ATP. *Aerobic* energy production refers to the need for oxygen to assist in the chemical reaction process. *Anaerobic*, by contrast, means that oxygen is not required for the chemical reactions to take place.

Anaerobic and aerobic energy production is concerned with creating the ATP necessary for every cell in the body to function. For adventure sports participation, however, we most commonly discuss the energy systems in relation to the energy requirements for a particular activity. The predominant energy system and energy requirements differ for adventure sports such as bouldering, mountain biking, canoe polo and sailing. Just as a car has gears, so the body needs a variety of systems to produce ATP: two low gears (1st and 2nd) for power and acceleration and a top gear (3rd or overdrive) for efficiency and long journeys. The 'human car' has three gears: 1st and 2nd gears are the anaerobic energy production systems while 3rd gear is the aerobic system. The purpose of all three systems is to create ATP to replace the stores in cells as they are used up when the cell is working.

The anaerobic energy systems

The two anaerobic sources of energy production are the phosphagen system (ATP-PCr) and glycolysis.

Phosphagen system

The most rapid source for re-synthesizing ATP is the phosphagen (ATP-PCr) system. Cells store phosphocreatine (PCr) within the cytosol which can be used, as shown in Figure 4.22, to re-synthesize ATP. The simple nature of the chemical reaction and the limited stores of PCr mean that this system is a rapidly available short-term energy system. Through the reaction depicted in Figure 4.22, phosphocreatine is initially broken down to its creatine and phosphate components. The energy released from this exergonic reaction enables phosphate to form a high-energy bond with ADP synthesizing ATP. The arrows in Figure 4.22 show that the reaction is reversible and PCr can also be re-synthesized to replenish cytosol stores. Adventure sports such as short bouldering problems of up to 10 s duration, will use this source of energy as the predominant fuel source.

Glycolysis

The second anaerobic energy production mechanism is glycolysis. In adventure sports activities

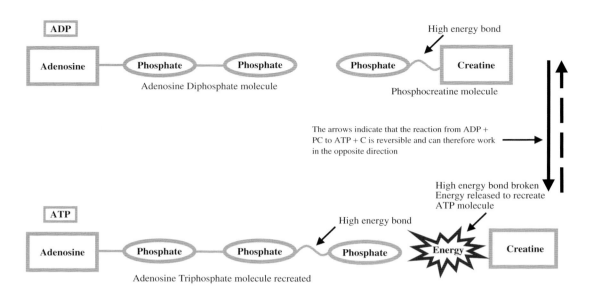

Figure 4.22 The recreation of ATP from the breakdown of phosphocreatine: the high-energy bond between creatine and phosphate (phosphocreatine) is broken to release the phosphate and energy to recreate ATP

lasting longer than 10 s and less than 90 s of maximal effort or for a sprint finish, such as competitive climbing, a ski or snowboard run and coffee-grinding in sailing, glycolysis represents the main ATP production mechanism. Glycolysis is the initial breakdown system for carbohydrate which is found within the body as either glucose or glycogen (glucose is stored in the liver and muscles as glycogen, chains of glucose molecules linked together). As can be seen from Figure 4.23, glycolysis involves a ten-step breakdown of glucose (nine steps for glycogen) to produce pyruvate and ATP. The breakdown of glucose, or the storage form glycogen, takes place in the cytosol of each cell. The ten steps of glycolysis are described in more detail in Chapter 7. The important concept to note at this time from Figure 4.23 is that there are many more reactions to produce ATP through glycolysis than through the ATP-PCr system. It is for this reason that glycolysis represents the second system for energy production. It is, however, a faster ATP production mechanism than the aerobic system.

Glycolysis does not require oxygen for any of the chemical reactions in its ten steps and is therefore an anaerobic energy production mechanism. The end product of glycolysis is pyruvate. The fate of pyruvate is dependent upon the intensity of exercise. When exercising for a duration of longer than 90 s and at a lower intensity than maximum, pyruvate subsequently provides an energy source for aerobic metabolism. Some textbooks refer to this as aerobic glycolysis. However, this is a confusing term as by the very nature of the chemical reactions involved glycolysis is always anaerobic. Perhaps a better term would be slow or *slower rate glycolysis*. When exercising at a lower intensity and long duration (slower rate) pyruvate is subsequently utilized (in processes requiring oxygen) within the aerobic system for further energy production.

When exercising near to or at maximal levels for over 10–20 s, the fate of pyruvate is altered. This rate of energy production is sometimes referred to as anaerobic glycolysis; however, the textbooks that use this term are not implying anything about a change in the nature of glycolysis but rather the fate of pyruvate at the end of glycolysis. During high-intensity short to medium-duration exercise, the aerobic system cannot utilize

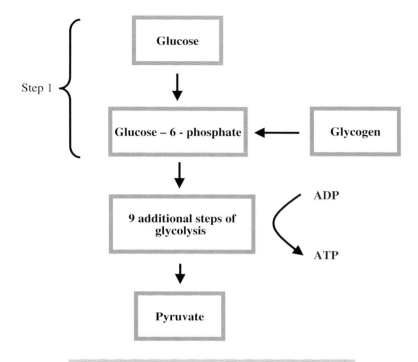

Figure 4.23 Glycolysis and the production of ATP

pyruvate at the rate at which it is produced. As a consequence, through a further reaction, pyruvate is converted to lactic acid. Perhaps a better term to describe the second fate of pyruvate at times of intense exercise is *fast rate glycolysis*.

The formation of lactic acid which creates acidity in cells and pain in muscles during high-intensity exercise (such as during a competitive downhill ski run or rock climb) limits the duration for which you can continue to exercise at that intensity. The lactic acid produced through high-intensity exercise, using glycolysis to form ATP, causes a rise in acidity in the cells. The enzymes (catalysts) that assist the reactions of glycolysis can only work within a narrow pH range. As lactic acid levels rise and change the pH within the cell, the enzymes are inhibited in their work. This impairs our ability to continue to produce ATP through glycolysis, and leads to fatigue.

Key point

The end product of glycolysis, the breakdown mechanism for carbohydrate in the form of glucose or glycogen, is ATP and pyruvate. The process of glycolysis is unchanged by the intensity of exercise; however, the fate of pyruvate is altered. In lower rate glycolysis, pyruvate can be utilized to produce further ATP through the aerobic system. In fast rate glycolysis, however, pyruvate is converted to lactic acid.

The aerobic energy system

Energy production through glycolysis can only uti-lize carbohydrate, whereas during the production of ATP through aerobic metabolism cells can also use fat and protein. The production of ATP within each cell for anaerobic purposes takes place within the cytoplasm of the cell (see Figure 3.1). Aerobic metabolism takes place within the cytoplasm and mitochondria of each cell. Depending on their function, each cell can have from a few hundred to several thousand mitochondria. The mechanisms through which fat, protein and carbohydrate can be used by cells to create ATP, in the presence of oxygen (supplied to each cell from the lungs via the bloodstream), are shown in Figure 4.24. Although all three nutrients can be used for energy production, it is thought that protein is responsible for supplying only 10–15 per cent of our energy needs. The main aerobic fuel sources are therefore fat and carbohydrate.

Without looking further than Figure 4.24, it becomes clear that the process of ATP production through aerobic metabolism is slower than anaerobic mechanisms. Aerobic energy production requires two stages: an initial breakdown within

the cytoplasm which is different for each nutrient, and a common secondary breakdown within the mitochondria of the cell. The initial breakdown processes for fat, carbohydrate and protein are through β-oxidation, glycolysis and deamination, respectively. After this stage, all three nutrients enter the secondary breakdown process. In the mitochondria, the three nutrients are further chemically altered through the Krebs cycle (also known as the citric acid cycle or tricarboxylic acid cycle) and the electron transport chain to produce ATP. Energy production through aerobic metabolism represents the largest supplier of ATP and as such can provide, with relatively inexhaustible supplies, the energy required for a sea kayak expedition, mountain marathon or a day's scrambling.

The duration and intensity of an adventure sport will determine the predominant energy system. Figure 4.25 shows the relationship between the time duration and predominant energy system for adventure sports.

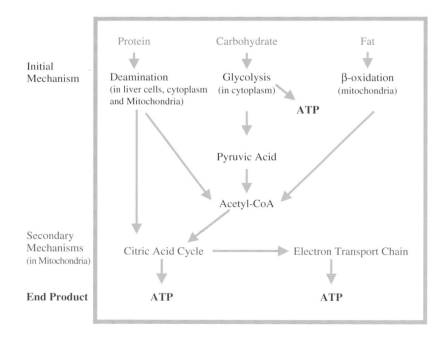

Figure 4.24 The production of ATP through aerobic energy production

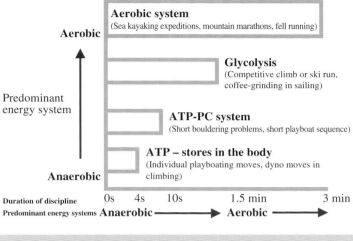

Figure 4.25 The energy systems in relation to adventure sports

Key point

The aerobic system has two phases: the initial breakdown of carbohydrate, fat and protein and the secondary mechanisms of Krebs cycle and electron transport chain to produce ATP. The initial breakdown of carbohydrate is through glycolysis (slower rate glycolysis), fat through β-oxidation and protein through deamination. Aerobic metabolism results in the highest net gain in ATP. However, due to the number of reactions involved, it is a slower process than either of the anaerobic ATP production systems.

4.5 Summary and study questions

The muscular system

The muscular system is responsible for movement, stabilization of posture, heat production and the storage of both oxygen and nutrients. There are three types of muscle tissue in the human body. Skeletal muscle is physically connected to the skeletal system by tendons that attach to the skeletal periosteum sheath. Skeletal muscle is the only muscle type that can be voluntarily contracted. Smooth muscle and cardiac muscle are under the control of the autonomic nervous system. Smooth muscle is found in the digestive tract and surrounds blood vessels and the airways. Cardiac muscle forms the heart. As such both blood and food are moved within the body by the muscular system and, together with the skeletal system, the muscular system allows the movement of the body as a whole.

Muscle cell membranes are referred to as sarcolemma and contain a special kind of cytoplasm known as sarcoplasm. Oxygen is stored within the sarcoplasm by myoglobin. Muscle fibres are grouped together into fascicles which comprise muscles. Inside each muscle fibre are tiny fibres called myofibrils (the contractile fibres of the cell), constructed of proteins arranged in sarcomeres.

Sarcomeres comprise a number of proteins of which actin and myosin are the most common. Together with troponin and tropomyosin, actin and myosin perform the contractile function of the muscle fibre. Actin fibres provide a framework that the myosin fibres slide within. Myosin heads bind to sites on the actin forming crossbridges. The breakdown of ATP provides energy that the myosin head stores and then uses to perform a powerstroke which pulls the actin filament towards the middle of the sarcomere, thus shortening the overall length. The binding of a new ATP molecule to the myosin head allows it to release the actin filament and become 'cocked'. The myosin head is then ready to bind again and perform another powerstroke. Essentially, the filaments slide along each other to shorten or lengthen the sarcomere, referred to as Sliding Filament Theory.

Muscle fibres come in two main types: Type I (slow twitch) and Type II (fast twitch). Type I fibres have higher aerobic capabilities while Type II fibres have greater anaerobic potential and are able to generate greater force. Skeletal muscles are made up from both types of fibres and individuals differ as to the percentages of each their bodies contain. As a result each of us is genetically more suited some adventure sports rather than others.

Muscles can contract in three different ways. When a muscle applying force shortens, the contraction is said to be concentric. In an eccentric contraction, the muscle applying force lengthens (for instance in lowering or braking actions). Finally, if the muscle applying force remains at a constant length the contraction is described as isometric. In any contraction, the working muscle is referred to as the prime mover or agonist. Normally muscles are arranged in pairs that have opposite effects and the non-working muscle is referred to as the antagonist. In fact, the antagonist muscle will often be providing a controlling function. Skeletal muscles normally allow movement by operating across joints and the particular arrangement of each muscle and joint will mean that one of three types of lever is created (first, second or third class). Each of these levers have different implications for force production and movement speed.

During any resistance training, the exercises chosen will require a particular kind of contraction from the muscles involved. The most common kind of contraction will be against a constant load. These kinds of contractions are isotonic. A resistance machine with cams, however, might be designed to provide a load that varies throughout the movement so that force production within the muscle is constant. This type of contraction is described as isokinetic. Some exercises require a bouncing action in order to pre-tension a muscle before a concentric contraction and these are referred to as plyometric contractions.

The respiratory system

The main function of the respiratory system is to facilitate the exchange of oxygen and carbon dioxide between the surrounding air and the body's cells. Oxygen reaches the cells of the body through three stages: pulmonary ventilation (breathing), alveolar ventilation (oxygen diffusion into the bloodstream) and finally cellular ventilation (the diffusion of oxygen across the plasma membrane into the cell). The respiratory tract is generally divided into two zones: the conducting zone and the respiratory zone.

The depth and rate of pulmonary ventilation (breathing) is highly adaptable with the volume of air moved through the lungs in 1 min rising from 6 L at rest to 180 L in a highly-trained adventure sports participant working maximally. At rest, the contraction of the intercostals and diaphragm results in inspiration and relaxation of these muscles leads to expiration. During exercise, active expiration is required and this is achieved through the contraction of the internal intercostals and abdominals. Pulmonary gas exchange takes place between the air in the alveoli and the capillary network that surrounds them. The flow of gases across the gaseous exchange membrane takes place by diffusion. Oxygen and carbon dioxide move from areas of high concentration to low concentration.

The volumes and ratios of the gases in expired air are particularly interesting for physiologists. The amount and type of work that an adventure sports participant is performing can be inferred from measurements carried out during or after gas collection. Maximal oxygen uptake or $\dot{V}O_{2\,max}$ is generally viewed as the definitive measure of aerobic fitness and is an assessment of the body's ability to consume oxygen. Another commonly used fitness measure is the anaerobic threshold, an activity intensity level at which the transition from aerobic to anaerobic exercise takes place. The respiratory exchange ratio RER = $\dot{V}CO_2 / \dot{V}O_2$ allows physiologists to assess which fuels (lipids or carbohydrates) are being utilized for energy production.

The cardiovascular system

The cardiovascular system is the main transport system in the body. It comprises three components: the vascular system, the heart and blood. The cardiovascular system performs three functions: it is a transport system for nutrients, waste products and hormones; it plays a key role in temperature regulation; and it contributes to the functioning of the immune system.

Blood

Blood consists of plasma and cellular components including red and white blood cells and platelets. Plasma is water based but also has proteins and other water-soluble components dissolved into it. Plasma also acts as a fluid that the cellular components of blood can be suspended in. Red blood cells (erythrocytes) contain haemoglobin and provide the main oxygen transport facility of the blood. White blood cells (leukocytes) are part of the immune system and are responsible for fighting viral, bacterial and fungal infections. There are five types of white blood cell, three of which have granules and two of which have none. The lifespan of a red blood cell is around 120 days, whereas white blood cells can have a lifespan of years in a healthy person or only a few days if someone is ill. Platelets (thrombocytes), together with the protein fibrinogen, provide the blood with its clotting ability.

The heart

The heart is constructed from a unique kind of muscle fibre called cardiac muscle which is able to function continuously without fatigue. It consists of two separate pumps that contract together. There are four chambers making up the structure of the heart: the atrium and ventricle on the right side of the heart are responsible for collecting and pumping blood to the lungs (pulmonary circulation) and the atrium and ventricle on the left supply the rest of the body (systemic circulation). The total volume of blood ejected from the left ventricle in one minute is referred to as cardiac output, determined by the stroke volume and heart rate (both of which increase during exercise). Heart rate is therefore closely related to exercise and intensity, which is why physiologists and athletes often use heart rate monitors during testing and training.

The vascular system contains smooth muscle which aids the heart's function in moving blood around the body. The main vessels of the vascular system are arteries, arterioles, capillaries, venules and veins. Arteries and arterioles carry blood away from the heart and to other tissues. Capillary networks are the areas where gaseous and nutrient exchange takes place. Venules and veins carry blood back from the tissues to the heart. Although arteries and veins appear to have similar functions i.e. the transport of large volumes of blood, they have quite different constructions. Arteries have thicker walls and a smaller diameter 'bore' and veins have one-way valves which help the skeletal muscles pump blood back to the heart. Arterioles help control blood flow by actively increasing in bore size (vasodilation) or decreasing in bore size (vasoconstriction).

The energy systems

The 'energy systems' are not really true systems in the same sense as other systems in the body such as the muscular or nervous system. Energy systems are actually processes required to provide the cells of the body with the energy required to sustain life and perform work during activities such as adventure sports. Adenosine triphosphate (ATP) is used as a fuel by every cell in the body and glucose, fatty acids and amino acids can be converted into ATP for use by cells. Cells can produce ATP through three mechanisms or energy systems. There are two anaerobic (without oxygen) energy systems: the phosphagen system (which allows very rapid but short-term synthesis of ATP) and glycolysis (needed to support anaerobic activity longer than 10 s). The aerobic energy system is the slowest producer of ATP but allows much longer production of ATP than anaerobic systems. The aerobic system is quite capable of producing sustained energy over a full day of adventure activities.

Study questions

1. What is the Sliding Filament Theory? Describe how the Sliding Filament Theory explains muscular contraction.

2. How might the composition of an adventure sports performer's muscle fibres impact on the types of activity they perform best in?

3. How do isometric, isotonic and plyometric muscular contractions differ? Describe an adventure sports example of each of these types of contraction.

4. During respiration, what is meant by the term tidal volume? Describe the changes in respiration that occur when exercising hard.

5. Why do physiologists convert the volume of gases collected to standard temperature and pressure dry (STPD)? What is the standard temperature ($^\circ$C) and pressure (mm Hg) for gases?

6. Why are there two sounds every time the human heart beats? What processes occur during each of these sounds?

7. The heart's natural rhythm, dictated by the pacemaker, is around 100 bts min^{-1}. How is the heart rate decreased or increased during rest or exercise?

8. Why is adenosine triphosphate (ATP) so important in cell metabolism? What is the most common breakdown of ATP?

5

Training fundamentals

If an adventure sports performer undertakes an exercise programme it will impact upon their fitness. This chapter concerns fitness for adventure sports and is divided into two sections. The first section provides an introduction to training: understanding fitness, what it is and how to measure it. The second section describes the general principles for designing a training programme and how to develop a specific plan for any adventure sport. This section includes the longer-term aspects of programme development as well as specific strength and conditioning exercises for individual training sessions.

5.1 Introduction to training methodology

Physiology is the study of how our bodies function – how we produce the energy and strength we need for our adventure sport. Knowledge of the physiological or physical principles that apply to each adventure sport will further improve the athlete or coach's ability to interpret performance. Our performance is related to the interaction of the skill, physiological and psychological elements involved in our chosen sport, as shown in Figure 5.1. This and subsequent chapters are concerned with the physiological principles

that underlie performance and how they can be addressed through training and preparation.

Key point

Performance is determined by the interaction of skill, physiological and psychological factors. As performers we need to identify which factor is the rate-limiter at any stage in our development.

The relationship between and the relative importance of the individual components within this performance model varies between adventure sports. For sports such as marathon paddling, hill walking and cross-country skiing, the physiological element at any performance level might well comprise the largest component, whereas for freestyle kayaking, sailing and climbing the skill components is predominant. Sports differ as to the relative importance of the physiological, skill and psychological components.

The relationship between the physiological skill and psychological elements for any one of us must be viewed as a dynamic. At any one time, an individual performer can be limited by physical, mental or skill-based constraints. The relative balance between the three components will change

Adventure Sport Physiology   Nick Draper and Chris Hodgson
© 2008 John Wiley & Sons, Ltd

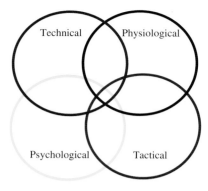

Figure 5.1 The components of adventure sports performance

over time, between sports and individuals. For example, the importance of the mental, physical and skill components may differ for climbers, windsurfers and marathon paddlers. The trick for us as athletes, seeking to improve our performance, is to identify which of the components is our 'rate-limiter' at any particular stage in our development. For example, in river kayaking, the inability to cross an eddy line and enter the downstream current could be due to technical (skill) errors (such as not having all the blade in the water), not being physically powerful enough to punch across the eddy line or due to anxiety created by the thought of capsize.

A major part of the physiological aspects of adventure sports concerns fitness. To develop any sort of training programme for adventure sport we must ask some fundamental questions about the nature of the activity. One of the key questions relates to fitness and how it improves performance. By being fitter we can improve our performance as measured by quicker times, improved technical quality or improved decision making due to lower overall stress on the body.

At its most basic level, to be fit is to be healthy or in good athletic condition. Fitness for adventure sports, however, is specific. Fitness for mountaineering expeditions means something very different to fitness for canoe polo and windsurfing. As an athlete or coach you must be able to identify the components of fitness that are the most

important to your sport. This section identifies the major components of fitness that will impact to a greater or lesser extent on each performer depending on their sport. In developing a training programme, you will need to identify which of these components are fundamental to improving your performance.

Key point

Fitness is specific to each adventure sport; a performer or their coach must be able to identify the key components of fitness for their activity.

The components of fitness

The components of fitness for adventure sports can be divided into those that are trainable in fitness session and those that are more appropriately developed in skill-based training.

Anaerobic power and endurance

Anaerobic relates to energy produced without the need for oxygen from the lungs. The body has two main mechanisms through which to produce energy anaerobically: the phosphagen system and fast rate glycolysis. These mechanisms were introduced in Chapter 4 and are further developed in Chapters 6 and 7, respectively. For sporting activities lasting less than 1.5–2 min, such as sprint events, for a sprint finish or for intermittent events (with sprints and recovery periods) such as canoe polo, bouldering competitions or a surf kayak heat, anaerobic systems provide a major fuel source.

Anaerobic mechanisms provide the 1st and 2nd gears for energy production. The 1st gear provides the energy required for *power* and *power endurance* adventure sports. *Power* is strength expressed at speed (strength × speed) and represents very short explosive movements such as a dyno (releasing one or more holds to catch a higher hold) move in climbing. *Power endurance* involves the use of a maximal expression of force over a more sustained period, up to around 10 s, such as seen during a short overhanging bouldering problem. *Anaerobic endurance* activities require the body's 2nd gear and this involves fast rate glycolysis. In events lasting longer than 10 s and up to around 90 s, glycolysis becomes the predominant mechanism for energy production. As was introduced in Chapter 4, fast rate glycolysis results in the production of lactic acid and provides a limitation to the length of time for which high intensity exercise can be maintained. Examples of anaerobic endurance adventure sports include short wildwater kayak races and competition climbing routes.

Strength and speed

These components are normally described as separate aspects of fitness, as shown in Table 5.1. In reality, however, strength and speed are very closely related to, and incorporated within, anaerobic performance. *Strength* represents the maximum force a muscle or group of muscles can generate. This is most often associated with resistance training, such as that for a one repetition maximum bench press. Strength in isolation is of a lesser immediate importance in adventure sports; however, as a component of power it is of great importance. *Speed* represents moving from one point to another as quickly as possible. Power, or high-speed strength, has a great application to many forms of outdoor activity. The ability to

make a dyno in climbing, the 4–5 strokes made to catch a wave when surfing, breaking in to the faster flow of a river from an eddy and the start in sprint events all rely on power. Power represents the functional application of both strength and speed.

The maximal expression strength or force required for bench press in strength training can only be maintained for a matter of seconds. In a similar way maximal speed, as demonstrated in sprint running, is achieved in the first few seconds of maximal exercise and can be maintained for a very short period of time. As a sprinter continues to run their speed decreases. As such, the time duration for the expression of maximal strength and speed, along with their clear link with power, mean these aspects of fitness are closely related to anaerobic power and power endurance. The immediate source of anaerobic energy (the 1st gear) is discussed in detail in Chapter 6 and glycolysis (2nd gear) is explained in Chapter 7. Figure 5.2 provides an example of anaerobic performance: a kayak surf run.

Aerobic endurance

While muscular endurance relates to the ability of a muscle or group of muscles to maintain work without fatiguing, aerobic power or capacity refers to the ability of the body to produce energy for exercise involving the whole body over a longer period of time. The energy for adventure sports

Table 5.1 The components of fitness

Physically trainable components of fitness	Skill-related fitness components	
Anaerobic power and endurance	Agility	
Speed	Coordination	Dynamic Balance
Strength	Balance	
Aerobic endurance	Reaction time	
Muscular endurance		
Flexibility		
Body composition		

Figure 5.2 Anaerobic performance: Simon Hammond, former surf kayak world champion carving a bottom turn (photograph courtesy of Sam Ponting)

that last for longer than 90 s is mainly provided through aerobic mechanisms. The term 'aerobic' means in the presence of oxygen. Figures 5.3 and 5.4 provide illustrations of aerobic adventure sports performance. The body has a variety of systems or gears that are used to provide the energy we need for performance and the aerobic system, the 3rd gear or 'overdrive', uses oxygen to assist in the final conversion of the food we eat into fuel for exercise. A mountaineer or Munroist will require oxygen to produce energy for the several hours duration of an ascent. Aerobic endurance is often measured using maximal oxygen uptake $\dot{V}O_{2\,max}$. This represents the maximum amount of oxygen the body can use each minute to provide energy. There are a variety of ways this can be measured or estimated such as through the multi-stage fitness test (Bleep test) or in a laboratory on a treadmill.

Figure 5.3 Aerobic performance: Marathon paddler Anna Hemmings competing for Great Britain (photograph courtesy of Mark Watson)

Figure 5.4   Aerobic performance: Tackling a Munro Nordic style (photograph courtesy of Bob Sharp)

In a physiology laboratory, gas analysis equipment can be used to examine how much oxygen an athlete consumes per minute.

Key point

Aerobic endurance refers to the body's ability to produce energy with the use of oxygen. Many adventure sports rely upon a sound level of aerobic endurance for performance.

Expensive equipment is not needed to estimate aerobic fitness. The following section on monitoring fitness provides details of a number of aerobic fitness tests that could easily be used or developed for any outdoor activity. In addition, aerobic metabolism is the focus of Chapters 9 and 10.

Muscular endurance

Muscular endurance relates to the ability of a certain muscle or muscle group to maintain repeated muscular contractions against a given resistance. This resistance could represent a small percentage of your maximal contraction (see strength above) or a higher percentage. The closer a load is to your maximum, the shorter is the time for which the effort can be maintained. For adventure sports, there are many examples where muscular endurance can impact upon performance. The ability to perform the repeated contractions necessary to paddle for a day whitewater paddling, pumping the sail in windsurfing and poling in cross-country skiing all draw upon muscular endurance. The load for each individual paddle stroke, sail pump or poling action are small relative to maximum possible contraction, thus enabling you to maintain the action for long periods of time. If your fatigue before the end of the activity, one aspect you might need to develop would be your muscular endurance.

Flexibility

The terms flexibility and mobility are often used interchangeably to refer to the range of motion about the joints of the body. *Mobility* does indeed refer to the range of motion about a joint; *flexibility*, however, is more than this and includes the stretch in the soft tissues that surround the joint and enable the range of motion. There are two forms of flexibility that can be identified: passive and active flexibility. *Active flexibility* refers to the ability of the muscles around a joint to be able to move a limb through a range of motion. For example in climbing, the ability to raise a leg high for a rock-over move requires active flexibility; the greater your active flexibility the higher you would be able to place your foot. *Passive flexibility* refers to the range of motion through which a muscle can be placed with the aid of other muscles or gravity i.e. stretching. Sitting and stretching to touch your toes would be an example of static stretching when the arms and the weight of the body (gravity) can be used to increase the range of motion.

Often linked with warm-up and cool-down, flexibility has a significant role to play for adventure sports performers. For example, good flexibility in the wrists, shoulders, trunk and hips is important for the canoeist. Trunk rotation for a variety of kayak strokes, cross-deck paddling for open boaters and flexion and extension onto front and back decks for rolling all rely on good levels of mobility and are improved by flexibility work carried out as part of a training programme. Flexibility is an excellent illustration of the close relationship between the physical and the skill-based elements of adventure sports. A lack of trunk rotation flexibility can easily become a rate-limiter for improvements with, for example, cartwheeling for playboaters. Hyper-mobility is also a concern for canoeists where too much shoulder flexibility can create problems with shoulder dislocations so there is a balance to be maintained. A balance between strength training to protect joints as well as flexibility training to maintain and improve range of motion is a useful concern for all adventure sport participants.

Body composition

Body composition refers to the physical make up of the body. In its broadest sense, this refers to the size and shape of an individual. For physiologists this aspect most commonly refers to the relationship between two components: fat mass and fat-free mass. Research into adventure sports and body composition has concentrated on the body fat percentages for elite performers. For elite athletes in slalom canoe, alpine skiing, cross-country skiing and sport climbing, the mean body fat percentages were between 5–10 per cent for males and 10–20 per cent for females. The lowest recorded body

fat percentages were found for climbers and cross-country skiers. Body composition has health implications for all athletes. However, for paddlers, surfers, sailors and windsurfers taking part in a non-weight bearing and non-weight categorized sports, percentage fat mass has fewer implications for performance than in other sports. For all adventure sports, providing diet remains unchanged, increased training volumes will normally lead to a positive shift in body composition (i.e. a decrease in fat percentage in relation to lean mass).

Dynamic balance

Physiologists describe three different balance-related components that are more meaningfully combined for adventure sports and can be termed dynamic balance. Balance is the ability to hold a static position such as a handstand for a length of time under control. Agility relates to a moving form of balance e.g. a series of movements that an agile athlete can make while remaining in balance. Co-ordination is the ability to effect efficient movements to achieve a goal. As individual components they have a varying relationship to all adventure sports depending upon the activity, but are perhaps more meaningful when combined to comprise dynamic balance. In an adventure sport context, this refers to the ability to use skilful coordinated movements to maintain position or make progress on the water or land. Balance and agility as part of this movement are fundamental to the skilful aspects of any particular sport. There are similarities between the dynamic balance demands between some adventure sports such as inland kayaking, surf kayaking, surfing, skiing and snowboarding. As a skill-related component of fitness, dynamic balance retains a uniqueness to it that requires practice of that specific activity to develop performance.

Reaction time

Reaction time is the interval between the introduction of a stimulus and the beginning of a response. For example, on paddling a new river how quickly does a kayaker react after spotting a feature and, if necessary, take evasive action e.g. catching an eddy before a waterfall. It is a key aspect of sports performance and as such is defined by physiologists as one of the components of fitness. However, reaction time is much less trainable in a physiological context than components such as strength or aerobic endurance and is best developed as part of skills training for any adventure sport. It is discipline-specific, in that reaction time training for paddlers is best carried out in a boat.

Adventure sports are generally highly skilful activities and as such any fitness training needs to be balanced with skill development. Improvements in fitness can help skill development and lead to improved performance. With the highly varied nature of activities that are the subject of this textbook it would be impossible to examine all the components of fitness important to each adventure sport. With knowledge of your sport and the knowledge developed from using this textbook, however, you should be able to identify the main components. It is these components that should form the basis of your fitness assessment and subsequent training programme.

Monitoring training: fitness programme assessment

There are a wide variety of assessment tools available that can be employed to measure fitness and training in adventure sports. Each of the tests to be discussed relates to one of the components of fitness described in the previous section. After deciding which of the components of fitness are fundamental to your adventure sport, it is necessary to decide which assessment tools should be used to monitor a training programme.

A key question is: why are assessment tools necessary? There are three main reasons why assessing fitness is important to an athlete or their coach. Firstly, if you do not test you cannot know the level at which to set the training programme for a performer or similarly, if you do not carry out testing at the end of a phase of training you will be unable to decide if the programme is working. Secondly,

if you conduct a test it can help to give you or your performer some positive feedback that the hard work in training is working (therefore improving motivation) and can also help to set realistic training goals for the next phase of training. Lastly, if you do not carry out testing for a programme which an adventure sports athlete is following, you will have an incomplete picture or profile of the performer. Using video to analyse techniques and skill development, good communication with the athlete about the psychological preparation and physiological assessments of fitness improvement allow the performer or their coach to have a more complete picture in the search for the performance rate-limiter at any particular stage.

As mentioned above, the battery of tests decided upon should be based on the central components of fitness for your sport. These fitness components are different for each adventure sport and the exact schedule of tests you decide to use should reflect this. This section will be split into three main parts: aspects of assessment that can be common across sports, discipline-specific assessment tools and finally a section with guidelines for organizing a testing schedule.

Key point

Keeping race results, video clips, height, weight, nutrition and health history measurements for a performer can form the basis of monitoring long-term athlete development.

Generic performance assessment tools

For all athletes, especially growing ones, it is useful to keep a record of height and weight. A record of the health history of an athlete can help to monitor the general well-being of a performer as well as keep a check of any indicators of overtraining. Monitoring nutrition for a performer can help them to keep a focus on what they put into their bodies, helping them with repair and recovery from training and competition. Keeping previous race, performance and test records is a really useful way of monitoring development for a performer. Although most often used as a coaching tool, a video clip diary kept over time can additionally provide an excellent record of performance development in skill-based tests.

Discipline-specific assessment tools

The development of a battery of tests should be specific to the components of fitness relevant to each activity. When designing a battery of tests it is really useful to be inventive. It would be impossible in a chapter such as this to decide on a battery of tests for each adventure sport. The tests described are therefore examples that you should modify and adapt as you feel is most appropriate for your sport, making decisions based on the key components of fitness.

Key point

In order to evaluate whether a training programme is working, you need to assess its impact i.e. check whether the performer is getting fitter. Tests should be specific to the sport, the components of fitness and carried out before and after each phase of training.

Anaerobic power and endurance

Anaerobic power – the application of strength at speed – is explosive by definition and tests to examine this aspect of fitness should reflect this. If possible, these should be related to the adventure sport. However, there are a number of generic anaerobic power tests that could be employed as part of a monitoring programme for a performer.

Vertical jump The vertical jump test provides a measure of anaerobic power. To complete the test, the performer stands sideways to the measuring board (see Figure 5.5) 15 cm from the board and first of all chalks their nearest hand fingertips. The participant then reaches to the highest point they can reach, making a mark on the board. The

Figure 5.5 Athlete performing vertical jump next to a jump board

performer then re-chalks their fingertips and jumps two-footed from standing to touch the measuring board as high as possible.

The best of three trials is recorded. The vertical jump height is determined by measuring the difference between the jump height and the standing height (cm).

Margaria step test The Margaria step test provides an alternative anaerobic power test, although it requires a very accurate timing device, staircase with at least 12 steps 17.5 cm high and a flat lead-up area.

The performer starts 2 m away from the start of the steps and sprints as fast as possible to the steps. In climbing the steps, the participant must only touch the 2nd, 4th, 6th, 8th, 10th and 12th steps (taking two steps at a time). The time is recorded for the best of three trials. The following formula is then used to calculate power P (watts):

$$P = \frac{9.807 \times M \times H}{t}$$

where M is the participant's mass (kg), H is the vertical height from step 8 to step 12 (m), 9.807 represents normal acceleration due to gravity (m s^{-2}) and t is the time taken to complete the climb (s).

Margaria step test worked example:

A participant aged 19 years weighing 73 kg takes of 0.826 s to complete the Margaria step test, where the height of stairs (vertical height between the 8th and 12th step) was 1.05 m.

$$P = \frac{9.807 \times M \times H}{t}$$
$$= \frac{9.807 \times 73 \times 1.05}{0.826}$$
$$= 910$$

i.e. Anaerobic power of participant is 910 W.

Wingate anaerobic test The Wingate anaerobic test (WAnT) is a 30 s maximal sprint test performed on a cycle ergometer. Developed at the Wingate Institute and first proposed in 1974, the WAnT has been used to assess anaerobic fitness for a wide variety of sports. When using a cycle ergometer for the test, the height of the saddle needs to be adjusted to the correct height for the participant. This can be achieved by the participant placing a heel on a pedal at its lowest point. The saddle height is adjusted correctly when the leg is almost, but not quite, straightened.

The participant should complete a 5 min warm-up with a resistance (load) of 1 N for females and 1.5 N for males at a pedal rate of 60 rpm. Once completed, the participant makes his or herself ready for the test and sets the pedal on the lead leg at about 10 o'clock ready to commence the test. The load for the test is usually set at 7.5 per cent of bodyweight, although this can be adjusted to a higher loading for trained athletes. A photocell on the flywheel is used to count revolutions, from which a computer calculates power output for the test. On the word 'go', the participant cycles as hard as possible for the duration of the test while remaining in their seat.

A typical power curve for a WAnT can be seen in Figure 5.6. The higher the peak power output, and the longer and higher the performer can maintain the power output, the better the anaerobic performance. The fatigue index can be used to

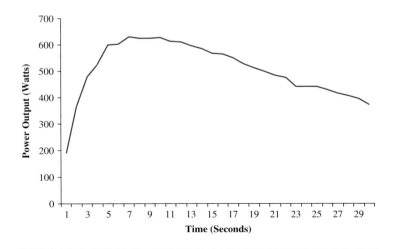

Figure 5.6 A typical power curve for the Wingate anaerobic test

provide further information on the rate of fatigue for the performer. The peak power output can be used as a measure of anaerobic power and the fatigue index a gauge of anaerobic capacity.

Wingate anaerobic test example:

The fatigue index for a participant with a peak (maximum) power output (PPO) of 648 W, a mean power output of 508 W and an end (minimum) power output (EPO) of 355 W can be calculated from:

$$\text{Fatigue Index } (\%) = \frac{(PPO - EPO)}{PPO} \times 100$$

Therefore,

$$\text{Fatigue Index } (\%) = \frac{(648 - 355)}{648} \times 100$$

$$= 0.4522 \times 100$$

$$= 45.22\%$$

Sport-specific tests

There are a variety of adventure sport specific tests that could be developed for monitoring a performer's fitness. In canoeing and kayaking, short courses could be set to assess take-off power or sprint speed. For flat water paddlers this would be very close to boat-based skill work. Power assessment for surf kayakers would most logically be based around the four or five power strokes needed to pick up a wave and could be measured on flat or moving water, assessing the speed or distance. In climbing the use of an adjustable dyno could be employed such that the greater the height achieved, the greater the power of the participant.

Anaerobic capacity assessment would need to be based around the anaerobic demands of the activity. As mentioned in the fitness components section above, anaerobic energy production relates to the '2nd gear' of energy production for events such as sprints, sprint finishes in distance races

or for intermittent events such as canoe polo or match sailing. For anaerobic capacity sports, performances would be of high intensity and lasting from 10 s to 90 s. The duration or nature of the test should reflect the sport for which the assessment is being devised. There are a wide variety of tests that can be used to assess anaerobic capacity. Although not developed for adventure sports specifically, these tests could be modified to fit a specific sport by an inventive performer or coach. The demands of the adventure sport should dictate the test to be employed by a performer. The following tests are those that could be modified to suit any particular adventure sport, making them sport-specific, and are described from shortest to longest duration.

Key point

There are a variety of anaerobic tests that can be used to assess a training programme. Choose or develop tests that are most specific to your adventure activity and the type of training being undertaken.

T-test   The T-test is a timed run on a T-shaped course. The layout for the course is shown in Figure 5.7. The distances and exact nature of the course could be altered to suit a particular

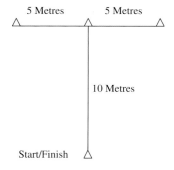

Figure 5.7   Course set-up for the T-test (adapted from Semenick, 1990)

adventure sport. For example, this test might be adapted very usefully for a sport such as canoe polo. For this test, the performer runs from the start line to the top of the T then sidesteps to the edge of the top of the T, sidesteps back to other top edge of the T then back to the middle where they turn and sprint to the finish line. The best time from three trials is recorded.

Repeat sprints   Variable-distance repeat sprints could involve either a straight course or out and back course. The distance and nature of the course would be specific to the adventure sport. The sprint could be repeated a number of times with recovery time between sprints varied to suit the sport. The time taken for each sprint could be recorded. The better the individual's anaerobic performance, the lower their first sprint time and drop-off between repetitions.

6 × 150 m rowing test   Similar in format to the repeat sprints, the 6 × 150 m rowing anaerobic test on a Concept II indoor rower offers an alternative repeat anaerobic test that can be used with adventure sports performers from a variety of activities. The Concept II rower can be programmed for the 150 m intervals with 30 s recoveries. Each sprint takes approximately 25–35 s and the time and average power should be recorded for each sprint. Results for such a test are listed in Table 5.2. A power curve can be constructed using power outputs. The higher and more horizontal the power curve, the higher the individual's anaerobic power and their endurance (Figure 5.8).

Basketball sprint test   The basketball sprint test involves four out and back increasing distance sprints. The layout for the test is shown in Figure 5.9. The performer completes the test four times with a 2 min rest between each repetition. The test could be adapted to increase the specificity for a particular land-based adventure activity or re-designed as a boat test for a paddler or sailor. The distances would need to be adapted to the activity. The intervals and numbers of repetitions could also be modified to fit a specific adventure sport with the time taken for each repetition being recorded.

Strength and speed

Assessment of *strength* development needs to be based around the most appropriate exercises for the adventure sport. The methods of assessment could include one repetition maximum lifts e.g. the maximum amount lifted for one repetition of the bench press, or a 5–10 repetition maximum could be utilized. The number of repetitions should be based upon the sport or the age of the athletes. Exercises, depending on the age and experience of the athletes, could be carried out as free weight core lifts or on multi-gym machines. The advantage with free weights is that they tend to require groups of muscles to work together e.g. lifting the bar or controlling sideways movement. They therefore more closely match adventure sports where performers use muscles working in functional groups e.g. dynos

Table 5.2   Results for a 6 × 150 m test

Repetition No.	Time (s)	Mean Power (W)	Heart rate (bts min^{-1})
Row 1	34.1	238	132
Row 2	34.3	234	140
Row 3	35.5	211	161
Row 4	35	219	165
Row 5	36.1	200	174
Row 6	35.7	208	179

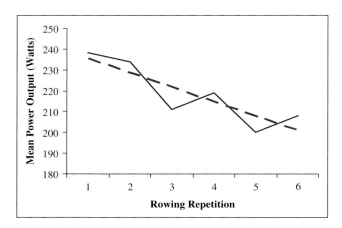

Figure 5.8 Results for a 6 × 150 m test shown in a table and then graphically. The best fit line (red dotted line) indicates the general trend across the test. The higher the initial power output and the more horizontal the trendline, the better the performance

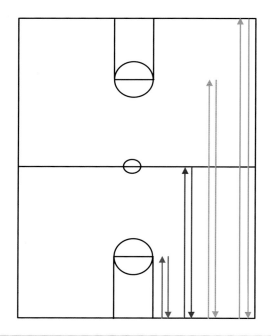

Figure 5.9 Basketball sprint test: the participant sprints to (1) the freethrow line and back, (2) the halfway line and back, (3) the far freethrow line and back, and (4) the far baseline and back (adapted from Semenick, 1984)

in rock climbing require coordinated whole body movement and muscle group integration. Later in this chapter, a wide range of weight training exercises for adventure sports are described and these could easily form the basis for a strength assessment.

Speed is assessed very simply as time taken to cover a specific distance over land or water. Example distances for land-based tests could be 10, 20 or 30 m etc. up to a maximum of around 100 m, depending on the individual athlete. The accuracy of timing would be very important for speed tests. The electronic sensors used in the Margaria step test enable a greater accuracy with results. With these sensors, an infra-red beam is cut off as the athlete passes the first sensor, starting the timing clock. As the performer passes the second sensor, the timing device is stopped.

Aerobic capacity

The traditional gold standard (seen as the most accurate measure) for assessment of aerobic capacity is the maximal oxygen uptake test or $\dot{V}O_{2\,max}$ (the maximum volume of O_2 consumed per minute) as developed by Archibald Hill

(Chapter 1). In a laboratory setting, the test is carried out using a treadmill, cycle ergometer (literally a work-measuring device), rowing machine or kayak ergometer. The testing device to be used should be based on the athlete's sport due to the specificity of fitness. $\dot{V}O_{2\,max}$ is reported in absolute terms ($L\,min^{-1}$) or relative to mass ($mL\,kg^{-1}\,min^{-1}$).

To convert from absolute to relative values you simply multiply by 1000 (to change from litres to millilitres) and then divide by mass. For example, a $\dot{V}O_{2\,max}$ of 4.2 L min^{-1} for a 70 kg male would represent a relative $\dot{V}O_{2\,max}$ of 60 mL kg^{-1} min^{-1} ($4.2 \times 1000/70$). For comparative purposes, it is more useful to record data in relative form rather than absolute figures. By doing this, you take into account the body mass of the participants. For instance, females tend to have lower absolute $\dot{V}O_{2\,max}$ but when converted to a relative value male and female $\dot{V}O_{2\,max}$ tend to be much closer.

After a warm-up, the athlete completes an 8–12 min protocol designed to take them to their maximum work rate with increases in the workload at regular increments throughout the test. The test time is designed to allow the aerobic or oxygen transport system to get to its maximum. During the test, the amount of oxygen consumed for each stage up to the athlete's maximum is recorded through direct pulmonary gas exchange measurement. Pulmonary gas exchange measurement involves the analysis of expired air from which the percentage of oxygen consumed and carbon dioxide produced can be calculated. The aerobically fitter the athlete, the longer they will be able to carry on the test and the greater the volume of oxygen they will be able to use to produce energy. Figure 5.10 provides an illustration of $\dot{V}O_{2\,max}$ tests being conducted on cycle and kayak ergometers, as well as a treadmill.

One of the physiological relationships of this test that makes heart rate (HR) monitors so useful is that HR and O_2 consumption have a very close linear relationship. This is unsurprising as it is the heart that is required to pump the oxygen we need for exercise to the working muscles. The harder you work at each stage in the test, the more oxygen you need to use and the harder the heart needs to work to supply the required oxygen.

Due to the relationship between HR and O_2 consumption and the expense and availability of laboratory testing, many maximal and sub-maximal (tests that do not take an athlete to maximum but use a level below to estimate maximum) aerobic tests based on HR have been developed for predicting aerobic fitness. Table 5.3 lists a variety of tests that can be used based upon their method for assessment. Researchers, clinicians, sports physiologists and coaches use a wide range of these tests according to their appropriateness for the athletes to be tested.

Where a high degree of accuracy is required, a laboratory-based $\dot{V}O_{2\,max}$ with gas analysis may be the most appropriate measure. For example in 2005/2006 the authors, as part of an athlete support programme for an endurance athlete training for a race to the North Pole, the performer was tested at regular intervals using a treadmill-based gas analysis with the use of Douglas bags to collect the expired air samples. The results of the tests were then used to adjust the next phase of training for the athlete.

When a group of performers are to be tested, time and financial constraints may mean that the most appropriate test would be one such as the multi-stage fitness test (bleep test – MSFT). In the MSFT, a number of athletes can be tested at the same time and from their finishing point in the test, a $\dot{V}O_{2\,max}$ can be predicted. In a clinical setting, or where taking participants to their maximum could be a problem, sub-maximal tests such as the Rockport walk, the Queen's College step test or the Åstrand–Åstrand nomogram test can be used to predict $\dot{V}O_{2\,max}$.

Key point

There are a wide variety of measures of aerobic capacity. Using knowledge of the different tests it is possible to select one that is specific to any adventure sport.

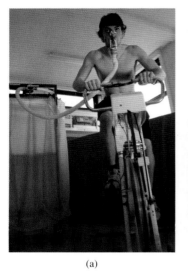
(a)

(b)

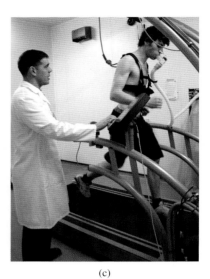
(c)

Figure 5.10 $\dot{V}O_{2\,max}$ tests being conducted on (a) cycle and (b) kayak ergometers and (c) on a treadmill; in each case the expired air sample is being collected using Douglas bags

Table 5.3 Maximal and sub-maximal assessment instruments that could be employed in adventure sports to predict or measure $\dot{V}O_{2\,max}$

Maximal assessments of $\dot{V}O_{2\,max}$	Maximal prediction tests of $\dot{V}O_{2\,max}$	Sub-maximal predictions of $\dot{V}O_{2\,max}$
Laboratory-based tests	*Laboratory-based tests: Cycle ergometer*	*Laboratory-based tests: Cycle ergometer*
Gas analysis $\dot{V}O_{2\,max}$ with Douglas Bags (off-line)	Åstrand maximal test	Åstrand-Åstrand nomogram
On-line gas analysis	Fox maximal test	
Common modes of analysis	*Laboratory-based tests: Treadmill*	*Laboratory-based tests: Treadmill*
Cycle Ergometer	Balke Protocol	Balke Protocol: single-stage test
Treadmill	Bruce Protocol	Bruce Protocol: multi-stage test
Rowing Ergometer		
Kayak Ergometer		
Common protocols	*Field-based tests*	*Field-based tests*
Bruce Protocol	Multi-Stage Fitness Test	Queen's College step test
Balk Protocol	Aero Test	Rockport walk
Athlete Led Protocol		Cooper's 12 min run
		1.5 mile run test

Athlete led protocol: Treadmill for measurement of $\dot{V}O_{2\,max}$ Laboratory-based maximal tests with direct measurement of pulmonary gas exchange can be conducted off-line, through the use of Douglas bags to collect expired air samples for later analysis, or by on-line breath-by-breath (b^2) analysis conducted during the test. The advantage of the Douglas bag system is accuracy, although you have to wait until after testing before you obtain results and cannot use the system in the field easily. Portable on-line gas analysis systems have an advantage over the Douglas bag system because they can be relatively easily used in the field (a great advantage for adventure sports) and provide oxygen consumption data during the test. The disadvantage to the on-line systems is their lower accuracy compared to Douglas bag gas analysis, although their accuracy is improving with each generation of system that becomes available.

Key point

In direct pulmonary gas exchange measurement the percentage of oxygen and carbon dioxide in collected expired air are used to calculate the volumes of oxygen consumed and carbon dioxide produced at rest or during exercise.

The athlete-led protocol (ALP) can be used with either on-line or Douglas bag gas collection. This $\dot{V}O_{2\,max}$ assessment is conducted using a treadmill following normal laboratory health and safety regulations. The test protocol is summarized in Figure 5.11. In the ALP, the athlete to be tested is familiarized with the test which starts with a 6–8 kph initial running speed and a 0 per cent gradient on the treadmill.

The ALP derives its name from the fact that the athlete controls the progression of the test. In the initial phase or the test, the speed on the treadmill is increased by 1 kph at the end of each minute. This continues until the athlete points up with their index finger (rather than talking, due to the gas collection equipment) while continuing to run. At this stage the athlete is indicating that the maximal cadence (running speed) at which they can cope has been reached and they now need gradient to be added each minute as the test continues. This commences phase two of the test, where the treadmill gradient is increased by 1 per cent at the end of each minute. This phase continues until exhaustion and the performer being no longer able to continue the test.

Multi-stage fitness test This test was developed as a convenient maximal field test to predict $\dot{V}O_{2\,max}$ and can be used to assess one or more subjects simultaneously. A commercial package which includes a compact disc (CD) to conduct the test, instructions and $\dot{V}O_{2\,max}$ prediction tables is readily available and probably is a test familiar to many adventure sports participants. It has been used by many sports teams and athletes including English rugby and British judo. The MSFT was introduced in 1988 after research into the development of progressive shuttle run tests was published in 1982. In the MSFT, participants run back and forth between two lines marked 20 m apart following progressively quicker bleeps on the MSFT CD. The level at which at maximum is reached is recorded. The shuttle test result can be used in conjunction with the prediction table to identify the participant's estimated $\dot{V}O_{2\,max}$. Figure 5.12 shows two athletes taking part in the MSFT.

Åstrand–Åstrand nomogram This is a laboratory-based sub-maximal assessment of aerobic fitness from which $\dot{V}O_{2\,max}$ can be predicted. The test is conducted using a cycle ergometer where the resistance against which the participant pedals and the cadence (pedal speed) can be monitored. Once the seat height and handlebars have been adjusted for the participant, they begin to pedal at 50 rpm. The initial work load should be set at between 75–100 W for females and 100–150 W for men. The workload is then manipulated such that the subject elicits a heart rate of between 130–170 bts min^{-1}. Once the heart rate reaches the target zone, the athlete continues to pedal for 6 min at this intensity. The heart rate should be recorded at the end of the 5th and 6th minute. If the heart rates at these two

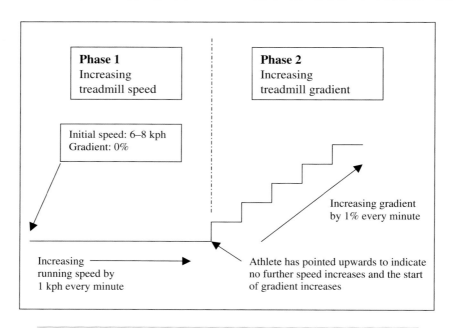

Figure 5.11 Athlete led protocol for treadmill assessment of $\dot{V}O_{2\,max}$

Figure 5.12 Athletes taking part in the multi-stage fitness test

times differ by less than 5 bts min^{-1} the test is complete. If the difference is greater than this, the test would be continued until the heart rate differs by less than 5 bts min^{-1} over two consecutive minutes.

An Åstrand–Åstrand nomogram is then consulted to identify the participants estimated $\dot{V}O_{2\,max}$. Figure 5.13 shows a nomogram where a line is drawn between the pulse rate (mean heart rate for the final 2 min of the test) and the work

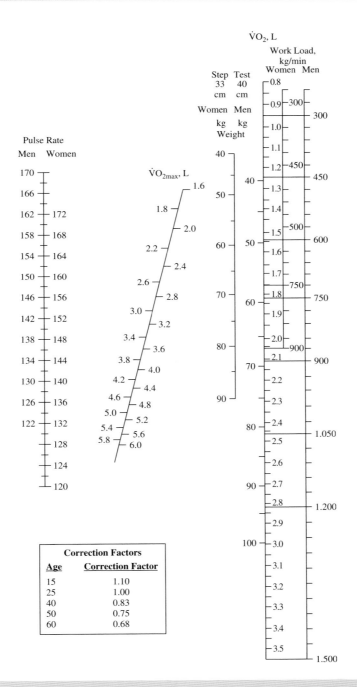

Figure 5.13 Åstrand-Åstrand nomogram: a line is drawn between the workload and heart rate (different for male and female) to reveal a predicted absolute $\dot{V}O_{2\,max}$ value. Aerobic Capacity in Men and Women with Special Reference to Age by I Åstrand 1960 – Acta Physiologica Scandinavica (suppl. 169) p51. Reproduced with permission from Wiley-Blackwell

rate at which the performer was pedalling. The predicted $\dot{V}O_{2\,max}$ is read from the centre line. As maximal heart rate has been demonstrated to decrease with age, there is a correction factor included in the nomogram to account for differences in age. The score read from the centre line would be multiplied by the age correction factor to arrive at the age adjusted $\dot{V}O_{2\,max}$.

Rockport walk The Rockport walk is a submaximal field test, the results of which can be used to predict $\dot{V}O_{2\,max}$. In this simple test, participants complete a 1 mile walk as quickly as possible. At the end of the test, the performer counts their pulse for 15 s. This figure is then multiplied by 4 to give a minute value (bts min^{-1}). The following equation can then be used to provide a predicted $\dot{V}O_{2\,max}$.

$$\dot{V}O_{2\,max} = 6.9652 + 0.02006095\,M - 0.0257\,A$$
$$+ 0.5955\,S - 0.224\,T_1 - 0.0115\,HR$$

where M is mass (kg), A is age (years), S is sex ($S = 1$ for male and $S = 0$ for female), T_1 is the time to walk 1 mile (minutes) and HR is heart rate (bts min^{-1}).

Rockport walk worked example:

What is $\dot{V}O_{2\,max}$ for a male aged 40 years, weighing 85 kg, who took 13 mins to walk 1 mile with a heart rate of 88 bts min^{-1} after doing so?

$$\dot{V}O_{2\,max} = 6.9652 + 0.02006095\,M$$
$$- 0.0257\,A + 0.5955\,S - 0.224\,T_1$$
$$- 0.0115\,HR$$
$$= 6.9652 + (1.705) - (0.0257 \times 40)$$
$$+ (0.5955 \times 1) - (0.224 \times 13)$$
$$- (0.0115 \times 88)$$
$$= 6.9652 + 0.7735 - 1.028$$

$$+ 0.5955 - 2.912 - 1.012$$
$$= 4.31388$$

i.e. $\dot{V}O_{2\,max}$ is therefore 4.31 L min^{-1} or 50.71 ($= 4.31 \times 1000/M$) mL kg^{-1} min^{-1}

As well as these measures of $\dot{V}O_{2\,max}$ to provide an assessment of aerobic fitness, a number of performance-related tests have been developed for work with rowers and paddlers. Time to complete a test is the parameter measured, including 1000, 2000 and 6000 m timed distance tests. The key concept with all these tests is to match as closely as possible the requirements of the activity to the test. Research has shown that kayakers being tested for aerobic fitness score relatively higher on a kayak ergometer than they do on a treadmill (running) or a cycle ergometer and higher still when they are tested in their own boats. Specificity, which is discussed in more detail in the next section, as a principle of training, is as important for testing as it is for training. The use of HR monitors as a training tool will be discussed further in the section on conditioning training.

Muscular endurance

A variety of body weight resistance circuit training exercises or sport-specific drills can be used to assess muscular endurance ability. The key with these exercises is to maintain the quality of the movements throughout the test time. Exercises can be carried out on a singular basis, such as how many sit-ups can be completed in a minute, or for example in a 30 s mini-circuit with press-ups, sit-ups, triceps dips, star jumps with 30 s being completed. The type of exercises would be determined by the activity and could include appropriate sport-specific drills as well as generic exercises. Examples of core muscular endurance tests can be found at www.wiley.com/go/draper although it is beneficial to develop adventure sport-specific exercises, such as coffee grinding style exercises for match-racing crew members in sailing.

Flexibility

There are a variety of generic flexibility assessment tools that can be used to measure flexibility. One of the most commonly used as a general measure of flexibility is the sit and reach test. The sit and reach test must be conducted after a participant has thoroughly warmed up and is ready to make a maximal stretch. The test assesses low back and hip flexibility. The participant must remove shoes and socks and places their heels against the testing apparatus. In one smooth movement they should reach forward, keeping their legs against the floor as far as possible. The best of three trials would be recorded. The advantage with this test is that it is simple to administer and relatively quick to complete. As a result, it has proved a popular estimate of overall body flexibility. However, research indicates that it does not have a good correlation with hamstring flexibility and has a poor agreement with low back flexibility as directly measured.

When carrying out flexibility assessments, it is best if they are based around the movements and range of motion required for an adventure sport. For example, trunk rotation and flexion/extension of lower back would be relevant to kayakers. An example of flexibility specificity can be seen through the Brentoflex, which is under-going further development at the University of Canterbury (New Zealand), and was designed to provide four climbing-specific flexibility tests including a functional climbing flexibility measure. Shown in Figure 5.14 is a demonstration of the load, reach and hold (LRH) test. Developing specific adventure sport based flexibility tests that are valid and reliable for use in each assessment would provide the performer with useful information about his or her specific flexibility for their activity. This information could then inform future training for that athlete.

Body composition

There is a wide variety of instruments and calculations that can be used to assess body composition. A simple measurement of body

(a)   (b)

Figure 5.14 Brentoflex LRH flexibility test: the participant starts on the lower holds and makes a rock-over move at their maximal stretch (hip flexion) to reach a higher hold

uld be the body mass index in kg m^2. This is calculated from t (mass) data using the following equation:

$$BMI = \frac{M}{H^2}$$

where M is mass (kg) and H is height (m).

Body mass index worked example

Calculate the BMI of a participant with a mass of 63 kg and height of 1.67 m.

$$BMI = \frac{M}{H^2} = \frac{63}{1.67 \times 1.67} = \frac{63}{2.789} = 22.560$$

i.e. BMI of the participant is 22.6 kg m^2

The advantage of using BMI as a measure of body composition is that the components can be measured quickly and relatively unobtrusively; the index can also be calculated quickly. The main disadvantage of using BMI is that the mass element of the equation takes no account of the fat or fat-free mass. This means that two participants could have the same BMI score although true body compositions could be very different: one muscular and the other with a higher percentage of body fat. It is for this reason that a number of measures have been developed to more accurately assess body composition.

Two commonly used measures of body composition (body fat percent) are skin fold callipers and body composition analysers. Skin fold callipers are used to measure a fold of skin at a variety of body sites. The results of these are used with prediction equations to provide a measure of body fat percentage. The intra-tester (test-retest) and inter-tester (agreement between all testers) reliability must be high to ensure accurate results. Body composition analysers uses bioelectrical impedance analysis (BIA) to assess body composition. The participant is connected by electrodes to the analyser. A low level electrical current is passed through the body and the impedance to flow is measured.

Body water is high in electrolytes so is an excellent conductor of electrical current. Fat-free mass has a relatively high water content compared to fat mass, and consequently the differences in resistance to conductivity can be used to assess body fat percentage. Body composition analysers will provide an estimate (using prediction equations based on bioelectrical impedance results) of total fat mass, fat-free mass and the fat mass for each limb and the trunk. As the calculation of the results is based on water content in the body and its conductivity, the hydration status of an individual will directly affect the results. When a participant last ate, drank or exercised before a test therefore impacts upon the results and has to be controlled to maintain the integrity of the analysis.

The key concerns with this aspect of fitness assessment are the reliability of measurements (see below) and the way in which results are used. As a society, we have become obsessed with our body sizes and shapes. We therefore need to be sensitive when handling or assessing body composition results. Even when dealing with professional athletes, who when compared with the majority of society have excellent body compositions, results of a body composition analysis can cause distress. Balancing data sensitivity, high margins of error and the most common and easily accessible methods are important.

Advanced assessment tools

There are a number of advanced assessment tools that can be employed to measure the fitness of an adventure sports performer. The $\dot{V}O_{2\,max}$ test has traditionally been the standard measure or predictor of aerobic fitness. More recently, however, research has shown that the lactate threshold (LT) is a more accurate predictor of success in aerobic endurance activities. The measurement of blood lactate concentration is based on research which has shown that as an athlete increases their effort towards maximum, they make a transition towards increasingly anaerobic exercise and lactate begins to accumulate in the muscle fibre and blood.

As described in Chapter 4, lactate is continuously produced and removed during metabolism.

At rest, and during lower intensity exercise, lactate removal matches lactate production and blood lactate levels remain consistent. As an athlete moves towards maximal effort, such as for the finish of a race, lactate is produced at a faster rate than it can be removed. As a consequence, lactate and hydrogen ions (H^+) begin to accumulate in the muscles and blood.

Key point

Lactic acid is produced during fast rate glycolysis when there is insufficient oxygen to enable the aerobic system to continue the breakdown of carbohydrate. Once lactic acid is produced, it quickly dissociates to form lactate and a free hydrogen ion which results in a decrease in pH within the muscle fibre and a rise of acidity in the muscle. High levels of lactate and H^+ have been associated with fatigue in aerobic endurance activities. The terms lactic acid and lactate are often used interchangeably.

The lactate threshold represents the level of work above which an athlete's lactate removal mechanisms cannot keep pace with production.

Above this threshold, lactate and hydrogen ions begin to accumulate in the blood, contributing to fatigue in aerobic endurance adventure sports. The higher percentage of an athlete's $\dot{V}O_{2\,max}$ at which this occurs, the better for performance in endurance events. In other words, if two fell runners had the same $\dot{V}O_{2\,max}$ but one could operate at 80 per cent before passing their LT while the other could only operate at 70 per cent, the former would be able to run at a higher speed during a race. An illustration of this can be seen in Figure 5.15.

Research indicates that knowledge of the level at which lactate threshold occurs is a better predictor of performance than $\dot{V}O_{2\,max}$ alone. As a result, many sports scientists are now using a test of lactate threshold as a measure of aerobic endurance capability for elite athletes. The concept of lactate threshold and the transition to increasing anaerobic metabolism is covered in more detail in Chapter 9, where the focus is upon high-intensity aerobic endurance adventure sports. In these activities, performers attempt to race close to the LT and train to improve the work rate or exercise intensity at which it occurs.

Guidelines for administration of testing

There are a number of useful principles that can be followed to assist with designing the tests to be used.

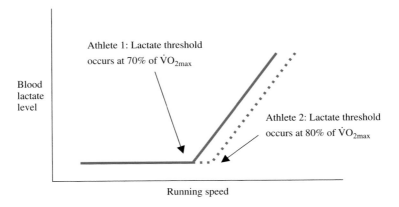

Figure 5.15   How the percentage of $\dot{V}O_{2\,max}$ at which lactate threshold is met affects performance

Validity of the test   Does the test measure what it really should? For example, does the distance run, paddled, climbed or biked assess the aerobic system, the anaerobic system or a combination of both? How does the test match the requirements of your adventure sport? There are many tests that can be employed for each component of fitness and you need to select the most valid. Are they sufficiently specific to your activity?

Reliability of the test   There are two aspects of reliability that must be considered: intra-tester (test-retest) reliability and inter-tester reliability. If someone is tested one day and then retested again the next couple of days before any changes in their fitness have taken place, the same or a very similar score should be achieved if the test is reliable. If scores are different, this may be due to a lack of test-retest reliability. In the same vein, there may be situations where two different people carry out the testing for an athlete. In a reliable test, the performer would achieve the same score regardless of who was testing. If so, the test is said to have inter-tester reliability.

Key point

Tests must be valid and reliable measures of performance. An analogy for this can be drawn from archery. The validity of a test refers to the ability of the arrow to hit the target and the reliability refers to the consistency with which the target is hit with each arrow.

Ordering of tests   Two factors that are highly specific to adventure sports need to be considered in carrying out any fitness assessment. If conducted in the field (as opposed to in the laboratory), the environmental conditions and equipment used by an athlete may impact upon results and should be taken into account by the assessor. Due to the nature of our sports, changes in weather, the terrain or the equipment being used, etc. can have a significant influence on test results. That is not to say testing should be abandoned if there is a change in environmental conditions, but these factors should be recorded and taken into account when evaluating results.

The period before testing can also have an influence on results. For instance, a really heavy training session the night before testing can influence the following day's assessment. When administering a test battery, you should try to mimic as closely as possible the lead-in to the previous tests and, where possible, allow athletes to take tests in a rested state.

Each time a fitness assessment is conducted, the battery of tests should be completed in the same order as the previous assessment. The order in which a performer completes a series of tests will also influence the levels they achieve. For example, scores for strength tests will be negatively influenced by any maximal aerobic test carried out before the strength test, whereas aerobic results, after an appropriate rest, are not detrimentally affected by strength assessment. The following assessment schedule is therefore proposed:

- collection of biographical data and non-exercising tests (i.e. age, height, weight, body composition);
- maximal strength tests;
- flexibility and agility tests;
- sprint tests;
- maximal aerobic or anaerobic tests; and
- muscular endurance tests.

If you are concerned about the influence of one test on the score achieved in a later test, you could decide to complete testing over two days rather than one.

5.2   Developing and designing training programmes

To improve fitness for an adventure sport, a coach or performer needs to consider developing a training programme based upon the results of a sport-specific fitness assessment. This section covers the key components to be considered in the development of a training programme.

Principles of training

The key principles of training have been identified over time by physiologists and should serve as guidelines for any performer or coach developing a programme for an adventure sport.

Overload principle

This principle is based upon research which has shown that the body will respond to training loads above those that it normally meets. If a sedentary person begins training, all the exercises would be new and would therefore be overloading the person's body. This will bring about physiological adaptations for that individual i.e. improvements in their fitness. An adventure sports performer must apply this same principle when designing a training programme. To improve fitness, we must train with loads that are above those we would normally meet. To do this, we can adjust the frequency, intensity or time (FIT) for which we train: frequency being the number of times per week that we train, intensity being the level at which we work (for example at what percentage of our heart rate maximum we work) and time being the duration of each training session. Each of these parameters can be manipulated in programme design to create a training overload through which to achieve an improvement in fitness.

Key point

The principles of fitness should be used as guidelines for the development of any training programme.

Progression principle

The principle of progression is very closely linked to that of overload. As physiological adaptation

takes place (as we becomes fitter), a training intensity that was previously an overload become less difficult. To continue to progress in our training, the FIT of training needs to be adjusted to cater for improvements in fitness. We must progressively increase the training overload.

An example of this can be made with strength training. You decide that a strength (weight) training programme would improve your sports performance and design a programme that you carry out three times per week at a local gym. The load for each exercise in the initial weeks will represent an overload. However, as you become fitter you will need to increase the weight lifted to maintain progress. A simple way to do this is to use the two-for-two rule. If you can complete two repetitions more than your planned repetition goal in the last set of an exercise over two consecutive sessions, you should increase the weight lifted. For example, if for biceps curls you are performing 3 sets of 10 repetitions but in the last set for two sessions in a row you can complete 12 repetitions, it is time to increase the load. By adjusting the FIT you can manipulate any aspect of your training to maintain a progressive overload.

Key point

Maintaining a progressive overload in training enables you to continue to make gains with your training. The two-for-two principles can be used to ensure that you are maintaining a training overload.

Specificity principle

Research into physiological adaptations has provided us with a number of guidelines regarding specificity of training. The specificity of training principle refers to the need for training to be particular to the desired goals.

Strength training specificity   Strength training is commonly employed by athletes from many sports to help improve performance. Research tells us that resistance training can help improve strength, power, speed and muscular endurance, all of which can be beneficial to our performance (the amount depending on our adventure sport). The goals of strength training can be manipulated and should therefore be made specific to the discipline. Firstly, the exercises selected should be specific to the activity chosen. For example, upper body exercises for a kayaker should take priority over lower body exercises and movements made with weights should match as closely as possible movements made in the activity. Secondly, strength gains have been found to match the speed at which training was carried out. For example if training was carried out at slow speed to develop strength, the maximal expression of strength would be found moving at that slow speed. If you are training for an explosive event that requires power, then training speeds should reflect this to maximize possible gains.

Conditioning training specificity Training results are specific to the muscles trained and there is often little transfer to other muscles even between events such as running and swimming. A good illustration of this is a study which involved 15 healthy but sedentary subjects swim training for 1 hr per day, 3 days per week, for 10 weeks. All subjects carried out swimming and running tests at the start and end of the study. Not surprisingly, the results found an 11 per cent improvement in the swimming test scores for the group, but only a 1.5 per cent change in the running test results.

When training for specific activities such as sailing, climbing, canoeing and mountain biking the training overload should engage the specific muscles and energy systems used for the discipline. In other words, if you wish to get fit for paddling you need to train in your boat as a regular part of your programme. In a similar fashion to the speed aspect of strength training discussed above, the conditioning aspects of training should match the activity for which you are training e.g. sprint training for anaerobic sprint events, endurance training for aerobic events and a mixed protocol for intermittent events such as canoe polo or surf kayaking.

Programme specificity There are a number of aspects with regard to programme design where specificity needs to be addressed. In developing a training schedule, the focus of training should move from generalized to more specific training as time comes closer to competition. This will be addressed further in the section below on periodization. The time at which training is carried out should match competition or performance times as closely as possible. Research has shown that adaptations to training are specific to the time of day during which the training took place. If training for an event which will take place in the morning, it is better if some of the training occurs at this time of day. In a similar way, research has found that modelling competition in training can help make adaptations specific to the event. It is important for you as a performer or coach to be aware of the aerobic and anaerobic contributions to energy production for your discipline. Training should mimic or model these relative contributions.

Key point

The strength training and conditioning aspects of a training programme should be specific to the relevant adventure sport. Matching the time at which training takes place, the energy demands of training and the resistance exercises to the adventure sport will further improve gains in performance.

Regression principle

When an athlete stops training they will lose any gains they have made relatively rapidly. The effects of detraining can be seen very clearly in anyone who has broken a leg or arm and had it kept in plaster for 6 weeks. When the plaster is removed, the resulting atrophy (loss in muscle size) is due to detraining or regression. The regression principle states that ceasing training, dependent upon the period and degree of detraining (ranging from normal active life but no training to complete bed-rest), will result in a loss of fitness gains made through previous training. This has clear implications for breaks in training or off-season phases and the need to maintain some level of fitness. It is for this reason that many top athletes have maintenance programmes to retain a basic level of fitness before pre-season training. The same principle can guide planning the training programme for any level of adventure sports performer.

Individualization principle

If you set two very similar canoe polo forwards on a training programme, the response of each to training would vary for each aspect of training. Individualization, while making sound physiological sense, is a fundamental coaching principle. Just as skill development sessions should be devised on an individual basis, training programmes should be developed in response to each performer's fitness assessment results and their training goals. Research tells us that optimal training benefits are obtained through individually developed programmes. This does not, however, mean that all training should be carried out individually. During training, adventure sports performers will benefit from having sessions conducted individually, in pairs, and in groups of various sizes throughout the programme.

Variation principle

Adding variety such as by changing the number of paddlers training or by having changes in a training programme can help with motivation and enable a paddler to avoid any boredom with training. Dependent upon the level of competition, the training volume for an adventure sports performer can be up to and over 1000 hr per year; as such the athlete or their coach needs to add variety to the programme to maintain motivation and enthusiasm. Variation in training loads can help not only with motivation but also assist in avoiding overtraining illnesses. Training close to your maximum in every session can overstress the body. Adding variety to sessions, for instance by having hard and easy training days, can help an athlete avoid such illnesses or injuries. This can be achieved by manipulating the FIT in a training programme, a central aspect of the following section dealing with long-term training programme development i.e. periodization.

Periodization

Periodization is based upon the General Adaptation Syndrome (GAS). The body responds to a situation of stress by first showing *alarm reaction* i.e. shock to body and muscle soreness, then *resistance* (*adaptation*) which leads to improvement of performance and finally *exhaustion* when a body is able to make no further improvement to the training. In response to strength and conditioning, the stress arises from the different forms of exercise encountered during training. However, in its broadest sense, stress is the reaction of the body to any demand. Figure 5.16 shows a model of the GAS. If training is not reduced at this stage, an overtrained state may occur. Periodization is a system of training programme design through which a performer can best develop their training overtime to peak for a major event or for a competition season, without going into the exhaustion phase.

Since the original model of periodization was proposed by Matveyev in 1966, a number of variations in training programmes have been developed to suit athletes from different sports. Athletes from any adventure sport and level of performance now have a wide range of periodization options from which they can develop an appropriate training programme. As well as avoiding overtraining, periodizing training can help the performer avoid boredom. Research has shown that having a phased programme and knowledge of what is coming next in preparation for an event helps an athlete cope with the training volumes to be completed. The athlete can see clearly why they are training and how they are progressing to their goal.

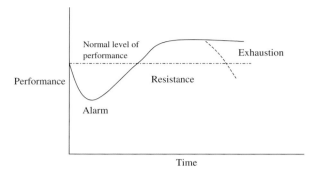

Figure 5.16   General adaptation syndrome (GAS) (adapted from Seyle, 1956)

Key point

Periodization, dividing the training year into manageable and progressive sections, is an excellent way to set realistic goals for the performer. It enables him or her to identify exactly where they are in the progress towards a main competition or event.

The concept behind periodization is to divide the training year into distinct phases. These phases build upon one another to prepare an athlete for a specific event or season of competition. Programmes can be designed over a short or long period of time and be designed individually for each performer. The most common lengths of time for a periodized programme would be between several weeks and one year, depending on the level of performance and age of the athlete. Within periodization, these phases of training have specific names as listed in Table 5.4. A macrocycle is the largest unit of a programme such as a training year, divided into a number of mesocycles which last for one to several months. Mesocycles are divided into microcycles, typically one week long. By dividing the programme in this way, it is possible for an adventure sports performer to put what they are doing for the next training session into context with the whole week's training, see how that links to the next mesocycle and understand how it leads to the event for which they are training.

Programme design   The success of periodized training is the variation in the volume (quantity) and intensity (quality) of training. Figure 5.17

Table 5.4   Possible training units within a periodized programme

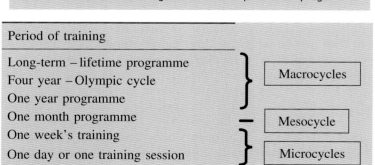

Period of training	
Long-term – lifetime programme	
Four year – Olympic cycle	Macrocycles
One year programme	
One month programme	Mesocycle
One week's training	
One day or one training session	Microcycles

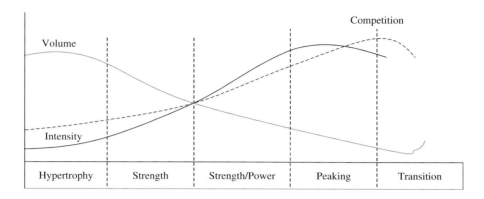

Figure 5.17 General periodization model performance curve

provides a model of how the relationship between volume and intensity of training is managed through a periodized programme. The performer or their coach needs to manage the volume and intensity through the programme in order to avoid overtraining but attain a peak in performance. During the early part of training, the volume of training is kept high but at a lower intensity; as time comes closer to the main event, the intensity is designed to rise to competition level and as a result, the volume of training decreases dramatically. As can be seen, the model training period (macrocycle) is divided into four mesocycles (each divided into a series of microcycles). The hypertrophy phase is a preparatory phase and is designed to get the body ready to train, concentrating on increasing muscle mass and endurance. The end of the strength phase sees the cross-over between intensity and volume of training with quantity decreasing to allow a greater focus on the quality of the sessions. The strength and the strength/power phases represent the key phases of training when the hard work is completed by the performer, ready for the final peaking phase that tapers training for competition.

Key point

The phases in a training programme are built one on top of the other. The hypertrophy phase is the foundation stage to get the body ready for the hard training to come. The strength phase is a physically hard stage for any athlete and should provide the strength base for the speed and strength development in the strength/power phase. In the peaking phase, the quantity or training is reduced but the quality (intensity) is maintained as the performer nears the event for which they are training.

An example programme with fitness assessment and periodization can be found on the web site www.wiley.com/go/draper.

Long-term athlete development

If a performer starts their sport at a young age or a coach works with younger athletes, are there some fundamental differences that should be taken into account? The short answer to this, unsurprisingly, is yes. The individualization principle of training is a key concept for all performers, including younger athletes. Children are not simply mini-adults and we must take notice of differences between them and adults and between males as well as females. Growth, development and maturation have implications for long-term athlete development.

Growth refers to increases in size, both of the body as a whole and of individual components. For example, the heart increases by between 10–20 times in size between birth and adulthood. Development refers to the gaining of abilities and behaviours such as social, emotional and sporting skills. Maturation, a term often associated with growth and development, refers to progress towards the mature biological functioning of the body. Growth and development and gender differences have implications for all athletes. This section examines these differences, provides a way to assess growth and development and suggests a model for long-term athlete development.

Measurement of age in growth and development

There are many different ways by which age can be reported. Traditionally for teaching and coaching, we group by chronological age (years, months, days). However, the biological age of a year group may show a wide variation. The range in biological age for a group of 13–14 year olds can be as wide as from 9–16 years old. Research tells us that while we follow a similar pattern of development from birth to adulthood, the timing and extent of changes is highly individual. In addition, girls tend to mature physically about 2–2.5 years before boys. Coaches should take these differences into account when working with groups.

Height and weight (mass)

Two of the most widely studied and easily recorded measures of progress to maturity are height and weight. There is a variety of points at which height and weight measures in childhood can be used as a predictor of adult values; for instance, on average 50 per cent of adult height is reached at the age of 2 years old. From birth to puberty, weight gain is constant and similar for males and females. During puberty there is a significant rise in weight for boys and girls. For males this rise is mainly in muscle mass and for females it results from an increase in bone mass, fat mass and muscle mass.

Parameters of great interest to physiologists and coaches working with young athletes found by keeping longitudinal (long-term) records of height and weight are the individual's peak height velocities (PHV) and peak weight velocities (PWV). These are the periods in growth where height and weight gain are at their maximum rates. For each of us, there are two peaks in height and weight velocities: the first between birth and two years of age and the second during puberty. The start of the second increase in height gain starts before puberty with PHV occurring at about 12 years old for girls and 14 years old for boys. For girls, PWV is reached on average at the same time as PHV while for boys it is reached after PHV (14.5 years old). Research indicates that knowledge of the timing of PHV for an individual is an important indicator for long-term athlete development.

Key point

Peak height and peak weight velocities are critical stages in the development for each adventure sports performer. Monitoring height/weight changes on a regular basis for younger athletes can enable a coach to time the introduction of new aspects of training.

Motor development

A major part of childhood is spent learning basic movement patterns and skills such as walking, running, jumping, hopping, skipping, throwing, kicking, catching and striking. These movement patterns are the foundations for developing sport-specific skills learned in later life. Research has shown that a rounded and fulfilled motor development in childhood is vital to future sporting performance. Skills that are not developed during this vital stage are not compensated for later in life and will leave deficits in global motor performance that can impact on sports performance. Researchers indicate that for successful long-term development, an adventure sports performer and his/her coach should address these core motor movement patterns at the appropriate stages in development. Gender differences in these basic motor abilities are most clearly evident for throwing and striking and may reflect societal differences in programmes for boys and girls. Such differences can be addressed through exposure to a full range of activities that are non-gender specific for all children.

A model for long-term athlete development

In recent years, sport governing bodies around the world have advocated a longer-term approach to training than normally considered as part of periodization. The implementation of long-term athlete development strategies for British sport, for example, have led to changes to the development programmes for adventure sports such as canoesport, sailing, mountain biking, climbing and orienteering. The implementation of the LTAD programme in the UK has been conducted in conjunction with the researcher Istvan Balyi who proposed a five-stage development programme for all athletes. The identified stages relate to growth, development and gender differences and have implications for long-term athlete programmes. This model has physical considerations that need to be addressed at each level.

Key point

Taking a long-term approach to training can help all adventure sports athletes in their development. For younger performers, there are certain critical periods that can be addressed to maximize sporting (physical) development. Awareness of these can help the coach plan progressions in training.

Pre-training or fundamental stage (5–8 years old) This stage is focused on the development of fundamental movement abilities mentioned in the growth and development section above. It is concerned with the development of physical literacy, fundamental movement and the following skill areas: walking, striking, coordination, running, buoyancy, speed, jumping, gliding, throwing, agility, catching and balance. Istvan Balyi refers to this stage in his publications as FUNdamentals: indicating the emphasis that should be the basis for all sessions.

These fundamental abilities, which should continue to be developed as children move into the learning to train stage, also need to be combined with basic sport specific skills such as edge control, boat awareness, etc. in kayaking. The emphasis at this stage should be upon fun and children should follow a wide variety of activities.

Learning to train stage (8–12 years old) During this stage, the general principles of training along with sport-specific skills should be taught. The fundamental motor abilities should continue to be a focus as research indicates that this stage represents a peak for motor coordination development. The main focus here should also be on training and fun as opposed to competition. It is more important to continue to develop physical literacy than to win competitions.

Training to train stage (12–16 years old) It is during this stage that performers will begin their adolescent growth spurts and achieve PHV and PWV. The onset of PHV can be found by measuring young paddlers' height monthly. When they get closer to the PHV point, height can be measured daily to plot the actual timing of PHV. Research indicates that this is a critical time for developing aerobic and speed training. The timing of the PWV should also be used to introduce bodyweight and foundation skills in strength training. This stage should also see a gradual and progressive move towards more adventure sport-specific skill training.

Training to compete stage (16–20 years) During this stage, periodized programmes and higher volume training programmes should be further developed. The emphasis with regard to the fitness aspects of training should be upon maximizing strength through additional weight training. Competitions become more important, with programmes geared towards two or three key events.

Training to win stage (18+ years old) If a performer chooses to continue with competition development, this stage would represent the final in athletic development. The training work would be highly specialized and focused on key competitions. The training development from the fundamental stage to this level should have provided the educational tools for the adventure sports performer to be a largely independent athlete who has a coach with them to guide their future development rather than to cover the basics of training.

Lifelong paddling fun stage (12+ years old)

The model proposed by Istvan Balyi is a five-stage model devised as a guide for performer development following a competition route. Adventure sports are fairly unique activities in that they are often pursued as lifelong passions. They can be followed at a recreational or competitive level. For adventure sports there exists, therefore, six stages of development. An athlete could enter this sixth stage at any age from 12 onwards. In addition, many adults come to adventure sports later in life and can pursue their discipline as a lifelong recreational activity. This could include personal sports development, master's level events, first descents, expeditions and coaching. Any athlete or their coach can usefully adopt a long-term approach to athlete development.

Strength training and conditioning

So far we have examined the components of fitness that may apply to any particular adventure sport, general principles for designing a training programme, how to divide up a training year to peak for an event and how to evaluate the success of the programme. Having done this, and with knowledge of the energy systems that were introduced in Chapter 4, the next step is to examine the range of strength (also called resistance or weight) training and conditioning exercises that can form the basis of a training programme for an adventure sports performer. From a physiological context to develop fitness, training is based upon

the manipulation of FIT for strength and conditioning training. Strength training refers to the development of the muscular system and conditioning training for improvements in the functioning of energy systems. The development of a strength and conditioning programme needs to be specific to the individual and the adventure sport.

Strength training

In its simplest form, strength training involves adding a resistance to a muscular contraction so that an overload is created. Exercises conducted with an overload will result in increases of strength. There are now a great number of gyms and sports centres in many cities worldwide where you can undertake a strength training programme. Resistance training can be employed to specifically address deficits in strength, power or muscular endurance. Strength or resistance training can be carried out using free weights, machine weights and bodyweight or a combination.

With the choice of equipment available there is a vast range of exercises that can be performed for any muscle group, isolating the muscle or as part of functional group (compound lift). Carrying out a preacher curl would be an example of a muscle (the biceps) being trained in isolation whereas a power clean, a very important exercise for athletes, works a large number of muscles in coordinated movement pattern.

Time for weight training may be limited. As a consequence, compound exercises should form the basis for a strength programme as, by their use of muscles working in coordinated units, they are more specific to sporting performance. During a rehabilitation programme when recovering from an injury, an adventure sports performer may increase the number of isolated muscle group exercises in order to train specific muscle groups. For example, in canoeing and kayaking, boaters work their muscles to a large extent in functional groups (several muscle groups working together) to bring about propulsion. Therefore, following the specificity principle we should try to match this in the weight training programmes we follow to improve our strength, power or muscular endurance.

Key point

Strength training programmes can be designed to develop a wide range of muscular goals including pure muscular strength, power, muscle size and endurance. The repetitions, sets and weight can be manipulated to bring about desired results.

Strength training can be used to bring about a variety of results through manipulating the weight, repetitions and sets carried out for each particular exercise. Table 5.5 provides details of the variation in reps and sets that would be required for each particular outcome. The weight for each exercise should be decided by experimenting with the weights until the athlete finds a load at which, in the last set, he or she can just complete the repetitions required. Start with lower weights and build up. After you have been to the gym for a short period you will be able to find the appropriate weight quite quickly – do not sacrifice technique for extra weight. The two-for-two rule (see progression principle above) can be used to make decisions about increasing training loads as your body makes adaptations to training.

A strength training coach at a gym should know all of the exercises in this chapter and would be able to provide you with further advice on a choice of exercises and technique. A variety of general strength training exercises for adventure sports are provided in Table 5.6. For the compound lifts and using free weights, an adventure sports performer should take additional care with technique.

Research supports the use of free weights and the compound exercises for athletes; however, the technique employed is important. After coaching, free weights can become an essential part in a strength training programme.

Key point

Compound lifts especially require good technique for correct execution. Take a longer-term view to their inclusion in your programme. Learn correct technique before adding weight. Without good technique, they can result in injury. The use of a strength training coach is important in the development of safe technique. The use of wooden discs provides an excellent way to learn the core functional lifts. Once learned, the core functional lifts are highly specific to sports performance and a very efficient form of training for adventure sports performers.

Strength training exercises

There are a wide range of general exercises that can be usefully undertaken in a fitness programme for adventure sports, listed in Table 5.6. Photographs providing illustrations of the exercises in Table 5.6 can be found on www.wiley.com/go/draper. The number of exercises in a session should be between 8–12 and those selected should be those that you believe are the most specific to the performer's adventure sport. For example, in

Table 5.5 Suggested repetitions and sets for specific resistance training outcomes

Outcome	Strength/power	Strength	Hypertrophy	Muscular endurance
Repetition range	1–3	5–8	8–12	15+
Sets	3–5	4	2–4	3
Rest between sets (mins)	3–5	3–5	1–3	1

Table 5.6 General exercises that could be included in a strength training programme; the compound lifts should eventually become a central component of strength training sessions for adventure sports athletes

Body part	Exercise	Equipment	Photograph sequence no.
Shoulders	Lateral raise	Dumbells or machine	1a & 1b
	Shoulder press	Barbell, dumbells or machine	2a & 2b
	Shrugs	Dumbells	3a & 3b
	Upright row	Barbell	4a & 4b
Legs	Leg press	Machine	5a & 5b
	Leg extension	Machine	6a & 6b
	Leg curl	Machine	7a & 7b
	Lunge	Dumbells	8a & 8b
	Calf raise	Dumbell	9a & 9b
Upper back	Seated row	Cable pulley	10a & 10b
	Lat pulldown	Machine	11a & 11b
	Bent-arm pullover	Dumbell	12a & 12b
	One-arm dumbell row	Dumbell	13a & 13b
	Chin-ups	Chin bar or beam	14a & 14b
Lower back	Back extension	Machine or floor exercise	15a & 15b
	Prone cobra	Dumbells	16a & 16b
Chest	Dumbell flys	Dumbells	17a & 17b
	Press-ups	Floor	18a & 18b
	Bench press	Barbell or machine	19a & 19b
	Incline bench press	Barbell or machine	20a & 20b
Arms	Tricep press	Machine	21a & 21b
	Tricep extension	Machine	22a & 22b
	Bicep curl	Ez bar, barbell or dumbell	23a & 23b
	Reverse curl	Ez bar, barbell or dumbell	24a & 24b
Abdominals	Abdominal curls	Machine, floor or Swiss ball	25a & 25b
	V-sits	Floor	26a & 26b
	Hanging leg raise	Hanging from pull-up bar	27a & 27b
Compound lifts	Power clean	Olympic bar	28a, 28b, 28c & 28d
	Deadlift	Olympic bar	29a, 29b & 29c
	High pull	Olympic bar	30a & 30b
	Squat (front)	Olympic bar	31a, 31b & 31c
	Squat (back)	Olympic bar	32a & 32b
	Military press	Olympic bar	33a & 33b

paddlesports it is clear that the major requirement relates to upper body strength. However, the low back, abdominals and legs are important as assistor muscles and weaknesses in these aspects could impact negatively upon performance.

Table 5.7 provides a variety of adventure sport-specific exercises that could be added to an adventure sports performer's programme. Photograph sequences illustrating these exercises can be found at www.wiley.com/go/draper. Through the inclusion of general and sport-specific exercises, an athlete could develop a more comprehensive strength training programme addressing individual weaknesses as well as discipline-specific requirements.

Programme order

Having selected the exercises for a programme, there are a number of ways they can be ordered depending upon the desired outcomes. For maximal strength and power gain, exercises can be carried out from largest to smallest muscle groups (for example bench press before tricep extension), or changing from upper body to lower body exercises (for example Lat Pulldown to Squat) where maximizing rest is important. For hypertrophy and muscular endurance training you could consider moving from small to large muscles groups (triceps extension before bench press) or completing all exercises for one body part before moving on to the next. Weight training exercises can also be linked with matched plyometric exercises in complex training that can also impact upon the power or endurance capacity of muscles. An example of this would be bench press being matched with clap press-ups: a strength exercise to pre-load the system followed by a plyometric power exercise.

By altering the order of exercises or splitting routines into separate body parts, it is possible for an athlete to train in the gym up to six days per week. However, this degree of strength training specialization would be more appropriate to sports where the physical domain is larger than the skill component. This is not the case for adventure sports; as such a maximum of three weight training sessions per week would allow a more appropriate focus on conditioning and skill development.

As well as weight training, there are a number of alternative forms of resistance training that might prove beneficial to adventure sports. One of the most important of these is *plyometric training* that can be used in conjunction with strength training. Plyometric exercises use pre-tensioning of muscles to increase force generation. Research has shown that they can help with power development, especially when linked with weight training exercises. *Yoga* and *functional stability training* both offer forms of exercise that can help with core strength and flexibility for adventure sports and could be considered as part of strength and conditioning. Finally, *circuit training*, a form of training that combines strength and conditioning training, can provide an efficient form of training for adventure sports.

A carefully planned and balanced strength training programme can be a central part of injury prevention and rehabilitation for an adventure sports performer. The key to injury prevention is about keeping a balance between muscle groups around the main joints of the activity. For instance, in canoeing and kayaking, where pulling from the shoulder makes up a large part of the demands of the sport, to avoid muscle imbalances strength training programmes should include exercises that develop the pushing from the shoulder muscle movements. Having suitable rest and recovery periods between training, a good basic diet and including a warm-up and cool-down before and after training can also help with avoiding both overtraining illnesses and injuries.

For rehabilitation purposes, the use of a physiotherapist or similar medical professional can help with specific injury recovery strategies. In addition, they can provide specific strength training exercises that can be used during the later stages of injury recovery to assist with strengthening the muscles around joints. An example of this is the lower arm abduction and adduction exercises in Table 5.6 that can be incorporated in to a resistance training programme and can assist in both rehabilitation from, and prevention of, shoulder impingement injuries.

Table 5.7 Adventure sport-specific exercises that could be included in a strength training programme. For some exercises such as clock lunge (a suggested exercise for canoe and kayak), the basic movement is the same as for the photograph sequence in the general exercises so the reader is referred to the photograph sequences in Table 5.6

Adventure sport	Exercise	Equipment	Photograph sequence no.
Climbing	Plate pinch	Barbell plates	A1
	Typewriters	Chin bar or beam	B1, B2, & B3
	Chin and leg lift	Chin bar or beam	C1, C2, & C3
	Assisted one-arm chin-ups	Chin bar or beam	D1 & D2
	Dynamic chins	Chin bar or beam	E1 & E2
	Toe lift	Chin bar or beam	F1 & F2
	Body curl	Chin bar or beam	G1 & G2
Canoe and kayak	Swiss ball dumbell press	Dumbells and Swiss ball	H1 & H2
	Swiss ball lateral raise	Dumbells and Swiss ball	I1 & I2
	High pulley wood chop	Cable pulley	J1 & J2
	Low pulley wood chop	Cable pulley	K1 & K2
	Clock lunge	Dumbells	8a & 8b
	Sumo deadlift	Dumbell	L1 & L2
	One-arm Swiss ball press	Dumbell and Swiss ball	M1 & M2
Surfing and windsurfing	Oblique twist on Swiss ball	Swiss ball and bar	N1, N2, & N3
	High pulley wood chop	Cable pulley	J1 & J2
	Low pulley wood chop	Cable pulley	K1 & K2
	Toe touches (left and right)	Chin bar or beam	O1, O2, & O3
	Clap press-ups	Floor	P1, P2, & P3
	Dumbell curl and press	Dumbells	Q1, Q2, Q3, & Q4
	One-arm dumbell row	Dumbell	R1 & R2
	Squat 21s	Olympic bar or dumbbells	32a & 32b
	Side loaded squats	Dumbell	S1 & S2
Caving	Seated toe press	Dumbells	T1 & T2
	Plate pinch	Barbell plates	A1
	Dynamic chins	Chin bar or beam	E1 & E2
	One-arm dumbell row	Dumbell	13a & 13b
	Press and squat	Dumbell	U1, U2, & U3
	Press-up cross	Floor	V1, V2, & V3
	Reverse curls	Barbell	W1 & W2

Table 5.7 (continued)

Adventure sport	Exercise	Equipment	Photograph sequence no.
Mountain biking	Dumbell jump squat	Dumbells	S1 & S2
	Split jump lunges	Dumbells	X1
	Seated toe press	Dumbells	T1 & T2
	Tricep extension on a Swiss ball	Dumbell and Swiss ball	22a & 22b
	Dumbell curl and press	Dumbells	Q1, Q2, Q3, & Q4

Conditioning training

This aspect of a training programme involves anaerobic and aerobic training sessions, specific to the demands of the performer's adventure sport. Table 5.8 provides an overview of the various forms of conditioning that can be undertaken as part of a training programme. For sprint training and short-duration interval training, the method

for measuring each of the efforts is how close the athlete is to his or her maximum. For longer duration intervals and all other forms of training in Table 5.8, it is possible to use heart rate as an indicator of effort. The development of heart rate monitors has made it relatively easy and inexpensive to monitor training effort.

The more exact a coach or performer is in monitoring training, the more accurately he or she

Table 5.8 Anaerobic and aerobic conditioning training methods

Type of training	System trained	Intensity	Repetition	Sets	Duration of repetitions
Sprint training	Anaerobic	90–100 % of maximum effort	5–10	1–5	5–30 s
Interval training	Anaerobic/ Aerobic	85–100 % of maximum effort/heart rate	3–10	1–3	30 s to 2 min
Medium-paced continuous training	Aerobic	70–80 % of maximum heart rate	1–3	1	3–10 min
Fast-paced continuous training	Aerobic	80–90 % of maximum heart rate	1–3	1	1–5 min
Fartlek* (Speedplay)	Aerobic	70–85 % of maximum heart rate	1	1	40–60 min
Long slow distance training	Aerobic	50–70 % of maximum heart rate	1	1	40–60 min or more

*(Fartlek is a Swedish word meaning speedplay and refers to a form of training where the intensity is varied from easy to medium to high paced throughout the session)

can evaluate the success of the programme. The most accurate way to calculate maximum heart rate would be to design a test to take the subject to their maximum effort over a period of 8–12 min. Ideally, the test should be adventure sport-specific i.e. running-based for orienteers, fell runners and mountain marathoners. The maximum heart rate recorded can subsequently be used to calculate the individual training zones for conditioning training.

A very common and simple method through which to estimate heart maximum is to use the *220–age formula* which was first suggested by Fox *et al.* (1971). For example, if a performer was aged 23 years their predicted heart rate maximum would be 197 bts min^{-1}. This formula is based upon the principle that maximal heart rate is around 220 bts min^{-1} at birth and decreases with age at a rate of approximately 1 bt year^{-1}. Since this formula for estimation of heart rate maximum was first proposed, there have been many subsequent studies suggesting alternatives. One of the most useful for males and females across all age groups was proposed by Tanaka *et al.* (2001) after testing 18 000 participants. In estimating heart rate maximum, it appears be best to use the Tanaka *et al.* method as described below.

Calculation of estimated maximal heart rate as per Tanaka *et al.*, for a participant aged 23 years:

$$HR_{max} = 208 - (0.7 \times age)$$
$$= 208 - (0.7 \times 23) = 208 - 16.1$$
$$= 191.9$$

i.e. maximum heart rate is 192 bts min^{-1}

The data in Table 5.9 provide an illustration of the range in heart rate maximum that can occur through the use of different methods. The 188 HR_{max} target zones were calculated for the same 23 year old athlete to illustrate the Fox *et al.* and Tanaka *et al.* methods. The results for each method are different. To achieve the highest accuracy in determining HR_{max}, it is therefore appropriate to use an actual determination of the athlete's HR_{max} rather than an estimation of the value. In a similar way, there is a specificity of heart maximum that is relative to the method employed to determine

Table 5.9 Example target heart rate zones for the three methods of determining heart rate maximum (HR_{max}). Note: For a 23 year old participant, HR_{max} is 197 and 192 from the 220–age method and Tanaka *et al.* (2001) formula, respectively, and is measured as 188

Type of aerobic training	Intensity	Target HR zones 220–age	Target HR zones Tanaka	Target HR zones Max Test
Medium-paced continuous training	70–80% of maximum HR	138–158	164–154	132–150
Fast-paced continuous training	80–90% of maximum HR	158–177	154–173	150–169
Fartlek or Speedplay	70–85% of maximum HR	158–167	134–163	132–160
Long slow distance training	50–70% of maximum HR	99–138	96–134	94–132

the maximum. For an adventure racer or tri-athlete it is appropriate to determine heart rate maximum for all three or more disciplines as appropriate. For a kayaker, heart rate maximum is best determined in boat if target zones are sought that apply to paddlesport training.

Key point

Conditioning training programmes can be designed with a great deal of variety. One of the most versatile forms of conditioning training is interval work where the repetitions, set interval lengths and recovery can be manipulated to meet many training goals.

Further to this, the majority of the conditioning training for an adventure sport can and should match as closely as possible the demands of the activity. For instance, fitness for hill walkers, fell runners and orienteers should be carried out on natural terrain or on a treadmill rather than a cycle ergometer in the gym. For canoeists, conditioning training should ideally be conducted in a boat. However, training on rowing machines, such as the Concept II rower which is readily available in many gyms, is more specific to paddlesport than using any other gym equipment for developing aerobic and anaerobic fitness.

Key point

The key with all aspects of resistance and conditioning training is not to lose focus on the most important aspects of your sport. It would be very easy to have yoga, functional stability work, weight training and conditioning sessions dominating your training. Adventure sports are dominated by skill and therefore for most of our development, the sport should be at the centre of our training.

Warm-up and cool-down

The warm-up and cool-down are essential parts of training and performance that are often overlooked for all sports performers. For adventure sports performers, especially when the activity is performed in poor environmental conditions, this is particularly true. Depending on the level of performer, warm-ups may often only be carried out for national governing body assessments, rather than being seen as an essential part of the preparation for exercise. Cool-downs are seldom completed by adventure sports athletes, especially when it is cold and getting dark at the end of a long day. In these conditions performers want to put kit away and get changed and warm rather than think about completing a cool-down strategy. Research, however, indicates that both warm-up and cool-down can improve performance, help avoid injury and promote recovery.

Benefits of warm-up A number of research projects have identified the performance and injury avoidance benefits of including a warm-up as part of any training session or competition. Warm-ups generally include a pulse raiser, followed by a variety of stretches of mobilizing exercises and a further more intense pulse raiser. The warm-up can also be divided into general and sport-specific exercises should become more specific to the adventure sport to be undertaken as the warm-up progresses. The benefits of completing a warm-up include improvements in the following related performance aspects.

- Blood flow to muscles and muscle temperature: sudden bouts of strenuous exercise without warm-up have been shown to lead to abnormal heart performance through inadequate oxygen supply to the heart muscle.

- Oxygen utilization and the functioning of the energy systems, including reducing lactic acid build-up.

- Nerve transmission and reaction time have been shown to improve in the early stages of exercise as a result of warm-up.

- Strength and power at the start of exercise have been shown to increase as result of the pulse-raising aspects of warm-up.

- Flexibility is improved through warm-up, which may well assist in injury avoidance.

- Psychological readiness for the activity by focusing the mind and body on the session ahead, enabling a performer to ready him or herself mentally and physically before the main activity.

- Improved performance during the first minutes of strenuous exercise.

Benefits of cool-down   The way in which an athlete completes their training session or performance has implications for their recovery from the exercise. Most sports people are aware of the benefits of warm-up, but perhaps fewer consider the benefits of a cool-down at the end of activity. In the rush to get packed away at the end of the day a cool-down is often neglected. This has implications for the removal of metabolites produced during exercise, completing flexibility exercises at the optimal time and reducing the possible effects of delayed onset muscle soreness (DOMS).

The multi-day nature of many adventure sports makes attention to a cool-down an important aspect for all performers. Skiers participate in trips up to two weeks long, paddlers travel to the Alps for river descents over a number of days, big wall climbers are on the rock face for several days at a time and boulderers will visit Fontainebleau for several days at a time. Research indicates that, within the restrictions of the activity (i.e. jogging is not a feasible option after a day big-wall climbing), completing a cool-down at the end of each day's activity will improve recovery before the next. By day three or four of a trip to Font this could have significant impact on climbing performance.

The benefits of employing a cool-down strategy for adventure sports performers include:

- reducing the impact of DOMS that is often felt 1–2 days after strenuous training sessions, by gradually lowering pulse rate and completing stretching exercises;

- assisting the body with the removal of substances created during exercise such as lactic acid, that would otherwise tend to pool in the active muscles; and

- stretching to improve flexibility has been shown to be most effective for warm muscles and is therefore best carried out during cool-down.

Format for warm-up and cool-down   There are a number of general principles that can be followed in the development of a warm-up or cool-down session.

- Warm-ups should be specific, or become more specific as they progress, to the activity to be undertaken. Where possible warm-ups should include, for example, in-boat activities for paddlers and easy climbing for climbers.

- The warm-up should include a pulse-raising activity and a variety of mobilizing exercises centred on the key joints involved in the exercise. Stretching exercises can be included where necessary for the activity or the needs of the individual (when recovering from injury). However, the muscles will not be as warm at this stage in the session as they will be at the end of exercise.

- A cool-down included at the end of a session should include exercise to gradually lower heart rate, assisting with the removal of exercise metabolites, and flexibility exercises to promote recovery and improve flexibility.

Flexibility training   An essential component of any cool-down session is the inclusion of a range of sport-specific flexibility exercises. Developing an optimal level of flexibility specific to each adventure sport can help to improve performance. This optimal level of flexibility relates to the fact that a lack of flexibility (hypomobility) could cause soft tissue injury during performance and become a rate limiter for skill development in that activity. Conversely, too much flexibility (hypermobility) can result in joint and soft tissue injury. For example, hypermobility of the shoulder joint for a kayaker may be partly responsible for shoulder dislocation during a capsize or when caught in a stopper. A balance needs to be maintained between flexibility training and muscle development (through strength training). There are three main forms of exercise training that can be completed to improve flexibility: static stretching, ballistic stretching and proprioceptive neuromuscular facilitation (PNF).

Static stretches These involve a performer holding a position, usually for 10–30 s, in order to apply a stretch to the soft tissue surrounding a particular joint. The movement is made slowly and held in position for the length of the stretch. Static stretches have the advantage that they need only the athlete to perform them, they are simple to complete and are least likely to result in injury. Relative to other methods of stretching, they cause the least amount of pain during execution.

Ballistic stretches As the name suggests, these require motion and momentum to induce the stretch. This movement is most often completed in a bouncing rhythmic fashion. Ballistic stretches often use the same exercises or positions as static stretches but are completed with movement to push the stretch further. Martial arts exponents and hurdlers are athletes who often make very dynamic movements in extreme ranges of motion and have traditionally used the exercises in their warm-ups and cool-downs. The disadvantages with this form of stretching result from the movement and momentum that can be generated during the stretch. With the bouncing motion, there is an increased risk of injury and a resistance to the stretch from the antagonistic muscle to that being stretched. Unless employed for a sport-specific reason, static stretching has the same benefits as ballistic stretching, with a lower risk of soft tissue injury.

Proprioceptive neuromuscular facilitation PNF requires a partner with experience to complete this form of stretching. There are a variety of PNF forms; the most commonly used involves an isometric contraction of the muscle to be stretched just prior to the stretch (contract-relax-contract method). Research has found that there is an increase in muscle relaxation after a muscle

contraction. It is thought that this occurs due to a decrease in the muscle's protective mechanism (against the stretch) brought about through the muscle contraction prior to stretch. The golgi tendon organs, proprioceptors and muscle spindles (Chapter 3) normally inhibit the stretch of a muscle as it reaches its limits of range of motion to protect against injury. The isometric contraction of the muscle prior to each stretch appears to enable the protective response to be inhibited, allowing a greater stretch to be achieved.

Figure 5.18 shows a contract-relax-contract PNF stretch for the hamstrings. This involves the partner kneeling close to the athlete and moving their leg (hamstrings muscle) to its lengthened position. The athlete then contracts the hamstrings against the resistance of their partner, enabling an isometric contraction. The contraction is held for 5–6 s after which the muscle is relaxed. During the relaxation phase, the partner moves

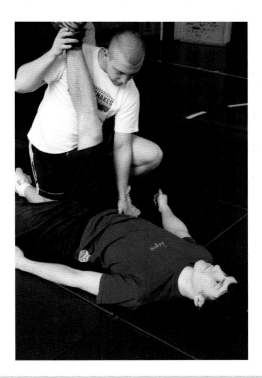

Figure 5.18 Contract-relax-contract PNF exercise for the hamstrings

the limb increasing the range of motion and the stretch. The limb is then contracted again and the process repeated three to four times for each muscle. This form of stretching has been shown to result in greater improvements in flexibility than for either ballistic or static stretching. It is very important, however, that the partner and the athlete are trained in using this form of flexibility training due to the additional stretch applied and the subsequent increased risk of injury. Examples of a range of general flexibility exercises for adventure sports performers can be found at www.wiley.com/go/draper.

Key point

Warm-ups are used more often as part of a training session for adventure sports performers. The most appropriate are those that closely match the demands of the discipline. Cool-downs are less frequently carried out at the end of a session, but can have a very positive impact upon recovery from training and competition. Flexibility training can most usefully be performed during a cool-down when muscles have been active over a longer period of time.

5.3 Summary and study questions

Physical fitness for any sports-based activity can be broken down into key areas or components. While these components are common across activities, every specific adventure activity will require a different emphasis on these components. Generic components are:

- anaerobic power and power endurance;
- speed;
- strength;
- aerobic endurance;

- muscular endurance;
- flexibility;
- body composition;
- dynamic balance (agility, coordination, balance); and
- reaction time.

Assessment

In order to improve an adventure sports athlete's performance, it is important to understand the requirements of the specific activity and be able to assess the current physical preparation of the athlete. An athlete's physical fitness can be assessed using generic assessment tools and discipline-specific tools. Tests for any particular athlete need to be chosen carefully in order to provide a profile of the athlete that is relevant to performance in their activity. While some common tests can be used off the shelf, others might need to be adapted to more closely resemble the demands of the specific adventure sport. The more sport-specific a test is, the more likely the results will relate to the performance in the actual activity.

Recording the progress of an adventure sports athlete can be a useful way to monitor progress, evaluate the effectiveness of training and motivate both coach and athlete. The tests that you, as a coach or athlete, choose to measure will depend on the resources that are available and the level of accuracy required. If you want to be able to compare preparation during, or after, a training programme then you will need access to the same resources at each testing point.

If you are a coach then remember that results of testing can be sensitive, especially when comparison to peers can be made or testing relates to body size and shape. Athletes might need to be given information with a high degree of sensitivity and results should remain confidential. Testing athletes in a group can be motivating but can also be very stressful depending on the particular circumstances. The coach should consider this when planning any fitness evaluations.

If you are going to employ any fitness test, then you should consider its validity and reliability. Does the test measure the relevant parameter and is it going to give you a trustworthy score? A good test will do both. You will probably want to carry out more than one test; does the order of testing allow for each test to be completed properly? Remember to record the exact order of tests so that you can repeat the sequence when the time comes to retest.

Training

The body will only adapt to activities that are more demanding than those in the athlete's everyday life. Setting activities that are above this level will result in overload and this is the stimulus for improvement. When the current training programme becomes an established activity the body has adjusted to, then the demand of the activities will need to be increased to maintain the rate of improvement. This can be through increases in frequency, intensity or time (FIT) of training activities. These principles are referred to as overload and progression.

Any training will result in very particular or specific gains. Training needs to be planned very carefully so that these fit the demands of the activity being trained for. Well-planned training will involve activities with similar movement patterns, speeds, times and loads to our adventure sport. Programmes also need to be tailored to an individual's needs.

Periodization means dividing the training period into segments and developing particular fitness components in each phase. This can prevent staleness and allow fitness components to build upon each other ready for a peak during the competitive or performance period. Fitness should be tested at the end of each phase to monitor improvements.

Long-term athlete development is an approach to training which tries to take into account the athlete's age and learning stage. This briefly means that you will need a different focus for adults or young people and for novices, intermediates and advance performers. There are windows of opportunity for certain types of training activity as well as the necessity to build sequentially on solid foundations.

Warming up and cooling down is essential not only for performance but also to obtain the maximum benefits from training. These also need to be specific to the adventure activity or training that the athlete is undertaking. Cool-downs are often neglected and can have real impacts on performance over consecutive days on an expedition or an activity trip.

Study questions

1. What is meant by aerobic and anaerobic capacity? How are they related to the duration of an adventure activity or element of an activity?

2. What is meant by the terms reliability and validity of a fitness test? How would you try to ensure reliability and validity if you were testing the aerobic capacity of a mountain biker?

3. How would you calculate Body Mass Index (BMI)? Why is it limited as a method of body composition assessment?

4. Why would a 30 min rowing session not be particularly good aerobic training for a downhill skier? What activities would you suggest instead and why?

5. Explain the two-for-two rule. How does it relate to overload and progression in resistance training?

6. What gains might we expect for an athlete engaged in plyometric training?

7. What is PNF stretching? Why is it thought to result in greater increases in flexibility when compared to static stretching?

Part II
The Physiological Demands of Adventure Sports

Chapters 6–10 cover activity-specific physiology for adventure sports. We examine the physiology of high-intensity short-duration power and power endurance activities (Chapter 6), anaerobic endurance (Chapter 7), intermittent aerobic and anaerobic metabolism (Chapter 8) and aerobic adventure sports (Chapters 9 and 10), in terms of specific sports and activities.

Adventure Sport Physiology   Nick Draper and Chris Hodgson
© 2008 John Wiley & Sons, Ltd

6

Power and power endurance: the explosive activities

6.1 Introduction

This chapter concerns the physiology of the power or explosive activities. The adventure sports that fall into this intensity and duration category include an individual playboat/freestyle move, a rock climbing dyno, a snowboard/ski move or a short, powerful bouldering move. The time duration for sustaining such an effort is relatively short, up to around 10 s, but this can result in the highest power outputs humans can achieve. Figure 6.1 provides a representation of the duration range for the activities covered in this chapter.

Maximal power outputs can be measured in a variety of ways and many of these were described and illustrated in Chapter 5. The peak power output achieved in a vertical jump, standing board jump, a Margaria step test or in the first few seconds of a Wingate anaerobic test (WAnT) all assess very short duration maximal efforts. With regard to adventure sports, compared to other chapters in Part II, there are relatively few activities that fall into the all-out power short-duration category. Perhaps the most appropriate illustration of a short-duration power endurance adventure sport would be an ascent of a short, steep, powerful bouldering problem which is at

your limit. There are, however, many adventure sports that involve powerful moves and when carried out in isolation, the energy required for these efforts would be supplied through short-term anaerobic metabolism. The 3–4 strokes to drive a kayak or canoe across an eddy line, the double pump (powerful paddlestroke) to set up a flatwater cartwheel in a playboat and dyno moves in climbing (where the climber releases all four limbs from the rock to drive up and re-catch/deadpoint at a higher point) are examples of power or explosive movements within adventure sports. To improve power, perhaps the most effective tools are a combination of body weight plyometric exercises adopted in conjunction with a strength training programme. Ideally, the exercises adopted should match, as closely as possible, the movements made in the adventure sport for which you are training. The training section in Chapter 5 provided a wide range of exercises that can be used as the basis for a programme to improve your power performance.

This chapter examines the physiology of power and power endurance activities. Due to the obvious importance of the muscular system to powerful movements this chapter begins with a further examination of muscle fibre types, building on information provided on the muscular system in

Adventure Sport Physiology Nick Draper and Chris Hodgson
© 2008 John Wiley & Sons, Ltd

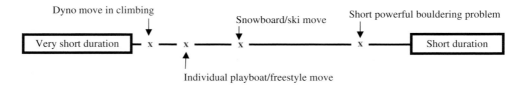

Figure 6.1   Duration continuum for the power and power endurance adventure sports

Chapter 4. This is followed by a more detailed explanation of energy production for explosive activities, as well as an examination of fatigue in response to power and power endurance activities and how the body adapts to specific power development training. In addition, a brief account of the delayed onset of muscle soreness (DOMS) and muscular cramp are provided. Finally, a brief review of possible dietary manipulations for power activities is provided. Naomi Buys, a member of the Great Britain Bouldering Team, describes the physical demands of bouldering. A journal article by Vandenberghe and co-workers on the use of creatine as a nutritional supplement is included, illustrating the research that has been conducted in this area of physiology.

6.2   Muscle fibre types

In Chapter 4, the structure and functioning of the muscular system was described. The different types of muscle fibre which human muscles comprise were briefly mentioned. The two main muscle fibre Types, I and II, were described and their basic physiology explained. There are, however, further sub-divisions of fibre type that can be made and important distinctions that can be identified. The nature of these sub-divisions is important to muscle recruitment for power or explosive activities, and will also have implications for glycolytic and aerobic energy production.

The identification of two main divisions of muscle fibre type originated from early research into the structure and function of muscles. In the 19th century, researchers identified differences in muscle fibre types visible to the naked eye. Louis Ranvier (1835–1922), who was the first to identify the presence of myelin around nerve fibres and the gaps (nodes of Ranvier) in the sheath, published a paper in 1873 detailing structural differences in the white and red muscle fibres of rabbits and rays. These fibres can be seen in chicken meat, the red fibres of the legs being Type I fibres and the white meat of the breast and wings being Type II fibres. The difference in colour between the two types of fibres has been identified as being due to the presence of high levels of myoglobin and the denser capillarity of the aerobic Type I fibres.

Since this early research, new techniques have been developed which have enabled researchers to identify further muscle fibre sub-divisions. Technical developments in the identification of muscle fibre protein and enzyme presence through histochemistry, the study of the microscopic structure and chemistry of cells and tissues, led to the identification of Type II sub-divisions. Histochemical research has shown differences in the properties of myosin adenosine triphosphatase (ATPase) between Type I and Type II fibres. Discussed in more detail in Chapter 4, ATPase is the enzyme associated with the thick myosin filaments responsible for the hydrolysis (catabolism) of ATP attached to the myosin head. The breakdown of myosin ATP supplies the energy for the power-stroke in contraction and the release of the myosin head from the actin binding site. The difference in ATPase properties between muscle fibres types has enabled researchers to stain sections of muscle tissue and identify further sub-divisions. Figure 6.2 provides a photomicrograph of a micro-section of rat muscle, revealing the Type I and II fibres.

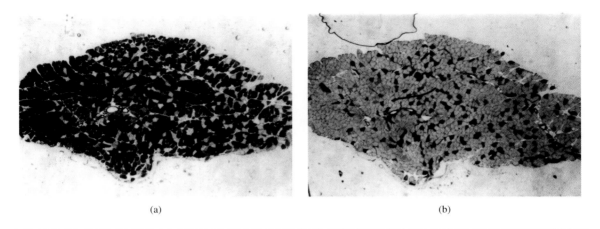

(a)   (b)

Figure 6.2   Photomicrographs incubated at (a) pH 4.35 and (b) pH 10.2 revealing fibre types in rat muscle (photograph courtesy of Chris Wilkinson)

Sections of muscle tissue can be taken from humans or animals. Rat muscle tissue has been frequently studied in physiological research as its general properties are similar to human muscle. The muscle fibre types identified in humans can all be found in the muscle fibres of rats. A muscle biopsy is used in humans to sample a small section of muscle tissue.

By histochemical staining, three sub-divisions of Type II muscle fibres have been identified. These are commonly referred to as Types II_a, II_b and II_c. Figure 6.2(a) shows Type I fibres which are stained black, Type II_a grey and Type II_b white. Figure 6.2(b) depicts Type I fibres which are stained white, Type II_a black and II_b grey. Type II_c are far less commonly identified in humans and are thought to account for only 1–2 per cent of muscle fibre types. Due to their rare occurrence as well as limits in current knowledge of their function, the discussion in this section concerns the structure and function of Type I, Type II_a and Type II_b fibres.

It is worth noting two points at this stage. Firstly, the advent of *polyacrylamide gel electrophoresis* in the study of the histochemistry has enabled even further sub-divisions of muscle fibre to be identified. Gel electrophoresis, which is commonly carried out in biochemistry laboratories such as that at the University of Canterbury, enables the separation of individual myosin filaments and further understanding of their different properties. Through this method, up to nine muscle fibre types have been identified in the muscle tissue of rats and rabbits. Secondly, although commonly called Type II_b fibres in humans, research has indicated that the majority of these fibres found are in fact Type II_x fibres. Examination of rat muscle shows the presence of Type II_x and II_b fibres which have different properties including speed of contraction. Type II_b are slightly faster contracting than Type II_x fibres. Comparison of human and rat muscle fibres has revealed that human Type II_b fibres are actually structurally and functionally the same as Type II_x fibres found in rat muscle. Consequently, they should be more correctly referred to as Type II_x fibres. In this textbook, II_b will subsequently be referred to as $II_{b(x)}$ to concur with the nomenclature in other textbooks but to remind the reader that the term II_x is more appropriate when discussing human muscle tissue. Table 6.1 provides details of the variety of classifications developed for Type I and II muscle fibres. The details provided here are intended to help the reader when referring to other textbooks.

Table 6.1 The varied nomenclature for muscle fibre types

Type I	Type II_a	Type $II_{b(x)}$	← Terms used in this textbook
Type I	Type II_a	Type II_x	
Type I	Type II_a	Type II_b	
Slow twitch	Fast twitch a	Fast twitch b	
Slow oxidative (SO)	Fast oxidative glycolytic (FOG)	Fast glycolytic (FG)	Terms used in other textbooks
Slow contraction (S)	Fast contraction, fatigue resistant (FR)	Fast contraction, fast fatigue (FF)	

The descriptors in Table 6.1 provide a useful insight to the properties of each type of fibre. Type I, the slow oxidative fibres, contain slow rate myosin ATPase and rely on oxidative (aerobic) mechanisms for energy production. As a consequence, although Type I fibres take twice as long to reach peak contraction (around 100 ms) and generate less force than Type II fibres, they are much more resistant to fatigue. Conversely Type $II_{b(x)}$ fibres which contain fast rate myosin ATPase, reach peak contraction quickly. They can generate more force than Type I fibres using anaerobic glycolytic metabolism for energy production, meaning they fatigue relatively quickly. Type II_a, also called fast oxidative glycolytic fibres, fall between Type I and Type $II_{b(x)}$ fibres with regard to many of their properties and functions. They are more fatigue resistant than Type $II_{b(x)}$ fibres, although less so than Type I fibres as they can utilize both oxidative and glycolytic means for energy metabolism. Table 6.2 provides a summary of the properties of each fibre type commonly found in human skeletal muscle.

The recruitment of muscle fibres during exercise is dependent upon the intensity and duration. When standing, where muscles have only to contract against the forces of gravity, or during low intensity exercise such as walking, paddling very slowly or climbing on large holds, Type I fibres are recruited to provide the necessary muscular contractions. When the intensity of exercise is higher, for instance during activities such as jogging or steady-rate paddling, Type I and II_a fibres are recruited. When all-out effort is required, for the finish of a marathon paddle or for a maximal bouldering problem, all muscle fibre types are called upon to provide the necessary muscular contractions and force generation. The body, however, has protective mechanisms to avoid recruiting all motor units during maximal effort to avoid damage to muscle attachments and tendons. In a maximal contraction we therefore normally recruit up to 70 per cent of the available muscle fibres. In situations of grave danger, there have been incidents of people being able to lift heavy weights to free a trapped person inside. It is thought that such capabilities, which often lead to major injury to the person as a result of the effort, are enabled through an emotionally driven maximal recruitment that overrides the normal protective mechanisms.

With any individual, it is possible to identify all human muscle fibre types including Type II_c. An average muscles contain around 50 per cent

Table 6.2 Properties of Type I and II muscle fibres

	Type I fibres	Type II$_a$ fibres	Type II$_{b(x)}$ fibres
Properties			
Appearance	Red	Pinkish	White
ATPase hydrolysis rate	Slow	Fast	Fast
Contraction speed	Slow	Fast	Fast
Contraction force	Low	Medium	High
Fatigue resistance	High	Medium	Low
Oxidative capacity			
Myoglobin content	High	High	Low
Capillarization	High	High	Low
Number of mitochondria	High	High	Low
Oxidative enzyme concentration	High	Medium	Low
Triglyceride content	High	Medium	Low
Glycolytic capacity			
Glycogen stores	Low	Medium	High
Glycolytic enzyme activity	Low	High	High
Phosphagen system capacity			
Phosphocreatine stores	Low	High	High
Creatine kinase stores	Low	Medium	High
Recruitment pattern			
Order	First recruited	Second recruited	Third recruited
Exercise intensity for recruitment	Low	Medium–high	High–maximal

Type I, 30 per cent Type II$_a$, 19 per cent Type II$_{b(x)}$ and 1 per cent Type II$_c$ fibres. The distribution of muscle fibre types varies with their role within the body. The soleus, which is an antigravity or postural muscle that requires low force contractions over long periods of time, comprises around 90 per cent Type I fibres. The muscles of the lower limbs tend to have relatively more Type II$_a$ then the muscles of the upper limbs, which tend to have greater numbers of Type II$_{b(x)}$ fibres.

Muscle fibre distribution is determined initially by individual genetic endowment and then by training which mimics the nature versus nurture hypothesis. Are athletes drawn to a sport because of their genetic muscle fibre type endowment or does their muscle fibre type alter as they train for a particular sport? The most likely answer, as with the nature versus nurture argument, lies somewhere in the middle. A certain level of genetic endowment is essential for high level performance. However, the effects of training for a particular activity have been shown to alter the properties of fibre types. Sedentary individuals tend to have a muscle fibre distribution of around 50 per cent Type I and 50 per cent Type II fibres. Those involved in power events tend to develop higher percentage of Type II fibres and those involved in endurance events such as mountain marathon or fell running have high levels of Type I fibres in the trained muscles (in the lower limbs particularly). The largest muscle fibres in trained individuals tend to be those which are most recruited during exercise. Endurance training has been shown to lead to an increase in the size of Type I fibres and a gradual shift in the properties of Type $II_{b(x)}$ fibres, towards those of Type II_a fibres. Strength and power training, due to the all-out nature of muscle recruitment, has been shown to increase the size of Type I and II fibres and result in improvements in the anaerobic capacity of the trained muscle fibres.

Key point

Three main types of muscle fibre can be identified in humans. Type I contract more slowly than Type II fibres and produce lower relative force levels, but are more fatigue resistant. Type II_a have a fast contraction speed with medium force production but fatigue more quickly than Type I fibres. Type $II_{b(x)}$ are capable of contracting at the highest speed, producing the highest levels of force but fatigue more quickly than Type I or II_a fibres. The speed of muscle fibre contraction or twitch is determined by the form of myosin ATPase present within the fibre. Type II fibres have faster-acting ATPase than Type I fibres which results in a halving of the time to peak contraction compared to slow twitch fibres. Muscle fibre distribution can be determined through muscle biopsy and analysis of the cross-sectional fibre content. Athletes in endurance events such as marathon paddling and fell running tend to have a higher distribution of Type I fibres compared to those involved in strength and power sports.

6.3 The phosphagen system

The phosphagen system provides the main energy source for explosive activities. The primary mechanisms for energy production for power-based sports include the initial breakdown of ATP and the ATP-PCr reaction which were introduced in Chapter 4. This section provides a more in-depth discussion of the reactions that take place within the phosphagen system during the immediate response to maximal efforts. The phosphagen system, sometimes called the ATP-PCr, alactic acid or immediate energy system, is the key mechanism for energy production during the first 10 s of exercise.

Adenosine triphosphate (ATP)

Every cell in the body needs energy to perform its role. The body's main energy source is adenosine triphosphate (ATP). Figure 6.3 shows the structure of ATP which comprises an adenosine molecule with three phosphate molecules bonded to it. The adenosine molecule is composed of a *base* (adenine) and a *sugar* (ribose) which together is called a *nucleoside*. There are several different forms of nucleoside, which bond with up to three phosphates, in the same way as adenosine, to form *nucleotides*.

Adenine is one of five bases which are fundamental to cell structure and functioning. The five bases, illustrated in Figure 6.4, are *adenine* and *guanine* (double-ringed *purines*), *thymine, cytosine,* and *uracil* (the latter three of which are single-ringed *pyrimidines*). Each of the five bases is a protein-based structure and therefore contains a nitrogen component which is associated with the basic structure of all proteins (Chapter 2). These five bases are common to every cell in the body as they form an essential component of our deoxyribonucleic acid (DNA) and ribonucleic acid (RNA) – the genetic coding inherited from our parents found in the nucleus of every cell and used as the blueprint for its creation, development and functioning. The letters *ACGT* in DNA coding and *ACGU* in RNA coding may be familiar: these come from the above-mentioned five bases,

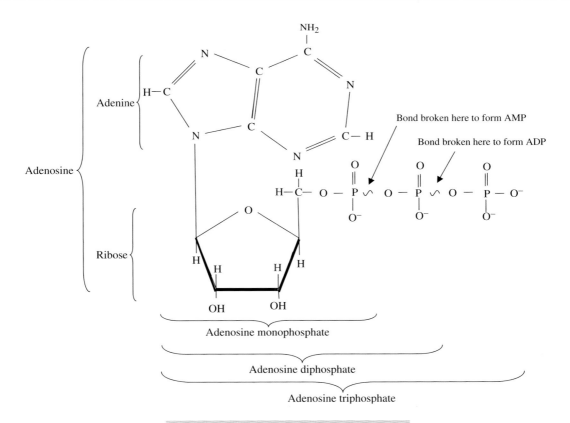

Figure 6.3 The chemical structure of ATP

of which adenine (A) is one. Adenine, cytosine, guanine, thymine and uracil form the basis for the structure of DNA and RNA.

Each of the five bases bonds with a sugar, either deoxyribose or ribose, to form a nucleoside. As previously mentioned, adenine bonds with ribose to form adenosine. Adenosine is the base molecule for ATP, and is an example of a nucleoside. Just as adenosine bonds with phosphate molecules to form AMP (adenosine monophosphate), ADP and ATP, the other nucleosides (for example guanosine and uridine) can bond with phosphates to create *nucleotides*. As with ATP, the final two phosphate bonds store energy that can be used to fuel the work of a cell. If ATP is the petrol for our bodies, then the other nucleotides represent diesel or liquid petroleum gas; alternative fuels for the cells of the body. For example, guanosine triphosphate

(GTP) is used as an energy source for protein synthesis and one molecule of GTP is produced during the Krebs cycle (Chapter 10). Uridine triphosphate (UTP), by way of a further example, is used by the liver and muscles to provide the energy necessary for the synthesis of glycogen. ATP, however, is the main fuel for human energy production and the muscular contractions that are central to movement in adventure sports and everyday life.

The bonds of the final two phosphates in ATP, as already mentioned, store energy. The stored energy is released when the bond is broken and the phosphate ion separated from the ATP molecule (creating adenosine diphosphate – an adenosine molecule with two phosphate molecules still attached). For instance, as was described in the section on muscular system, ATP is required to drive all muscle contractions. The energy from

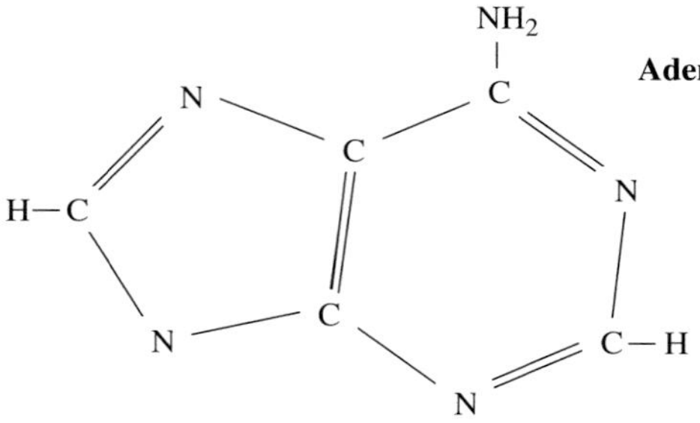

Adenine: Key component of ATP, DNA and RNA

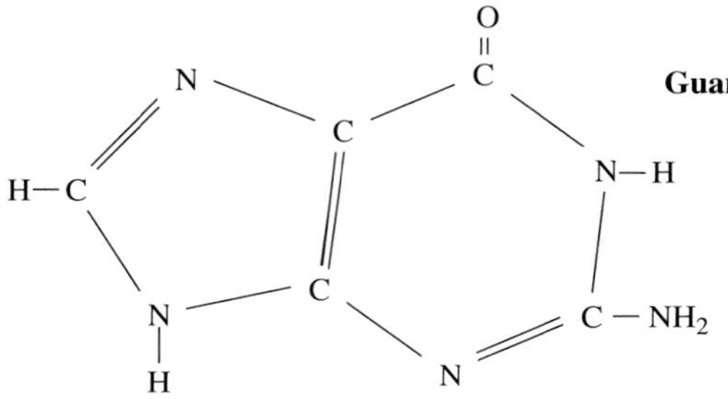

Guanine: Key component of DNA and RNA

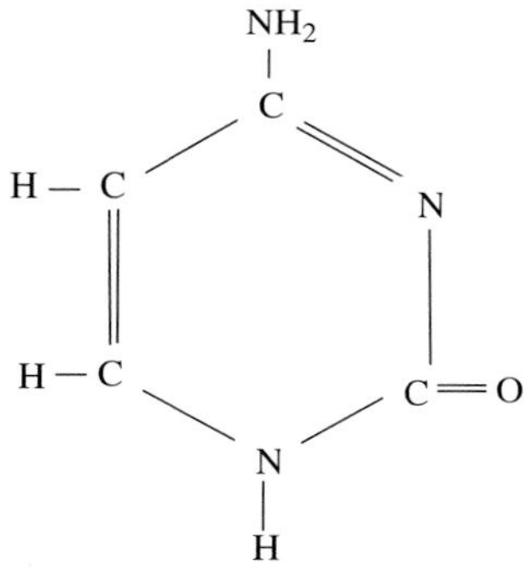

Cytosine: Key component of DNA and RNA

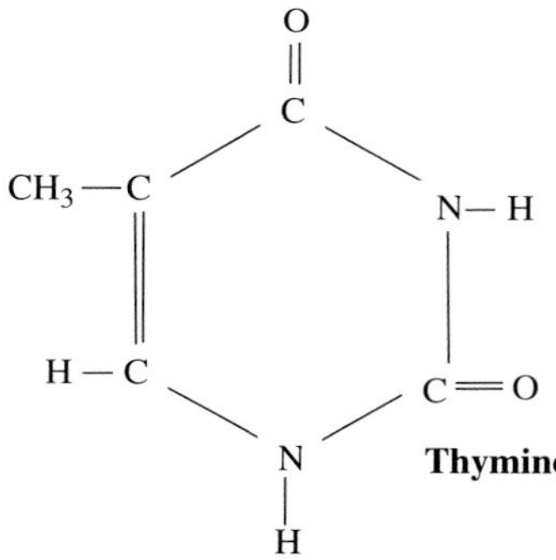

Thymine: Key component of DNA

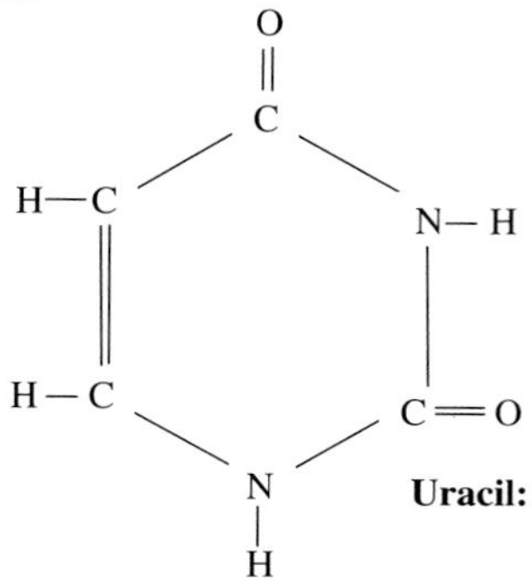

Uracil: Key component of RNA

Figure 6.4 The structure of the bases

the breakdown of ATP to adenosine diphosphate enables the powerstroke in muscle contraction and the release of each myosin head from its binding site on actin.

As was described in Chapter 4, when one phosphate is broken from ATP, energy is released and an ADP molecule remains. There is, however, further energy that can be gained by cleaving a second phosphate from ADP. This leaves the adenosine molecule bonded to only one phosphate which is called, unsurprisingly, adenosine monophosphate (AMP). Figure 6.3 shows diagrammatically the relationship between the number of phosphates attached to the adenosine molecule and its identification. Adenosine triphosphate stores provide the first 1–2 s of exercise during explosive activities. For power events that last for longer than this (up to 10 s of activity) the ATP stores must be replenished. The body has three phosphagen mechanisms which serve in different ways to maintain power output beyond the 1–2 s provided by the ATP stores. The most important of these for immediate ATP re-synthesis, the ATP-PCr reaction, was introduced in Chapter 4.

Key point

ATP, a nucleotide, is the main source of energy production within the body. Muscles store sufficient supplies of ATP to power the first 1–2 s of maximal effort during exercise, to complete activities such as a dyno move in rock climbing. ATP, along with the other four nucleotides GTP, UTP, TTP and CTP, provide potential energy sources and in their base form are key components to the structure of DNA and RNA, the blueprint for cellular functioning.

Phosphocreatine and ATP-PCr reaction

The structure of phosphocreatine (PCr) is shown in Figure 6.5. At rest the body stores between

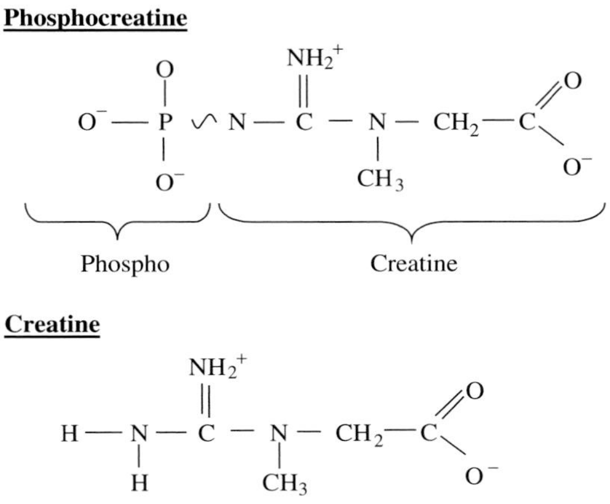

Figure 6.5 Structure of phosphocreatine and creatine

18–20 mmol kg^{-1} of PCr in the muscles, about four times the stores of ATP maintained by the body. During maximal exercise, ATP stores in the working muscle would be exhausted in 1–2 s of effort. However, ATP is reformed very quickly through the breakdown of the PCr stores in a reaction catalysed by creatine kinase.

The breakdown of ATP and the ATP-PCr reaction, or as it is sometimes called creatine kinase reaction, is shown in Figures 4.21 and 4.22 in Chapter 4. The stores of ATP and PCr are maintained within the sarcoplasm (the muscle cell body) which is also the location for the catabolism of ATP and the ATP-PCr reaction. From the onset of exercise, muscle ATP stores are immediately catabolized to form ADP and Pi (inorganic phosphate). The immediate proximity of PCr to the newly created ADP molecules, along with the presence of creatine kinase, enables the ATP-PCr reaction to proceed at a fast rate and maintain ATP supplies for muscular contraction. During exercise, the reversible ATP-PCr reaction favours the formation of ATP. At rest, however, aerobic metabolism enables the regeneration of PCr stores within the muscle.

As can be seen from Figure 6.5, PCr is created from the bonding of creatine and inorganic phosphate molecules. The bond creates stores of energy which are released when PCr is broken down during exercise, enabling the formation of ATP from ADP. Creatine is consumed within a diet containing meat or can be synthesized in the liver from several amino acids. The supplementation of dietary creatine has been suggested as an ergogenic aid for power-based sports and is sold commercially. The role of creatine in the improvement of performance in explosive activities is discussed later in this chapter in the nutrition and ergogenic aids section, and is the subject of the **Getting into research** paper for this chapter.

Key point

The ATP-PCr reaction enables the regeneration of ATP from ADP and represents the main energy source for 1–2 s of activity and up to around 10 s. The ATP-PCr system represents the power endurance system.

Adenosine diphosphate and the adenylate kinase reaction

During explosive activities, as an individual approaches 10 s of maximal effort, the stores of PCr will start to diminish and ADP will accumulate above resting levels. Phosphocreatine concentrations begin to fall because muscle fibre stores are limited and PCr cannot be resynthesized during high-intensity exercise. The rise in ADP levels trigger two parallel reactions that help to maintain all-out effort and increase activation of glycolysis. The first described here involves ADP and the second explained in the next section utilizes AMP, a product of this first reaction. The two reactions are coupled, the first being triggered when ADP levels begin to rise and the second begins as AMP begins to accumulate.

The first reaction, adenylate kinase reaction, is a supplementary phosphagen energy system mechanism which serves to produce further ATP to maintain maximal power output. The second reaction, the adenylate deaminase reaction, is not part of energy production for the phosphagen system. It is described here because it is triggered by increases in AMP, a product of the adenylate kinase reaction, and serves to activate energy production beyond the phosphagens: glycolysis. The first reaction, shown in Figure 6.6, is catalysed by the enzyme adenylate kinase (sometimes called *myokinase* or, when referring to the reaction in cells other than muscle tissue, *AMP kinase*). Adenylate kinase catalyses a reaction as ADP concentrations rise, splitting a further phosphate from one ADP and bonding it to another, creating ATP and AMP. The ATP molecule is subsequently utilized to maintain the maximal effort and the AMP molecules play an important role in the activation of glycolytic enzymes. In other words, the presence of increased AMP levels in the muscle fibre, the result of the adenylate kinase reaction, stimulates the rate of glycolysis to sustain maximal power output beyond 10 s.

Key point

As maximal effort continues towards 10 s, PCr stores begin to decline and ADP begins to accumulate. The adenylate kinase reaction enables two ADP molecules to be utilized to produce further ATP to maintain phosphagen system power output. The end

Figure 6.6 The adenylate kinase reaction

product of the adenylate kinase reaction includes an ATP and an AMP. If maximal effort continues beyond 10 s AMP, which serves as a chemical trigger to increase the rate of glycolysis, is further degraded to IMP (inosine monophosphate) as part of the deaminase kinase reaction.

Adenosine monophosphate and the adenylate deaminase reaction

In the section on protein metabolism in Chapter 2, the process of amino acid deamination was described. As a result of deamination, the amine or nitrogen component of an amino acid is removed. This process enables the remaining part of an amino acid, the α-keto acid, to be catabolized to produce energy. In the adenylate deaminase reaction, with the addition of water, the amine component of AMP is deaminated to leave inosine monophosphate (IMP) and the amine group which forms ammonia (NH_3). The adenylate deaminase reaction occurs to alleviate a build-up of AMP within the sarcoplasm which would inhibit the adenylate kinase reaction and interfere with sustained ATP production. This reaction is irreversible and during recovery IMP is removed from the muscle fibre. Ammonia (NH_3), a base, quickly accepts a hydrogen ion to become an ammonia ion (NH_4^+) which serves as a further activator for glycolysis. Ammonia is subsequently excreted from the body via the urea cycle. The effect of increasing acidity during exercise within muscle fibres as a result of sustained energy production is discussed in more detail in Chapter 7. The adenylate deaminase reaction is shown in Figure 6.7.

The complete degradation of ATP during maximal effort exercise could see an ATP molecule broken down to ADP, then to AMP and finally irreversibly to IMP. The degradation of ATP to IMP only occurs when a maximal effort moves beyond 10 s towards 45–90 s duration, such as during a sprint or slalom kayak race. Figure 6.8 provides an overview of the reactions within the phosphagen system that can lead to the degradation of ATP to IMP.

The reactions within the phosphagen system, the immediate source of energy for muscular contraction, can be thought of in two phases. The hydrolysis of ATP provides the explosive energy for the first 1–2 s of activity or the power required for a very short-duration maximal power movement such as a dyno move in climbing. Beyond the first 1–2 s of activity, the ATP-PCr reaction subsequently supported by the adenylate kinase reaction provide the sustained power required for a short duration (up to 10 s) power sequence of movement such as the completion of a steep bouldering route.

Initial explosive efforts can be thought of as *power* activities and those requiring sustained maximal effort as *power endurance*. Beyond 10 s, to maintain maximal power output, glycolysis (further stimulated by the presence of AMP) increasingly becomes the main fuel source. The utilization of glycolysis to provide all-out power for activities of duration longer than 10 s can be thought of as *anaerobic endurance* and this process is described in Chapter 7. The adenylate deaminase reaction occurs as a natural effect of maintaining maximal power output beyond 10 s as part of anaerobic endurance.

Beyond 90 s, the aerobic system increasingly becomes the fuel source to maintain maximal energy output. *Aerobic endurance* is covered in Chapters 9 and 10. Absolute peak power output is attained by human effort within the first 3–4 s

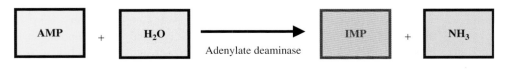

Figure 6.7 The adenylate deaminase reaction

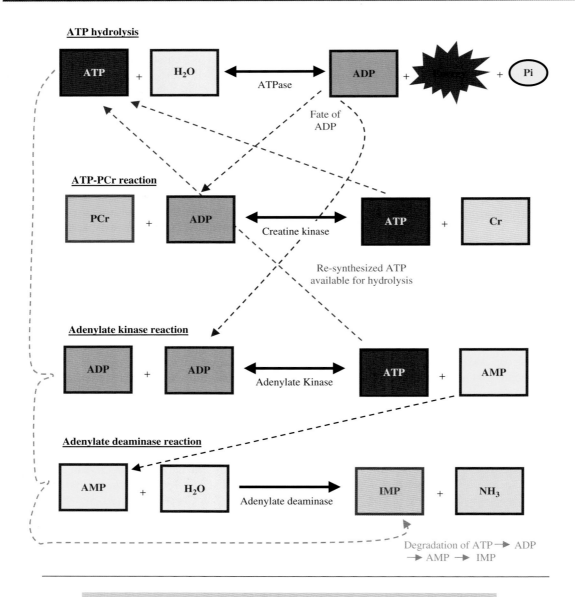

Figure 6.8   Reactions of the Phosphagen system: the possible fates of ATP

of exercise (somewhere between power and the very start of power endurance). Beyond this, power output drops so that attaining peak power for power endurance, anaerobic endurance and aerobic endurance is relative to that energy production system. Maximal effort, however, is a key component across the energy systems for many adventure sports. The phosphagen system is responsible for maximal power output for very short and short-duration power and power endurance activities. Maximal power output through glycolysis, anaerobic endurance and through the aerobic system, aerobic power, are covered in the following relevant chapters.

Key point

The phosphagen system provides the key energy production mechanism for power (1–2 s activity) and power endurance (up to 10 s of all-out effort). The catabolism of ATP stores within the muscle provides the initial energy for muscular contraction and movement. The regeneration of ATP through the ATP-PCr and adenylate kinase reactions provides the main energy source during power endurance activities. Beyond 10 s of activity, glycolysis and subsequently aerobic metabolism become the main fuel sources for the regeneration of ATP and continuation of exercise.

6.4 Physiological response to power and power endurance activities and fatigue

This section examines the physiological response to exercise and the impact on the body of power and power endurance activities. The causes of fatigue in explosive adventure sports are also discussed. At the start of exercise, regardless of intensity and duration, the nervous and endocrine systems alter the functioning of the respiratory and cardiovascular systems to meet the demands of exercise. The general response of these systems to exercise has an application in all adventure sports. The specific response after a general increase in functioning at the onset of exercise is determined by the intensity and duration of exercise. The nature of fatigue, in a similar way, is specific to the energy demands placed upon the body, such that the causes of fatigue in explosive activities are different to those for high-intensity aerobic endurance adventure sports (Chapter 9).

The sympathetic nervous system is the key driver of the physiological changes within the body at the start of exercise. Sympathetic stimulation causes a relaxation of the airway in the lungs, an increase in heart rate, breathing rate and depth, an inhibition of digestion, stimulation of endocrine function, release of exercise-related hormones and vasoconstriction of blood vessels supplying the organs of the abdomen, such as the stomach, kidneys and intestine. The sympathetic nervous and endocrine responses to exercise are concerned with two main functions: (1) increasing transport and exchange of metabolites, and (2) increasing the fuel substrate availability and, where possible, glucose sparing.

Hormonal response to exercise is mainly associated with increases in cardiovascular response, metabolism and substrate availability. The endocrine response to exercise provides a less immediate, but longer enduring, reinforcement of stimulation initiated by the sympathetic nervous system. The hormones thyroxine, adrenalin and noradrenalin have a general effect on cells and all serve to increase the rate of metabolism during increasing levels of exercise. In addition to this general effect, noradrenalin and adrenalin, released at exercise intensities above 50 and 60 per cent of $\dot{V}O_{2\,max}$ respectively, stimulate increases in cardiac output and alterations in blood flow during exercise. Glucagon, adrenalin, noradrenalin and cortisol increase the availability of glucose in the blood during exercise. Glucagon, adrenalin and noradrenalin stimulate the liver to catabolize glycogen back to glucose through the process of *glycogenolysis* and to deaminate amino acids to create glucose through the process of *gluconeogenesis*. Cortisol enhances glucose availability through its stimulation of amino acid gluconeogenesis.

Carbohydrate stores in the form of glucose and glycogen, unlike the body's fat stores, are limited in supply and therefore a number of hormones stimulate the catabolism of fats and amino acids to promote glycogen sparing. These hormones also stimulate the catabolism of fat as an energy source when glucose levels are low, such as during endurance activities. The hormones adrenalin, noradrenalin, cortisol, human growth hormone and glucagon all increase the activation of the enzyme *hormone sensitive lipase* which is the key enzyme in the mobilization of fats for metabolism during the process of lipolysis (lipolysis and fat catabolism are discussed in more detail in Chapter 10). Through their general stimulatory effect on metabolic rate thyroxine, adrenalin and noradrenalin serve to enhance the catabolism of fats through β oxidation.

The general respiratory response to exercise was discussed in detail in Chapter 4. The onset of exercise stimulates an increase in the depth and rate of breathing, a response that is controlled by the medulla oblongata and the pons (each located in the brain stem). The onset of exercise increases the levels of CO_2 and H^+ and decreases the level of O_2 in the blood. Chemoreceptors in the brain, aorta and common carotid arteries detect these changes and relay this information to the respiratory driver. The primary driver for increases in respiration is the level of CO_2 and when higher levels are detected it creates a strong stimulus for increases in depth and rate of breathing. Increases in breathing rate and depth create increased availability of oxygen for diffusion into the bloodstream and the removal from the body of the increased concentrations of carbon dioxide arising from exercise.

The sympathetic nervous system is the primary driver for the increases in respiration. With regard to respiration for power and power endurance events, the maximal intensity of the activities, along with their very short duration, can create interruptions in the normal increase in respiration rate during the exercise. Research regarding bouldering has revealed the incidence of periodic breath-holding during ascent of power endurance bouldering problems. The breath-holding appears to be linked to fixation of the core in a Valsalva manoeuvre style maximal muscle recruitment, to make movements between holds. The nature of respiration during power and power endurance

Biography: Naomi Buys

Adventure sport: Rock climbing/Bouldering

Twenty-four year old Naomi is from Burnley, Lancashire, and began climbing 12 years ago at a local school which had a climbing wall. Naomi was immediately fascinated with climbing and learned to lead on at a variety of Yorkshire gritstone crags. Hooked on climbing, she has climbed at many Lancashire quarries, normally climbing around three times a week. When she married Jordon, who is also a climber, the two went on to improve each other's climbing grades. Her sport leading grade is F7c and she came fifth in her first attempt at the British Bouldering Championships. She is now a member of the British Bouldering Team and was featured in

Figure 6.9 Naomi Buys (photograph courtesy of Alex Messenger)

the August 2006 edition of Climber magazine. Each week, Naomi mixes two–three outdoor climbing sessions with indoor training, fingerboard work, circuit training, swimming and cycling.

Clever Beaver

Grade: V7/8/Font 7a+/7b
Location: Parisella's Cave, the Great Orme, Llandudno, Wales
Rock type: Limestone
Completed: June 2006

Parisella's Cave is one of the weirdest places I have ever climbed. It is ultra-steep with a compact goat dung flooring. There are beautiful views out to sea over Llandudno Bay and yet, such is the fascination of the shiny limestone holds, you don't often stop to look. Occasionally, you may have the pleasure of climbing to the sounds of the carousel on the nearby pier. Usually you can just hear the waves crashing on the rocky beach below. The other strange feature of the cave is that it is inhabited by an incredibly strong beast called Jamie Cassidy, who can often be found randomly aping between improbable holds.

Clever Beaver takes a line at the right hand side of the cave, starting with both hands in a juggy break. A brutal move off an undercut with the left hand leads to hitting a sloper that you can only move off if you utilize a foot lock and keep good body tension. You need to bring your left hand into an undercut in order to rearrange your right on the sloper to an unhelpful crimp. Next, you need to swap your feet into a toe-hook. Once you have brought your left hand up to the next edge you can release the toe-hook, paste your foot back on and campus to the finishing jug. This was definitely the hardest move for me as it required a lot of power in the arms, with not much help available from the foot holds. Clever Beaver has seven moves and requires five foot changes.

The first time I tried Clever Beaver I couldn't do any of the moves. A year later, I returned with much more confidence and worked the problem except for the last move. It was difficult for me to reach the last set of holds from the floor so I had to build a 'cheat stone' with the pads. No matter how hard I tried I could not latch that last hold so I grudgingly backed off the problem.

Figure 6.10 Naomi Buys on Clever Beaver (Font 7a+/7b), Parisella's Cave, the Great Orme, Llandudno, Wales (photograph courtesy of Alex Messenger)

Living in Burnley, I was facing a three-hour drive to the cave for this problem. I wanted to make sure I ticked it next visit. We set aside a day to return and I trained specifically for that last campus move by doing pull-ups on my fingerboard and practising dynamic moves at Boulder UK (my local climbing wall). The day came and I was so tired I couldn't lift my arms above my head! Thankfully Jordan completed his project so it wasn't a wasted trip.

Next opportunity to attempt the problem came in a more relaxed setting. We went to the International Meet, hosted at Plas y Brenin in North Wales, and were rained off on the first day. I jumped at the chance to sneak to the shelter of Parisella's. With no pressure, I did the whole problem to the last move on my first attempt. I then worked the last move and did it! All I had to do now was link it. Unfortunately, the day was very damp and the holds became too slick with condensation. I had to leave the problem for yet another day.

I didn't get a chance to try the problem again for over a month as I was busy training for a World Cup bouldering competition in Italy (this involved working hard boulder problems at indoor walls, mostly about 8–12 moves long). I finally got the chance to go in June 2006. All the training paid off and I completed the problem – it did take me about 20 attempts though! Altogether, it took 30 seconds to climb Clever Beaver, but many hours were involved in making it happen.

activities requires further investigation, although it appears that activities with the highest intensity levels and the shortest durations increase the likelihood of breath-holding during exercise.

Cardiovascular response to exercise begins before the exercise commences with an anticipatory rise in heart rate and blood flow. Research suggests that the rate of the anticipatory rise in heart rate is related to the intensity of the exercise ahead. The greater the intensity of the exercise, the higher the anticipatory heart rate response. With regard to adventure sports, the anticipatory rise in heart rate for power and power endurance activities would be higher than that for aerobic endurance sports. The cardiovascular responses to exercise involve alterations in cardiac output and blood flow.

The heart's output, introduced in Chapter 4, is the sum of stroke volume and heart rate. In other words, the cardiac output is determined by how much blood is ejected from the heart with each contraction and how many times the heart beats per minute. Increases in heart rate are initially stimulated by the sympathetic division of the central nervous system and subsequently by the joint action of the CNS and the hormones adrenalin and noradrenalin. Before exercise intensity reaches around 50 per cent of $\dot{V}O_{2\,max}$, increases in cardiac output are brought about through rises in the heart rate and stroke volume. Increases in the contractility (force of contraction) of the myocardium, as well as increased filling and preload brought about by the Frank–Starling mechanism, results in rises in stroke volume (Chapter 4). Adrenalin and noradrenalin reinforce the CNS-driven increases in stroke volume by stimulating the increases in myocardial contractility. Above 50 per cent of $\dot{V}O_{2\,max}$, further increases in cardiac output occur through rises in heart rate to a maximum of around 220–age for each individual.

Increases in blood flow to the muscles are brought about through an increased venous return and by alteration in blood circulation. At rest, muscles receive around 15 per cent of the total blood flow; during exercise this can increase to 80 per cent of total volume. The onset of exercise increases the skeletal pump, thereby increasing venous return and total blood flow. Central nervous and hormonal stimulation jointly bring about vasoconstriction and vasodilation to redirect blood flow to the working muscles. Vasodilation is stimulated in the arteries supplying the working muscles, the brain and the heart during exercise, whereas vasoconstriction is increased in the visceral arteries (supplying the organs of the abdomen). Reduced blood flow to the stomach and the intestine inhibits digestion, while the reduction to the kidneys results in a drop in urine production, which helps to maintain hydration and consequently blood volume.

Fatigue during exercise concerns a drop in expected or potential power output for any given activity. There are a variety of mechanisms which bring about fatigue during exercise. The nature of fatigue for any adventure sport is determined by the intensity and duration of the activity: the energy demands it places on the body. Due to the very short 1–2 s duration of power adventure sports or movements, such as a dyno in climbing, fatigue is not a reason for failure during execution. In these types of activities it is a lack of muscular strength, power or skill that causes failure rather than the onset of fatigue.

Maximal power outputs are normally reached within 3–5 s of all-out effort. As the duration of an explosive power activity nears 10 s, the intensity of exercise will begin to diminish. The performer, unable to sustain the intensity of exercise, will begin to fatigue. During maximal power endurance activities lasting from up to 10 s and beyond 30 s, a number of factors either acting individually or in combination may be responsible.

In power endurance activities, the exercise intensity necessitates a high rate of ATP hydrolysis and re-synthesis primarily through the ATP-PCr reaction, but also through the adenylate kinase reaction which was described above. As the duration of maximal exercise passes peak power output, at around 5 s of activity, the PCr stores begin to diminish, Pi begin to accumulate and, despite the positive contribution of the adenylate kinase reaction, ADP concentrations increase. In addition, it is believed that the high intensity of exercise during power endurance activities leads to a breakdown of intracellular Ca^{2+} transport within the muscle fibre. The accumulation of Pi and breakdown of Ca^{2+} transport have been linked to a disruption in the excitation of muscle fibres required for muscular contraction. Accumulation of ADP within a muscle fibre has been shown to inhibit ATP hydrolysis and the decrease in PCr stores associated with maximal activity results in a drop in the rate of ATP re-synthesis. The presence of all these potentially fatiguing metabolites within the muscle fibre makes it difficult to pin-point the exact mechanisms behind fatigue. Currently, however, the decrease in PCr stores is thought to be the primary mechanism for fatigue in power endurance activities.

Key point

Even before the start of exercise, the body makes a number of alterations in functioning to maintain homeostasis. The alterations in function, driven by sympathetic nervous stimulation, result in enhanced transport and exchange of metabolites, along with increased fuel substrate availability. The increase in CO_2 and H^+, resulting from the commencement of exercise, are detected by chemoreceptors within the body and bring about increases in the rate and depth of breathing. There are a number of cardio-vascular responses to exercise that serve to improve circulation. Heart rate and cardiac output increase in response to nervous system stimulation, resulting in an increase in blood flow to the working muscles. A combination of vasoconstriction and vasodilation serve to direct blood away from areas such as the digestive tract towards the active muscles. Fatigue in power and power endurance activities is believed to be associated with decreases in PCr concentration and disruption to Ca^{2+} transport during exercise.

Delayed onset of muscle soreness (DOMS)

The exercise adage of 'no pain, no gain' can be evidenced immediately after activity or as a delayed response. The immediate feeling of fatigue and pain after exercise results from the accumulation of exercise metabolites such as hydrogen ions in the muscles and, in the case of rock climbing, sail pumping and intense weight-training, the pump caused by blood pooling in the working muscles. These feelings normally subside within minutes or an hour or two after exercise. Intense exercise, particularly the eccentric contractions associated with running downhill in fell running or a mountain marathon, or resulting from weight training where a relatively heavy load is taken through a full range of motion (as with bench press or squats) can, however, result in a time-delayed muscle soreness known as delayed onset of muscle soreness (DOMS). DOMS normally occurs with 1–2 days of the exercise and lasts for up to a week post-exercise. During an occurrence of DOMS, an adventure sports performer might experience the acute muscle soreness normally associated with exercise which would dissipate in the normal time frame (1–2 hours after exercise). However, after an asymptomatic period (often the day following exercise) the effects of DOMS appear.

As already mentioned, DOMS appears to occur more frequently following eccentric muscular contractions, but is also associated with an increase in training volume and intensity, returning to training after a break or the commencement of a weight training programme. The symptoms of DOMS include muscle stiffness, soreness and often leave the areas affected feeling tender to touch. These symptoms tend to peak at onset, dissipating over the following days. The exact cause of DOMS is undergoing further investigation; however, the two prime mechanisms appear to be muscle tissue damage and inflammation around the affected areas.

Evidence of DOMS can be assessed through microscopic examination of tissue samples or through assessment of creatine kinase and the presence of myoglobin in the blood. DOMS can result in the rupture of the sarcolemma (the muscle cell membrane) causing the contents to leak, which is why myoglobin and creatine kinase can be detected in the blood. In cases of DOMS, microscopic muscle tissue examination reveals sarcolemma rupture as well as possible tearing of the Z-discs – the links between neighbouring sarcomeres and damage to the myofibrils.

The muscle fibre micro-trauma occurring in DOMS induces an inflammatory response in the area of tissue damage. Blood analysis during DOMS reveals elevated leukocyte (white blood cell) levels, an immune system response to the trauma and subsequent inflammation. Granular neutrophils and macrophages (which form as a result of monocytes i.e. agranular leukocytes) move into the area of tissue trauma. Neutrophils release oxidants (free radicals) that attack invading micro-organisms and release proteins which have an antibiotic effect in the damaged cells, while macrophages consume the cellular debris resulting from the trauma.

Key point

Free radicals (described in more detail in Chapter 2) can be detrimental to our health. However, those released by neutrophils as part of immune response serve a beneficial function. It is excess levels of free radicals, particularly those introduced into the body through smoking and air pollution, that can have a negative effect on our health.

Research indicates that the muscle micro-trauma and inflammatory response associated with DOMS has a negative effect on maximal strength while its effects remain with the body. These effects are thought to occur as a direct consequence of the tissue trauma and interference with the excitation-contraction coupling for the motor unit within the traumatized area. Research also indicates that fast twitch muscle fibres, closely associated with maximal contractions and expressions of power and strength, are more susceptible to the effects of DOMS. For sub-maximal exercise however, there appears to be no negative physiological effect on performance, although the work is perceived as harder by athletes due to the soreness and stiffness.

Antioxidant supplementation, particularly vitamin E, has been suggested as beneficial in recovery from DOMS, avoiding tissue structure damage associated with excess free radicals released by the neutrophils at the trauma site. The main recommendation for recovery from DOMS, however, is light exercise and stretching of the affected area. Once the muscle cells have repaired themselves, there appear to be some benefits from an episode of DOMS. Research indicates that an occurrence of DOMS is a precursor to subsequent gains in strength associated with muscular hypertrophy (increase in muscle size). In addition, a single incidence of DOMS appears to protect the body from subsequent occurrences during a particular phase of training. Although the effects of DOMS can be reduced through training programme management, the commencement of a new programme, new phase of training or new form of exercise often appears to result in an episode of DOMS.

Muscular cramps

Muscular cramps, painful abnormal muscular contraction of a single muscle group, are not fully understood by researchers. For adventure sports performers where the legs are either highly active during the activity (walking, climbing, etc.) or held stationary for long periods (kayaking and canoeing), the muscle groups in the upper and lower leg are particularly prone to muscular cramps. The original theory as to the cause of muscular cramps was based around fluid and electrolyte imbalances within the body, associated with exercise in the heat for a long duration. The incidence of muscular cramps during periods of rest and when fluid and electrolyte levels were balanced has led to further research in the area.

Muscular cramps are though to be triggered by an irritation to a muscle group. The irritation results in the hyperexcitability of the motor neuron(s), innervating that motor unit and bringing about an uncontrolled muscular contraction (cramp). Irritations that can trigger cramp include muscle overuse associated with fatigue, fluid imbalances, electrolyte imbalances (particularly potassium), holding a position for too long (such as sitting in a kayak or kneeling in a canoe) and the associated lack of blood flow to the muscle group. The immediate treatment for muscular cramp is to stretch the muscle and, if appropriate, the replacement of fluids and electrolytes. The most successful preventative measures in the incidence of cramp appear to be the implementation of a regular stretching programme as part of a cool-down after training, activity or competition and maintenance of electrolyte levels, particularly potassium. Regular stretching of the commonly affected muscle groups does appear to be effective in the prevention of cramp.

Key point

Delayed onset of muscle soreness occurs 1–2 days after training and symptoms remain up to a week post-exercise. The stiffness, soreness and tenderness of muscles brought about through DOMS is thought to occur due to muscle tissue damage and inflammation around the area of trauma. Light exercise and vitamin E supplementation are thought to be beneficial in recovery from DOMS. Muscle cramps, sudden painful muscular contractions are thought to occur due to irritation of muscle fibres. Sources of irritation include dehydration, electrolyte imbalances, remaining in one position for too long or a lack of blood flow to the affected area. Following a regular stretching regime, along with maintaining electrolyte balance particularly through potassium intake, appear to be the most beneficial methods for decreasing the incidence of muscular cramp.

6.5   Physiological adaptations to power training

Training to improve power and power endurance requires a combination of sprint (high-intensity short-duration activity) and strength training. Research clearly indicates that these forms of training result in improvements in performance, as shown through an increase in speed, power or strength following this type of training. There is generally a common agreement between physiologists regarding the adaptations to endurance training; however, research findings in the lesser-studied field of anaerobic (including power and power endurance activities) metabolism are more equivocal. The physiological adaptations in response to sprint and strength training described in this section represent those that are more widely supported in the literature.

As with other aspects of anaerobic physiology, there is a greater divergence in views as to the mechanisms behind anaerobic adaptations to training. The reason behind the divergence perhaps relates to the reliance on animal studies and the reliability of making inferences about human metabolism from animal studies, greater variation in the types of training undertaken and the use of a wider variety of assessment instruments than are commonly employed for aerobic adaptation studies. An example of this can be found in the uncertainty that exists regarding the notion of human muscle fibre *hyperplasia*. Hyperplasia refers to an increase in the number of muscle fibres in response to sprint and resistance training. Research with animals has identified incidences of both hypertrophy (increase in muscle fibre size) and hyperplasia in response to training. The application of this finding to humans is yet to find common acceptance among physiologists, for whom muscle hypertrophy is the most widely-accepted consequence of sprint and resistance training.

The more commonly agreed adaptations to sprint and resistance training include increases in; muscular strength and size, ATP, PCr and creatine stores, ATP-PCr enzyme activity, neuromuscular functioning, efficiency of movement, ligament and tendon strength, bone mineral content and fat-free mass as well as an associated decrease in body fat percentage. Sprint and resistance exercise leads to a greater recruitment of Type II muscle fibres than aerobic exercise and, as a consequence, this type of training results in an increase in the cross-sectional area of Type II muscle fibres. Type I fibres also increase in size through power and power endurance training, but to a lesser extent. It is thought that the increases in muscle fibre cross-sectional area relate to increases in the size and number of myofibrils along with associated increases in the number of myosin and actin filaments along each sarcomere. In addition, research suggests that there is a gradual shift of a small percentage of Type I fibres to Type II fibres, which serves to improve performance.

There appears to be a general agreement regarding the increase in activity of the phosphagen system enzymes, creatine kinase and adenylate kinase in response to sprint and strength training. As was described earlier in this chapter, creatine kinase catalyses the reaction to catabolize PCr to form ATP and adenylate kinase (sometimes referred to as myokinase) is the enzyme that catalyzes the formation of ATP and AMP from two ADP molecules. Theses mechanisms serve to maintain the supply of ATP during power endurance activities. Sprint training particularly has been found to enhance creatine kinase and adenylate kinase activity, resulting in improved power endurance performance. Results regarding increased ATP and PCr stores in response to training remain equivocal, although research identifying differences between endurance and power athletes revealed higher resting muscle stores of these substrates in power activity performers.

Strength gains through resistance training can be up to 25 per cent or more, dependent upon the training status of the individual and the type of programme followed. With regard to strength training, the use of low repetitions and higher relative loads is associated with greater gains in strength, while higher repetitions and lower relative loads results in greater incidences of muscle fibre hypertrophy. In the early stages of a strength training programme, research indicates that gains are more related to improvements in neuromuscular functioning (coordination of movement) than to hypertrophy. Beyond 8–12 weeks, the strength gains appear to result from improvements in contractile force and muscle fibre size. The neuromuscular adaptation associated with sprint and resistance training results in an improvement in muscle fibre recruitment and improved efficiency of movement. Neuromuscular adaptation occurs more quickly for simple one-limb movements such as that encountered in a bicep curl than for more complex and technical movements such as a power clean, sprint action or a 'windmill dyno' in climbing.

Key point

Anaerobic training for power and power endurance result in physiological adaptations which chiefly concern muscle and energy systems. Anaerobic training, primarily sprint and resistance training have been shown to increase muscle fibre size (mainly in Type II fibres) and muscular strength. Resting stores of ATP, PCr and creatine along with improved ATP-PCr system enzyme activity (creatine kinase and adenylate kinase), have been shown to improve in response to anaerobic training.

6.6 Nutritional ergogenic aids to power performance

In its broadest sense, an ergogenic aid describes any substance taken into the body that improves performance. In a sporting context, this has led to individuals taking supplements that are illegal and despite the short-term benefits for performance are damaging to long-term health. This section in each chapter will focus on those ergogenic aids that are not harmful to long-term health, legal

for sports performance and have well-conducted research supporting their ergogenic properties. There are far fewer nutritional supplements that do have an ergogenic effect on performance than the supplements industry would have us believe.

The basic nutritional strategy for performance in any adventure sport, regardless of the intensity and duration of the activity, should be to follow a healthy balanced diet. Details of the components of a healthy diet are included in Chapter 2. The key aspect is to plan your diet to maximize the *naturalness* of your food (minimizing processing and cooking as much as possible), have *variety* and eat with *moderation*. For power and power endurance adventure sports performers, it is also beneficial to monitor protein intake and increase this above the levels for normal population. As described in Chapter 2, protein intake for adventure sports performers should be increased to between 1.0–2.0 g per kg of bodyweight. For those involved in the power adventure sports, or for those wishing to develop their power for performance, daily intake should be closer to the 2.0 g per kg of bodyweight per day. With a balanced healthy diet, it should be possible to supply all the required protein within your normal meals.

There are many thousands of reportedly ergogenic aids that athletes from any sport can include in their diet to improve performance. Some of these work; however, there are far fewer products available that have good quality research behind them to justify the claims made by the manufacturers. The nutritional supplement industry is worth billions of pounds worldwide, yet in most cases improvements in our basic diet can outstrip any manufactured supplement. There are some ergogenic aids, the focus of this section in each chapter in Part II, that do appear to work and through carefully controlled research have been shown to improve performance. For power, power endurance, anaerobic endurance and indeed high-intensity intermittent exercise performance (Chapters 7 and 8) supplementation with creatine does appear to have an ergogenic effect on performance.

Creatine

Creatine is a naturally occurring compound contained in meat and fish or synthesized within the body from the amino acids glycine, arginine and methionine. The majority (over 90 per cent) of the body's stores of creatine are held within muscle fibres. The body has a natural turnover rate of about $2 \, g \, day^{-1}$. A normal diet provides around $2 \, g \, day^{-1}$ and 2 g, in the catabolized form of creatinine, are excreted from the body within urine. During normal conditions, the body's levels of creatine therefore remain constant. Vegans and vegetarians take in much smaller volumes of creatine in their diet, so rely on endogenous synthesis of creatine and excrete less creatinine in their urine.

Once inside the muscle fibre, creatine readily combines with phosphate to create phosphocreatine (PCr). The importance of PCr for high-intensity exercise is described in detail in the section on the phosphagen system earlier in this chapter. Phosphocreatine represents the body's immediate source of re-synthesis of ATP during high-intensity exercise. Consequently, increasing stores of PCr has been linked to improvements in performance in strength, power, power endurance, anaerobic endurance and high-intensity intermittent activities. Research by Harris *et al.* (1992) revealed that PCr and creatine stores within the muscle can be increased through supplementation of dietary creatine. A loading dose of $20 \, g \, day^{-1}$ for 6 days resulted in an increase of 20 per cent in total muscle creatine, of which one-third was converted to PCr. It is believed that the increased PCr stores result in an increase in stores for high-intensity exercise, thereby delaying reliance on fast rate glycolysis and delaying fatigue in power endurance activities.

Since these findings were published, and in light of the success of British athletes such as Sally Gunnell and Linford Christie at the 1992 Olympic Barcelona Games who reportedly trained using creatine, the supplement has been the focus of numerous subsequent studies. Research findings have indicated improvements in tests relating to strength, power, anaerobic endurance and repeat sprint laboratory test performance. Studies by Balsom *et al.* (1993) and Casey *et al.* (1996) reported improvements in performance after creatinine supplementation for $10 \times 6 \, s$ sprints with a 30 s recovery (delayed effects of fatigue) and $2 \times 30 \, s$ sprints with 4 min recovery (increased peak power and total work) respectively. Although some studies have not found supportive results, the majority report improvements during laboratory-based tests. Less clear-cut results have been found in performance-based tests. It appears that the closer the required power output is to an individual's maximum, or when high-intensity repeat activity is required, the better the impact on performance. In swimming for instance, even in short (25 m) sprints, perhaps due to the highly technical nature of the activity, some studies have reported no improvement in performance through creatine supplementation. In repeat sprint swimming ($10 \times 50 \, m$), however, creatine supplementation has been reported to result in an improvement in performance as measured through swimming velocity. This perhaps has important implications for the high-skill sports that are the subject of this textbook.

In terms of loading doses for creatine supplementation, a study by Hultman *et al.* (1996) indicated that either a low dose sustained approach or initially high loading dose followed by a lower maintenance regime can be followed with equal success. A loading regime of $20 \, g \, day^{-1}$ for 6 days followed by $2 \, g \, day^{-1}$ for the remaining 22 days resulted in a very similar increase in total muscle creatine stores when compared to a $3 \, g \, day^{-1}$ regimen.

In recent years, research has begun to indicate that the benefits and increased storage decrease over time. In response to this, Harris *et al.* (1992) suggested interruptions in use to better maintain the increased creatine storage benefits of supplementation. Currently, after 12 weeks use of creatine during training, a wash-out period of 4 weeks (non-creatine supplementation phase) appears to be beneficial for maintenance of above-baseline creatine stores. The effect of such a regime, along with the effectiveness of creatine for high-intensity adventure sports, requires further research.

Creatine is a legal supplement that appears to improve performance in power, power endurance and anaerobic endurance. There is, however, an ethical or moral if not legal decision which an adventure sports performer must take in deciding to supplement with creatine. Depending on the views of the individual athlete, creatine supplementation goes against the spirit of the World Anti-Doping Agency (WADA). Anti-doping policy is defined as a substance that is taken in abnormal quantities with the aim of improving performance. The choices an athlete has to make with regard to any supplementation are very personal and can be illustrated by the case of one British athlete who won a bronze medal in the 1999 Judo World Championships. The British under 63 kg player, Karen Roberts, made the decision not to take creatine despite its ergogenic effects due to a personal ethical standpoint. For Roberts, in a competitive sport context, the use of creatine was cheating. Each athlete has to make up their own mind about the use of creatine or any other ergogenic aid. Research appears to suggest that there are no long-term health effects of taking creatine. To date, however, there has not been a longitudinal study regarding the long-term effects and it is worth bearing this in mind when making a decision to use creatine as a supplement. The use of a wash-out phase, as was described above, would appear to lessen any potentially harmful effects.

Key point

The most important dietary strategy for any adventure sport performer is to adopt a healthy balanced diet. This means maximizing the naturalness and variety in your diet as well as following a principle of moderation. For power and power endurance activities, the use of creatine supplementation could be considered to improve performance. Creatine has been shown to improve strength, power and power endurance. If using creatine, it is important to include

a wash-out phase in any supplementation strategy.

Getting into research Long-term creatine intake (Vandenberghe *et al.*, 1997)

Creatine appears to provide a nutritional supplement that has an ergogenic effect for power and power endurance adventure sports. In addition, creatine has been linked with increases in fat-free mass. Vandenberghe *et al.* investigated the effects of long-term creatine supplementation on resistance (strength) training and intermittent exercise. The study was conducted with 19 female sedentary volunteers, to examine the effects of long-term creatine usage on performance and body composition.

The study had three phases: an initial 4 day loading phase where only creatine was supplemented; a 10 week training phase where strength training was conducted in addition to low dose creatine supplementation; and finally a 10 week detraining phase where the participants continued to take creatine after ceasing strength training. The study was double-blind and the participants were randomly assigned to either creatine or placebo group.

In the loading phase, participants in the creatine group took 5 g creatine four times per day (20 g per day) for four days. The placebo group followed the same regime, but took maltodextrine tablets in place of creatine. The doses were taken in tablet form and were designed to look and taste the same. During the 10 week training phase, the dosage for both groups was dropped to 2.5 g twice a day (5 g per day) for the course of the study. During this phase, all participants followed a strength training programme that involved six exercises; leg press, bench press, leg curl, leg extension, squat and shoulder press. The load for each session was calculated as 70 per cent 1 RM values, with 1 RM reassessments after 5 and 10 weeks of this phase.

In the detraining phase, the participants (of which there were 13 who continued in the study: seven creatine and six placebo) stopped training but continued taking the lower dose of creatine (or placebo) for the additional 10 weeks of detraining. Creatine and ATP concentrations were determined by use of nuclear magnetic resonance (NMR) spectroscopy of the right gastrocnemius muscle. The assessment involved an intermittent exercise protocol of five sets of 30 arm flexions with a 2 min rest period between each set. The torque during each repetition was recorded. Body composition was assessed using underwater (hydrostatic) weighing at the start of the study and 5 weeks and 10 weeks after training.

The participants reported no side-effects and felt they could not guess whether they were taking creatine or the placebo. The 4 day loading phase led to a 6 per cent increase (from 22.5–24.2 mmol kg wet wt) in PCr stores which remained elevated to a similar level during the training and detraining phases. The 1 RM lifts total for the placebo group was 465 kg at the start of the study and rose to 553 kg at 5 weeks and 606 kg by the end of the training phase (10 weeks). This represented an initial increase (from start to week 5) of 88 kg and

a total increase (from start to week 10) of 141 kg i.e. a 30 per cent increase. The total lifts for the creatine group rose from 481–612 kg and finally to 711 kg during the same time phases, representing a 48 per cent increase over baseline levels.

In response to the intermittent exercise testing, the 10 week creatine and exercise programme led to 11, 18, 20, 21 and 25 per cent higher scores for the five sets of lifts when compared with the placebo group. The fat-free mass of the participants rose for both groups but was higher for the creatine group. At 5 weeks there was a 2.5 per cent increase in fat-free mass for the placebo group, and a 4.5 per cent rise for the creatine supplement participants. At 10 weeks these values were 3.7 and 5.8 per cent, respectively.

These results indicate that long-term low dose creatine supplementation maintained PCr levels about 6 per cent above baseline levels after being established by a higher dose loading phase. The low dose and training phase led to improvement in performance (1 RM value and intermittent arm-flexion torque) for both creatine and placebo groups. However, the rise was significantly higher for the creatine group. The detraining phase led to similar losses in strength. However, the PCr levels remained elevated for the creatine supplementation group. The higher PCr levels returned to baseline within 4 weeks of cessation of creatine supplementation. The study presented evidence that creatine supplementation enhanced muscle strength development during strength training for sedentary females.

6.7 Summary and study questions

Muscle fibre types

Very few adventure activities fall into the absolute power category, but many activities will include power during part of a more sustained activity. To understand the differences and adaptations to training in power activities, it is essential to understand the kinds of muscle fibre that make up human skeletal muscles. While these are often divided into Type I and Type II, there are in fact important subdivisions. Type II fibres can be divided into three further types designated by physiologists as Type II_a, Type II_b and Type II_c. In fact, in human muscle tissue, the vast majority of Type II muscle fibres are Type II_a and Type II_b (referred to in this textbook as Type $II_{b(x)}$, as most textbooks refer to these fibres as Type II_b even though they are more correctly Type II_x).

Type I fibres are described as slow oxidative fibres and rely on aerobic energy pathways. Type $II_{b(x)}$ fibres generate greater forces than Type I fibres and use the anaerobic glycolytic energy pathway which means they fatigue more quickly. Type II_a fibres are an intermediate fibre, also referred to as fast oxidative glycolytic fibres. Type II_a fibres use both oxidative or glycolytic pathways for energy production, falling between Type I and Type $II_{b(x)}$ fibres for speed and force of contraction.

When muscles are required during exercise, fibres are recruited in order. Type I fibres are recruited first. If the intensity is too high for the Type I fibres to cope alone, then Type II_a fibres are also recruited. For efforts close to or at maximum, Type $II_{b(x)}$ fibres are also recruited. Generally, no more than 70 per cent of the available muscle fibres will be recruited even for maximal efforts. This reserve is believed to be a protective mechanism that prevents injury during maximal contractions.

While all human beings contain all muscle fibre types, the exact proportion of each type of fibre differs both between individuals and between muscles in different parts of the body. Postural muscles tend to be high in Type I fibres and muscles in the upper limbs contain proportionally more Type $II_{b(x)}$ fibres. Athletes tend to have a distribution of muscle fibres appropriate to the demands of their activities. This is probably due both to individuals selecting activities to which they are naturally physically suited and also due to changes within muscles that are exposed to particular activities and training.

The phosphagen system

The main energy source for power-based explosive activities is the phosphagen system. The main energy source for cells in the body is adenosine triphosphate (ATP) which can be broken down to release energy. Adenosine is a nucleoside, and bonds with phosphates to become adenosine monophosphate (AMP), adenosine diphosphate (ADP) and adenosine triphosphate (ATP). The second and third phosphate bonds store the energy required for cell metabolism. In fact, there are several other nucleosides that are also capable of bonding with phosphate and therefore storing energy; however, adenosine is the nucleoside responsible for most cell metabolism including muscular contraction.

ATP is degraded to ADP when one phosphate bond is broken. It is still possible at this stage to separate another phosphate ion to produce more energy and AMP. For activities that last longer than 1–2 s, it is necessary to replenish the ATP stores and the body has two ways of achieving this. Phosphocreatine (PCr) stores provide the most important and immediate mechanism for the re-synthesis of ATP. ATP can be replenished very quickly from the breakdown of PCr stores in a reaction catalysed by the enzyme creatine kinase. The creatine stores are approximately four times the size of the ATP stores, allowing athletes to perform for up to 10 s before this source of energy is depleted. Creatine can be synthesized from amino acids and be consumed in the diets of those who eat meat or fish. Additional creatine supplementation is common among power athletes.

As the PCr stores become depleted, the levels of ADP present in the muscle cells will begin to rise. This rise in ADP triggers two further energy system responses. Firstly, as ADP levels rise, the enzyme adenylate kinase splits a second phosphate ion from ADP and bonds it to another ADP molecule, creating one AMP molecule and one replenished ATP molecule available for energy production. The increasing presence of AMP acts as a trigger which stimulates an increase in the rate of glycolysis. The second reaction, while not providing further energy, is coupled with the adenylate kinase reaction. As levels of AMP increase through the adenylate kinase reaction, the amine component of AMP is catabolized with water to produce inosine monophosphate (IMP) and an amine group which forms ammonia (NH_3), quickly becoming an ammonia ion (NH_4^+). This reaction, catalysed by the enzyme adenylate deaminase, is irreversible so catabolized AMP cannot be replenished and is removed from the muscle fibre. It occurs to reduce the build-up of AMP within the cell. This process will normally take place when a maximal effort moves toward 45–90 s.

Physiological responses to power and power endurance activities

Regardless of the intensity or duration of activities, the nervous and endocrine systems alter the functioning of the respiratory and cardiovascular systems to meet the general demands of exercise. The sympathetic nervous system will increase the rate and depth of respiration as well as increasing heart rate and triggering the release of exercise-related hormones alongside an inhibition of the organs of the digestive system and kidneys. The endocrine system provides a longer-term effect on metabolism, blood flow, the availability of glucose and the catabolism of fats and amino acids.

During the onset of exercise, the levels of carbon dioxide (CO_2) and hydrogen ions (H^+) increase and the level of oxygen (O_2) decreases. These changes are detected by chemoreceptors in the brain, aorta and carotid arteries and relayed to the respiratory driver. The increase in CO_2 is the main driver for increases in respiration. Cardiovascular responses to exercise begin prior to the exercise itself in an anticipatory response which is related to the intensity of the expected exercise. Increases in cardiac output up to exercise intensity levels of 50 per cent of $\dot{V}O_{2\,max}$ are due to increases in heart rate and stroke volume. Once 50 per cent of $\dot{V}O_{2\,max}$ has been reached, further increases in cardiac output are due to increasing heart rate.

The increases in cardiac output during exercise are also accompanied by alterations in the blood circulation that favour the muscles. At rest, muscles receive 15 per cent of the blood flow but this can increase to 80 per cent during exercise. The working muscles also act as a pump, increasing venous return. Vasodilation and vasoconstriction are used to target blood flow to the active muscles. The reduced blood flow to the kidneys means that hydration and therefore blood volumes are maintained during prolonged exercise.

Fatigue

Fatigue during exercise occurs when a drop in expected or potential power occurs during an activity and fatigue during adventure sports can arise through a number of mechanisms, depending on the nature of the activity. In efforts of less than 2 s, fatigue will not normally be a factor in failure. However, as the effort approaches 10 s and beyond then fatigue can become a factor in failure. The decrease in PCr stores and accumulation of specific breakdown products are believed to disrupt muscular contraction during power endurance activities.

The immediate feeling of fatigue and pain after exercise is due to the accumulation of exercise metabolites and is often referred to as a 'pump'. Further pain and discomfort can occur later (within 1–2 days), particularly if eccentric contractions have been involved in an activity. This time-delayed muscle soreness is known as delayed onset muscle soreness (DOMS) and can last up to a week. DOMS is due to both inflammation and muscle tissue damage at a cellular level. While peak strength is reduced during DOMS, there seems to be an adaptive response and strength levels can increase after recovery.

Cramp

The mechanism that results in muscular cramps is not fully understood. Originally it was believed that imbalances in fluid and electrolyte levels were responsible, but it is now thought that a number of sources of irritation with a muscle group can trigger cramp. The irritation results in hyperexcitability of the motor neurons and uncontrolled contraction. Stretching and replacement of fluid and electrolytes is the most common treatment.

Adaptations to power training

Power and power endurance performance are best improved through a combination of high-intensity, short-duration activity and strength training. While performance improvements are clear, the mechanisms behind these improvements are often debated. While physiologists agree that hypertrophy, an increase in muscle fibre size, occurs as a result of power training there is much debate about the possibility of hyperplasia, an increase in the number of muscle fibres. Muscle size, strength, energy stores, enzyme driven activity and efficiency are known to increase in response to sprint and strength training. Muscle fibres can also, to a limited extent, change their characteristics to suit the demands of training. Neuromuscular adaptation and increased efficiency are two of the earliest changes to take place when a training programme is adopted; improvements in muscle fibre contractile force and size occur later.

Ergogenic aids

Nutritional and ergogenic aids are commonly available for strength and power training. However, the number of substances shown to have a beneficial effect is actually much smaller than you might expect, despite the claim of many commercial supplement suppliers. The basis of nutrition for performance activities should come from a balanced and healthy diet.

Creatine is one of the few legal supplements that has been shown to improve performance during power and power endurance activities. Many studies have shown improvements in controlled laboratory settings, but the usefulness of creatine for many real-world activities is still equivocal. Performance in activities involving power and power endurance close to the individual's limits is most likely to improve through creatine supplementation.

Study questions

1. How would you define power and power endurance? Describe an adventure sport example of each of these.

2. Why do we generally consider human muscles to consist of three main types of muscle fibre? Describe the characteristics of these three types of human muscle fibre.

3. What is the controversy over designating the third common group of fibres in human muscle 'Type II_b'?

4. Describe the mechanisms responsible for energy production in activities of up to 10 s duration.

5. Why is the breakdown of ATP to ADP very different from the breakdown of AMP to IMP?

6. List the hormones released by the endocrine system at the onset of exercise. What are the key roles of these hormones during exercise?

7. Why are blood CO_2 levels important during exercise? What is the role of the respiratory driver?

8. What are the most likely explanations for fatigue of the muscle fibre during power endurance activity?

9. What is DOMS? Explain why DOMS occurs and its significance for physical activity and training.

10. Describe the muscular adaptations to sprint and resistance training. What is meant by hypertrophy and hyperplasia?

7

Anaerobic endurance: the lactate tolerance and management activities

7.1 Introduction

The specific physiological demands of sports that predominantly rely upon glycolytic energy production are described in this chapter. Examples of the sports within the anaerobic endurance category are shown in Table 7.1. The relative duration of each upon a duration continuum is shown in Figure 7.1. The relative intensity of each of the adventure sports in this chapter would be near the maximum level possible, given the duration of each activity.

When an adventure sports performer exercises at maximal levels the initial source of energy comes from ATP stores in the muscles and then from energy production via the phosphagen system, described in Chapter 6. If these two sources of energy describe what was referred to in the introduction to the energy systems (Chapter 4) as the 1st gear for human energy production, then glycolysis represents the 2nd gear. The reactions within glycolysis commence with the onset of exercise. However, it takes several seconds for ATP to be realized through the glycolytic pathway. From this point onwards, glycolysis becomes increasingly important and soon the predominant source of energy for continued high-intensity exercise. For events lasting between 10–90 s, glycolysis provides the key mechanism for energy production. The role of glycolysis is not, however, limited to the first 90 s of high-intensity activity. Glycolytic metabolism continues to make a significant contribution to total energy supply during events lasting up to 10 min and for the high-intensity bursts during intermittent adventure activities such as those described in Chapter 8. The glycolytic pathway provides a fundamental system for energy production for adventure sports.

During high-intensity exercise exceeding 10 s, a by-product of glycolytic metabolism is lactic acid. Lactic acid, or lactate as it is often referred to, changes the pH of the cell in which it is produced. The enzymes of glycolysis – the catalysts of the reactions in glycolysis that help to drive the process – work best within a specific pH band (discussed in more detail later in this chapter). As the lactate levels increase, they impact on the pH levels in the cells and these can inhibit the continued functioning of glycolysis. For the adventure sports covered in this chapter, such as climbing routes, competitive climbing, deep water solo, skiing or snowboarding runs, short wildwater kayak races, short kayak/canoe slalom races and the 200 m and 500 m sprint kayak/canoe races, the

Adventure Sport Physiology   Nick Draper and Chris Hodgson
© 2008 John Wiley & Sons, Ltd

Table 7.1   The anaerobic endurance adventure sports

Water-based activities	Land-based activities
Short wildwater kayak races	Longer powerful bouldering routes
200 m and 500 m kayak/canoe sprints	Competition climbing
Slalom kayak/canoe races	Sport climbing routes
	Deep water solo climbing routes
	Ski or snowboard runs

management of and ultimately tolerance of lactate is a key aspect of performance. The time duration of these activities means that glycolysis provides the predominant energy source for performance and the high intensity will result in significant rises in lactate above resting levels.

This chapter begins with a more detailed description of energy production through the glycolytic pathway and discusses the nature of and impact of lactate on performance. The nature of fatigue in anaerobic endurance activities is described, with the response of the body to this type of exercise. We review training and testing for anaerobic endurance activities. Specific nutritional and ergogenic aids for anaerobic endurance activities are also considered. In addition, there are performer contributions from Nick Smith and Katherine Schirrmacher on slalom canoeing and sport climbing and a summary of a research paper

by Booth *et al.* regarding the energy cost of sport climbing. There are a small number of adventure sports (rock climbing, kayaking, skiing and sailing) where there have been a number of studies on the physiology of each activity. Where such research is available, a review is made within the most relevant chapter. Specific physiological findings for rock climbing and snowsports are presented here.

7.2   Glycolysis

Glycolysis provides the predominant fuel source for events lasting 10–90 s. Beyond 90 s, glycolysis continues to play a significant role in energy production for high-intensity activities lasting up to 10 min. The glycolytic pathway forms the second of the body's anaerobic energy systems, meaning that it does not require oxygen for any of the

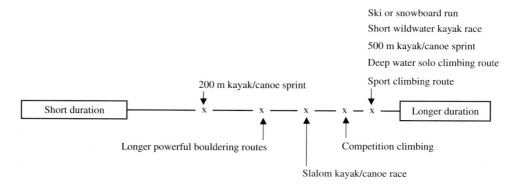

Figure 7.1   The duration continuum for anaerobic endurance adventure sports

reactions that take place during glycolysis. Glycolysis serves two functions: to produce *ATP* to maintain supplies for exercise and to produce *pyruvate* for aerobic metabolism. The reactions or steps in glycolysis are shown in Figure 7.2 and the chemical structure of each of the reaction products is illustrated in Figure 7.3.

As a simple consequence of the number of reactions, it takes time for the glycolysis to become the predominant energy system during high-intensity activity. After the first few seconds of activity, ATP becomes available through glycolysis and it begins to replace the phosphagen system as the predominant energy production mechanism. The pathway in glycolysis was first described by Gustav Embden and subsequently demonstrated by Otto Meyerhof (two German physiologists – Chapter 1) and consequently is sometimes referred to as the Embden–Meyerhof pathway. Glycolysis describes the degradation of the carbohydrates *glucose* and *glycogen* (glycogen initially through glycogenolysis which is described below) to pyruvate through a series of ten steps or reactions (eleven for glycogen) to produce a net gain of two ATP molecules for each glucose molecule. The reactions for glycolysis, and indeed glycogenolysis, take place within the cytosol, the gel-like mass that surrounds the organelles within the cytoplasm (Chapter 3), where all the reagents are stored for use during exercise. Textbooks vary as to whether they refer to the cytosol or the cytoplasm as the site for glycolysis. However, the authors are referring to the same location as the cytosol forms part of the cytoplasm.

In Chapter 4, there was a brief introduction to, and overview of, the energy systems for adventure sports. Two forms of glycolysis were referred to – *slower rate* and *fast rate* glycolysis – the rate being dependent upon the intensity of exercise. These two rates of glycolysis are sometimes referred to as *anaerobic* and *aerobic* glycolysis. However, by employing this system of reference to describe these differential rates of glycolysis, there is a possibility of misinterpretation. Glycolysis is an anaerobic process; it does not require oxygen for any of the ten reactions involved in the production of ATP. The reference to anaerobic and aerobic refers to the fate of pyruvate, the end product of glycolysis (in addition to ATP), and this is determined by the intensity of exercise being performed. During high-intensity or fast rate glycolysis the fate of pyruvate is lactic acid – a process that again does not require oxygen and is therefore anaerobic – hence anaerobic glycolysis. During lower intensity exercise or slower rate glycolysis, pyruvate and hydrogen ions can be utilized within the aerobic system for further energy production which requires oxygen and has therefore been referred to as aerobic glycolysis. The possible fates of pyruvate are shown in Figures 7.2 and 7.4. To avoid this possible confusion the terms slower rate and fast rate glycolysis have been used throughout this textbook.

Key point

Glycolysis involves the degradation of glucose and glycogen to form pyruvate with the net gain of two (three for glycogen) ATP molecules. Glycolysis has two functions: to produce ATP to continue anaerobic metabolism and to produce pyruvate for aerobic metabolism. The glycolytic pathway involves ten reactions which take place within the cytoplasm. During high-intensity exercise, the fate of pyruvate is lactic acid which ultimately contributes to the limitations in duration of exercise. When exercising at a lower intensity, pyruvate can be utilized for aerobic metabolism and lactate levels do not rise to levels that limit the duration of performance.

Glycogenolysis

As was described in the section on carbohydrates in Chapter 2, glucose is stored in long highly branched glucose chains called glycogen. Glycogen is formed through the process of *glycogenesis*. Glycogen is stored in the liver and muscles with the body retaining around 500 g during rest, 100 g in the liver and 400 g in the muscles. During

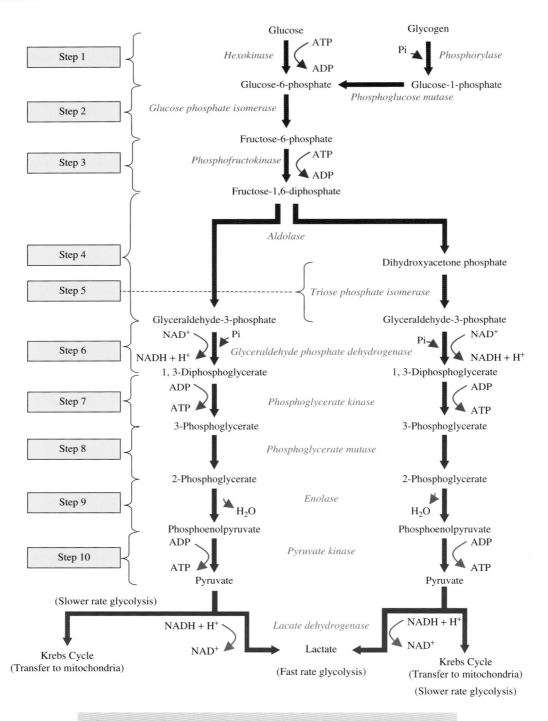

Figure 7.2   The steps and reactions in glycolysis and glycogenolysis

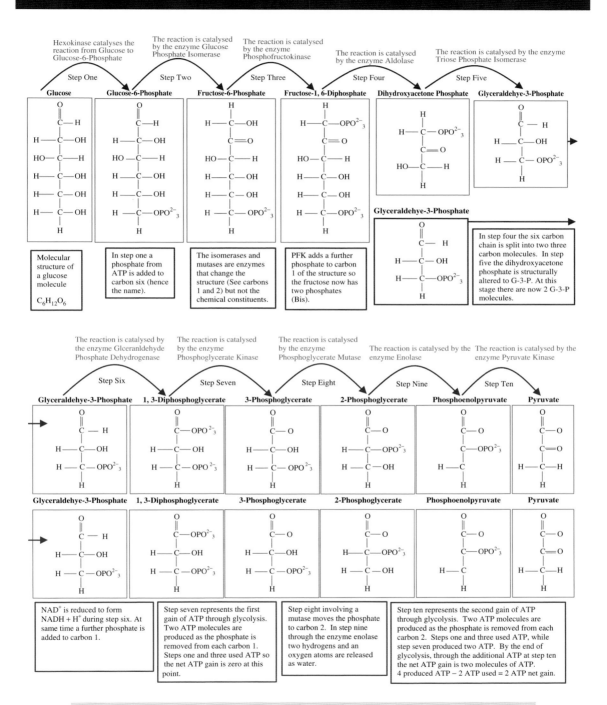

Figure 7.3 Chemical composition and structure of the components within glycolysis

exercise, and specifically for glycolysis, glycogen has to be catabolized to release the stored glucose molecules. The process of glycogen breakdown is called *glycogenolysis* and involves two reactions to convert glycogen to glucose-6-phosphate which can enter the glycolytic pathway. The reactions within glycogenolysis are shown in Figure 7.2. In the first step of glycogenolysis, glycogen is converted to glucose-1-phosphate through a reaction catalysed by phosphorylase during which a glucose molecule is detached from the glycogen chain and a phosphate added to the glucose-based structure. In the second step, catalysed by the enzyme phosphoglucose mutase, the phosphate is moved from the first carbon to the sixth carbon resulting in the formation of glucose-6-phosphate which can then enter at step two of glycolysis.

Key point

Glycogen, the storage form of glucose has to be broken down into single carbohydrate units before it can be utilized within glycolysis. Glycogenolysis describes the two reactions necessary to convert glycogen to glucose-6-phosphate which can enter the glycolytic pathway.

The glycolytic pathway

The ten steps of glycolysis begin with glucose and end with the formation of two pyruvate molecules for each one of glucose, along with a net gain of two ATP molecules. Although not as rapid as the phosphagen system, glycolysis can provide energy for exercise at a much faster rate than through aerobic means, which is essential for the continuation of high-intensity exercise after PCr stores begin to diminish. The reactions or steps within glycolysis are shown in Figure 7.2. In addition, the chemical composition of the resultants of each reaction is shown in Figure 7.3. It is worthwhile, for a clear understanding of

glycolysis, to look at each step described here and shown in Figure 7.2 and then to refer to what happened chemically to the structure in Figure 7.3.

The language used to describe each of the resultants can appear complex. However, when reference is made to Figure 7.3, the names are generally very logical. For instance the first step in glycolysis (from glucose to glucose-6-phosphate), catalysed by the enzyme hexokinase, involves the addition of a phosphate group to the sixth carbon of glucose. The name of the product of the reaction is therefore very appropriate. In addition, it is worth noting at this point that the enzymes that end in *mutase* and *isomerase* catalyse reactions that change the structure of the substance without adding anything new, whereas those ending in *lase* or *nase* such as aldo*lase* and phosphofructoki*nase* (PFK) are responsible for catalysing reactions that add or remove a chemical component to the structure. For example, in the reaction catalysed hexoki*nase*, a phosphate group is added to glucose in the first step of glycolysis and phosphoglucose *mutase* catalyses the movement of phosphate from carbon 1 to carbon 6 during the formation of glucose-6-phosphate in the second reaction of glycogenolysis. With this knowledge, and referring to both the text and Figures 7.2 and 7.3, it should be possible to achieve a clear understanding of each of the steps in glycolysis.

The ten steps of glycolysis are as follows.

1. The conversion of glucose to glucose-6-phosphate, is catalysed by the enzyme *hexokinase*. Through the breakdown of ATP to ADP, a phosphate is added to carbon 6 of glucose. This results in a net loss of ATP as a result of the first step (net ATP = -1). The second step in glycolysis rearranges the chemical structure of glucose-6-phosphate and is consequently catalysed by an isomerase.

2. Glucose-6-phosphate is altered to fructose-6-phosphate, an isomer (i.e. has the same chemical composition but a different structure), catalysed by the enzyme glucose phosphate isomerase.

Fate of pyruvate during slower rate glycolysis:

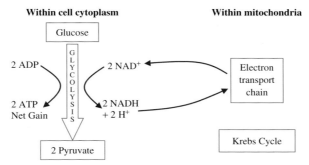

Fate of pyruvate during fast rate glycolysis:

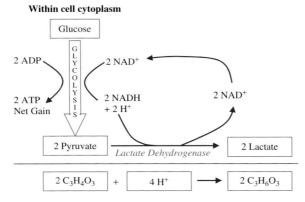

Figure 7.4 Fate of pyruvate during fast and slow rate glycolysis

3. A second phosphate is added to the structure from the breakdown of a further ATP. This reaction is catalysed by the enzyme phosphofructokinase (PFK) which is also an important agent in the control of the rate of glycolysis. The reaction results in the phosphorylation of fructose-6-phosphate to fructose-1, 6-diphosphate (Figure 7.3) which, as the names suggests, has a phosphate group attached to carbon 1 and 6. The further use of ATP in the phosphorylation of the original glucose molecule means that, at this step, the process of glycolysis results in a net loss of two ATP (net ATP = -2). So far, glycolysis has cost energy and not created it.

4. The enzyme aldolase catalyses the splitting of the hexose (six carbon chain) into two triose (three carbon chain) molecules, each of which contains a phosphate. The two trioses are structurally different due to the locations of the phosphates. In fructose-1, 6-diphosphate the phosphates were attached to carbons 1 and 6. The splitting of fructose-1, 6-phosphate leads to a difference in the location of the phosphate for the two trioses. Dihydroxyacetone phosphate, formed from the upper portion of the fructose-1, 6-phosphate, has the phosphate group attached to its first carbon. However, glyceraldehyde-3-phosphate has the phosphate located on the third carbon.

5. Dihydroxyacetone phosphate is restructured to form another glyceraldehyde-3-phosphate. The enzyme triose phosphate isomerase

catalyses the fifth reaction in glycolysis, the conversion of dihydroxyacetone to glyceraldehyde-3-phosphate. This reaction is necessary because only glyceraldehyde-3-phosphate can participate in the subsequent reactions of glycolysis. Once two molecules of glyceraldehyde-3-phosphate have been created, the sixth step of glycolysis takes place.

6. A further phosphate is bonded to the first carbon of each triose to form 1,3-diphosphoglycerate, catalysed by the enzyme glyceraldehyde phosphate dehydrogenase. In addition, two hydrogen atoms are detached from the structure and reduce nicotinamide adenine dinucleotide (NAD^+) to $NADH + H^+$. Nicotinamide adenine dinucleotide is a co-enzyme carrier (a derivative of niacin – one of the B vitamins) which transports hydrogen to either the mitochondria for use within aerobic metabolism or deposits them with pyruvate to form lactic acid (see the section on lactate below and Chapter 9). In summary, step 6 involves the addition of a further phosphate and the removal of two hydrogen atoms from each of the trioses, resulting in the formation of 1,3-diphosphoglycerate, a reaction that is catalysed by glyceraldehyde phosphate dehydrogenase. Up until the sixth step of glycolysis, there continues to be a net loss of two ATP (net ATP = −2).

7. The first production of ATP from glycolysis occurs at this step. Phosphoglycerate kinase catalyses the phosphorylation of the phosphate from carbon 1 to ADP. This results in the creation of an ATP from each of the trioses. The energy balance for glycolysis after this step is therefore two ATP consumed (during steps 1 and 3) and two ATP produced (net ATP gain = 0).

8. The remaining phosphate in 3-phosphoglycerate (the product of step 7) is transferred, in a reaction catalysed by phosphoglycerate mutase, from carbon 3 to carbon 2. The product of this reaction is unsurprisingly 2-phosphoglycerate.

9. In the penultimate step, 2-phosphoglycerate is dehydrated (an H_2O molecule is removed from each of the trioses) in a reaction catalysed by enolase to form phosphoenolpyruvate.

10. Pyruvate kinase catalyses the phosphorylation of a further ADP for each of the trioses, resulting in the production of two further ATP and two pyruvate molecules. This results in a net gain of two ATP for glycolysis (net ATP gain = +2). The fate of pyruvate at the end of glycolysis is determined by the exercise intensity. During fast rate glycolysis, pyruvate is converted to lactate which will ultimately limit the duration of exercise. During slower rate glycolysis, pyruvate becomes a substrate for aerobic metabolism.

Key point

The glycolytic pathway involves 10 steps in the degradation of glucose to pyruvate. The ten steps are illustrated in Figures 7.2 and 7.3. Steps 1 and 3 each use one ATP to form the products of their reactions. After the split of the hexose (6 carbon chain) glucose derivative at step 4, two triose (3 carbon chain) molecules are formed. As a result of the reactions in steps 7 and 10, two ATP are produced for each triose resulting in a gain of four ATP. The net gain of ATP through glycolysis is two ATP. The second product of glycolysis, pyruvate, will either be used as a substrate for aerobic metabolism or be converted to lactate. The fate of pyruvate is determined by the intensity of exercise.

Control of the rate of glycolysis

The rate of glycolysis can be inhibited or increased through the presence of a variety of substances within the cytoplasm. Control over the rate of glycolysis is essential to ensure ATP production via the glycolytic pathway meets the needs of the cell. As the duration of high-intensity exercise nears and then passes 10 s of activity, the need for ATP through glycolysis is increased. As was discussed in Chapter 6, at around 10 s or so the PCr concentrations drop and ADP levels rise. In response to this, the adenylate kinase reaction and adenylate deaminase reactions are stimulated. The adenylate kinase reaction creates an ATP and an AMP molecule from two ADP molecules. If high-intensity exercise is maintained for a longer duration (up to 90 s of activity, such as for the duration of a kayak sprint race), the rises in AMP stimulates the adenylate deaminase reaction which results in the conversion of AMP to IMP and NH_4^+. The rises in ADP and Pi concentration through the catabolism of ATP during exercise, the increases in AMP though the adenylate kinase reaction and the rises in NH_4^+ through the adenylate deaminase reaction all stimulate the rate of glycolysis. High levels of ATP, PCr, citrate and glucose-6-phosphate inhibit the rate of glycolysis.

The mechanism for the control of glycolysis is through the functioning of key glycolytic enzymes. The key rate limiting enzymes in glycolysis are hexokinase and PFK. The hexokinase reaction (the first step in glycolysis) is slowed in response to an increase in glucose-6-phosphate levels (the product of the first step). The major enzyme limiting the rate of glycolysis is PFK. Specifically, increases the presence of ADP, AMP and Pi serve to stimulate PFK activity and increase the rate of glycolysis. The presence of ATP, citrate and H^+ serve to inhibit the functioning of PFK and slow the rate of glycolysis. Citrate, produced in the mitochondria (Chapter 10), leaks in to the cytoplasm when aerobic metabolism is meeting the demands of exercise. The presence of citrate in the cytoplasm decreases the rate of glycolysis by inhibiting PFK activity.

Key point

The rate of glycolysis can be increased and decreased according to the needs of the cell. The stimulation or inhibition of glycolysis is brought about through the presence of key substances on the enzymes that catalyse the reactions in glycolysis. The presence of ATP, PCr and citrate slow the rate of glycolysis, while increased concentrations of ADP, AMP, Pi and NH_4^+ stimulate glycolytic energy production.

Lactate

During lower-intensity exercise, NAD^+, the hydrogen carrier for glycolysis, can keep pace with the release of hydrogen atoms removed from glyceraldehydes-3-phosphate in reaction 6. As can be seen from Figure 7.4, under conditions of slow rate glycolysis, mitochondrial oxygen supply is sufficient to enable NAD^+ to shuttle hydrogen atoms to the mitochondria for aerobic metabolic consumption. During high-intensity exercise, ATP phosphorylated through glycolysis is required at a faster rate than oxygen can be supplied to the mitochondria. As a consequence, the hydrogen atoms are transferred to pyruvate in a reaction catalysed by lactate dehydrogenase to form lactic acid. Although the formation of lactic acid and the concomitant rise in H^+ results in a drop in cytoplasmic pH, a major component of fatigue during high-intensity exercise, without this mechanism the rate of glycolysis would be impaired more quickly. In other words, the formation of lactate during high-intensity exercise enables NAD^+ to continue to function as a hydrogen atom carrier for glycolysis to sustain anaerobic performance.

As can be seen from Figure 7.4 during slower rate glycolysis, the end product of the breakdown of glucose is pyruvate. At lower intensities of exercise, the pyruvate can be utilized in aerobic metabolism to form additional ATP molecules. The fate of pyruvate during high-intensity exercise, as shown in Figure 7.4, is lactate. The terms *lactic acid* and *lactate* are normally used interchangeably, although it is lactic acid that is produced when the hydrogen atoms are bonded with pyruvate. The chemical formula for lactic acid is

$C_3H_6O_3$, but once formed lactic acid quickly dissociates to lactate, a salt of lactic acid. Lactate is formed when lactic acid releases a hydrogen ion to form $C_3H_5O_3 + H^+$. As lactic acid very quickly dissociates to lactate, both terms are used to describe the product of fast rate glycolysis.

Lactate concentration can be measured relatively simply in the laboratory and the field from a capillary blood sample. The processes for lactate sampling in the laboratory and field are illustrated in Figure 7.5. Research indicates that beyond an initial threshold (Chapter 9), lactate concentration

(a)   (b)   (c)

(d)   (e)

Figure 7.5 Laboratory and field measures of capillary lactate concentration: **laboratory analysis** depicting (a) blood sampling, (b) introduction of the sample for analysis and (c) printout of result; **field analysis** showing (d) blood sample being taken from a climber and (e) readout obtained in the field

levels rise linearly with increasing exercise intensity. The rises in lactate concentration can be used to indicate the glycolytic contribution to energy production. Information regarding lactate concentration levels can be useful for training and performance in adventure sports where fast rate and slower rate glycolysis make a significant contribution to energy production. Examples would include kayak and canoe sprints, kayak and canoe slalom, sport climbing and high-intensity endurance activities such completing mountain or kayak marathons.

The fate of lactate

Frederick Hopkins and Walter Fletcher provided evidence in 1907 of the relationship between muscle activity and lactate production (see also Chapter 9). They also demonstrated that skeletal muscles were able to remove lactate from the muscle. Later Carl and Gerty Cori, who worked with Otto Meyerhof on the identification of the pathway in glycolysis, demonstrated one fate of lactate: its use in the re-synthesis of glycogen in the liver. The husband and wife team demonstrated that lactic acid is not an end product of glycolytic metabolism but an energy intermediate for the re-synthesis of glucose and glycogen.

Since the determination of lactate as an energy rich by-product of glycolysis, researchers have gone on to identify a number of routes through which the energy stores in lactate can be used for metabolic purpose. A major fate of lactate is conversion to glucose and subsequently glycogen in the liver. This process, known as the *Cori cycle* after Carl and Gerty Cori who first proposed it, is shown in Figure 7.6.

The process of glucose re-synthesis from lactate which occurs in the Cori cycle is one form of *gluconeogenesis* i.e. the creation of glucose and glycogen from non-direct dietary carbohydrate. The bodily mechanisms for glucose creation are described in the next section. Through the Cori cycle, lactate and indeed pyruvate produced in the muscles can be transported to the liver, taken

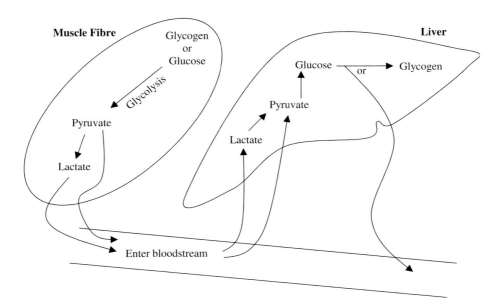

Figure 7.6 The Cori cycle: pyruvate or lactate are diffused into the bloodstream and transported to the liver, where they can both be converted to glucose to re-enter the bloodstream for transport to other cells in the body, or into glycogen to replenish liver glycogen stores

up and converted to glucose and then stored as glycogen. Research has shown that the Cori cycle provides a major substrate for glycogen creation in the liver, second only to the absorption of dietary blood glucose.

The Cori cycle provides a mechanism for lactate conversion to glycogen in the liver. However, lactate is also processed for use within muscle fibres themselves. Through *intercellular shuttling*, lactate is transported from Type $II_{b(x)}$ (fast twitch) muscle fibres, which are major producers of lactate during high-intensity exercise, to the highly oxidative Type I fibres (slow twitch). Within Type I fibres, catalysed by the enzyme lactate dehydrogenase (LDH, the same enzyme involved in the reversible reaction to create lactate), lactate is oxidized to pyruvate and subsequently converted to acetyl CoA for use within aerobic metabolism.

More recently, a second form of *intracellular shuttling* has been suggested within muscle fibres. Researchers have identified the presence of LDH in the mitochondria of skeletal and cardiac muscle, suggesting the ability of muscles to convert lactate to acetyl CoA within their own mitochondria. These two mechanisms establish the importance of muscle fibres as not only producers but also consumers of lactate. Within the body, lactate can be shuttled to mitochondria within the producing fibres and shuttled to neighbouring muscle fibres or to the liver for conversion to an energy substrate. In addition, *cardiac muscle* (the heart) can utilize lactate directly as a substrate for glycolysis. This mechanism is particularly important during sustained high-intensity exercise when glucose concentrations may fall as a result of the high consumption rate.

Key point

Lactic acid, a product of fast rate glycolysis, quickly dissociates in the cytosol to lactate. Lactate is not a waste product of glycolytic metabolism. There are four key fates of lactic acid: (1) the heart can use lactate as a direct substrate for metabolism, otherwise it can be (2) shuttled to neighbouring Type I muscle fibres within the muscle where it was produced or via the blood to other non-working muscles, (3) possibly oxidized within the mitochondria of a producing muscle or (4) transported to the liver for conversion to pyruvate or glucose.

Gluconeogenesis

Gluconeogenesis, the creation of glucose from non-dietary carbohydrate sources, provides an important series of mechanisms for glucose recreation, particularly during long-duration activities or when dietary carbohydrate intake is reduced. *Neo* (new) in gluconeogenesis can be helpful in distinguishing between this and other similar terms such as glycogenesis, glycogenolysis and indeed glycolysis. (Recall that *glycogenesis* involves the formation of glycogen from glucose, *glycogenolysis*, as in hydrolysis, refers to the breakdown of glycogen to glucose and *glycolysis* is the degradation of glucose – and glycogen – to produce pyruvate and ATP).

Non-dietary sources of glucose for gluconeogenesis include: *lactate* and *pyruvate* converted to glucose through the Cori cycle as described above; *amino acids* through the alanine-glucose cycle described in Chapter 2 and through the glutamine-glucose cycle, a similar deamination process; or *glycerol*, the carbon-based backbone to which fatty acids bond to form triglycerides. The liver plays a key role in gluconeogenesis and is the site for the Cori cycle, alanine-glucose cycle and conversion of glycerol to glucose. Additionally, a small proportion of total gluconeogenesis takes place within the kidneys (around 10 per cent). The glutamine-glucose shuttle is an example of a gluconeogenic mechanism that takes place within the cortex of the kidneys. Gluconeogenesis is stimulated by the adrenal cortex hormone cortisol, the adrenal medulla hormones adrenalin and noradrenalin and by the pancreatic hormone glucagon.

7.3 Physiological response to anaerobic endurance activities and fatigue

The general metabolic, cardiovascular and respiratory response at the onset of exercise was described in Chapter 6. In response to the commencement of anaerobic endurance activity, stroke volume, heart rate and respiratory rate rise to meet the demands of exercise. The predominant energy system for adventure activities lasting up to 90 s is glycolysis. The high-intensity level of exercise in the anaerobic endurance activities covered in this chapter means that aerobic metabolism will play a lesser, but increasing role in energy production. As a consequence, the active muscles will rely on the re-synthesis of ATP through fast rate glycolysis. The physiological consequence of fast rate glycolysis is the production of lactic acid, which quickly dissociates to lactate and free hydrogen ions. The high rate of ATP hydrolysis and reliance on fast rate glycolysis in short-duration high-intensity adventure sports results in an accumulation of lactate and H^+. The accumulation of lactate and H^+ ultimately leads to fatigue in anaerobic endurance events. However, the muscle fibres have a number of intermediary mechanisms through which the onset of fatigue can be delayed.

In Chapter 9 the focus is upon high-intensity aerobic endurance activities, which due to their high intensity and relatively short duration continue to receive a significant proportion of the energy for ATP re-synthesis through fast rate glycolysis. Examples of such activities include 1000 m kayak sprint races, downhill mountain bike races and short-duration wildwater races. Indeed, some of the highest lactate levels recorded for adventure sports are found in these types of activity. As a result, the implications of lactate on performance and its role as a mechanism for fatigue in some aerobic endurance events is as relevant to those involved in high-intensity short-duration aerobic activities as it is to those involved in anaerobic endurance events.

The cut-off point between aerobic and anaerobic endurance activities in this book is 90 s as from this point onwards, the aerobic system is increasingly predominant for ATP re-synthesis. However, it is important to recognize that this is an artificial division and the transition to predominantly aerobic metabolism exists on a continuum. The predominant fuel for ATP regeneration in high-intensity activities over 10 s increasingly becomes fast rate glycolysis. In anaerobic endurance events and short-duration aerobic events, the accumulation of lactate and hydrogen ions plays a significant role in fatigue.

In anaerobic endurance activities, the relatively short duration means that performers are focused primarily on maintaining power output and therefore *lactate tolerance*: how much lactate are they prepared to accept? In the relatively shorter duration aerobic endurance activities, the focus becomes *lactate management*: push too hard too early and performance becomes detrimentally affected. An optimal level of exercise intensity must be achieved; one which will result in high lactate levels but allow successful completion of the activity. In longer duration high-intensity aerobic activities, beyond 8–10 min, the focus becomes how hard the athlete can push before they cross the *lactate threshold* and begin to accumulate lactate. Longer duration high-intensity endurance adventure sports performers attempt to stay as close to the lactate threshold as possible to achieve optimal performance. The lactate threshold, as it applies to high-intensity aerobic endurance events, is discussed in Chapter 9.

As was described in the above section on the fate of lactate, it is not a waste product of metabolism. Lactate is an intermediary energy substrate that can be used directly by the cardiac muscles of the heart and may, to a limited extent, provide an immediate energy source within the mitochondria (intracellular shuttle). Lactate can be oxidized to regenerate glucose through the Cori cycle and intercellular shuttling. During exercise, the fast rate of glycolysis necessitates an immediate system for managing the lactate and H^+ produced during metabolism. The body's buffering mechanisms provide a temporary measure for neutralizing the acidic effects of lactic acid produced during fast rate glycolysis.

The scale for measuring the relative acidity or alkalinity of a solution is pH. The pH scale was established by the Danish biochemist Søren Sørensen (1868–1939) where the p relates to the power or concentration of H ions. The scale ranges from very acidic solutions such as hydrochloric acid (found in the gastric fluids of the stomach, Chapter 2) which has a pH of 1.0, to highly alkali solutions such as sodium hydroxide (found in drain cleaner) with a pH 14.0. Figure 7.7 provides an illustration of the pH scale. Blood, with a normal resting pH of 7.35, has a narrow band of tolerable limits in pH from 6.9 to a maximum pH 7.5. The muscles, the lactic acid producers during exercise, can tolerate lower pH levels down to 6.6 from a normal resting pH of around 7.1.

Acids, such as lactic acid produced during fast rate glycolysis, dissociate in solution to release their H^+ ions increasing acidity. *Bases* such as ammonia (NH_3) produced during the adenylate kinase reaction (Chapter 6) accept H^+ ions in solution to form hydroxide (OH^-) ions, thereby increasing alkalinity. The acid-base balance of solution depends on the concentration of hydrogen and hydroxide ions present. When $H^+ > OH^-$, the solution is acidic, $H^+ = OH^-$ the solution is neutral, and $H^+ < OH^-$ the solution is alkaline. The maintenance of blood pH within tolerable limits is essential to life. Acidosis arising in diseases such as diabetes and alkalosis in response to the low PO_2 at altitude are potentially fatal if left untreated. The body's buffering mechanisms are responsible for maintaining pH at rest and, importantly, during exercise. The buffers operate within all the body fluids including intracellular fluid (inside cells) and extracellular fluid (around cells and plasma, the fluid portion of blood).

In the body's cells, including muscle fibres, inorganic phosphate (Pi) form the primary buffer for lactic acid. The hydrolysis of ATP molecules for muscular contraction liberates inorganic phosphate which are then available to accept the H ions resulting from the dissociation of lactic acid to lactate and H^+. As can be seen from Figure 6.8, the catabolism of the final phosphate from ATP creates a HPO_4^{2-} i.e. a phosphate ion. The phosphate ion generated combines with an H^+ to form $H_2PO_4^-$ (dihydrogen phosphate) which prevents the pH of the intracellular fluid from rising. The phosphate ion has buffered the H^+ and decreasing its acidic affect. The equation for this reaction is:

$$HPO_4^{2-} + H^+ \rightarrow H_2PO_4^-.$$

Lactic acid produced during fast rate glycolysis also diffuses from the muscle fibres to the bloodstream where the primary buffer is carbonate. In this buffering reaction, the H^+ ions from lactic acid bind with bicarbonate ions to produce water and carbon dioxide. The equation for this reaction is:

$$HCO_3^- + H^+ \rightarrow CO_2 + H_2O.$$

The CO_2 produced during carbonate buffering is expelled from the body via the lungs through respiration. It is of interest to note that the CO_2

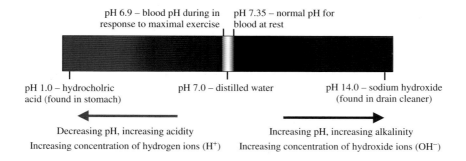

pH 6.9 – blood pH during in response to maximal exercise

pH 7.35 – normal pH for blood at rest

pH 1.0 – hydrocholric acid (found in stomach)

pH 7.0 – distilled water

pH 14.0 – sodium hydroxide (found in drain cleaner)

Decreasing pH, increasing acidity
Increasing concentration of hydrogen ions (H^+)

Increasing pH, increasing alkalinity
Increasing concentration of hydroxide ions (OH^-)

Figure 7.7   pH scale, demonstrating the narrow range of blood pH

produced through lactic acid buffering is the reason that RER values rise above 1.0 during maximal exercise, such as during a $\dot{V}O_{2\,max}$ test. The respiratory exchange ratio (Chapter 4) provides the proportion of the oxygen consumed to carbon dioxide produced during the catabolism of fat and carbohydrate. At rest and during low intensity exercise, fat provides the main fuel for ATP generation. As exercise intensity increases towards maximum, carbohydrate (through glycolysis) becomes the predominant fuel for ATP re-synthesis. The RER value for metabolism of fat is 0.70 while the RER for glucose is 1.0, indicating that one molecule of oxygen is used for every molecule of carbon dioxide produced. This represents the highest level of RER for fuel substrate catabolism. During a maximal test, when carbonate buffering is in operation to neutralize H^+ produced through fast rate glycolysis, additional CO_2 is produced which results in a rise of RER to $+1.0$. One indicator used by physiologists to show that a $\dot{V}O_{2\,max}$ test represents a true maximum for an athlete is to check the RER value to see if it has risen to a value of 1.15 or over. If so, carbonate buffering is active and a true maximum may well have been reached.

The kidneys play an important role in the maintenance of the acid-base balance within the body. The kidneys can filter H^+ ions from the blood and either excrete them from the body via the urea cycle or use them to create further bicarbonate ions for blood buffering. It is the kidneys that are responsible for the regulation of bicarbonate levels within the blood. Blood proteins and haemoglobin provide alternative buffers for hydrogen ions when the production of lactate exceeds the carbonate buffering mechanism capacity. During the latter stages of maximal intensity exercise, which lasts around 1–2 min (falling between anaerobic endurance and high-intensity short-duration aerobic endurance), the adenylate deaminase reaction (Chapter 6) stimulated by the increase in adenosine monophosphate (AMP) molecules begins to break down AMP to form inosine monophosphate (IMP) and the base ammonia (NH_3). The ammonia created also accepts hydrogen ions, binding with them before removal from the body via the kidneys. The equation for this reaction is:

$$NH_3 + H^+ \rightarrow NH_4^+.$$

Fatigue in anaerobic endurance activities occurs primarily through increases in H^+ concentration within the muscle fibre, inhibiting glycolysis and interfering with the excitation coupling mechanism during muscular contraction. Despite the effects of the body's buffering mechanisms during maximal intensity exercise, the levels of lactate and H^+ rise dramatically. It is currently thought that it is not the presence of lactate in the muscle fibre which causes fatigue, rather the presence of its dissociate partner H^+ which alters the pH within the muscle fibre. Research indicates that enzymes function optimally within narrow pH bands. The increasing acidosis within the muscle fibre appears to inhibit the functioning of enzymes such as PFK and therefore hinder the rate of glycolysis. In addition, the decreasing pH is thought to result in interruptions in the contraction of individual sarcomeres, thereby weakening muscular contraction. The feelings of fatigue that follow maximal anaerobic endurance exercise are also thought to result from the change in cellular pH. Chemoreceptors and nocioceptors within the muscle fibres detect the fall in pH and relay these as feelings of pain to the brain.

Although not thought to be directly responsible for fatigue in anaerobic endurance activities, the rise in blood lactate concentration that is concomitant with the drop in cellular pH is used by physiologists as a measure of exercise intensity. Resting blood lactate concentration is around $1.0\,mmol\,L^{-1}$ (produced by red blood cells whose only mechanism for energy production is through glycolysis). During maximal exercise, lactate concentrations can reach as high as $25–30\,mmol\,L^{-1}$ in motivated athletes. A typical lactate concentration for university physical education and adventure education students in the Wingate anaerobic test (30 s duration) is around $10–12\,mmol\,L^{-1}$. For dinghy sailing, lactate concentrations of between $4–8\,mmol\,L^{-1}$ during a demanding 40 min sailing test have been reported. In response to maximal kayak ergometer assessments, elite slalom, sprint and whitewater paddlers have recorded lactate concentrations from $5.3-13.3$ mmol L^{-1}. In competitive rock climbing, post-climb lactate concentrations in climbers at the 1993 World championships ranged from $6.4-7.4$ mmol L^{-1}.

Key point

The predominant energy system during the adventure sports which form the focus of this chapter is fast rate glycolysis. Fast rate glycolysis results in the accumulation of lactate and hydrogen ions which can alter cellular and blood pH. The presence of the cellular buffers phosphate and ammonia (produced during maximal exercise lasting over $1-1.5$ min through the adenylate deaminase reaction) and the blood buffer bicarbonate help to maintain pH within the body's internal environment.

Biography: Nick Smith

Adventure sport: C2 Slalom

Nick Smith started canoeing at the age of 12 with a local Scout group. He then joined Stratford-on-Avon Kayaks and moved into slalom kayak competition. At the age of 15 he tried his hand at C1 and C2 and quickly dropped kayaking altogether. Nick proceeded to move up the slalom divisions and after a successful summer season in Bourg St Maurice, moved to Nottingham to train more seriously for C1 and C2. He made the British World Cup Team the following year and then in 1996 he teamed up with Stuart Bowman in C2. The pair immediately hit it off, having previously been rivals in C1. In the spring of 1997 the pair made the decision to concentrate on C2 paddling and prepare for the 2000 Sydney Olympic Games.

Figure 7.8   Nick Smith (courtesy of Owen Hughes)

The pair qualified and finished in 4th place, just 0.54 seconds away from a bronze medal. Nick and Stuart have gone on to take medals at international level, finishing 9th at the Athens 2004 Olympic Games and 3rd in the 2006 World Cup at Augsburg, Germany.

The 2002 Canoe Slalom World Championships

The Canoe Slalom World Championships at Bourg St Maurice in the French Alps was always going to be one of the toughest races. Feeding from Val d'Isere and Tignes, the glacier-fed river creates one of the world's most exciting and demanding stretches of whitewater for canoe slalom competition.

Negotiating 20 gates on the rapids, a typical competition run takes around 120 s with extra time penalties added for gates touched. During qualification each boat has two runs, and these times and penalties are added to give a two-run total. Some boats are eliminated before racing the semi-finals on the next day, on a modified course. After this single run, only the top ten proceed to take their second run in the finals. I compete in double canoe (C2) with Stuart Bowman, but there are three other classes: ladies and men's kayak (K1) and single canoe (C1).

In recent years, artificial channels have been taking over natural rivers as prime sites for canoe slalom competition. Being purpose built, the rapids can be designed and tweaked with concrete rocks and moveable plastic obstacles to create the desired waves, eddies, stoppers and chutes of water. In addition, the banks, footpaths, changing rooms, car parking, etc. create a full purpose-built facility.

Racing at Bourg St Maurice, however, puts you out of your comfort zone! There's an element of the unknown on this grade 4 river. The rocks can be sharp and hazardous and not always visible, so there is a real danger of damaging your boat or paddle. I was offline once in training, and hit an underwater rock. It ripped a 10 inch hole in the hull and we just made it to the side before sinking!

Also there's the risk of being turned upside down and having to perform an Eskimo roll in the fast currents or, even worse, breaking or losing your paddle and being forced to swim out of your boat. Experimenting with one move in training resulted in a wave flipping our boat over. I was awkwardly positioned and it seemed to take me ages to get my paddle round to perform an Eskimo roll. C2 requires a coordinated rolling effort and when we righted the boat after two attempts, both of us were gasping for air and with the shock of the cold water. But this upped our confidence because we had survived the inevitable ducking and forged a greater relationship with the river gods!

The water flow on an artificial course can be controlled exactly. However, early summer snowmelt, despite being regulated with HEP schemes, meant variable water flow during training. Sometimes we trained on 50 cumecs (cubic metres per second), and at this level the river was seriously fast and pushy. Some people opted out! But the plus side was that back on the usual 35 cumecs, things seemed a lot easier. The race was held on 30 cumecs as the organizers feared too many couldn't cope.

The sheer power of the water means that it's all too easy to lose control, miss a turn or an eddy and be carried off downstream. This requires a significant change in technique and skill to judge the speed of the water to stay ahead of the game (interestingly, boat manufacturers had designed bigger volume boats this year specifically for this river). Many of us have spent previous summers here because of the beautiful Alpine environment and because nothing else compares to this slalom course! In the run up to the Worlds it was important to get enough time on the river so two comprehensive training camps were planned before the championships.

Bourg St Maurice is a very hard place to paddle, requiring excellent boat positioning to make maximum use of the power of the water. Brute strength and ignorance is not the way! Keeping the boat on line and with a smooth technique helps to conserve energy (mental and physical) for the latter part of the run, where aching arms and burning lungs can lead to costly mistakes.

The course that we race on (i.e. position and sequence of the gates) is designed and set only the day before the race. There is no practice so we must plan our race strategy by looking at the course from the bank and using our experience from training. The two camps allowed us plenty of water time to become accustomed to the rapids, get a 'feel' of the water and try as many moves as possible. Training sessions would involve progressing down the river in sections, concentrating just on one section and walking back to repeat, trying different techniques, as well as full-length competition simulation runs. Video analysis is used both for feedback during training and dissection and discussion afterwards.

There are additional considerations specific to this race site. The volume and speed of the whitewater makes itself known physically. The extra adrenalin, fast reactions, speed and power of paddlestrokes, combined with an altitude affect (850 m) result in extremely fatiguing paddling! We included running and mountain biking higher up in the ski resorts to help boost specific fitness and at home we incorporated repeated 4 min maximum efforts into our training to increase our $\dot{V}O_{2\,max}$.

The river water is icy cold, yet air temperature can easily be around 30°. We wanted to wear our normal summer short-sleeved cags but found we lost some feeling in our arms and there was a more rapid onset of lactic acid accumulation as they were getting continually wet. Our sponsor, PeakUK, designed some very thin long-sleeved neoprene tops that provided perfect insulation and maximum flexibility.

In relation to the heat, we were careful to keep well hydrated, using isotonic drinks during and after training. The team sports scientist kept check of our hydration levels by measuring body mass immediately before and after training, plus urine tests each morning.

There is a certain nervous exhilaration of canoeing here, whether training or racing, but here I was, on the start line of the World Championships. The river roared on, as usual, but crowds lined the banks, the commentary blared out and the TV cameras focused in. I could feel the heat of the sun and the cold chill of the river water, that same river that we had been technically and psychologically trying to master. On that day, we had the confidence and feel for the whitewater, were in a good physical state and had a good race strategy: the keys to performing at one's best.

Figure 7.9 Nick Smith (left) and Stuart Bowman competing at the 2005 World Championships in Sydney, where they finished seventh (photograph courtesy of Owen Hughes)

7.4 Physiological adaptations to anaerobic endurance training

Anaerobic training leads to improvements in glycolytic metabolic functioning and fatigue resistance. Interval training is perhaps the single most useful method for improving anaerobic endurance capacity. Through such training, the timing of exercise and recovery periods can be closely monitored and altered as training progresses. For anaerobic endurance performance, high-intensity exercise bouts with short recovery periods (not long enough to enable lactate clearance to near resting levels) are typically employed. The intensity and duration of exercise would be increased and the recovery duration decreased to maintain the training stress overtime. Anaerobic training, due to the recruitment of Type II fibres, will also lead to improvements in strength and fibre cross-sectional area (Chapter 6).

Anaerobic endurance training leads to greater glucose uptake during recovery and increased resting glycogen stores. The increase in glycogen stores will result in an increased anaerobic ability during exercise. A key mechanism for improvement in anaerobic endurance capacity relates to improved glycolytic enzyme activity. Just as sprint and strength training lead to improvements in phosphagen enzyme activity, anaerobic training such as interval training leads to increased enzyme activity such as phosphofructokinase (PFK) and hexokinase (HK). HK catalyses the first reaction in glycolysis where glucose is altered to glucose-6-phosphate and PFK catalyses the third step in glycolysis (fructose-6-phosphate to fructose-1,6-diphosphate). Both these reactions involve the addition of one phosphate (from the breakdown of ATP) to the resultant of the step.

Research indicates that improvement in enzyme activity for PFK and HK can be as high as 50 per cent above pre-training levels. Some studies have found additional improvements in the functioning of phosphorylase and lactate dehydrogenase (LDH), but these findings are less consistent. Phosphorylase is the catalyst for the initial reaction in glycogen breakdown in glycogenolysis, the equivalent of HK in glycolysis, while LDH is the enzyme responsible for catalysing the conversion of pyruvate to lactate at the end of glycolysis. Although the increases in glycolytic enzyme activity are not as substantial or as influential as realized through aerobic training, the increased glycolytic enzyme activity still results in improved force generation and the ability to sustain contractions during anaerobic endurance exercise.

Anaerobic endurance ability is improved after training through improved buffering capabilities within the blood and active muscles. Research indicates that buffering capabilities are improved by 25 per cent in response to anaerobic training. The net effects of the increased buffering capability will be to delay the onset of fatigue, thereby improving anaerobic performance. In addition, anaerobic training has been shown to lead to an increased tolerance of H^+ produced during fast rate glycolysis. Individuals who engage in anaerobic training develop an improved tolerance of H^+ accumulation as lactic acid dissociates during high-intensity exercise, thought to occur due to improved motivation or an ability to cope with the pain associated with decreasing pH. Lastly, and perhaps surprisingly, anaerobic training has been shown to result in small improvements in aerobic metabolism that help to delay fatigue. Anaerobic endurance training, such as interval sessions, has been shown to result in improved activity in Krebs cycle enzymes such as citrate synthase, succinate dehydrogenase and malate dehydrogenase (that catalyse steps 1, 7 and 9, respectively). Although these improvements are not as substantial as those realized through aerobic training, they still contribute to a reduction in the anaerobic load and increase time to fatigue during high-intensity endurance activities.

Key point

Anaerobic endurance training results in a number of physiological adaptations that serve to improve performance. Anaerobic training leads to improvements in glucose uptake during recovery, leading to improved glycogen storage and availability during exercise. For glycolysis and Krebs cycle enzyme activity are improved through anaerobic training. Research indicates that the increased activity of enzymes such as HK and PFK result in improved anaerobic performance. Although increased to a lesser extent than through aerobic training, improvements in Kreb cycle enzyme activity appear to reduce the anaerobic load, thereby improving performance.

7.5   Nutritional ergogenic aids to anaerobic performance

This section explains nutritional strategies or ergogenic supplements beneficial to performance in the adventure sports that are the focus of this chapter. In common with Chapter 6, the basic starting point of any attempt at dietary manipulation to improve performance should begin by examining normal dietary intake. Where possible, a *normal healthy balanced diet* including attention to moderation, variety and naturalness of foods will provide an excellent starting point for any adventure sports performer.

The main focus of ergogenic aids that have been developed to improve sports performance is upon maintaining energy delivery and delaying fatigue. For example, in Chapter 6 the use of creatine supplementation has been cited as an ergogenic aid due to its positive effect on total body creatine stores. Creatine offers a possible mechanism for improved performance in anaerobic endurance events and research findings are generally supportive of this potential benefit. Creatine, therefore, offers one possible supplement for improved performance in anaerobic endurance activities. With regard to delaying fatigue and maintaining performance, in light of the high levels of lactate and hydrogen ions produced during the high-intensity endurance activities that are the focus of this chapter, the potential benefits of a variety of pH buffering supplements are reviewed here.

Bicarbonate, phosphate and more recently β-alanine have been used in a variety of studies to increase cellular and blood pH and decrease the effects of hydrogen ion accumulation, the result of high-intensity exercise, fast rate glycolysis and the dissociation of lactic acid to lactate.

As described earlier, phosphate and bicarbonate buffering within the muscle fibres and blood present a short-term buffer to the decreasing pH associated with increasing hydrogen ion concentration during high-intensity short-duration exercise. If the concentrations of bicarbonate and phosphate within the body can be increased this should, in theory, result in improved buffering capability and improved performance. Researchers have investigated the effects of bicarbonate and phosphate loading on fatigue and performance in anaerobic endurance activities over many years. More recently, researchers have begun to examine the potential of β-alanine as a supplement to assist in pH buffering.

Sodium bicarbonate

Bicarbonate ingestion, in the form of sodium bicarbonate (baking powder), appears to have an ergogenic effect for anaerobic endurance, short-duration intermittent and high-intensity aerobic endurance activities. The use of sodium bicarbonate appears to assist performance by increasing extracellular pH and, as a result of the rise in blood alkalinity, stimulating an increase in shuttling of lactate and hydrogen ions from muscle fibres to the bloodstream. Research generally indicates that the ingestion of sodium bicarbonate leads to an increase in pH which serves to counteract the decreases in pH that occur during maximal exercise. With regard to improved lactate and hydrogen ion transfer from active muscle fibres, there appear to be pH reactive transport proteins that, in response to the rises in blood pH associated with bicarbonate ingestion, increase the rate of transport of lactate and H^+ into the bloodstream. Transport proteins, originally identified by Roth and Brooks in 1990, react to the difference in pH between the active muscle fibres and extracellular fluid which occurs during high-intensity exercise and move lactate and hydrogen ions from the areas of low pH (in the muscle fibres) to areas of higher pH (extracellular fluid including blood plasma).

The dual action of extracellular buffering and increased rate of lactate and H^+ transfer to the bloodstream is believed to assist in maintaining pH in muscle fibres and the blood, thereby delaying fatigue. The pH beneficial effects of bicarbonate buffering are also thought to lead to a decrease in feelings of fatigue (central fatigue), appearing to have an additional positive effect on time to fatigue. Research indicates that the effects of bicarbonate buffering are best observed in high-intensity activities lasting from 1–7 min including intermittent exercise. This has application in adventure sports that are the focus of Chapters 7–9.

The loading dose for best response needs to be 300 mg per kg bodyweight, which should be ingested 1–2 hr before exercise. There have been incidences of gastrointestinal upset (bloating, abdominal pain, cramps and diarrhoea) reported in response to sodium bicarbonate ingestion, but these effects can be lessened by taking the dose in 10 min intervals starting 2 hr before exercise. Used in this way, sodium bicarbonate supplement should present no major health risks and, indeed, is commonly used as a stomach antacid. Excessive or long-term use can present harmful effects including alkalosis and muscle contraction interference and should therefore be avoided. Further research for its application in adventure sports such as bouldering or climbing competition require further research although, as for creatine supplementation, its use is contrary to the spirit of the WADA regulations.

Phosphorus

The mineral *phosphorus* is found in the body as the salt phosphate. Most of the body's phosphate is combined with calcium to form calcium phosphate for use in teeth and bones, as described in Chapter 2. Phosphate can exit in its inorganic (Pi) form or combined with organic matter to form the organic phosphates found in ATP, PCr or the phospholipids that form the basis for cell membrane structure. *Phosphate loading* has been postulated as a mechanism for improving performance and decreasing fatigue through pH buffering, although the results of studies are far less convincing for phosphate than they are for bicarbonate. It is also suggested that increasing extracellular and intracellular phosphate concentration through supplementation increases the phosphagen available for PCr and ATP synthesis. There is, however, a lack of research evidence to support this theory.

The most likely ergogenic effect of phosphate to date appears to be its impact on 2,3-diphosphoglycerate (2,3-DPG). Produced in red blood cells, 2,3-DPG enhances the release of oxygen from erythrocytes. Research indicates that phosphate supplementation increases the synthesis of 2,3-DPG which results in an increase in oxygen release from the erythrocytes to the muscle fibres. The increase in oxygen release has been measured in improved oxygen uptake results, leading to decreased lactate accumulation in response to a given workload. Despite its role in cellular buffering described above, phosphate loading appears to have a minimal effect of fatigue delay in anaerobic endurance events. If phosphate loading does have an ergogenic effect, it is most likely connected with improvements in aerobic performance.

Carnosine

In recent years the buffering capacity of carnosine, a dipeptide derivative of the amino acids alanine and histidine, has been investigated. Carnosine, found in high concentration in muscle fibres, represents an effective H^+ buffer. The availability of carnosine within the muscle appears to be limited by the availability of β-alanine. As a consequence, Harris *et al.* (2006) have investigated the effects of β-alanine supplementation on lactate buffering and performance. In their research, β-alanine supplementation was found to increase carnosine stores within the muscle by between 42–80 per cent dependent on the loading regimen.

In a study of maximal isometric contraction endurance and increased carnosine stores resulted in 11–14 per cent improvements in time to exhaustion. In a cycle ergometer study, designed to elicit exhaustion in 150 s, times to exhaustion were increased by 12 per cent and 15 per cent after a 4

and 10 week supplementation period, respectively. It does appear that β-alanine supplementation offers a buffer to H^+, thereby delaying fatigue and improving performance in anaerobic endurance and short-term high-intensity aerobic endurance exercise. Further research is required in applied sport and adventure sports contexts, outside the laboratory, to further illuminate the potential performance benefits.

Key point

Beyond a normal healthy balanced diet, the employment of a buffering supplement strategy may serve to improve anaerobic performance. Supplementation of dietary intake of buffering substrates can help to alleviate the deleterious effects of increasing H^+ by neutralizing the acidity. Bicarbonate and phosphate, along with β-alanine ingestion, have been demonstrated to improve H^+ buffering and anaerobic performance.

7.6   Sport-specific physiology

The physiology of rock climbing

Rock climbing incorporates a very diverse range of climbing-based activities from bouldering to big wall climbing. The diversity in climbing and the physiological challenge associated with this range of disciplines is perhaps best reflected in the fact that one or more climbing activities are found within every chapter in Part II of this textbook. Rock climbing in its various forms can present a challenge across a range of exercise intensities and durations. It is this diversity that makes study of the physiology of rock climbing both fascinating and complex. In recent years the popularity of, and access to, rock climbing has increased as a result of the development of artificial climbing walls.

Biography: Katherine Schirrmacher

Adventure sport: Rock climbing

Katherine has been climbing for 13 years and during that time has climbed many different rock types and across disciplines including bouldering, sport climbing and trad climbing. She has climbed at crags all over the world but began her career on the Welsh sea cliffs, Lake District crags and the Italian Dolomites. In the past four years, Katherine has found herself drawn to the more physical nature of bouldering, sport climbing and competition climbing. Last year she gave up her NHS job to concentrate full time on climbing. Katherine has climbed bouldering problems up to V9, led 8a on bolts and E6 on trad

Figure 7.10 Katherine Schirrmacher (photograph courtesy of Alex Messenger)

routes. Since entering climbing competitions, she has made the finals in three World Cup bouldering events, with her best finish so far being 11th place. Katherine lives in Sheffield with her husband Nic and coaches climbing in her spare time.

Raindogs

Raindogs is like the 100 m sprint but is only 12 m high and takes 1 min 20 s to climb. Requiring a very specific type of fitness, almost only applicable to this route, Raindogs is the ultimate test of power endurance. There is no section of desperately hard moves but the climbing is relentless and resting is almost impossible. After all the difficulties below, the final move of grabbing the lower-off chain at the top needs one more burst of power. People are known to spend days falling off here. A short bit of white rock at Malham Cove, it is swamped by the 100 m sweep of cliff above it. This is one of Britain's classic bolted routes and at the magic grade of 8a, many people aspire to climb it. It's a benchmark which I wanted to test myself against.

The climbing is intricate and purposeful. Every hand movement seems to involve about three foot changes. There are many different and subtle possibilities depending on your body shape. In fact, despite high climbing standards today, it is so complicated no one has yet succeeded on the route on their first go with no prior knowledge. Once you get past the first bolt the movements are long, sometimes powerful and you have to climb fast. The clock starts ticking right away and it's a race against time before your arms drain and the route spits you off.

From the very first moment, I was drawn to the route. None of the moves seemed especially hard, it was simply a question of putting it all together. Simple? The journey began; both in the steps I had to take to get myself physically in shape and mentally as time and time again I fell and I wondered whether I would ever reach the top.

I spent a period of time on it trying 'links', climbing from the second bolt to the top, then first bolt to the top and finally ground to the top. Within a matter of about three days I found myself climbing from the first bolt to the top so I tried from the ground. However hard I tried, nothing would get me to the top. Within a couple of metres of the chain and within a matter of seconds my arms would give out and gravity got the better of me.

Conditions were vital: too cold and I wouldn't be able to feel my fingers, too

Figure 7.11 Katherine Schirrmacher on Raindogs (8a), Malham Cove, UK (photograph courtesy of Alex Messenger)

warm and my fingers would slip on warm rock, wet rock was a no-go. I battled with hot weather and became obsessed with the forecast and the temp gauge in my car. At the bottom, one of the pockets stays stubbornly damp even when the sun is shining hard and needs a rag stuffing in to dry it, in order to stop fingers slipping. Clutching at straws I stuffed newspaper cuttings of Kelly Holmes' Olympic success into the pocket, hoping that some of her success would rub off on me.

A re-think was needed and the following year I geared myself properly to the route convinced that nothing was going to get between me and the top. I practiced circuits in my local climbing wall lasting 1 min 20 s and kept up a dose of powerful climbing. Pull-ups, push-ups and other core body exercises took over my normal 'having fun at the climbing wall'. It was time to get serious. Within a short time I was actually climbing the best I had ever climbed and when I went on the route I was already better than I had been the previous year.

I can clearly remember the day I did it. I smiled when I saw 12°C on the temp gauge as I parked my car. The previous day on the route, I had arrived within one hand movement of the top but for some reason my body seemed to fold in at the knees flat like an envelope bringing me downwards dangling again on the rope. However the moment I finally grabbed the chain, my whole body was working as one, strong from the toes to the knees, and up through my arms. The tip of my middle finger crept through the chain. For an instant my body wasn't sure if it could make it but my mind took over as the rest of my finger curled through. Letting out a cheer, the job was done.

The research interest in rock climbing has paralleled the growth in popularity of the sport. At present, perhaps due to the complexity of the sport, the available research describes performance and performers adequately. However, research is limited within the area of underlying mechanisms critical to success in climbing and the methods for improving performance. For this information, an interested climber needs to turn to a growing body of texts written by high level performers describing training methods they have developed. This section will briefly review the available research before examining key concepts for training and preparation in rock climbing.

A variety of studies have reported the physiological profiles of high and elite level climbers. Results suggest that top level climbers have a relatively low body mass, low body fat percentage, good flexibility, a high upper body to strength ratio, good muscular endurance and a high level of upper body power. There has been some debate as to the relative importance of aerobic fitness for rock climbing. Some studies have reported relatively high aerobic endurance results for rock climbers with $\dot{V}O_{2\,max}$ of between $50-55$ mL kg^{-1} min^{-1} being reported for climbers in treadmill-based tests. Other researchers, such as Billat *et al.* (1995), have suggested that for rock climbing the overall percentage of $\dot{V}O_{2\,max}$ required is relatively small. Sheel *et al.* (2003) found that the percentage of $\dot{V}O_{2\,max}$ required increases with the difficulty of climb. Despite this, Mermier *et al.* (2000) identified that there maybe a breakdown in the linear relation between heart rate and oxygen consumption. They suggested this

was the result of a reduced importance of a aerobic endurance. Sheel (2004) suggests that this breakdown is perhaps the result of the isometric forearm contractions associated with rock climbing, leading to a disproportionate rise in heart rate and blood pressure. In addition, an anxiety-induced rise in heart rate disproportionate to exercise intensity may result from the possibility of falling. For elite level climbers, however, one might expect such anxiety to be less pronounced than for recreational climbers.

Blood lactate concentrations in response to rock climbing have been analysed in a number of studies with values from $3-8.5$ mmol L^{-1} being reported. In competition (UIAA 1993 World Championships), Gerbert and Werner (1999) reported that lactate concentrations rose with difficulty of climbing and height attained. The relative blood lactate concentrations in response to climbing appear to be lower than might be expected for the intensity of exercise when compared with treadmill exercise. Watts (2004) suggests that this may relate to the relatively small lactate-producing muscle mass involved in exercise compared to treadmill running. In agreement, our research has led us to suggest the possibility of the arms as net producers and legs as net consumers of lactate during climbing. Grant *et al.* (1996, 2001) found that elite (higher than 6a technical grade) climbers had significantly higher 90° arm-hang times and pull-up scores compared to recreational climbers. Using a high-step measure of leg and hip flexibility, Grant *et al.* (1996, 2001) found higher flexibility scores for the elite climbers in their study compared to recreational and non-climbers.

The conflicting results regarding the relative contribution of aerobic endurance to rock climbing reveals the difficulty in investigating the physiology of rock climbing. Not only do climbers take part in a diverse range of activities, but the physiological implications of top-roping are perhaps different to those experienced in bouldering or lead climbing. In addition, the surface upon which climbing takes place varies considerably, from slab to steep overhanging climbing and from gritstone to slate. The nature of the moves will impact upon the energy demands. Some climbs have one technical move with easier climbing for the rest of the ascent, while others are more sustained at the grade. The presence of a crux move at different heights could also impact on the physiological demands. It would perhaps be beneficial to future research if studies examined in a logical progression a manipulation of single variables to more clearly identify the physiological differences between climbing and E1 5b on grit and limestone, on top-rope or lead or as a steep climb rather than a slab climb. An agreed reference for climbers would be beneficial such as the use of a 90° wall with a repeated movement throughout the climb, making it more cyclic and therefore closer to cycling or running. From such a starting point, comparison between studies would become more meaningful. The difficulty with such an approach is that it would require a large number of studies to move the climbing knowledge base forward.

Perhaps the most comprehensive model for performance in rock climbing was presented by Goddard and Neumann (1993) and included six main factors: coordination and technique, tactics, physical fitness, psychology, background conditions and external conditions. With regard to physical fitness, Goddard and Neumann (1993) described the terms power, power-endurance, local endurance and general endurance. Along with these aspects of climbing, they stressed the importance of flexibility for climbing. Other training guides have used a variety of terms to describe similar aspects of fitness for climbing which can result in confusion for the reader. Table 7.2 provides a summary of the different terminology used in common training guides for climbing and links them with the terminology more common to physiology.

Power refers to all-out effort over a short period of time. Goddard and Neumann (1993) refer to power as requiring between 80–100 per cent of maximal strength over a short time period for up to 3–5 moves. *Power-endurance* involves 50–80 per cent of maximum strength for 8–15 moves. Power and power-endurance climbing are anaerobic in nature, the predominant energy source being ATP stores and the ATP-PCr system, respectively. Referred to as *local endurance* by Goddard and Neumann (1993), this aspect of energy production is typically known as muscular endurance by physiologists. The term local endurance was perhaps selected to reflect the specific local loading placed upon the forearms, designed to highlight the unique nature of muscular endurance required in rock climbing. Muscular endurance for rock climbing can be thought of as requiring between 25–50 per cent of maximal strength over long sustained pitches.

Goddard and Neumann (1993) believe local (muscular) endurance is the limiting endurance factor in climbing. A climber will fall from a route due to muscular fatigue of the forearms long before they reach their $\dot{V}O_{2\,max}$. Referred to as *general endurance* by Goddard and Neumann, this aspect of fitness for climbing is commonly referred to as aerobic endurance and is the focus of Chapters 9 and 10 in this textbook. As with the research regarding this aspect of fitness for climbing, climbers assign different levels of importance to a good oxygen consumption level. Goddard and Neumann advocate aerobic training at a level between 65–75 per cent of heart rate maximum in order to relieve stress and as a method for fat burning. Aerobic training can be undertaken through a variety of mediums such as running, cycling swimming or rowing. Most authors agree that in climbing, aerobic endurance is not likely to be a limiting factor to performance.

Climbing training zones, regardless of the terminology used to describe them, have a physiological application to the climber. When making decisions about the type of training to undertake, a climber needs to start with their own personal goals and current level of performance. To improve performance, a climber must identify their specific rate limiters and the type of climbing for which they

Table 7.2 Physiological terminology prevalent in rock climbing

General physiology terminology	Goddard and Neumann (Performance rock climbing)	Eric Hörst (Training for climbing)	Hague and Hunter (self coached climber)	Heather Reynolds Sagar (Climbing your best)	Michelle Hurni (Coaching climbing)	McClure and Binney (Climb magazine)
Power	Power	Strength and power	Strength and power	Power	Power	Strength
Power-endurance	Power-endurance	Anaerobic	Anaerobic endurance	Power-endurance (high)		Strength endurance
Muscular-endurance	Local endurance		Endurance	Power-endurance (low) → Endurance, Stamina, Recovery	Endurance	Local endurance
Aerobic endurance	General endurance	Aerobic capacity				General endurance

(General physiology terminology: Power and Power-endurance bracketed as "Anaerobic"; Eric Hörst: Strength and power and Anaerobic bracketed as "Anaerobic"; Hague and Hunter: Anaerobic endurance and Endurance bracketed as "Endurance"; Michelle Hurni: Power-endurance and Endurance bracketed as "Endurance") are training. Training for a boulderer is different to that of a primarily trad climber. The complication comes for climbers who take part in more than one climbing discipline. It is important for a climber to be realistic about the aspects of their climbing that are holding them back. The rate limiter for a climber could be physiological, a lack of muscular endurance or from a different source such as a psychological or technical issue. To progress, a climber must address their rate limiter at any specific time in training.

There are a variety of training tools, specific to rock climbing, that have been advocated for performance improvement. This section finishes with a short review of a number of climbing-specific training tools for rock climbing. On a climbing wall or outside (indoor climbing walls generally provide a good range of routes and types of climbing within a relatively smaller area) a variety of types of training can be undertaken to improve performance. Training methods can include; *continuous climbing* for a set time period (often involving traversing); *interval training* where the length of climb, intensity and recovery duration can be manipulated to alter the training goal (power, power-endurance or muscular endurance); *up and down climbing* (staying on the holds in both directions); *pyramid sessions* where you move up and down grades to your maximum; and *limit training* where you spend a session climbing routes that are at or just below your technical limit.

Campus board training, mainly developed for higher-level climbers, can provide a specific training tool. Campus board training has most application to bouldering and sport climbing where power and power-endurance are the primary concerns. Campus board training is physically demanding and should be entered into with care, after a thorough warm-up and with knowledge of correct technique. Campus boards can be used for laddering and dynos with feet on a supporting hold or with no foot contact. *Laddering* involves moving up the campus board rungs in a hand-over-hand or matching fashion.

Campus boards generally have a variety of rungs sizes, the smaller of which place greater strain on the fingers. Climbers number the rungs from the bottom to the top. It is possible to ladder using every rung (1 – 2 – 3 – 4 – 5 – 6 – 7) or to miss rungs (1 – 3 – 5 – 7), which increases the difficulty. In *dynos*, or double dynos as they are often known, the climber starts on a rung near the middle and then drops down a rung catching it double-handed before immediately springing back up to catch the starting rung or a higher rung (rung 2 – 1 – 2 or rung 2 – 1 – 3 – 2 – 4 – 3 – 5). A variety of *pull-ups* can be performed to improve power-endurance e.g. leading to one-arm pull-ups. When linked with lock-offs, they become more specific to climbing and can be of benefit to all levels of climber.

Systems board training falls between campus board training and wall/rock climbing. A systems board enables a climber to train a specific technical aspect of climbing, such as crimping, while completing laps ensures physiological development. Systems board routes can be specifically designed to focus on one aspect of climbing and then changed to meet another need. The specific repetitions, rest periods and duration of a session should be developed for each individual with their own training goals, level of fitness and experience taken into account. Although a highly technical adventure sport, rock climbing has a clear physiological component that can be improved through training.

For examples of campus board use, laddering, pull-ups and systems board training please visit www.wiley.com/go/draper.

Key point

There is an increasing body of research relating to rock climbing. Typically elite rock climbers have high power to mass ratios and low body fat percentage. The development of training methods for rock climbing has come from the programmes followed by elite climbers. The different methods suggested by elite climbers have led to a diversity in terminology to describe the training and performance zones for rock climbing. The terms power, power-endurance, muscular endurance and aerobic endurance are more commonly recognized by physiologists. Training tools to assist rock climbers include indoor climbing walls, campus boards and systems boards. Any training programme for a climber should be individually developed and focused on the individual's rate limiter.

Getting into research: energy cost of sport climbing (Booth et al., 1999)

Studies of rock climbing had previously been focused on the physiological attributes of various levels of climbers. Booth et al. set out to measure oxygen uptake, blood lactate concentration and heart rate in response to sport climbing. Prior to this study, the research consensus was that rock climbing was primarily anaerobic and did not require an oxygen consumption rate above 50 per cent of $\dot{V}O_{2 max}$. This study appears to represent the first published study of oxygen consumption in response to climbing on natural rock, made possible by the development of portable on-line gas analysis systems.

Seven climbers (six male and one female) with a mean experience of 8.9 years agreed to take part in the study. The climbers had a maximum on-site lead of between 6b–7a and mean age 25 years, height 175 cm and mass 62.6 kg. This study was conducted in two parts: the first utilizing a climbing treadmill to establish a climbing specific $\dot{V}O_{2climbpeak}$ for each climber and the second to measure blood oxygen consumption, heart rate and lactate concentration in response to outdoor rock climbing.

In the first test the climbers performed an incremental $\dot{V}O_{2 max}$ test, in this study specifically referred to as $\dot{V}O_{2climbpeak}$, on a climbing treadmill. The test had five increments split into three trials. Each of the trials was separated by a 20 min recovery due to the relatively high loading on the arms created during rock climbing. Trials one and two lasted for 5 min; trial three for 7 min. In the first trial the treadmill speed was set at 8 m min^{-1}, in the second trial at 10 m min^{-1} and in the third trial the first 5 min were conducted at 12 m min^{-1} and for the 6th and 7th minutes the speed was increased to 14 m min^{-1} and 16 m min^{-1}, respectively. Participants climbed to volitional fatigue which occurred within the third trial for all climbers. Oxygen consumption and heart rate were recorded using a K2 gas analyser and blood lactate concentration was determined for samples taken before the test and after each trial. In the second test, which was conducted on a separate occasion, the climbers ascended a 24.4 m outdoor climb on natural rock. The climb was constantly over-hanging (4 m) and graded at 5c. Climbers ascended with a slack top rope attached. A K2 gas analyser was worn to determine oxygen consumption and heart rate during the climb and blood lactate samples were collected before and after the climb.

In the climbing treadmill test, the participants achieved a mean $\dot{V}O_{2climbpeak}$ of 43.8 mL kg^{-1} min^{-1} and a mean heart rate maximum of 190 bts min^{-1}. Resting mean lactate concentration was 1.42 mmol L^{-1}, rising to 4.54, 6.50 and 10.2 mmol L^{-1} at the end of trials one to three, respectively. The mean time for the 24.4 m climb in the second part of the test was

7 min 36 s (range 6 min 28 s to 9 min 54 s) with no climber falling at any stage, representing a mean climbing velocity of 3.2 m min^{-1}. During the climb, mean heart rates were between 145–157 bts min^{-1} and the mean oxygen consumption was between 26–32.8 mL kg^{-1} min^{-1}. Blood lactate values rose from 1.3 mmol L^{-1} pre-climb to 4.5 mmol L^{-1} post-climb.

The results from this study indicate that mean oxygen consumption during an outdoor rock climb ranged from 60–75 per cent of $\dot{V}O_{2climbpeak}$. This represents a significant proportion of each climber's $\dot{V}O_{2climbpeak}$. As a result of this finding, the researchers concluded that aerobic metabolism perhaps represents a greater mechanism for energy production in rock climbing than had previously been thought. The length of climb, closeness to a climber's maximal grade, nature of the climb (rock type, degree of overhang, number and type of hand holds, etc.) will obviously greatly affect the result and require attention in future studies. Booth *et al.* conducted an important study, however, for it most likely represents the first conducted on natural rock where the energy costs of performance have been reported.

The physiology of snowsports

There has been a continued research interest, particularly within the Scandinavian countries, regarding the physiology of snowsports. The most developed research focus has been cross-country skiing, although the various forms of alpine skiing have received research attention. As yet, however, snowboarding has been the least explored.

Alpine skiing comprises four main disciplines: slalom, giant slalom, supergiant slalom and downhill, involving the descent of ski slopes at up to 145 kph (90 mph) in races of typically 45–60, 60–75, 75–90 and 90–140 s, respectively. The alpine skier is accelerated downhill by gravity and encounters centrifugal forces during turning. These forces, perhaps uniquely to skiing, require co-activation of the quadriceps and hamstring muscles to stabilize the hip, knee and ankle during descent and turning. The duration of exercise results in an alpine ski run requiring predominantly anaerobic metabolism with additional energy supplied through aerobic means.

Research findings by Karlsson *et al.* (1977) support this energy production balance as they identified post-race capillary blood lactate samples of 13 mmol L^{-1} for males to 10 mmol L^{-1} for females. There has been a debate in the literature regarding the importance of aerobic endurance for alpine skiers. Some Swedish researchers have emphasized the importance of aerobic fitness for alpine skiers, reporting $\dot{V}O_{2max}$ of 53 mL kg^{-1} min^{-1} for females and 68 mL kg^{-1} min^{-1} for males. More recently, researchers have stressed that aerobic fitness has a lesser role in successful performance in alpine skiing. Research indicates that alpine skiers tend to have a mixture of Type I and Type II muscle fibres, rather than one predominantly over another. Typically, downhill skiers tend to be heavier than slalom or giant slalom skiers. Comparative research suggests that alpine skiers have become heavier more recently, but without a rise in body fat percentage suggesting increases in muscle mass. Research has found that male alpine skiers have a body fat percentage of 6–11 per cent and females 13–23 per cent. The key components of fitness for alpine skiers appear to be anaerobic capacity, aerobic endurance and flexibility.

Cross-country ski races, over distances of 5–50 km, place substantial demands on aerobic fitness. The major research finding for cross-country regards the very high $\dot{V}O_{2max}$ value that have been reported for skiers. A major sports research focus in Scandinavian countries, specialist cross-country ergometers have been designed for the assessment of physiological functioning. Research with these ski ergometers and in the field has revealed that cross-country skiers have $\dot{V}O_{2\,max}$ value that are normally among the highest reported for endurance athletes, and are regularly over 80 mL kg^{-1} min^{-1} for males. Indeed, the highest reported $\dot{V}O_{2\,max}$ for an athlete, 94 mL kg^{-1} min^{-1} was from a cross-country skier. It is believed that the near maximal efforts encountered during uphill sections and the relative relaxation during a subsequent downhill are responsible for the maximization of $\dot{V}O_{2\,max}$ improvements. Research indicates that the intensity is very high during cross-country races, with performers typically working at between 80–90 per cent of $\dot{V}O_{2\,max}$. Not surprisingly, body fat percentage for cross-country skiers is typically low, between 5–12 per cent for males and 10–22 per cent for females.

The biathlon discipline adds a further challenge with the 3.5–5 kg rifle that must be carried, with the additional requirement of quickly lowering heart rate prior to shooting. To bring such a decrease in heart rate, cross-country skiers typically decrease the intensity of exercise about 60 s before arrival at a shooting station, consciously focusing on slowing heart rate prior to the first shot.

Key point

Alpine and cross-country skiing have been the focus of several research projects, particularly from Scandinavia. Recent research suggests that aerobic fitness has a lesser importance for alpine skiers than previously thought. In more recent years, alpine skiers appear to have gained muscle mass, perhaps from more thorough non-ski based physical preparation. Cross-country skiers have been reported as having the highest $\dot{V}O_{2\,max}$ of any adventure sports performers. Indeed, the highest $\dot{V}O_{2\,max}$ reported for any sport was for a cross-country skier.

7.7 Summary and study questions

Glycolysis

Although glycolysis commences with the onset of exercise, it takes several seconds before ATP can be realized through this pathway. For activities lasting 10–90 s, glycolysis is the key mechanism for energy production. Glycolysis makes a significant contribution to activities up to 10 min duration and for intermittent bursts of energy during longer activities. One of the most significant by-products of glycolysis is lactic acid (lactate), which can build up and inhibit the process of glycolysis itself. The management of lactate build-up and tolerance of increasing lactate levels is an important issue for many adventure activities.

Glycolysis is an anaerobic energy source and therefore does not require oxygen for its functioning. Glycolysis produces ATP supplies for exercise but also provides pyruvate for aerobic metabolism. During glycolysis, glycogen and glucose are broken down into pyruvate to produce a gain of two molecules of ATP for each glucose molecule. This process takes 10 chemical reactions for glucose and 11 for glycogen, which is why glycolysis is slower to produce ATP than the phosphagen system. There are two rates at which glycolysis can take place: during higher-intensity exercise pyruvate is converted to lactic acid, but when the exercise intensity is lower (slower rate glycolysis) pyruvate can be utilized within the aerobic energy pathway, thus avoiding the build-up of lactate that occurs with fast rate glycolysis.

Glycolysis, from glucose, is a 10-step process and actually requires ATP to be broken down in order to drive some of the stages. Reactions 1 and 2 each require 1 molecule of ATP and it is only at reaction 7 that 2 molecules of ATP are produced. The 10th reaction produces 2 more molecules of ATP to produce a net gain of 2 molecules.

The rate of glycolysis is adjusted to meet the needs of the cell. This process occurs through the action of the enzymes hexokinase and phosphofructokinase (PFK) and these respond to the concentrations of key substrates. During slower rate glycolysis, hydrogen atoms released in the 6th reaction of glycolysis are shuttled by NAD^+ to the mitochondria for aerobic metabolic consumption. During fast rate glycolysis, the supply of oxygen to the mitochondria cannot keep pace with the production of hydrogen ions and instead they are transferred to pyruvate to form lactic acid. Although increasing lactate and lactic acid levels limit the continuation of exercise, glycolysis would be slowed much sooner if the transfer of hydrogen to pyruvate did not occur. Lactic acid actually dissociates into lactate and H^+ but both terms are generally used interchangeably. Lactate concentration is relatively easy to measure through capillary blood sampling and can provide the physiologist with valuable information about exercise intensity levels and energy production.

Lactate

Lactate is often thought of as an end-product of metabolism. However, lactate is actually reused either as a direct fuel source or after it has been re-synthesized into glycogen. The process of converting lactate to glucose, gluconeogenesis, occurs through the Cori cycle. The Cori cycle takes place in the liver. Lactate is also shuttled to Type I muscle fibres where it is oxidized to form pyruvate and then converted into acetyl CoA for use within the aerobic pathway. It also seems likely that muscle fibres can re-synthesize lactate ready for use within their own mitochondria. Additionally, cardiac muscle is capable of using lactate directly as a substrate for glycolysis.

Gluconeogenesis

Gluconeogenesis is the creation of glucose from non-dietary sources and is particularly important during longer-duration activities or when the availability of dietary carbohydrates is limited. Both amino and fatty acids can also be used as fuels for gluconeogenesis.

In activities that take the performer longer than 90 s to complete, the aerobic pathway becomes increasingly important in relation to ATP re-synthesis. However, the transition from anaerobic to aerobic endurance activities is actually a gradual process. The length of activity is very important in terms of how a performer will deal with lactate. During short-duration anaerobic endurance events, the performer will be concerned with how much lactate they can cope with i.e. lactate tolerance. As the event becomes longer, the performers' focus must move towards lactate management and the activity intensity must be controlled to accommodate an optimal overall performance. In an activity lasting longer than 8–10 min, the performer must try to push as hard as possible without crossing the lactate threshold, the point after which lactate begins to accumulate and cannot be reprocessed as quickly as it is created.

Both the blood and muscles themselves have quite a narrow band of tolerance to acidity, meaning that the body needs a buffering system to cope with acidity. The primary buffer for lactic acid within cells is inorganic phosphate (Pi), while in the bloodstream it is bicarbonate. The buffering of lactate in the bloodstream releases CO_2 which is expelled through the lungs resulting in the RER values of more than 1.0 during maximal exercise. The kidneys also play an important role in eliminating H^+ from the blood, either excreting them through urine or using them to create bicarbonate ions for blood buffering. It is believed that excess H^+ in muscle fibres causes the increased acidity that results in fatigue and the sensation of pain during anaerobic endurance activity.

Adaptations to anaerobic endurance training

Interval training is considered one of the best training for anaerobic endurance. Anaerobic endurance training will increase glycogen uptake during recovery, increase resting glycogen stores and improved glycolytic enzyme activity. Buffering capability is also improved both within the muscles and blood and this will serve to delay the onset of fatigue. Additionally, training results in increased tolerance of H^+ and even small improvements in aerobic metabolism.

Ergogenic aids

Creatine offers the possibility of supplementation to increase performance in anaerobic endurance, as do supplements that may help H^+ buffering mechanisms such as bicarbonate, phosphate and β-alanine. Of these supplements, at present, bicarbonate has the most scientific support. β-alanine is a relatively new supplement and though early results are positive, there is still more work to do to establish whether it can provide a safe and effective supplement for real sports performance.

Research and advice in sports

At present, much of what we know about rock climbing physiology is from descriptive studies of performers who typically complete a battery of assessment tasks. There is less information about the demands on climbers in action. This has meant that training advice has been based on matching attributes accomplished climbers have developed or extrapolating from the assumed and measured demands of the activity such as the amount of time a climbing task might take.

Alpine ski racing and cross-country ski racing are activities with very different physiological demands. For alpine skiing, the necessity of high aerobic fitness is still debated but research and the duration of events suggest that the predominant energy system is anaerobic. The requirements for cross-country skiing, on the other hand, are far more unequivocal. Cross-country skiers have some of the highest aerobic capacities of any athlete and the highest ever $\dot{V}O_{2\,max}$ recorded was that of a cross-country skier.

Study questions

1. Why does it take longer to gain energy from glycolysis than the phosphocreatine system? What is the fuel for glycolysis?

2. Explain why fast rate glycolysis and slow rate glycolysis are more accurate terms than anaerobic glycolysis and aerobic glycolysis. What happens to pyruvate at the end of each of these processes?

3. How is the rate of glycolysis controlled? Which enzymes are involved in this process?

4. Explain the processes for dealing with lactate produced by glycolysis. How is cardiac muscle different to skeletal muscle in this respect?

5. What is gluconeogenesis? Where in the human body does this process take place?

6. Why is the length of an activity important in relation to the way a performer will need to deal with lactate? Explain the terms lactate tolerance, lactate management and lactate threshold.

7. In relation to anaerobic endurance what is meant by the term buffering? What is the difference between the buffering mechanisms of the muscle cell and the bloodstream?

8. How might improving buffering capacity help a rock climber improve their performance on a route?

8

Intermittent adventure activities

8.1 Introduction

The physiology of intermittent activities forms the focus of this chapter. Intermittent activities involve bursts of high-intensity exercise coupled with lower-intensity phases and/or periods of rest. The on-off nature of intermittent activities creates demands on the anaerobic and aerobic energy systems. The sole game-based adventure sport, one that comprises an intermittent activity, is canoe polo. Although played in a swimming pool or at a flatwater location, many of those involved in canoe polo are also involved in other forms of canoesport that fall more readily under the umbrella of adventure sports. Consequently, a chapter is included on intermittent activities to provide a resource for those with cross-discipline interests. In addition, a variety of adventure sport activities are included here which involve stop-start activity. The descent of a river by kayak or of a mountain on skis or snowboard is not normally completed in a single exercise block; instead kayakers and skiers descend in a series of lengths, regrouping and resting in eddies or landmarks before continuing. The rest phases may be to scout the next rapid or section of piste, however, the physiological implication is recovery and as such it makes the activity intermittent in nature. In addition, adventure sports such as surfing, kayak surfing, playboating, indoor climbing, single-pitch climbing in the Peak District, a bouldering session and dinghy sailing (single helms and doubles crews) involve intermittent activity.

The adventure sports discussed in this chapter are listed in Table 8.1 and are divided according to the nature of the intermittent activity. Three main forms of intermittent activity can be identified in adventure sports. Canoe polo, normally played in two halves, involves a game-related form of intermittent activity similar in physiology to that of football or hockey. The physiological demands in a playboating session or when surfing involves self-selected intermittent activity. During a surfing session after the paddle out, the surfer can rest outback before deciding which wave to pick-up to come in on, and as such the performer is making the decisions about the length of activity and rest periods. A similar pattern can be seen during a playboating session where the performer waits in an eddy (resting/recovering) before moving on to the wave/feature. In this situation, the performer (self) selects the length of rest and activity periods.

In contrast, the intermittent activity in coffee-grinding (winching) during match racing or that of doubles crews and singles helms in dinghy sailing is activity determined. That is, the physiological work required is determined by the nature of the course, wind conditions and sea state. A performer

Adventure Sport Physiology Nick Draper and Chris Hodgson
© 2008 John Wiley & Sons, Ltd

Table 8.1 The nature of intermittent activities in adventure sports

Adventure sport or discipline	Nature of intermittent activity
Canoe polo	Game-related intermittent activity
Surfing	
Surf kayaking	
Playboat/freestyle session	Self-selected intermittent activity
Bouldering session/competition	
Climbing session/competition	
Match racing in sailing (crew)	
Dinghy sailing (doubles crews)	Activity-determined intermittent activity
Dinghy sailing (singles helms)	

in these sailing-based adventure activities may have a long recovery period, such as that on a long-downwind leg, before having a period of frantic on-off activity during a tack. In this way, the performer has to recover while they can before the next bout of activity. The recovery periods are often largely determined by factors external to the crew.

We review the physiology of intermittent activity including demands and physiological response to exercise, fatigue and training. This is followed by sections on nutritional and ergogenic strategies for intermittent activities, the physiological implications of cold water immersion and physiology of water-based adventure sports. In addition, performer contributions by Ben Ainslie (match racing in sailing), Fiona Jarvie (freestyle kayaking), and Simon Hammond (kayak surfing) as well as a research summary on the physiology of intermittent exercise by Gaitanos *et al.* (1993) are featured in this chapter.

8.2 Physiology of intermittent activities

The physiology of intermittent activities, distinct from exercise of a continuous nature, has been

increasingly investigated since the 1960s. Intermittent exercise is defined as periods of exercise interspersed with rest or recovery periods. The past 50 years has seen a growth in the body of knowledge relating to the physiological demands, predominant energy systems and causes of fatigue in intermittent activity. Researchers have manipulated exercise and recovery periods to examine their effects on oxygen consumption and the various substrates of metabolism such as ATP, ADP, Pi and lactate.

With regard to exercise protocols, researchers have varied exercise intensity across three main levels. The exercise intensity has been set sub-maximally (below $\dot{V}O_{2\,max}$) at the same level as an individual's $\dot{V}O_{2\,max}$ (maximal), or supra-maximally, at an exercise intensity above the participant's $\dot{V}O_{2\,max}$. As discussed in Chapter 4, maximal human power outputs are achieved in the first 1–4 s of activities. The vertical jump or Wingate tests (Chapter 4) provide laboratory measures of maximal power output. Peak power outputs are achieved in the first 3–4 s of a WAnT and can be as high as 1000 W or more. The $\dot{V}O_{2\,max}$ test provides a measure of aerobic power. As an individual reaches their $\dot{V}O_{2\,max}$, power output tends to be around 25 per cent of that achieved in a power test such as the WAnT.

The available research regarding intermittent activities has been largely developed through laboratory-based studies with careful manipulation of specific variables. The manipulated variables in intermittent activity studies include the mode of exercise, exercise intensity, exercise duration, recovery duration and the number of work intervals (the duration of the test). Laboratory-based studies have normally been conducted on a cycle ergometer or a treadmill. The advantage with these laboratory-based studies is that control over many extraneous variables can be achieved.

Data collected has originated from regularized exercise protocols i.e. 10×6 s sprints with a 30 s recovery between each repetition. Through this approach, a great deal has been learned regarding the physiological response to these types of intermittent activity. The disadvantage of this approach, however, is that none of the adventure activities covered in this chapter (except in training) has regulated exercise periods. The exercise and recovery periods during a canoe polo match relate to events occurring during the progress of a game. A player may be involved in the game for a period of time before getting a recovery, and each recovery phase will vary in length. This non-regulated pattern is similar to the coffee-grinding required by crew members in match racing (sailing). A crew member may have to make several changes of sail in quick succession, before a longer rest period occurs. If this is taken into consideration, the research that has been conducted to date provides a great deal of useful information regarding the physiology of intermittent activities. In sports such as football, the use of video, heart rate monitors and computer software programmes have been employed in more recent research. The data collected provides further information regarding the distances covered, speeds of travel during a match and nature of fatigue within the natural environment of the activity. The availability of this technology for adventure sports, as well as the additional tool of GPS (global positioning system), has enabled the development of field-based research regarding the intermittent activities included in this chapter.

The duration of the activities has an impact upon the physiological demands and the exercise intensity maintained. Intermittent activities rely on energy production from anaerobic and aerobic metabolism. The predominant energy system for any specific activity is determined by the duration of the activity. For instance, the energy demands of a canoe polo game lasting 20 min (two 10 min halves) are different from those of a canoe polo tournament or the several rounds of a surf kayak competition. The predominant energy system and the drop in power output during a session will be determined by the duration of the activity. Research shows that intermittent activities rely on energy production from the phosphagen system, fast rate glycolysis and aerobic metabolism. This section goes on to describe research findings regarding the predominant energy system for specific intermittent activities. This knowledge can be applied to the adventure activities included within this chapter to better inform training methods for fitness development for each of the adventure sports.

One of the early studies of intermittent activities was conducted in 1960 by Christensen et al. (1960). The exercise intensity was set at a high intensity, such that exhaustion was reached within 5 min of continuous exercise. During 4 min of continuous running on a treadmill at that exercise intensity, the participants were able to complete a mean distance of 1300 m with a mean rise, in lactate to 16.85 mmol L^{-1} and a mean oxygen consumption of 5.6 L min^{-1}. On three subsequent days, the participants completed one of the following 30 min intermittent exercise protocols: (1) 10 s exercise matched with 5 s recovery (10/5); (2) 10 s exercise and 10 s recovery (10/10); and (3) 15 s exercise with 30 s recovery (15/30). The exercise intensity for the intermittent activity sessions was kept the same as for the 4 min continuous run. All three exercise protocols resulted in a lower lactate level (4.94, 2.25 and 1.80 mmol L^{-1}) at the end of exercise than for the continuous exercise, and the participants were able to complete the 30 min sessions at an intensity that would have caused exhaustion within 5 min of continuous running. The 10/5 intermittent protocol enabled the furthest distance to be covered within 30 min (mean distance run 6670 m), which included 120 exercise intervals resulting in a total work time of 20 min; note that this was at an intensity which would have caused fatigue within 5 min of continuous running.

Karlsson *et al.* (1967) conducted a study on the relative intensity of exercise. In this study, with 20 s exercise and 10 s recovery periods, the exercise intensity was manipulated to examine the effect on rate of fatigue during intermittent exercise. The influence of exercise intensity on exhaustion was easily demonstrated through the results achieved by one participant, who ran for 1 hr at a speed of 22 $km\,hr^{-1}$ but could only sustain exercise for 25 min when the intensity was raised to 22.75 $km\,hr^{-1}$. In a study by Margaria *et al.* (1969) with a supra-maximal exercise intensity that would cause fatigue in 30–40 s of continuous exercise, the recovery interval between the 10 s exercise periods was manipulated. When the interval between the exercise periods was 30 s, a much lower elevation of blood lactate concentration above resting levels was achieved compared to the 1:1 protocol (10 s exercise with 10 s recovery) where lactate concentrations rose progressively through the test. The duration of the exercise, as well as the exercise intensity and recovery duration, were shown to be important in intermittent exercise through research conducted by Saltin and Essén (1971). In their study, with a 1:2 recovery to exercise ratio, blood lactate concentrations rose only slightly above resting levels for the 10 and 20 s exercise intervals. When the exercise intervals were of 30 and 60 s duration, there were much higher rises in mean lactate concentrations.

Researchers have moved on to examine implications about substrate utilization during activity in more detail, extrapolating from findings. Essén *et al.* (1977) made a comparison regarding lactate concentrations, ATP and PCr stores in the muscle for continuous and intermittent exercise. After identification of the $\dot{V}O_{2\,max}$ for each of the five individuals in the study, participants completed two cycle ergometer protocols: one continuous and one intermittent. For the intermittent protocol, participants completed a 1:1 exercise to rest protocol for 1 hr. During the continuous protocol, the participants again cycled for 1 hr but at an exercise intensity of 55 per cent of $\dot{V}O_{2\,max}$ to maintain a matched mean intensity for the two tests. Results from the study revealed little difference in muscle lactate concentrations between the two exercise modalities.

However, large fluctuations in ATP and PCr stores between exercise and rest throughout the test were measured for the intermittent exercise protocol. After 5 min of exercise the PCr stores fell to 40 per cent of their resting levels, but during recovery were replenished to 70 per cent of their original level. Similar results were obtained throughout the hour for the intermittent exercise. During the continuous exercise, however, the levels of ATP and PCr were maintained at a level that was 50 per cent higher than during intermittent recovery. Further conclusions drawn by Essén, from this and later research, relate to differences in fat oxidation and muscle fibre type recruitment between intermittent and continuous exercise. Intermittent and continuous exercise tends to result in similar patterns of glycogen usage during exercise; however, fat oxidation appears to be elevated in intermittent exercise. During continuous exercise, mainly Type I fibres tend to be recruited whereas during intermittent exercise both Type I and Type II fibres are recruited. It is for this reason that lactate levels tend to be higher during intermittent exercise.

Gaitanos *et al.* (1993) published important findings regarding intermittent exercise (see **Getting into research** at end of chapter). In their study, the participants completed ten 6 s sprints on a cycle ergometer at a supra-maximal workload with a 30 s recovery between each sprint. Quantifying the fall in peak power across the sprints, they identified a mean drop of 33 per cent between the first and last sprint. Muscle biopsy analysis revealed distinct changes in energy system contribution to total energy production between the first and the tenth sprint. Hydrolysis of ATP stores represented 6.3 per cent and 3.8 per cent contribution to total energy utilization, which represented around a 2 per cent decrease between the first and the tenth sprint. This is indicative of an exercise-induced reduction of ATP stores across the sprints. The largest changes, however, were for breakdown of

PCr and glucose/glycogen through glycolysis. The relative energy contribution from glycolysis fell from 44.1–16.1 per cent and for PCr, breakdown rose from 49.6–80.1 per cent between the first and last sprint. These findings suggest that as duration of exercise increases, the relative contribution of glycolysis to energy production decreases. The relative decrease in glycolysis is reflective of the drop in peak power output during repeat sprints and places a greater relative demand on PCr catabolism and aerobic metabolism.

It is worth summarizing the findings of these studies as they reveal much about the physiology of intermittent exercise. It is possible for a performer to maintain a given work intensity for longer during intermittent than during continuous exercise. An individual can perform intermittent exercise for an hour or more at a maximal or supra-maximal exercise intensity that would cause fatigue within a few minutes if performed continuously. The exercise intensity, length of exercise bout and recovery duration will determine the time to exhaustion and blood lactate response in intermittent activities. The closer the exercise is to maximal power output and the shorter the recovery periods the greater will be the drop in peak power output over subsequent sprints. On the other hand, the duration of the intermittent activity, as well as impacting upon the intensity maintained, will alter the relative contribution of aerobic and anaerobic metabolism to performance.

In the early stages of exercise, glycolysis plays a greater relative role in energy production, reflective of initially higher lactate levels during intermittent exercise when compared to continuous exercise. As the duration of intermittent exercise continues, the relative contribution of glycolysis is decreased. The importance of PCr breakdown and aerobic metabolism, particularly of fat oxidation, is increased. Intermittent exercise has been shown to provide a greater utilization of fat oxidation than continuous exercise. It is believed that this occurs as a result of the recovery periods in intermittent exercise which enable replenishment of myoglobin and haemoglobin O_2 stores during each rest period, allowing for greater fat oxidation. This finding has led researchers to examine the possibilities

of intermittent exercise as an improved exercise mechanism for weight loss. The results of this research are as yet equivocal.

It appears that PCr and glycogen provide the main sources of energy during intermittent exercise, while fat oxidation and blood glucose provide the substrates for the recovery periods. With regard to muscle fibre type, research suggests that intermittent exercise results in the recruitment of Type I and II fibres as opposed to the mainly Type I fibres recruited during continuous exercise. Further research is required, particularly for adventure sports to determine the effects of intermittent exercise protocols for specific adventure activities. The key finding from the research to date is perhaps that for intermittent activities aerobic and anaerobic metabolism make a contribution to total energy production, whereas for the adventure sports that are the focus of the Chapters 6–7, 9 and 10, one energy system tends to be predominant. The exact contribution of each is determined by the duration and intensity of the exercise periods along with the length of recovery periods. The nature of energy production for a short bouldering session or a canoe polo game would be different to those for a series of heats in a surf kayak competition or a canoe polo tournament.

Key point

Research related to intermittent activities has primarily utilized treadmills and cycle ergometers as the exercise medium. During this research, which has developed over the past 50 years, the exercise intensity, duration and recovery duration have been manipulated to examine their effect on performance. Intermittent exercise enables a level of intensity to be maintained that would quickly cause fatigue during continuous exercise. The exercise intensity and recovery duration determine the length of time for which intermittent exercise can be maintained. During intermittent activity, there is a progressive decline in energy production from glycolysis and an increase in the relative reliance on the ATP-PCr system. Respiration during recovery periods in intermittent exercise appears to enable improved oxygen replenishment, enabling an increase in fat metabolism when compared to continuous exercise.

Fatigue in intermittent activities

The mechanisms for fatigue in intermittent adventure sports relate to the duration of the activities. In short-term higher-intensity intermittent activities, a variety of mechanisms for fatigue have been postulated that lead to an interruption in the contractile processes or an inhibition of excitation coupling. In prolonged intermittent activities of over 90 min duration, it is likely that glycogen depletion forms one of the major mechanisms behind fatigue. The exact mechanisms behind fatigue in intermittent activities, however, still require further research.

The role of increases in H^+ concentration and citrate leakage from the mitochondria in the inhibition of glycolysis were described in Chapter 7. Intermittent exercise research has indicated increased levels of H^+ and citrate in response to repeat sprints. In Chapter 6 Ca^{2+} transport disruption and PCr depletion have been suggested as possible mechanisms of fatigue in power endurance activities. All four mechanisms (increases in H^+ concentration, citrate leakage into the cytosol, Ca^{2+} transport disruption and PCr depletion) have been cited as possible causes of fatigue in short-term high-intensity intermittent activities.

In addition, the leakage of K^+ from within the sarcoplasm into the interstitial fluid (surrounding each muscle fibre) has been suggested in recent studies as a possible mechanism for fatigue in high-intensity activities. As described in Chapter 3, the Na^+-K^+ pump establishes a resting potential for nerve impulse conduction and muscle fibre contraction. The Na^+-K^+ pump, shown in Figure 3.5, moves sodium ions to the outside of the cell membrane into the interstitial fluid and transfers potassium ions into the cytosol. This movement of ions establishes the resting membrane potential. Research has identified increased interstitial K^+ concentrations at the point of fatigue, and consequently it has been suggested that these rises create a decrease in the resting membrane potential and disruption to excitation coupling (described in Chapter 4).

Glycogen stores in the muscles and liver represent a major substrate for ATP regeneration during exercise. During the high-intensity aerobic activities described in Chapter 9, the lower intensity but longer duration aerobic activities in Chapter 10 or prolonged intermittent activities (which by their nature include periods of high-intensity exercise), the body's glycogen stores can become depleted. The depletion of glycogen stores may be linked with fatigue in many high-intensity or long-duration adventure sports. Research indicates that during running, glycogen stores are used 40 times faster than during walking. It is for this reason, as described in Chapter 2, that an individual's glycogen stores can be sufficient to climb Ben Nevis, but can become depleted during a kayak or mountain marathon. Glycogen depletion leads to a decrease in substrate availability that is detrimental to performance and also associated with an increase in central fatigue. Research has shown that performers indicate increases in fatigue when operating in a glycogen-depleted state. The feeling of 'hitting the wall' in high-intensity endurance activities, such as in the later stages of a mountain or road marathon (at around 18–22 miles) or towards the end of a long day in the mountains, can largely be attributed to glycogen depletion. Glycogen depletion can have a significant negative effect on performance in intermittent activities and leads to an increase in feelings of fatigue for performers. There are a number of studies, however, that indicate that carbohydrate loading before exercise and glucose intake during exercise can delay the incidence of glycogen depletion. Strategies for delaying or avoiding glycogen depletion are discussed later in this chapter in the section on nutrition.

Key point

In shorter duration intermittent activities, the possible mechanisms for fatigue relate to factors that cause an inhibition of excitation coupling or interruption in the contractile processes. Factors that have been indicated in these mechanisms include increases in H^+, citrate leakage, Ca^{2+} transport disruption, PCr depletion and K^+ leakage. In longer-term intermittent activities, the key mechanism thought to be responsible for fatigue is glycogen depletion.

8.3 Training and recovery in intermittent activity

The adaptations to training for intermittent activities will be determined by the nature of the training undertaken. The development of a training programme for an intermittent activity should match the demands of the adventure sport. A time and motion analysis such as that carried out by Bussell (1995) for canoe polo (see the section below on the physiology of canoe polo) is essential to understand the demands of the activity and determine training programme structure. The type of training should based on the specific demands of the adventure sport. Events such as a bouldering competition are intermittent by nature, with enforced breaks before the next problem can be tackled. This type of event provides an example of another interesting aspect for those involved in intermittent activities to consider: the form of recovery strategy to employ after an exercise bout. This section examines the impact of active and passive recoveries upon subsequent performance in intermittent activities.

Research has shown that lactate uptake by skeletal muscle is increased when light exercise is performed in preference to complete rest. There appears to be a dynamic situation with regard to the exchange of lactate as a fuel source across muscle fibres within an exercising muscle and between working and non-working muscle groups. Glycolytic metabolism and the effects of lactate on performance have now been studied in relation to active and passive recovery for over 30 years. Research has predominately employed a single high-intensity short-duration exercise bout capable of producing high levels of lactate. In early studies, relatively long-duration recoveries were utilized to achieve significantly lower lactate levels. Hermansen and Stensvold (1972) and Belcastro and Bonen (1975), for example, employed periods of 30 min treadmill running at 60–70 per cent of $\dot{V}O_{2\,max}$ and recumbent cycling at 30–45 per cent of $\dot{V}O_{2\,max}$.

Later studies led by Ahmaidi *et al.* (1996), Bogdanis *et al.* (1996) and Corder *et al.* (2000) examined the effects of shorter-duration active recovery between repeated bouts of cycling or parallel squats. These recovery periods, which were of 4–5 min duration, were not designed to reduce lactate levels to pre-exercise conditions, but to maintain performance levels across subsequent trials and to optimize lactate removal when compared to passive recovery. Ahmaidi *et al.* and Corder *et al.* found significant reductions in lactate during short-duration low-intensity active recoveries when compared to passive recovery involving three or more bouts. In addition, performance and rating of perceived exertion (RPE) appeared to improve in the low intensity recovery condition.

Although research findings remain equivocal as to whether reductions in lactate lead to a resultant improvement in performance, it does appear that long periods of recovery such as those employed by Hermansen and Stensvold (1972) and Belcastro and Bonen (1975) may not be required in situations of repeated exercise. Our research regarding active and passive recovery in climbing employed a 3.5 min recovery between five 2 min bouts of exercise. The findings indicated that in a climbing context, a walking-based active recovery resulted in lower blood lactate concentrations at the end of each recovery period compared to a passive recovery. In addition, the climbers reported lower perceptions of exertion when following the active recovery strategy. These results suggest that active recovery strategies would be beneficial to climbers during the intermittent activity encountered in bouldering and sport climbing training sessions and competition. These findings may also have application to a range of adventure sports that involve intermittent activity.

Key point

Training for intermittent activities, as for any adventure sport, should be made as specific as possible. As an athlete comes closer to an event for which they are training the exercise intensity, duration and recovery intervals should match the demands in the adventure sport. Surfers should attempt short sharp sessions that match a heat in competition rather than continuing with long-duration practices. Research indicates that active recoveries can improve performance during intermittent exercise and this could have implications for adventure sports performers taking part in events such as bouldering competitions.

8.4   Nutrition and ergogenic aids for intermittent activities

Beyond the adoption of a normal healthy balanced diet for those involved in intermittent adventure sports, any dietary manipulations should be determined by the duration of the intermittent activity. For short-duration intermittent adventure sports, such as taking part in a canoe polo match, dietary manipulations could involve those covered in Chapters 6 and 7. Dependent upon the nature of the intermittent activity, the use of creatine supplementation (Chapter 6) or buffer (bicarbonate or β-alanine) loading (Chapter 7) could be beneficial to performance. In prolonged duration intermittent and high-intensity aerobic activities, dietary manipulation of carbohydrate intake has been shown to result in improvements in endurance performance. These manipulations can include pre-exercise carbohydrate loading and carbohydrate feeding during the activity through the use of sports drinks.

Carbohydrates

Carbohydrate loading, carbo-loading, glycogen loading or glycogen super-compensation involves the manipulation of diet to increase the glycogen stores prior to endurance exercise. As was discussed in the section on fatigue for intermittent and high-intensity aerobic endurance activities, glycogen depletion leads to 'hitting the wall' and is detrimental to performance. Research indicates that for prolonged (over 60 min) high-intensity activity (at or above 75 per cent of $\dot{V}O_{2\,max}$), carbohydrate loading can result in an improvement in performance, an increase in time to exhaustion and a delay in fatigue.

The classic studies into glycogen depletion and carbohydrate loading took place in the 1960s and form the basis from which modern forms of carbohydrate manipulation were developed. Research by Bergström et al. (1966) and Bergström and Hultman (1967) provided a major breakthrough in understanding about the effects of glycogen loading on performance. Bergström and Hultman (1967) carried out a one-legged cycle test to deplete muscle glycogen stores in the exercising leg. The participants exercised for several hours, after which muscle biopsies were taken to record the levels of glycogen stored within the leg muscle. After a 3 day period, during which a high-carbohydrate diet was followed, the glycogen levels were reassessed. In the exercised muscles, the glycogen stores were found to have nearly doubled compared to those in the non-exercising leg. This study indicated that a bout of glycogen depleting exercise prior to a carbohydrate rich diet would result in higher muscle storage of glycogen.

Bergström and Hultman (1967) then published further study findings regarding the effect of diet on performance. In this study the nine participants followed either a normal diet, a low-carbohydrate diet or a high-carbohydrate for three days, prior to a cycle ergometer ride to exhaustion at an exercise intensity equivalent to 75 per cent

$\dot{V}O_{2\,max}$. After following a normal diet, the mean exercise time was 115 min. This decreased to 60 min after following a low-carbohydrate diet. After three days with a high-carbohydrate diet, the time to exhaustion was increased to a mean 170 min. This experiment clearly identified that enhancing muscle glycogen stores above normal levels resulted in an improvement in performance and time to fatigue. These and subsequent studies have indicated that carbohydrate loading can result in a 40–100 per cent increase in glycogen stores, depending upon the regimen followed.

From the results of these and other related studies, a classical model for carbohydrate loading was created. This regime involved the athlete altering their diet and exercise pattern in the final week before a competition. The classical model for carbohydrate loading involved an athlete following a low-carbohydrate (fat and protein-based) diet for three days, performing a bout of exhaustive exercise to deplete glycogen stores, followed by the adoption of a high-carbohydrate diet for the remaining days before the event. In this way, glycogen stores were severely depleted before bouncing back above normal resting levels, hence the term glycogen super-compensation.

The use of this dietary manipulation undoubtedly led athletes to increase their pre-race glycogen stores, but had a number of drawbacks which led to the development of alternative strategies. The three days following a low-carbohydrate diet and exhaustive exercise caused feelings of fatigue, dizziness and mood change. These feelings have been linked to a decrease in self-confidence and self-belief prior to competition, which appeared to have a detrimental effect on subsequent performance. Further, following this regimen has been shown to interfere with the final taper an athlete needs to make for competition.

Sherman *et al.* (1981, 1982) provided a modified carbohydrate loading regimen that can be followed by an athlete as it avoids the drawbacks found with the classical model. The Sherman model indicated that a decrease in exercise duration in the week prior to competition and an increase in carbohydrate to constitute 70 per cent

of total calorific intake resulted in very similar pre-race gains in muscle glycogen storage. This model not only provides a simple method for increasing glycogen storage, but also presents a regimen that matches the alterations in training associated with the taper for competition.

Research indicates that a typical glycogen loading diet should contain around 500–600 g of carbohydrate per day. Table 8.2 provides an example menu for a carbohydrate intake within this range. As is shown in the Table, it is important that the modifications to carbohydrate intake are not supplied through simple sugars. Increases in total carbohydrate ingestion should come from increases in complex carbohydrates consumption found in foods such as wholewheat pasta, bread and cereals. The use of a high-carbohydrate diet prior to competition can improve performance and delay fatigue.

Research has identified one further aspect that should be considered by athletes before the adoption of a pre-race carbohydrate loading regimen. The improvements in performance for cycling have been better than those for running and the difference may well relate to the weight-bearing nature of running. Each gram of glycogen stored results in the storage of 2.7 g of water. Almost 4 g (2.7 g H_2O + 1 g glycogen) of weight added for each gram of glycogen stored creates a 0.5 kg to 2.0 kg increase in total body weight that must be carried for a fell race or mountain marathon. Research results indicate that the weight gain associated with carbohydrate loading would have a lesser impact for performance in kayak marathon and would in fact help to increase pre-race cellular hydration. It is very important to model any carbohydrate loading strategy during a practice or unimportant race, rather than employing a new diet before a major event.

A further aspect to consider for prolonged intermittent and high-intensity aerobic endurance activities is that of carbohydrate feeding during competition. In events lasting more than 1 hr, the consumption of a carbohydrate sports drink has been shown to result in improvements in performance and delaying of fatigue. The use of a carbohydrate sports drink can not only have a glycogen sparing effect, but can also help to

Table 8.2 Menu providing 500–600 g of carbohydrate as part of a carbohydrate-loading phase

Food item	Carbohydrate content (g)
Breakfast	
100 g muesli	70
100 g Crunchy Bran®	55
4 × Weetabix®	50
Skimmed milk	12
4 × wholewheat toast	60
1 × banana	30
Lunch	
Baked potato	65
Baked beans (1/2 tin)	30
Glass skimmed milk	12
1 × low fat yogurt	8
1 × apple	20
Dinner	
100 g wholewheat pasta	70
Pasta sauce (low fat)	5
1 × low fat yogurt	8
Snacks	
1 × apple	20
1 × banana	30
1 × cup of grapes	30
2 × bagels	60
Total	510–530

maintain hydration status and make fluid more palatable. Research indicates that a 1 per cent drop in bodyweight (through water loss during exercise) is associated with a 2 per cent drop in speed. The effects of dehydration can have a major impact on performance in endurance events. Study findings suggest that the use of a 6–8 per cent carbohydrate drink can result in improvements in performance. Use of higher concentrations of drink can affect gastric emptying and lead to discomfort, so ideally should be avoided. If employing a carbohydrate drinking schedule during a race, ingestion should begin within the first 30 min of exercise and continue for the duration of the race. Fluid should be taken on board at a rate of 500 mL water per hour. With a 6–8 per cent carbohydrate solution, this would represent an intake of about 30–50 g of carbohydrate per hour (about 6–8 g per 100 ml).

An additional advantage obtained through the use of a carbohydrate drink relates to water absorption rate. Water and sodium (useful for maintaining electrolyte balance, which is covered in Chapter 10) absorption rates appear to be enhanced when combined with carbohydrate. Increasing fluid intake during intermittent activities may be more straightforward than during a race such as a kayak marathon when hands are needed on the paddle. However, it is still important to practice a drink strategy before an important event. It is also important to work out the volume of fluid intake you can tolerate without gastric discomfort.

Research indicates that the use of Platypus® or Camelbak® style water bottles (when compared with a traditional water bottle approach) improves the frequency of fluid intake during exercise, the total fluid intake and performance in an adventure sports context. Platypus® or Camelbak® water bottles are worn on the back and have a mouthpiece that enables water supply on demand to the wearer. They are probably most widely used in mountain biking compared to other adventure sport settings. In longer kayak marathons, the main issue relates to the equation between the total volume of fluid required during a race, the time taken to change to another Platypus®, and the detrimental effect of not changing over to a new fluid supply. The decision taken regarding stopping paddling to change to a new Platypus® must be balanced against the duration of the race. Again, practice and the use of more than one Platypus® will establish the optimal solution for an individual performer. In multi-day trips, the issue of cleaning this type of water bottle to maintain hygiene and avoid bacteria build-up, particularly when using carbohydrate drinks, is an issue the user would need to address.

Key point

The nutritional strategy employed by an adventure sports performer should be determined by the duration of the intermittent activity. In shorter duration events, creatine supplementation and buffer loading could be considered. In longer duration intermittent activities, glycogen loading and feeding during the activity could be considered. Glycogen loading has been shown to increase muscle glycogen stores by between 40–100 per cent above normal resting levels and result in an improvement in time to fatigue during exercise.

8.5   Water immersion

In Chapters 8–10, the main environmental stresses that can impact upon adventure sports performance are described. Any individual involved in a water-based adventure sport has, at some stage, to deal with the additional demands placed on the body by water immersion. In adventure sports such as surfing, diving or freediving, water immersion is

Biography: Fiona Jarvie

Adventure sport: Freestyle kayak

Fiona started kayaking when she joined Reading University Canoe Club in 1994. She was immediately hooked on the sport and took up freestyle in 1997 after she got her first car, enabling her to travel to her local playspot at Hurley Weir. Fiona took part in her first competition at Hurley in 1998 and went on to win her first international medal at the 2002 European Championships in Thun, Switzerland. Since that time she has won medals at national and international levels, including 1st place at the 2004 White Nile Freestyle Festival. In January 2005, she placed 3rd at the Freestyle World Championships in Penrith, Australia. Fiona is a qualified primary school teacher. Her training for freestyle includes sessions in the gym as well as time spent on the water.

Figure 8.1   Fiona Jarvie (photograph courtesy of James Farquharson/Cam Sans)

2005 Freestyle World Championships

In November 2004, I had just returned from a month of paddling on the big waves of the White Nile, Uganda. That was good training, perhaps, for the World Championships in Australia in January 2005. I was certainly 'boat-fit': freestyle kayaking uses all sorts of different muscles: arms obviously, but legs and core stomach muscles are also essential to put the boat on edge, throw 15 kg of plastic around your body and slice the ends of the boat through the water. A month of intense paddling certainly strengthened all of the right muscles.

Competition Freestyle kayaking is very feature dependent; you need to ensure that you are training on a similar river feature to that of the competition so that you will be able to pull off the highest scoring moves on the day. The river features in Uganda were big waves, where it was easy to get the boat to bounce into the air. This was fantastic but not very good training for the World Championships in Australia, which were to be held in a small hole. Smaller features need smaller, more precise movements to get the boat vertical and to stay on the feature for the full run time of 45 s. It was time to practice at my local spot in the freezing December weather to enhance my ability to consistently pull off hole-moves (e.g. cartwheels, split-wheels and loops, equivalent to a front somersault where the boat comes right out of the water and over your head). In-between times I would visit the gym to keep up my aerobic fitness, doing 1 min interval-training reps to ensure that I would be able to last the 45 s competition run without tiring.

When I reached Australia in early January, 10 days before the competition, I felt physically fit. Thankfully it didn't take long to get over the jet-lag; in fact the jet-lag was beneficial as I was waking up at 6.00 am and our training sessions were at 7.30–8.00 am. However, I was concerned with the small amount of time I had in which to get used to the river feature. On different features, you need to work out where each move can be performed and you need to make small changes to an ingrained technique. You may have to wait a few seconds longer before you throw your arms to complete a loop, or you may have to speed up your cartwheels in order to stay in the hole and not 'flush' off (falling off the back of the wave i.e. meaning you have to paddle back up to the top of the eddy, something which could cost you the competition).

Organized practice times for each country were short, 1 hr a day for 30 people, approximately 2 min on the wave per person; not a lot of time to hone that technique! Video analysis with the team coach after the session really helped to see where on the wave the different moves needed to be set up, which moves were worth doing and which moves would definitely not score. Between official practice times, I was trying to get on the wave in the 'free' sessions at approximately the same time that my heats would be (midday), so that I could get used to being on the water at that time and work out when I would need to eat, etc.

When the day of the prelims arrived, I felt I was not fully prepared. I could have done with a day on the feature by myself to really get used to the way it fluctuated and to make sure I cound consistently pull off the moves I knew I was capable of. However, I had planned and practised what I was going to do in my 45 s ride which I had to repeat twice, the scores of which were added together. It was now a mental battle to keep myself calm enough to put what I knew into practice. Thankfully, it all came together and I got through to the semis in 2nd place.

I would have been more physically prepared for the semis if I'd had a full rest day between the heats and semis. However, to increase my confidence I felt I needed another short practice, just to be sure that my body was reacting to my brain at the right times on the wave. I felt I had achieved the biggest hurdle (getting through the heats), and perhaps I relaxed a little too much going into the semis as I only just made it through into the top 5 for the finals. Perhaps I would have benefited from that rest day.

Once again, I had to prepare myself mentally for the next round of competition, this time the finals. Freestyle finals take a knock-out format: all paddlers take a 45 s run and the lowest scoring paddler is 'knocked out'. This continues until the last two paddlers take their final run, and the highest scoring paddler is crowned Champion. This format of finals is demanding both physically and psychologically. If you reach the last two, you will have to do four runs of 45 s. This does not seem a long time, however it is difficult to fit the required moves into the time and it is extremely tiring as a majority of this time is spent in anaerobic exercise sprinting up the eddy or with your head underwater. Remembering to breathe when your head is out of the water is, obviously, extremely important but can be difficult when there are so many other things to think about! Holding my breath and forgetting to breathe has resulted in me being more exhausted than I should have been on a number of occasions.

In this knock-out final I was concerned whether I would make the moves that I knew I could do. One mistake would mean flushing off the back of the hole, and could cost the competition. Watching other people have bad rides gave me some confidence that I could do better (not the best way to compete).

My first ride was a high scoring one (I was in 2nd place) and thankfully I still had some energy left. My second ride was also good (although lower scoring than the last), but this time I was quite tired at the end. By my third ride, I knew I was in a medal position as there were only three of us left. I watched one girl have her ride and I knew exactly what I needed to do to get into 2nd place: achieve the moves I had completed on every ride so far. Unfortunately, I was so concerned about flushing off the back of the wave that I moved my boat around tentatively and didn't get the moves I needed

Figure 8.2 Fiona Jarvie competing at the 2005 Great Britain Team selections, Nottingham, June 2005 (photograph courtesy of Cam Sans)

to. In fact, it was my lowest scoring ride of the competition. When you get to a position where you are only 45 s away from being 2nd or even 1st in the World it becomes a very realistic prospect. I was very disappointed that I had been knocked out of the competition on a ride that was nowhere near my best. However, when I put what I had achieved over the whole competition into perspective, I was more than happy with the position of Freestyle World Championships Bronze Medallist!

prolonged and therefore a major issue regards the effects of water-induced hypothermia. In sports such as kayaking, canoeing and sailing, hypothermia presents a potential threat to homeostasis. In addition (perhaps with the exception of playboating/freestyle), a sudden and often unplanned water immersion can lead to cold shock. Gaining an understanding of the effects of short-term and long-term water immersion on physiological functioning is essential for those involved in water-based adventure activities, the focus of this chapter. Having an understanding of the threats to homeostasis presented by water immersion is important for coaches planning a day's river descent during the winter months when many rivers are at their best levels, for a group of sea kayakers planning a sea journey or for a sailor dealing with a sudden capsize during a race. For anyone involved in water-based adventure sports, if for no other reason than safety, knowledge of the effects of water immersion is essential.

In the summer months around Britain, mean sea temperature is around 15 °C. In the winter months this temperature falls to an average of 5 °C. In the Artic waters encountered by sea paddling expeditions the temperature is regularly close to freezing. Rivers in the UK have a temperature of over 20 °C in the summer to just above freezing in the winter months. Even in summer months, there can be a great variation in river temperatures. The Kaituna River in New Zealand benefits from geothermal warming and to a British paddler, feels like paddling in bath water. In contrast, a summer capsize when paddling glacier-fed alpine rivers, such as the Inn in Austria, can result in a sudden change

from a high (30 °C+) ambient temperature to a low (10 °C or less) water immersion temperature. The effects of a capsize and roll (or capsize and swim) creates an additional physiological stress over and above the physical demands. The body has a variety of mechanisms through which to attempt to overcome the stress induced by water immersion.

Weather changes or boat problems during ocean sailing races such as the Sydney to Hobart Race (described by Ben Ainslie in this chapter) or Vendee Globe single-handed, non-stop, round-the-world race can, in an instant, change the focus from racing to survival. An example of this can be seen through the survival of Tony Bullimore during the 1996 Vendee Globe race. Two months into the race (in January 1997) disaster struck for the 56 year old ex-Royal Marine from Bristol. In the middle of 60 knot winds and 15 m waves, his yacht the Excide Challenger lost its keel and capsized. Fortunately for Bullimore, who was below deck at the time of the capsize, the yacht had a series of compartments which trapped sufficient air for his survival. Nevertheless, he found himself without light and waist deep in water. Despite these problems, Bullimore was able to activate his Emergency Position Indicating Radio Beacon (EPIRB) and construct a hammock (from a cargo net) above the water line which enabled him to stay out of the water. It was five days later when, suffering from hypothermia, frostbite and dehydration, he was rescued by Australian Naval ship HMS Adelaide on responding to his EPIRB. He had sustained himself during his wait for rescue with limited water supplies and chocolate. His experience as a royal marine and sailor provided him with the knowledge necessary to survive the effects of his experience.

A number of common misconceptions had developed regarding the threat of water immersion to homeostasis. It is only recently that research (Golden and Tipton, 2002) has led to a clearer understanding of the physiological stress associated with water immersion. For instance, in the 1960s and 1970s it was regarded as beneficial to undress in water after an accidental immersion to avoid the clothing weighing you down and impeding swimming. In addition, victims were encouraged to swim around to keep warm. In survival training courses, trainees were encouraged to put their heads in the water as a method of drown-proofing. Research into the effects of water immersion have shown that contrary to these beliefs, clothes should be kept on to preserve heat and air trapped in clothing, movement should be kept to a minimum in order to avoid further heat and air loss and the head immersion for drown-proofing purposes accelerates the onset of hypothermia. The increased depth of understanding regarding the implications of cold water immersion presents an excellent aid for all water-based adventure sport performers and coaches.

Key point

Involvement in certain adventure sports, such as diving or kayaking, can involve planned or accidental water immersion. Developing an understanding of the effects of water immersion on the body is important for safety and improving performance.

Thermoregulation

Humans belong to a group of mammals and birds called homeotherms. Homeotherms, as a part of their homeostatic mechanisms, attempt to maintain a constant body temperature (around $37\,^{\circ}$C for humans). Thermoregulation requires humans to have a variety of mechanisms to either increase or reduce body temperature. Humans protect their body temperature to maintain an optimal internal environment for the cells of the body. Research indicates, for example, that when core body temperature falls below $37\,^{\circ}$C this leads to a slowing in brain cell activity and a decrease in our performance. In addition, the enzymes that catalyse metabolic reactions function optimally within a narrow temperature range.

Body temperature can be measured by a variety of methods. Commonly in the home, sublingual (under the tongue) temperature is used as an estimate of deep body or core temperature. However, the results of this type of measurement can be influenced by drinks taken prior to measurement. In addition, temperature at the body's surface is not always reflective of core temperature. For instance, in thermo-neutral conditions the skin temperature is normally around $33\,^{\circ}$C, lower by $4\,^{\circ}$C than core body temperature. As a consequence, in a clinical setting core body temperature is assessed using methods that provide a closer measure of core temperature. Core temperature T_c can be measured using a swallowed radio telemetry pill, T_{in} (intestinal) rectal probe (T_{re}), oesophageal probe (T_{os}) or through the use of a tympanic thermistor (T_{tym}) inserted into the ear canal. Most commonly in laboratory research T_{re} is used due to the expense of using telemetry pills (around £35 each), discomfort felt with T_{os} and the disagreement found between T_{tym} and other measures of T_c.

The body's thermoregulatory mechanisms maintain body temperature at around $37\,^{\circ}$C. The thermoneutral temperature for humans in water is around $35-35.5\,^{\circ}$C. When the water temperature is below this, humans will need to activate mechanisms to conserve existing heat and produce additional heat if necessary. Heat is lost from the body via radiation, convection, conduction and evaporation. These processes are discussed in more detail in Chapter 9. However, conduction and convection are introduced briefly here as they represent the key heat-loss methods in water.

In accordance with the second law of thermo-dynamics (see next section), heat will pass from the core to the tissues of the outer body. When immersed in water below body temperature, heat will move from the core and heat a boundary layer of water next to the skin. This boundary layer heating process results in heat loss from the core. When the water is moving or a person swims, the boundary layer will be disturbed and a new boundary layer (cooler than the old one) forms around the body. This movement of water against the skin in the sea (or a river) or water movement during swimming will enhance conductive cooling. It is partly for this reason that people, unless close enough to shore or a vessel, are now advised not to swim as the swimming movement will increase the rate at which the body cools. The properties of water make it far more effective as a medium through which heat transfer takes place compared to air. In other words, heat is lost much more quickly in water than in air. Due to its increased density, water has a greater thermal conductivity (about 24–25 times that of air) which results in humans cooling around four times faster in water. In addition, where normal clothing creates an insulative barrier to cooling in air, the body is almost 100 per cent exposed to cooling in water as the barrier is lost.

Key point

Humans are homeotherms and attempt to maintain core temperature at around 37 °C. There are a variety of methods through which core body temperature can be measured in the laboratory; the most commonly used is rectal temperature. Heat loss in water is primarily through conduction and convection.

Heat-production mechanisms

Two key laws of thermodynamics govern the transfer of heat as a form of energy. Firstly, as was described in Chapter 2, energy cannot be created or destroyed but is transferred from one form to another. Energy can be chemical, nuclear, mechanical, heat, light or electric (see Table 2.2). Secondly, heat flows along a thermal gradient from high to low. In a cold environment, such as that encountered during cold water immersion, the body has a variety of thermoregulatory mechanisms through which to maintain core temperature. Our metabolism represents the sum of all the reactions that take place within the body. As discussed in Chapter 10, human metabolic reactions are around 40 per cent efficient, meaning that 40 per cent of the energy released through the catabolism of ATP and other substrates provides energy for work. The other 60 per cent is liberated as heat. In a cold environment, heat produced through metabolism serves to maintain core temperature.

The body detects cold through receptors located in the skin and these provide sensory information to the brain to initiate heat loss and heat production strategies. The heat production strategies are designed to increase the rate of metabolism and therefore increase heat production. Firstly, when you enter water below the thermoneutral zone the muscles in the body will begin to stiffen. This stiffening of the muscles is the result of an increase in muscle tone. The increased muscle tone is not associated with any movement, but is the result of muscular contractions that generate heat. If the body remains in water, the next phase of heat production to protect temperature is initiated: shivering. Shivering represents a cyclic contraction of opposed muscle groups. Opposed or antagonistic pairs (see Chapter 4) of muscles are activated alternatively to avoid large muscle movements during the shivering process. Muscles contract at a rate of about 10–20 per second. The rate and intensity of shivering continues to increase the longer a person is in the water. At its peak, shivering can raise metabolism about five times above rest. The energy for shivering, like any muscular contraction, comes from carbohydrate and fat metabolism. Shivering, however, ceases when the glycogen stores are depleted. It is for this reason that in a continually cooling situation, someone can be seen to be shivering and then suddenly stop. They have not suddenly become warmer but have depleted their glycogen stores. Once shivering has stopped the rate of cooling will increase, exacerbating the situation.

As well as the stimulation of a shivering response to raise metabolism, the hypothalamus initiates an endocrine response to cold environments. The hypothalamus directly stimulates the adrenal medulla to release adrenalin and noradrenalin into the bloodstream. As described in Chapter 3, this affects all cells in the body and stimulates an increase the general rate of metabolism and an increased activation of glucose and fatty acid metabolism. To enhance the activation of metabolism the hypothalamus, through its effect on the anterior pituitary gland (APG), stimulates the thyroid gland to release thyroid hormone. In addition, again through its effect on the APG, the hypothalamus increases cortisol release from the adrenal cortex to further increase the supply of blood glucose.

Key point

Energy cannot be destroyed, but transfers from one form to another. For instance, electrical energy to a heater in the home will be converted to heat energy. Heat flows down a thermal gradient from low to high. There are a variety of mechanisms that can increase heat production when a body is exposed to a cold environment. The brain can stimulate an increase in metabolism (the chemical reactions in the body that liberate heat as a by-product). When introduced to the cold, muscles will begin by stiffening (increasing muscle tone) after which shivering is induced. Both of these mechanisms serve to increase heat production within the body.

Heat conservation mechanisms

The primary mechanism for heat conservation in the human body is vasoconstriction. The raising of goosebumps, or pilo erection, represents a further heat conservation in fur-covered animals. The raising of hair in cold environments is designed to trap air within the coat to create a larger insulative barrier to cold. This mechanism is still present in humans, despite not having sufficient hair for the establishment of a satisfactory insulative layer.

In the heat, around 4 L of blood is circulated to the skin during vasodilation for cooling purposes. In the cold, vasoconstriction results in a 99 per cent reduction of this flow. The flow of blood to the skin represents a major mechanism for core temperature regulation so in cold conditions, through the act of vasoconstriction, blood flow to the periphery is dramatically reduced and heat loss is minimized. By reducing blood flow to the periphery, the body establishes the skin and subcutaneous fat (the fat layer below the skin) as an insulative layer between the water and the body's core. The one exception to this is the scalp which maintains its blood flow to protect the functioning of the brain. The continued flow of blood to the scalp can result in the head being the main area of bodily heat loss. This is the main reason why drown-proofing is not taught in survival swimming courses.

The vasoconstriction associated with entering a cold environment leads to pressure receptors in the main arteries detecting an increase in blood volume. To counter this, some plasma is passed from the blood to the kidneys for excretion from the body as urine, thereby decreasing the volume of blood. This effect is called cold diuresis and is a familiar consequence of working in a cold environment. The cold diuresis sensation is heightened during full immersion in cold water. When standing in air, most blood pools in the legs (as was described in Chapter 4); as you enter water the pressure moves more blood into central flow, causing the blood volume receptors to detect an increase in blood volume thereby exacerbating cold diuresis.

If you kayak, surf, windsurf or sail in the winter months you may have found that at the start of activity your hands become cold quickly due to the effects of vasoconstriction but later appear to warm-up, meaning you can then operate for the day without painfully cold hands. This sensation, first described by Lewis (1930), was termed the *hunting reaction*. After about 5–10 min of cold exposure, blood vessels in the fingers dilate increasing blood flow and local temperature in the hands. An effect of this phenomenon, now commonly referred to as finger cold-induced vasodilation (CIVD), is an increase in manual dexterity which is normally beneficial to performance. It is thought that finger CIVD, which appears to serve the purpose of reducing the incidence of cold injuries to the hands, is brought about through a reduction in noradrenalin release during exposure to the cold. Research indicates that the finger CIVD is more pronounced when the body core temperature is maintained. Therefore, dressing to keep the core warm during water-based activities can maximize maintenance of manual dexterity and improve performance.

The body's responses to heat conservation have so far been described. Adventure sports performers can also employ clothing strategies that maximize heat retention during water-based activity participation. Research indicates that a normally dressed 70 kg male adult wearing light clothes and immersed in 12 °C water would cool at a rate of 2.3 °C per hour, representing a survival time of 65 min. Survival time would be extended to 4 hr if the individual was wearing a dry suit over the clothes, with a cooling rate of 1.1 °C per hour. The use of a 5 mm closed-cell neoprene wetsuit would decrease the cooling rate to 0.5 °C per hour and extend survival time to 10 hr. The use of a full immersion suit with insulative clothing beneath would result in a survival time of around 15 hours (cooling rate 0.3 °C per hour). It is for this reason that divers often wear fully-insulated dry suits. The enhanced thermal qualities of a wetsuit over a dry suit or dry trousers and cag combination (providing it has a dry seal) should be of interest to kayakers or canoeists, who often paddle in winter months.

Key point

To preserve heat in a cold environment, the body initiates pilo erection and vasoconstriction. In addition, the type of clothing worn by an adventure sports performer can help with insulation. During cold water immersion, the use of a wetsuit or dry suit with insulative clothing beneath can greatly extend survival time.

Effects of water immersion

In early research relating to the consequences of unplanned (cold) water immersion, drowning and hypothermia were alternatively highlighted as the key concerns for survival. Golden and Hervey (1981) established a series of stages in water immersion, accurately describing the risks associated with accidental water immersion, from statistics gathered on fatalities in open water swimming,. Research in the 1970s indicated that over half the people who died as a result of open-water immersion did so within 3 m of safety and two-thirds of these were reported as good swimmers. There was clearly an inherent risk in open water immersion, unrelated to hypothermia. As a consequence, Golden and Hervey suggested that the risks involved with accidental water immersion should be classified by the length of time of the immersion. Their risk classification system is shown in Table 8.3 with a summary of the predominant risk factors during each stage of immersion.

Initial response (Cold shock)

The initial response begins immediately on contact with water and peaks at around 30 s. Symptoms begin to subside as the immersion duration approaches 3 min. The initial or cold shock

Table 8.3 Stages in response to water immersion (adapted from Golden and Tipton, 2002)

Specific risk	Timing of response	Summary of risks
Initial response	0–3 min	Cold shock: immediate vasoconstriction, rise in BP and HR and breathing rate Associated risks: cardiac problems and arrhythmia, gasp response, dizziness, confusion, panic, reduced breath hold capacity, drowning
Short-term response	3–30 min	Associated risks: loss of manual dexterity, impaired co-ordination and swimming ability, decreasing strength, drowning
Long-term response	>30 min	Associated risks: hypothermia, drowning
Post-immersion response	On and post-rescue	Associated risks: afterdrop in temperature, cardiac problems

response does not occur on entry to warm water and is thought to be a response initiated by cold receptor stimulation. Research indicates that a greater response occurs in colder water and for those who have had fewer cold water immersions. The cold shock response is thought to be the primary cause of death in incidents when individuals drown within 3 m of safety. The cold receptors primarily trigger a CNS driven alterations in breathing and circulation. On entry to cold water (below 15 °C) there is an immediate increase in the respiratory drive. This results in an individual taking an initial large gasp of air (around 2–3 L) followed by the initiation of hyperventilation, where the rate of breathing is increased without an increase in depth of breathing. The incidence of hyperventilation can cause dizziness and confusion and is consequently thought to add to the panic associated with the immersion. Golden and Tipton (2002) believe that the most significant risk involved in the cold shock response relates to a decrease in breath-hold time. During the cold shock response, breath-hold time is reduced to around 10 s, whereas most individuals on the land can hold their breath for

around 60 s. The decrease in breath-hold time results in an increased risk of water inhalation and drowning.

Concomitant with the increases in breathing rate are CNS-driven alterations in circulation. Initial entry into cold water creates an immediate rise in blood pressure and heart rate and triggers vasoconstriction. These responses establish an increase in cardiac output that is thought to present a cardiac or stroke risk for those with cardiovascular disease. In addition, cold water immersion has been associated with the occurrence of cardiac arrhythmias created by an HR drive competition between the cold shock HR response (to increase HR) and the diving reflex (a drive to decrease HR). The diving reflex, common to many mammals, triggers vasoconstriction, breath-holding and slowing of the heart rate upon entry into water. Cardiac arrhythmia results from the competition between the cold shock response and diving reflex to control HR during cold water immersion.

Golden and Tipton (2002) have suggested a number of factors that appear to affect the cold shock response and could help during an accidental immersion. Firstly, safety measures should be put in place wherever possible to avoid the risk of accidental water immersion. The use of clothing appropriate to the risk of accidental immersion and the use of a buoyancy aid or lifejacket, as appropriate to the activity, can alleviate the cold shock response. Knowledge of the effects of initial cold water immersion for individuals for whom it is a risk can help the person to cope better with the effects. Where possible, an individual should hold on to something and remember that the symptoms will pass. Research indicates that the cold shock response is blunted in those who have a lower body fat percentage and are aerobically fitter. The use of habituation strategies, acclimatizing an individual to cold water immersions, has additionally been shown to reduce the cold shock response. This has implications for kayakers possibly using one or more habituation rolls in cold water before and during a river descent to lessen the cold shock response in a must-roll situation.

Key point

Golden and Hervey (1981) suggested a system of four stages regarding the effect of water immersion on the body. These four stages were cold-shock, short-term response, long-term response and post-immersion response. Cold shock (immediate response) lasts for the first 3 min of immersion and peaks around 30 s after water entry. The cold detected by cold receptors in the skin triggers a number of breathing and circulatory changes within the body, including an increase in breathing rate (hyperventilation) and competitive heart rate drives that can lead to cardiac arrhythmia. It is thought that cold shock is responsible for many of the fatalities that occur within 3 m of safety.

Short-term response

Beyond 3 min of immersion the short-term responses become the predominant concern for an individual. The short-term response to water immersion relate to cold-induced decreases in muscular strength, manual dexterity and coordination. Research relating to swimming ability in cold water has shown that performance is impaired as part of vicious circle of decreasing streamlining. Swimming in cold water, even for experienced swimmers, results in an increased breathing frequency. The increased breathing rate results in a less streamlined swimming position, which in turn leads to an increased oxygen demand creating a further rise in breathing rate and a worsening in body position in the water. The body becomes more vertical as the individual eventually becomes static in the water and effort becomes focused on breathing and keeping the head above water.

In a study with competent fully-clothed swimmers in 25 °C water, all 10 completed a 10 min swim. In water of 5 °C, only 3 managed to complete the 10 min swim. Increases in breathing rate and stroke rate and a decrease in swim steamlining were found to be higher in those that failed to complete the swim compared to those who succeeded in completing the 10 min swim at 5 °C. Golden and Tipton (2002) suggested that this knowledge should be made widely available to enable any individual to make an informed decision about swimming for shore in cold water. Where possible, try to complete manual tasks such as life jacket inflation as early as possible due to the decreasing manual dexterity and strength associated with short-term water immersion. Golden and Tipton (2002) recommend keeping swimming to a minimum and avoiding waving to rescuers to retain as much air trapped in clothing as possible.

Key point

The short-term responses to water immersion are those occurring between 3–30 min after entry into the water. Key aspects during this stage include decreases in strength, manual dexterity and coordination. These can lead to an impaired swimming ability as well as difficulties in tasks such as manipulating valves to add air manually to a life jacket, or releasing flares.

Long-term response

After 30 min of immersion, the primary concern becomes continued heat loss and hypothermia. Survival time during long-term water immersion is related to the rate of hypothermia and to the type of clothing worn. In response to continued water immersion, the body's key mechanism to generate heat and protect core body temperature is shivering. The shivering response is initiated soon after entry into cold water. In the early stages of the onset of hypothermia, shivering begins as an intermittent response. As temperature continues to decline, shivering becomes continuous until carbohydrate stores are exhausted. After this point, muscles go into a spasm pulling the body into a flexed foetal-like position and the individual becomes incapable of assisting in their rescue. At a temperature below 34 °C, brain function is impaired and this leads to unconsciousness. Heart failure follows as body temperature reaches around 28 °C. To protect against the progressive drop in core temperature the primary measures include clothing choice and, in the incidence of an accidental cold water immersion, removal from the water onto a floating object, for example. Although it may feel colder in the wind, the rate of cooling would be slowed. In light of the loss of strength and manual dexterity that is associated with short-term immersion, removing yourself from the water should be effected as quickly as possible.

Key point

Beyond 30 min of water immersion during the long-term response, the key concern is continued core temperature cooling.

Shivering, to protect core temperature, continues until glycogen depletion.

Post-immersion response

A number of risks associated with rescue have been identified and should be considered for individuals who have been involved in cold water immersion. First identified in the 1790s by physician James Currie, research has indicated that individuals who survive a water immersion episode often experience an 'afterdrop' in core body temperature following removal from the water. This afterdrop has been linked with post-immersion fatalities. It has been suggested that such deaths are caused by cold blood from the periphery entering central circulation leading to cardiac failure. Golden and Hervey suggested that post-immersion afterdrop is a consequence of the physics of heat transfer. Even after removal from water, an individual will experience a continued movement of heat from the warmer core to the periphery down the thermal gradient until the periphery is re-warmed. In the instance of an afterdrop fatality, Golden and Hervey suggested that the continued cooling post-immersion results in the individual's core temperature dropping to an irrecoverable level.

An important implication regarding post-immersion death has been identified from the change in pressure on the body that occurs on removal from the water. As was described earlier in this section, head-out water immersion leads to a hydrostatic squeeze on the body resulting in an increase in venous return, blood pressure, blood entering to the heart and cardiac output. In addition, the vasoconstriction associated with cold water immersion adds to this effect. To adjust to the increase in blood volume detected by blood vessel receptors, plasma is removed from the blood and passed from the body via the kidneys as urine. In this way, the blood volume is decreased.

When removed from the water, survivors were often lifted in a vertical position either into a boat or at the end of a helicopter strop. It is believed that vertical removal from the water results in the greatest challenge to a survivor's circulation. On removal from the water, a survivor faces a drop in blood pressure, blood volume and cardiac output due to the removal of the hydrostatic squeeze. The reduced blood volume (hypovolemia), a cold-induced impairment of baroreceptor response to adjust blood pressure and a drop in cardiac output may be the cause of post-immersion unconsciousness (lack of blood and oxygen to the brain) and cardiac problems. It is for this reason that many air-sea rescues are completed with horizontal recovery of the victim, which has been shown to decrease the post-immersion circulatory stress.

For anyone involved in adventure sports, especially those involved in water-based adventure sports, understanding the effects of water immersion is an essential component of an adequately prepared performer's knowledge. Recognition of the stages in cold water immersion could save the life of the adventure sports performer, a member of their group or a stranger they encountered during a day sailing or paddling.

Key point

The post-immersion stage can present a number of problems for a water immersion survivor. Post-water immersion, the body initially experiences a continued drop in temperature which is termed the afterdrop. Research indicates that horizontal evacuation from the water is safer than vertical (upright) rescue.

8.6   The physiology of water-based adventure sports

Sailing

Sailing is a very popular recreational activity and a highly competitive adventure sport. Events are held at European and World level and it is included in the Olympic schedule. Furthermore, some classes in sailing are among the most expensive and technically sophisticated of any sport. Until recently, however, the physiological demands in sailing – hiking, trapezing, sail pumping and coffee-grinding – were largely overlooked. Single-handed dinghy sailing, for instance, was considered a relatively static sport. The main physiological challenge was thought to be the isometric contractions of the legs and abdominal muscles required during hiking. Recent research has challenged the traditional view and provided information about the physiological demands of sailing.

Research suggests that sailing provides greater physiological demands than was previously thought, with the requirements for skipper and crew being dependent upon the class of boat sailed. This research suggests that the different roles played by helm and crews in each sailing class, along with the dramatic changes in physical stresses brought about by an accidental capsize, make it essential that the training is based around the physiological demands for the individual crew member. In most classes, the training programmes for the helm should be different to those for the crew to reflect these physiological differences.

The diversity of physiological demands in sailing is reflected by the fact that the roles for dinghy double handers, single handers and match racing crews can be considered as intermittent activity. In contrast, typically for dinghy double handers and match racing helms where technical and tactical consideration are paramount, the physiological demands are perhaps best categorized as lower intensity aerobic endurance (except in the event of a mid-race capsize!). Despite these groupings, the physiological demands and the development of a training programme for any sailor should be individually developed and based upon the current fitness level of the athlete and the demands of their role.

The main physiological challenges in sailing include hiking, trapezing, sail pumping and coffee-grinding. Table 8.4 provides a summary of the predominant demands of each class. During upwind sailing, hiking is required to counteract the heeling effect or capsizing movement created by the wind's effect on the sail. *Hiking* involves the helm, crew or both, moving to and then sitting out on the

Table 8.4 Predominant physiological demands by sailing class (adapted from Cunningham, 2004)

Class	Sailors	Predominant physiological demand
Laser (dinghy)	Single-handed	Hiking
Finn (dinghy)	Single-handed	Hiking
Europe (dinghy)	Single-handed	Hiking
470 (dinghy)	Double-handed	Helm: hiking, crew: trapezing
49er (dinghy)	Double-handed	Helm and crew: trapezing
Tornado (catamaran)	Double-handed	Helm and crew: trapezing
Mistral (windsurfer)	Single-handed	Sail pumping
Star (keelboat)	Double-handed	Helm: hiking, crew: hiking (harness assisted)
Yngling (keelboat)	Triple-handed	Helm: hiking, crew: hiking (hobble assisted)
Yachting (match racing)	Single/multi-handed	Helm: dependent on number of crew, crew: coffee-grinding

windward side of the boat to provide a counteraction to the force of the wind, thereby righting the boat. During hiking, the muscles of the legs and abdomen are primarily stressed. In double-hander boats, such as 470s and 49ers, the heeling effect is countered by the crew (and helm in 49ers) trapezing. *Trapezing* involves the crew or crew and helm suspending their body over the side of the boat. In order to achieve this position, a harness is worn which is attached to a wire from high on the mast. The roles of the helm and crew change during upwind and downwind sailing. For example, in the 470 class during an upwind leg, the helm will hike while the crew is relatively less physiologically stressed trapezing. On a downwind leg, the helm may be involved in pumping the mainsail (within legal limits) which is normally quite lightly loaded, while the crew is working relatively harder pumping the spinnaker which has a heavier loading.

The new Neil Pryde RSX class represents the only windsurfing class in the Olympic programme, having recently replaced the Mistral windsurfing board as the Olympic board. The Mistral class was adopted as the windsurfer for the 1996 Olympic Games and has been used in all subsequent games until Athens. After the Athens 2004 Olympic games, the Neil Pryde RSX board was adopted as the Olympic board. The Neil Pryde RSX class is unique to Olympic sailing as all the boards and sails are made in the same factory so all windsurfers are using the same basic craft to sail. Prior to the Atlanta Olympics (1993), a decision was taken to allow unrestricted sail pumping for windsurfing competition. *Sail pumping* involves drawing the sail to and from the windsurfer to use the sail as a 'wing', resulting in an increase in speed. Research indicates that sail pumping can result in substantial increases in speed, perhaps up to threefold in some wind speeds. The use of sail pumping creates a significant rise in the physiological demands placed upon the windsurfer.

In the Olympic classes, sailing competitions tend to have two races per day although the target time for race duration is different for each class. Typically, windsurfing races at the Olympics will take place over 35–45 min, Laser races over 60 min, Finns over 75 min and Star races over 90 min. The required wind speed is normally between 5–30 knots for racing to be allowed. With two races per day, it means that sailors can be on the water for up to 4–6 hr per day which increases the physical demands placed on each individual and has implications for nutritional strategies. In addition to the physical demands associated with each sailing class, environmental stresses of cold and hot ambient temperatures, the effects of wind exposure and possible water immersion can impact upon physiological functioning. All these issues must be addressed when planning a training programme for a sailor.

As mentioned in the introduction, many researchers believed that dinghy sailing placed relatively low physiological demands upon the sailor. Spurway and Burns (1993) suggested that the major demand rose through hiking and that isometric contraction work employing the muscles involved in hiking should predominate in training. Vogiatzis *et al.* (1993, 1994) suggested that aerobic training should not be emphasized for dinghy sailors (Laser class).

More recently, research has suggested that the movement required in hiking and trapezing places demands upon the aerobic capacity of the sailor. This concept is not new, however, as Harrison *et al.* (1988) suggest that hiking in a static position was a rarity due to changes in wind speed and direction that would necessitate continual movement on the part of the sailor to counteract changes in the heeling effect. As a result of their research, Devienne and Guezennec (2000) conclude that, in contrast to the majority of previous research, aerobic capacity should be considered as an important performance factor for sailors. Indeed, research by Cunningham (2007) with British Olympic sailors suggests that hiking on a specifically designed hiking ergometer can result in an oxygen consumption rate equivalent to 65–70 per cent of $\dot{V}O_{2\,max}$ during single-handed race simulations (Laser class). In agreement with this, Castagna and Brisswalter (2007) in an on-water study found very similar results, with hiking during a simulated race requiring up to 69 per cent of $\dot{V}O_{2\,max}$.

For the Mistral windsurfing class, research by Vogiatzis *et al.* (2002) suggests that sail pumping is a physically demanding activity that places significant demands upon the aerobic system. In their study, they compared the physiological cost of windsurfing with and without sail pumping (SP) and found that the mean oxygen cost more than doubled for both males and females (males $19.2\ mL\,kg^{-1}\,min^{-1}$ without SP compared to $48.4\ mL\,kg^{-1}\,min^{-1}$; females $15.7\ mL\,kg^{-1}\,min^{-1}$ compared to $40.2\ mL\,kg^{-1}\,min^{-1}$). These figures represent between 77–87 per cent of each sailor's $\dot{V}O_{2\,max}$. This result is in agreement with research by De Vito *et al.* (1997) who found windsurfing with sail pumping required 75 per cent of the $\dot{V}O_{2\,max}$ of the sailors in their study. It is thought that the new Olympic windsurfing class of the Neil Pryde RSX will show very similar physiological demands as the Mistral class. The results of recent research for both windsurfing and dinghy sailing have indicated the importance of aerobic capacity for sailors and led to changes in training programme emphasis.

Winches are used on the larger keel boats for hauling and adding tension to sails. In match racing pairs of cranks are attached to a pedestal and are used to power the winches. The cranks are moved in a bicycle action referred to as *coffee-grinding*. The winching systems on larger sailboats have a gearbox to provide multiple speeds for use with the large sails required to maximize speed in match racing and to allow the power to be transferred to alternative winch systems. A key role for matching racing crews is to provide the propulsive force required in coffee-grinding. Coffee-grinding on yachts during a race represents a high-intensity intermittent activity where upper body strength and a good aerobic base and anaerobic fitness are essential.

For helms, exercise which maintains a steady or decreased body mass along with hand-eye coordination and agility activities could be considered in addition to tactical and technical development. For dinghy and yacht racing crews, the intermittent activities chosen for training should match both the nature of the rest–exercise intervals during sailing and be as specific as possible to the exercise during sailing. For windsurfers, where the activity during racing can be classified as high-intensity aerobic endurance (the focus of Chapter 9), a major emphasis should be lactate threshold training.

In addition, work to develop strength (resistance training based around the Olympic lifts) and maintaining an aerobic base appear to be beneficial to performance.

Key point

Sailing in the past was considered a marginally active adventure sport. Recent research suggests that the physiological demands of the activity vary according to role in the boat, class of craft and race duration. The key physical demands in sailing are hiking, trapezing, sail pumping and coffee-grinding. In addition to these demands, the environmental stresses of heat, cold and water immersion can add to the challenge of sailing. When planning a training programme for a sailor the emphasis beyond skill development should reflect the key physiological demands of the class.

Biography: Ben Ainslie OBE

Adventure sport: Sailing

Ben, born in Macclesfield, began sailing at the age of 8 years at Restronguet near Falmouth in Cornwall. His father was a successful sailor and Ben quickly found success. At the age of 16 he became Laser Radial World Champion. Supported by his parents, Ben was picked for the 1996 Olympic Games at the age of 19 where he won a silver medal. He has continued his career with great success winning many international medals and Gold at the 2000 and 2004 Olympic Games in Laser and Finn classes, respectively. Ben has won eight European and World titles so far and recently won the 2005 Finn Gold Cup for a record fourth consecutive occasion. He plans to compete at the 2008 Beijing and 2012 London Olympic

Figure 8.3   Ben Ainslie (photograph courtesy of Richard Langdon)

Games as well as taking part in the America's Cup. He has been named ISAF World Sailor of the Year in 1998 and 2002 as well as being British Yachtsman of the Year on four occasions.

The 2005 Sydney–Hobart Race

The Sydney to Hobart yacht race is often referred to as the world's toughest yacht race. The history of the race is littered with stories of disaster or amazing feats of seamanship in extraordinary conditions. This 600 mile race is particularly tough due to the unique and extremely challenging conditions the boats encounter as they cross Bass straight, the body of ocean between Australia's southernmost tip and Hobart in Tasmania. Bass straight is notorious to yachtsman around the world as the southerly running currents run against the weather fronts that produce 40–80 mph winds. The waves produced by this phenomenon are quite literally mountainous and the combination of wind and water can be deadly. The most tragic of races was the 1998 Sydney–Hobart where the fleet of racers were caught in one of the biggest storms to hit Bass straight in the 20th century. Four yachtsmen were lost including Glyn Charles, a highly experienced Olympic yachtsman and a close friend.

I was invited to sail in the last Sydney–Hobart race on *Shockwave*, one of the biggest and most technically advanced racing yachts ever built. This boat is a far cry from the single handed Finn dinghy in which I raced at the 2004 Olympic Games in Athens. *Shockwave* is 100 ft long, 10 times the size of my Finn with a crew of 24. Up until this race, my offshore sailing experience was minimal and while I had proved myself at inshore racing, this was a big test! My Father skippered a boat in the first Whitbread around the World Race so I always felt that offshore sailing was in my genes, but I wasn't one hundred percent sure about how the race would play out.

The fleet of over 150 boats started in idyllic conditions on Sydney Harbour on Boxing Day, 2005. As we reached out of Sydney Heads, the spectator fleet and the number of helicopters in the air astounded me. I had been involved in some pretty big races but this was unbelievable!

Normally the prevailing wind is southerly and this gives the boats a beat into the wind all the way to Tasmania. However, the weather gods were on our side and as we headed out to sea we took the lead over our sister ship *Wild Oats* and set a spinnaker for a down wind sail. I was at the helm steering the boat and a cameraman came up to me to ask what I thought of my first Hobart race. 'Piece of piss, I don't know what all the fuss is about', I replied. As soon as I said it, I wished I had been a little more cautious since we still had over 500 miles to sail to Hobart and I knew the history of the race.

Figure 8.4 Ben Ainslie winning a history-making fourth Finn Gold Cup in Russia, 2005 (photograph courtesy of Peter Bentley)

My fears were realized on the second night at sea. The wind had built to 35 knots and the boat was screaming along at 30 knots in pitch-black conditions. I was off watch and below deck asleep as the boat flew into the back of a large wave, dropping the speed from 30 to 8 knots. The result was that the spinnaker broke into two and the bow frame was ripped away from the boat! The next thing I knew I was on the bow of the boat in my thermal underwear, a place I really shouldn't have been but I wanted to help. We had no lifelines or any of the equipment we should have had but it was one of those situations you just have to deal with. We eventually recovered the spinnaker and got the boat back on its feet. As we all got back to the safety of the cockpit someone started laughing and eventually the whole crew joined in, with sheer relief. The wipe-out cost us badly and we ended up loosing the race record and line honours to the very impressive *Wild Oats*.

Some time later I was watching a movie called the 'The Worlds Fastest Indian'. In the movie, the character played by Anthony Hopkins talks about risking everything for the buzz of being on his rusty old bike doing crazy speeds and really living, just for one minute! As I sat there I knew exactly what he meant. The experience of being on the edge like that will live with me forever.

Canoe polo

Canoe polo, rapidly increasing in popularity, is a fast and dynamic game played by two teams. Each team comprises eight players with five allowed on the pitch at any one time. Substitutions are made on a rolling on-off basis. Canoe polo is a classic intermittent activity with sprints of varying distance and intensity interspersed with variable duration recoveries. Many of the aspects important to performance in canoe polo can be derived from the preceding sections that have described the physiology, most likely causes of fatigue, nutritional strategies and training methods for intermittent activities. To date, however, there has been less research regarding the specific physiology of canoe polo.

One of the few studies in this area was a motion analysis completed by Chris Bussell (1995). In this study, covert filming of UK players taking part in National League Division One canoe polo matches was carried out. Five matches for the goalkeeper and ten matches for three defenders and four attackers were recorded during the completion of the study. The filming was completed covertly in order to minimize influence on the playing patterns for any of the participants.

From the video footage, a time and motion analysis was conducted beginning with velocity calculations for each of the participants. Movement was categorized according to whether it was of high intensity (sprinting forwards or backwards), intermediate intensity (tackling, turning the boat or restraining a player), low intensity (slow paddling) or stationary/drifting (little or no exercise being completed by the performer). The results of the study indicated that attackers travelled at a higher velocity than defenders, however, for players in both positions there was a general decline in intensity from the first to second half of the match. The defenders in the study appeared to have more balanced performance across the whole game, with just over 10 per cent of their time in both halves spent engaged in high-intensity exercise. The attackers, on the other hand, completed the first half of the game with close to 20 per cent of their time spent at high intensity, but in the second half this percentage fell to just over 6 per cent. The goalkeeper spent 15 per cent of the time in the first half and 12.5 per cent of u.. .. ond half engaged in high-intensity exercise. For players in all positions, around 70 per cent of their time was spent within the low and stationary/drifting categories. In the second half, all players spent a slightly increased amount of time in intermediate and low-intensity exercise or stationary/drifting. These results suggest that attention in training may require a focus on the maintenance of physiological performance during the second half of a match, particularly for the attackers.

Further research, perhaps additionally utilizing heart rate monitors and studying players from other countries, would be beneficial prior to decision-making regarding fitness programme development for canoe polo players. Nevertheless, in the time before such findings are available, research from other similar intermittent activities could be employed to inform training strategies for attackers and defenders in canoe polo. There are a number of canoe polo specific factors that would need to be addressed in making decisions regarding the physiology of canoe polo. National league and international competitions are typically organized such that several games are played on the same day. For instance, international competitions may consist of players taking part in 4–5 games in a day for up to 3–4 days and, as such, any training programmes should reflect the demands not just of one match but of a whole day's matches. Some competitions and many internationals take place on open water rather than in a swimming pool, which can place additional environmental stresses (wind and ambient temperature primarily) on canoe polo players. Further research regarding the physiology of canoe polo would be beneficial to improve understanding of the physical demands of the sport.

Key point

Canoe polo matches are played in two halves, often with several matches on the same day. Four intensities of exercise can be observed in a canoe polo match: high, intermediate, low-intensity exercise and stationary/drifting periods. Players typically spend more time involved in high-intensity exercise in the first half of a game. Few studies have been conducted regarding the physiology of canoe polo. Future research could further develop the work initiated by Bussell (1995) regarding the specific physiological demands of both a match and a competition in canoe polo.

Surfing

Developed many centuries ago, surfing has seen a dramatic rise in its popularity in more recent years although there is a limited availability of research regarding the physiology of the sport. Since the 1980s, studies have been conducted regarding the physiological characteristics of surfers, time and motion and common injuries during surfing. Research indicates that male and female surfers are generally shorter and lighter than athletes involved in other water-based activities such as swimming and water polo. Elite level male surfers have been found to have a mean height of 173–175 cm and females between 162–166 cm compared to means of 183–186 cm and 171–172 cm for male and female swimmers and polo players. With regard to mass, elite male and female surfers have been found to have means of 68 kg and 57–59 kg, while those for swimmers and water polo players have been recorded as between 78–86 kg for males and 63–64 kg for females. Body fat percentages of 10.5 per cent for males and around 21 per cent for females have been reported.

Research data suggests that, not surprisingly, surfers have higher recorded oxygen uptakes compared to untrained individuals. Data for oxygen uptake results have been collected using arm-crank ergometers (similar to a cycle ergometer as shown in Chapter 4, but moved with the arms rather than the legs), tethered board paddling and using a prone swim bench (a form of dry-land swimming ergometer where front crawl is the assessed stroke). Values of 40–54 mL kg^{-1} min^{-1} have been recorded for surfers with maximal heart rates during testing of around 180 bts min^{-1}. Maximal on-water heart rate values of around 171 bts min^{-1} have been reported, with an average of about 135 bts min^{-1} during surfing. This provides evidence for the inclusion of surfing as an alternative method for improving cardiovascular endurance. Indeed, some schools, particularly in the southwest of England, New Zealand and Australia, now include surfing within the Physical Education curriculum. A suggested model for the demands of surfing including physiological, psychological and skill-related factors and possible environmental issues is shown in Figure 8.5.

A number of investigators have reported the relative times spent paddling out (sometimes referred to as arm paddling), stationary and wave riding. These time and motion studies have used analysis of video footage to assess the time spent in each type of activity. Similar values for elite and recreational surfers have been identified. A study of recreational surfers reported that, of the time spent surfing, 44 per cent was spent paddling out, 35 per cent stationary and 5 per cent wave riding. A study of 42 male surfers in a professional surf competition recorded values of 51 per cent, 42 per cent and 3.8 per cent respectively. Taken together, these suggest that during a typical surf session perhaps as much as 50 per cent of the time is spent paddling out, 40 per cent stationary and 5 per cent actually surfing (none of these add to 100 per cent as each includes a miscellaneous category e.g. turning the board to check incoming wave sets and setting position ready to take-off for a wave). Of a typical 3–5 hr session, these values indicate that surfers spend somewhere around 9–15 min actually riding a wave. The most common injuries reported in surfing are cuts and grazes occurring through contact with other surfers and their boards, the beach or groins. With regard to overuse injuries, shoulder, low back and neck injuries represent those most commonly reported and appear to be closely linked with paddling out and departure from the board.

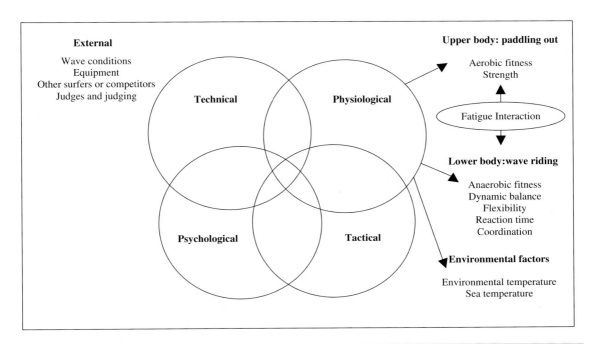

Figure 8.5 The components of surfing performance (adapted from Mendez-Villanueva and Bishop, 2005)

Key point

Surfing is a physiologically demanding adventure sport that appears to be beneficial for improving cardiovascular fitness. With as much as 50 per cent of the time spent paddling out, surfing provides an alternative fitness development sport for schools. Surfers have been found to have lower masses and body fat percentage than untrained individuals. Generally, surfers tend to be shorter than those involved in other water-based sports such as swimming or water polo.

Biography: Simon Hammond

Adventure sport: Surf kayak

Simon started surf kayaking with his local scout group in his home town of Bude, Cornwall. Simon recalls that he spent more time swimming than kayaking in those days. His first surf craft was an Australian-designed Raider Surf Ski. After qualifying as Physical Education teacher, he returned to Bude to run the family outdoor pursuits business with his wife Nicola. He has competed in many national and international surf kayaking events which culminated in him becoming World Surf Kayak Champion at the 2003 World Championships at Easkey, Ireland. Simon is a BCU level five coach and continues to live in Bude where he jointly runs Shoreline Outdoor Pursuits.

Figure 8.6 Simon Hammond (photograph courtesy of Sam Ponting)

2003 Surf World Championships

Before I went to Ireland for the 2003 Surf Kayak World Championships, I had no real expectation of coming home as the world champion. I did, however, allow myself the odd daydream of what it would feel like to win and I did go to Ireland with a plan!

Daydreaming and planning are a good start but, to be honest, my actual preparation for this event had been a disaster. In the spring I'd foolishly been persuaded into skateboarding down Bude's new half-pipe and on my very first attempt had caught one foot off the edge of the drop. As I dropped down the face, my legs were torn apart and I severely damaged my hamstring. I couldn't even sit in a kayak for two months let alone surf one! Then just as I started to recover from that injury, I developed housemaid's knee from building too many sandcastles on the beach with my two young sons. With less than two weeks to go before the competition, I had to completely re-pad the inside of my kayak to take away any pressure from my knee cap. This resulted in the kayak feeling completely different and I had no time to let my movements adjust to this new fit. I could still daydream!

Once over in Ireland I gained as much experience of the competition break as was possible. As it turned out, the first couple of days before the competition were pretty big so it was good to work out any nerves and at the same time start to learn how the reef at Easkey made the wave break. The competition was due to run over two weeks with the team event in the first week followed by the individual events in the second. This allowed everyone to get used to what the judges were impressed by, and luckily for me they seemed to be impressed by the way I surfed: not many tricks but deep in the pocket with big late bottom turns and even bigger off the lip top turns.

To get through this two week competition and stay in good shape meant getting organized back at the self-catering house with good quality and quantity meals. A good mixed and healthy diet and a big high-carbohydrate meal every night was required, the best being the roast chicken meal cooked by Bryce Barr half-way through the fortnight. Each day we'd pack up sandwiches, bananas and plenty of water to keep us going through the heats.

I spent an hour preparing for each heat, in which time I'd run down the road to be by myself and go through a warm up and joint mobilization routine, but most importantly switch off from all possible distractions. I'd have a drink of water, then get into my kit and spend 15 min watching the surf. I'd check my kayak in, get my bib on and then get onto the water to finish focusing for the event. I got into the habit of singing to myself while waiting for my heat to start, not any old song but one that kept me in my own world, gave me a buzz and stirred some aggression that I'd need during the heat.

With one minute to go I was off, sprinting to be the first to the peak and catch the first wave I could. Not that that's always a good thing, but I figured it might just put pressure on the other competitors, forcing them to play catch-up during our 20 min at the break. Others told me that you could tell who was going to win a heat just by their body language and degree of commitment as the heat began. If I could create the impression that I was the one to beat then I was already half-way to winning.

The Americans were very impressed with how aggressive the English team were during our paddle outs. The truth is that most English paddlers surf on beach breaks where you have to be aggressive in order to paddle out right through the breaking waves. Those who surf off reefs and points can be a little more laid back as they can paddle out around the on-coming waves. I think this background gave us a little edge as we were generally faster and fitter on the water than paddlers from some other nations.

I won my first heat and then had nothing to do until the next day when, by luck, I was drawn against one of the strongest paddlers in the competition. Someone asked if I'd seen the draw for tomorrow. 'No', I lied. 'You've got a tough heat', he was keen to tell me. What do you say to that? I guess it depends on how confident you're feeling and I felt confident.

I stayed confident and relaxed for my heat the next day, focused in and won and then had to wait another day for the quarters, semi's and finals. Whether you call it 'focused' or 'in-the-zone' it doesn't matter, the fact is: you can't stay like that for a whole week or you'd go crazy and this is where my friends Bryce, Paul and Eggy were so brilliant. Away from the beach, there was no talk of competition but instead tales of their own surfing adventures.

The final day was bound to be the most intense and some adaptations to my plans had to be made. Preparation time was cut and I concentrated on refuelling with honey sandwiches and re-hydrating with as much water as was comfortable. During my semi-final heat drawn against Pete Blenkinsop and John Grossman I had a terrible start without catching a single wave for the first 7 min. I still won and progressed to the final. In the final, I was up against Darren and Pete from Cornwall together with ex-world champion Dick Wold from California. With Dick just pleased to have made the final I told myself that all I needed to do was go out there and do my normal thing one more time.

Figure 8.7   Former surf kayak World Champion Simon Hammond catching some air (photograph courtesy of Sam Ponting)

The surf was dropping in size, so no point waiting for the big one! Remember, I told myself, if you think you're surfing badly the chances are that everyone else in the heat is thinking the same and don't bother looking or even thinking about how anyone else is doing. With hindsight, this last bit of self-advice saved me as with 5 min to go I was in third position with my very last wave giving me the points I needed to win.

Back at the beach the results weren't yet out; faces remained blank as we paddled back to shore. A crowd around the judge's stand waited for the results to be posted, no-one knew or dared to guess. Then the results were up, and Eggy was racing back to me with his finger in the air: 'You've won it!' he was shouting. The next few minutes were crazy, hugging people, whooping with joy, a telephone call home, an interview and the start of life with the biggest smile you'll ever see. Dave Johnson, another ex-world champion said to me that evening: 'It feels good doesn't it?' 'You bet it does!' was my reply.

Getting into research: intermittent maximal exercise (Gaitanos et al., 1993)

Many people take part in intermittent exercise sports. Intermittent sports are characterized by short bursts of maximal effort punctuated by rest or sub-maximal exercise. Gaitanos et al. (1993) made the point that more people engage in intermittent sports (football, tennis, basketball, hockey, etc.) than are involved in prolonged continuous exercises (running, cross-country skiing or distance swimming). In adventure sports, there are many examples of intermittent activities or forms of training that include intermittent exercise. Sport rock climbing and bouldering sessions, canoe polo matches, surf kayak, freestyle kayak and river running (canoe or kayak) are examples of adventure sports that inherently involve intermittent exercise. Despite such participation numbers, less is known about the physiology of intermittent sports compared to prolonged exercise. The aim of the study by Gaitanos et al. (1993) was to describe the metabolic changes that arise in response to repeat sprint exercise.

The eight male participants took part in 10 repeat sprints of 6 s duration with a 30 s recovery between each one. The mean age, height and mass of the participants was 26.7 years, 175.6 cm and 71.8 kg. Prior to the test, participants were familiarized with the testing procedure and instructed to refrain from exercise the day before testing. The protocol for testing included a rest period for pre-test blood and muscle biopsy sampling, standardized warm-up, the repeat sprints and a 10 min post-exercise recovery period.

The peak power output (PPO) and mean power output (MPO) for each sprint were recorded. Venous blood samples were taken pre- and post-sprint 1, post-sprint 5, post-sprint 9, post sprint-10 and at 3, 5, 10 min during the recovery. Blood samples were analysed for blood lactate concentration, glucose, adrenalin and noradrenalin. Muscle biopsies were taken from the vastus lateralis of the quadriceps muscle pre-, post-sprint 1, post-sprint 5 along with pre- and post-sprint 10. The muscle samples were analysed for a wide variety of metabolites which included ATP, ADP, AMP, PCr, creatine, pyruvate and lactate.

The key results for the study are shown in Table 8.5. The PPO and MPO for the first sprint were 1253 W and 870 W. The PPO fell by 16 per cent by the 5th sprint and 33 per cent by the 10th sprint. The MPO fell by 13 per cent and 27 per cent at the same points. The PCr concentration fell by 57 per cent after the first sprint, while glycogen and ATP fell by 13 and 14 per cent, respectively. By the end of sprint 9, the PCr concentration was 51 per cent of the resting value and decreased a further 16 per cent post-sprint 10. At the same stages, ATP and glycogen concentrations were 32 per cent and 30 per cent of resting values, respectively, but while glycogen concentration fell by almost a further 9 per cent, there was no change in the ATP concentration at the end of sprint 10.

From these results, Gaitanos *et al.* calculated that the percentage contribution from PCr degradation rose from 49.6 per cent in sprint 1 to 80.1 per cent in sprint 10, while the percentage contributions from glycolysis and ATP catabolism fell from 44.1–16.1 per cent and 6.3–3.8 per cent respectively.

The results of this study show a progressive decline in power output (peak and mean) across the 10 sprints, decreasing by 33 and 27 per cent respectively. As Gaitanos *et al.* point out, however, the MPO for sprint 10 still represented 73 per cent of the value for sprint 1. The slight decrease in lactate production between sprints 9 and 10 is suggestive of an inhibition of fast rate glycolysis. The identified decline in fast rate glycolysis in sprint 10 placed an increased relative reliance on PCr degradation and increased role for aerobic metabolism. The findings of Gaitanos *et al.* provide evidence to suggest that for intermittent exercise there is an increased reliance upon PCr catabolism and aerobic metabolism with increasing duration.

Table 8.5   Mean values for power output, blood lactate, plasma catecholamine and key muscle metabolites

	Pre	Sprint 1	Sprint 5	Sprint 10
PPO (W)	N/A	1253.3	1054.03	834.7
MPO (W)	N/A	870.1	760.47	638.65
BLa (mmol L^{-1})	0.6	1.9	9.2	pre 12.6/post 12
Adrenalin (nmol L^{-1})	0.4	1.9	4.2	5.1
Noradrenalin (nmol L^{-1})	1.7	3.3	15.7	22.3
ATP (mmol kg^{-1} dry wt)	24	20.9	N/A	pre 16.4/post 16.4
PCr (mmol kg^{-1} dry wt)	76.5	32.9	N/A	pre 37.5/post 12.2
Glycogen (mmol glucosyl units per kg dry wt muscle)	316.8	273.3	N/A	pre 221/post 201

8.7 Summary and study questions

Intermittent activity

A number of adventure sports are by necessity intermittent activities, meaning that they involve periods of high-intensity exercise coupled with periods of partial or complete rest. The approach of many adventure sports participants to their activities means that these also become intermittent activities. Some of these activities can be self-paced where the participant can choose how to structure the activity, while for others the demands of the activity will dictate the timing and duration of the activity and rest blocks.

There has been much research into intermittent activity and researchers have manipulated the timing, duration and proportion of exercise to rest as well as the level of exercise intensity. Different types of exercise have also been investigated, as well as overall duration and number of work/rest intervals. One problem with applying the findings of research to the real world of adventure sports is that no activity is likely to have the highly regulated structure that physiologists tend to employ in their laboratory-based studies. Rest intervals and exercise intensities are far more dynamic and variable in adventure sports. Recently sports such as football have tried to model the demands on athletes using technology such as GPS tracking and heart rate monitors so that studies can try to replicate more accurately the behaviour of a performer.

Research has consistently shown that during intermittent exercise, performers are able to achieve a much greater volume of work before exhaustion than during continuous exercise of the same intensity. Increasing the length of rests also leads to lower lactate accumulation. Fat oxidation seems to be elevated during intermittent exercise and both Type I and Type II fibres are recruited. In continuous activity, mainly Type I fibres contribute.

Fatigue

The mechanisms behind fatigue in intermittent activities are related to the duration and structure of the activities. In prolonged intermittent activity of 90 min or more, it is likely that glycogen depletion has an important role. In high-intensity short-duration intermittent activities, H^+ increases, citrate leakage, Ca^{2+} transport disruption and PCr depletion have all been identified as possible causes of fatigue. Recent studies have also suggested that K^+ leaking from the sarcoplasm into the interstitial fluid could be a mechanism for fatigue.

Glycogen depletion can also be a major problem during high-intensity aerobic endurance exercise. Since intermittent exercise includes periods of high-intensity exercise, it is quite possible to deplete glycogen stores much more quickly than would occur during a long, but low intensity, activity. Glycogen depletion can lead to central fatigue and the rapid deterioration in performance known as 'hitting the wall'.

Training and recovery

Not all forms of rest have been found to offer the same benefits in terms of recovery. Light exercise rather than complete rest seems to be more effective in reducing lactate levels during recovery periods. This is believed to be due to the enhanced ability of muscle fibres to take up and process lactate during light exercise.

During training it is always important that demands on an athlete replicate the demands of the event or activity. This is equally true with intermittent activities and performers and coaches should try to ensure a match between their adventure sport and the training they undertake.

Nutrition and ergogenic aids

Nutritional manipulation and ergogenic aids for intermittent activity will depend upon the specific nature and structure of the intermittent activity itself. For some activities, creatine supplementation or buffer loading could be beneficial. Carbohydrate loading protocols are a commonly used nutritional strategy for increasing glycogen stores prior to endurance exercise and this could be effective for some intermittent activities. Research has shown that carbohydrate loading can improve performance and delay fatigue in activities lasting over 60 min at or above 75 per cent of $\dot{V}O_{2\,max}$.

The classic method of carbohydrate loading was to deliberately deplete glycogen stores by eating a low-carbohydrate diet coupled with exhaustive exercise followed by a high-carbohydrate diet that would result in super compensation – higher glycogen stores going into an event. Problems with this approach were related to the low-carbohydrate and exhaustive exercise phase, which was detrimental to confidence and mood. More recent models for carbohydrate loading involve increasing carbohydrate intake and tapering exercise and this approach has been found to be effective in increasing glycogen storage without the drawbacks of the classical approach. It is, however, worth realizing that carbohydrate loading can also result in some weight gain as increased glycogen storage requires more fluid storage. As a result, it is best to practice carbohydrate loading to establish whether the overall effects are beneficial to performance.

Carbohydrate feeding during activity can be effective and sports drinks that supply carbohydrates have been shown to improve performance and delay fatigue. The benefit of using drink-based carbohydrate feeding is that both hydration and glycogen sparing can be achieved at the same time. It has also been shown that water absorption rates are higher when consuming a sports drink of the correct concentration than for pure water. Again, it is important to practice a drink/feeding strategy before trying to use it in an event.

Water immersion

Water immersion can cause a number of physiological problems for performers. Immersion in cold water can lead to hypothermia though increased cooling. Our understanding of the impact of water immersion has improved considerably in recent years. For instance, the dangers of hypothermia are now recognized to be of far greater consequence than allowed for in older advice on survival techniques. Minimizing heat loss is one of the priorities that adventure sports performers should bear in mind when formulating emergency procedures and planning. Human beings have a variety of thermoregulation mechanisms that they use to try and maintain a constant body temperature. The enzymes that catalyse metabolic reactions have a narrow band of temperature for optimal functioning, and performance drops when core body temperature falls below $37\,°C$.

Water is a far more effective conductor of heat than air and cooling during cold water immersion happens at a much faster rate than in the same temperature of air. Metabolic processes result in around 60 per cent of the available energy being converted into heat and therefore on immersion in water below thermoneutral temperatures ($35–35.5\,°C$), one of the initial responses is to increase muscle tone, stiffening the muscles, generating heat. If greater heat generation is needed, then shivering is initiated which can boost metabolism to about five times resting levels. The endocrine system also responds by releasing hormones that boost metabolism throughout the body.

As well as heat generation, the human body also employs heat conservation strategies in order to maintain a constant core temperature. Vasoconstriction is the main mechanism for heat conservation in human beings, through which the flow of blood to skin is reduced. Vasoconstriction effectively turns the skin and subcutaneous fat into insulation for the body's core. In contrast, blood flow to the scalp is maintained which protects the brain from cold but also means that the head becomes the major area for heat loss in many outdoor situations. The increase in the volume of blood in the core results in the kidneys working to excrete blood plasma as urine. The increased pressure on the body from water immersion adds to this effect by squeezing blood that would otherwise pool in the legs into the core. After 5–10 min of cold exposure, it is not uncommon for blood flow to the hands to be restored which protects the hands from cold injury and maintains dexterity. This response has been described as the hunting reflex, and the effect is more pronounced if the core temperature is maintained.

Cold shock can be a problem where a performer suffers an unprepared immersion in cold water. The cold shock response begins immediately on entry to cold water, peaks after 30 s and subsides in around 3 min. The cold shock response involves an increase in the respiratory demand and the individual will take a large gulp of air followed by hyperventilation. This hyperventilation can cause dizziness and confusion but also reduces breath-hold time which can result in inhalation of water and drowning. There is evidence that there can also be a battle for control of heart rate and respiration between the cold shock response and the mammalian dive reflex, which would normally result in breath-holding and slowing heart rate. This can trigger cardiac problems or stroke for those susceptible.

As immersion time increases, hypothermia becomes the biggest threat. Below 34 °C brain function is impaired and unconsciousness will follow. As the body temperature drops to 28 °C or below, heart failure will result in death. Adventure participants should make careful clothing choices that will offer increased protection but also make every effort to get out of the water, even onto a floating object. Although it may feel colder, heat loss will actually be slower.

Deterioration or even death can occur post immersion as body temperature can drop through continued cooling down the thermal gradient from the core to the periphery. The drop of pressure on the body can also lower venous return and cause circulation problems and vertical removal can compound this problem.

Sailing

Recent research into sailing suggests that the physiological demands on the individual are very much dictated by their specific role. The demands on different crew members for multi-handed sailing can be very different and depend upon the general direction of sailing for each leg of an event or journey. One of the key changes in our understanding of the physiological stresses during sailing is the recognition that sailing can have high aerobic demands; earlier assessments had suggested that isometric muscular endurance was the main demand on sailors. Training protocols used by sailors should mimic the demands of their role in terms of duration, intensity and structure.

Canoe polo

Canoe polo is relatively unexplored in terms of scientific research but is particularly interesting as it is a classic intermittent activity. The game requires players to perform sprints of varying distance and intensity punctuated with recovery periods of variable duration. Research has shown that even elite players often show a decline in the intensity of effort during the game, suggesting that training focusing on the maintenance of performance throughout the match may be beneficial.

Surfing

Much of the limited research into surfing has looked at the physiological characteristics of surfers. Surfers tend to be shorter and lighter than swimmers or water polo players. Surfers have recorded higher $\dot{V}O_{2\,max}$ scores than untrained individuals, which suggest swimming has a significant cardiovascular component. About half of a surfer's time during activity sessions is spent paddling out, a third or more is spent resting and 5 per cent or less is spent actually riding waves. This clearly makes surfing an intermittent activity.

Study questions

1. Which factors in relation to intermittent activities will have an impact on the physiological processes that underlie performance?

2. Why is fat oxidization likely to be greater during intermittent exercise than during continuous exercise? What role does fat oxidization play in sustaining performance?

3. Why might increased interstitial K^+ concentrations contribute to fatigue in intermittent activities?

4. Why is it important to match the demands of training to the specific demands of a performer's adventure sport? What might the first stage of this process be?

5. Why might light exercise be more beneficial than complete rest in some intermittent activities?

How have physiologists defined 'light exercise' during studies?

6. What advantages does a tapering protocol have over a classical carbohydrate loading protocol prior to an event? How much carbohydrate should an athlete consume during the tapering phase?

7. Describe the effects of cold shock and the diving reflex during initial immersion to cold water? What implication do these responses have on heart rate?

8. How does the human body cope with heat loss during prolonged immersion in cold water? Why is swimming in an attempt to reach a watercraft or the shore not recommended unless you are very close to safety?

9

Aerobic endurance part 1: high-intensity activities

9.1 Introduction

This chapter presents the first of the aerobic endurance chapters, based around high-intensity adventure sports. Chapter 10 examines the physiology of lower-intensity aerobic endurance adventure sports. All of the sports included in this chapter involve an element of competition which provides the driving force for the relatively high intensity. The sports covered in this chapter nine are shown in Table 9.1. Due directly to the competitive aspects within the adventure sports in this chapter it might easily have been called the racing or endurance racing chapter.

As a consequence of their high relative intensity, the duration of the activities in this chapter tends to be shorter than those which will be covered in Chapter 10. Despite this, the high-intensity endurance activities covered here have a relatively wide range in duration e.g. from events such as the 1000 m sprint kayak or canoe (around 4 min duration) which just fall into this category to mountain running and kayak marathons (2–3 hr). The relative duration of the adventure sports covered in this chapter is shown in the duration continuum in Figure 9.1. The duration of an individual aerobic endurance adventure sport has a direct impact

upon the relative intensity of performance. Clearly, it would not be possible for a sprint kayaker to maintain the physiological intensity they achieve in a 1000 m race over the distance required for a kayak marathon. When planning training for an individual taking part in an aerobic endurance adventure sport, the duration of the event should be central to decisions made about the preparation programme for that athlete. A high degree of physiological specificity, as for all the adventure sports covered in Part II of this book, is required for performance in high-intensity aerobic endurance adventure sports.

Since lactate threshold is important to high-intensity aerobic endurance, this chapter begins with a review of this topic. Performance in high-intensity aerobic endurance events is often affected by climatic conditions and so a review of the impact of thermal stress, heat and cold, is presented. This is followed by response to high-intensity endurance exercise, fatigue, training adaptations and nutritional strategies for performance. Specialist sections included in this chapter cover the physiology of mountain biking, orienteering and fell running, and kayaking and canoeing. Performer contributions from Tim Brabants (sprint kayaking), Jan Sleigh (wave sailing), Neil

Adventure Sport Physiology Nick Draper and Chris Hodgson
© 2008 John Wiley & Sons, Ltd

Table 9.1   The high-intensity aerobic endurance adventure sports

Water-based activities	Land-based activities
Wildwater kayak race	Climbing pitches near maximum
1000 m kayak/canoe sprint	Mountain marathon (26 mile)
Marathon kayaking/canoeing	Fell or mountain running
Windsurfing: Olympic racing	Orienteering: classic, short distance, sprint
Windsurfing: wavesailing	Mountain bike cross-country
	Mountain bike downhill

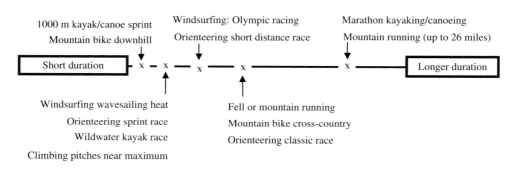

Figure 9.1   The duration continuum for high-intensity aerobic endurance adventure sports

Gresham (trad rock climbing), Anna Hemmings (kayak marathon) and Ian Tordoff (record-breaking kayak channel crossing) are included to highlight the physiological demands of the adventure sports within this category. In addition, a research summary on wash riding by Pérez-Landaluce *et al.* (1998) is included.

9.2   Aerobic-anaerobic transition and lactate threshold

The concept of lactate threshold (LT) was introduced briefly in Chapter 5 as an advanced measure of performance in the aerobic endurance adventure sports which are the focus of this chapter. In this section LT and, more generally, the transition from aerobic to increasingly anaerobic metabolism, is described in more detail. A clear understanding

of the aerobic-anaerobic transition (AAT) is very important for those interested in high-intensity aerobic endurance performance.

Research concerning the existence of an aerobic-anaerobic transition developed from the early work regarding the nature and production of lactic acid. Swedish physiologist Carl Scheele (Chapter 1) first isolated lactic acid in samples of sour milk in 1780. It is through this link with milk, a product of lactation, that lactic acid and lactate gained their names. In 1807 another Swedish scientist, Jöns Berzelius, isolated lactic acid in muscle calling it *sarcolactic acid*.

Berzelius (1779–1848) began his career as a doctor but later became fascinated with chemistry. His research included the discovery and isolation of several new elements such as silicon and selenium. It was Berzelius who developed the modern notation system for chemical formulae, such as the letter O for oxygen and H for hydrogen. The only difference from the system used today was that his notation included the numbers in formulae as superscripts rather than subscripts i.e. H_2O was H^2O in his system. Berzelius went on to determine that the concentration of sarcolactic acid in a muscle was associated with prior exercise. Concentration of sarcolactic acid was shown to increase as the intensity of exercise increased.

By 1833, the chemical composition and formula for lactic acid ($C_3H_6O_3$) had been determined. In 1907 Frederick Hopkins and Walter Fletcher, in a study on exercise to fatigue with amphibian muscle, determined that lactic acid concentration increased during anoxic (low or no oxygen) conditions. Later Hill and Lupton (1923) went on to state that the increase in lactic acid during exercise was the consequence of an insufficient oxygen supply.

The terms *aerobic* and *anaerobic* glycolysis arose from the conclusions drawn by Lupton and Hill regarding the formation of lactic acid during exercise. Aerobic glycolysis was used to describe slower rate glycolysis where the presence of oxygen in sufficient concentration enabled pyruvate to be completely oxidized or, more correctly, a pattern of glycolysis where lactate removal is equal to lactate production (resulting in no or only a small net gain lactate with increasing exercise). Anaerobic glycolysis described the fate of pyruvate during fast rate glycolysis where lactate production exceeds lactate removal, resulting in an accumulation of lactate within the muscle fibres and blood.

The use of these terms has led to some confusion in teaching physiology, due to the fact that glycolysis is an anaerobic (not requiring oxygen) metabolic process. It is the fate of pyruvate, the end product of glycolysis, which is determined by the intensity of exercise and the supply of oxygen for further aerobic metabolism. This is why the terms *slower rate* and *fast rate glycolysis* have been used throughout this textbook. The aerobic and anaerobic confusion is not helped by the common use of the terms aerobic and anaerobic to describe the metabolic processes for short-term high-intensity exercise (anaerobic) and longer duration exercise (aerobic).

Anaerobic metabolic processes described in Chapters 4, 6 and 7 refer to the phosphagen system and fast rate glycolysis. The aerobic system refers to the complete oxidation of carbohydrate, fat and protein to provide the energy necessary for lower intensity but longer duration activities (aerobic metabolism was outlined briefly in Chapter 4 and is described in detail in Chapter 10). The use of the terms aerobic and anaerobic in this context is far more useful within a physiological context and consequently has been adopted within this textbook.

Traditionally, the aerobic and anaerobic metabolic processes are described by beginning with short-term high-intensity exercise and then moving on to consider endurance activities. In this way, textbooks move from anaerobic to aerobic metabolism. For some activities, this matches the transition in energy production. However, for the high-intensity aerobic endurance activities that are the focus of this chapter, this is the reverse of the pattern of energy system contribution. Adventure sports such as marathon kayaking, classic orienteering or cross-country mountain bike racing have an aerobic base but draw upon anaerobic metabolism for short in-race bursts or for the finishing 'kick'.

An understanding of this conceptual difference is important when considering the transition from aerobic to anaerobic metabolism, the key concept of this section. During an endurance race, adventure sport athletes attempt to maintain an exercise level that is as high as possible, thereby maximizing aerobic metabolism, while avoiding substantial reliance on anaerobic (fast rate glycolysis or phosphagen system) metabolic processes that would lead to early fatigue. This point, maximizing aerobic metabolism while remaining just below substantial reliance on anaerobic metabolism, represents the AAT. The term AAT has been adopted for this section as an umbrella term encompassing the wide number of terms that have been used previously to describe this transition. For want of a better term, the anaerobic threshold has most commonly been used to describe the transition from aerobic to increasingly anaerobic exercise. The term anaerobic threshold, however, has a specific meaning which, as was clearly argued by Brooks (1985) and Davis (1985), precludes its use as an umbrella term. The AAT is an important physiological transition point for high-intensity aerobic endurance athletes. As a consequence, the various terms and concepts related to the AAT will be discussed before a review of the range of techniques used to identify this transition point.

Interest in the identification of an individual's AAT arose from a clinical rather than a sport-related concern for cardiac and similarly cardio-vascularly compromised patients. It was thought that the ATT provided a sub-maximal method for the assessment of exercise capacity for cardiac patients. Since then, interest in the AAT has spread to a sports performance context. The AAT is traditionally identified using invasive blood lactate analysis techniques or through the use of non-invasive gas analysis. The gas analysis methods were originally developed in the 1960s for a clinical setting where non-invasive techniques for ATT might be preferred. The use of gas analysis to identify the AAT was originally termed the *anaerobic threshold* (AnT) by Wasserman and McIlroy (1964) but is sometimes referred to as the *ventilatory threshold*. Wasserman and McIlroy (1964) suggested that a breakaway in pulmonary ventilation equated to the rise in lactate accumulation associated with increasing anaerobic exercise.

In Chapter 7 the concept of lactate and H^+ buffering was explained. During exercise, $\dot{V}_E$ initially increases linearly with exercise. However, at a specific point for each individual, a breakaway in ventilation occurs. This point was identified by Wasserman and McIlroy as the AnT and was said to coincide with the rise in CO_2 produced during lactate and H^+ buffering. It therefore indicated non-invasively the point at which the body begins to accumulate lactate i.e. the lactate threshold. Later research highlighted difficulties in the identification of ventilatory breakpoint. Consequently, other aspects of gas kinetics were examined to assess their possibility for a more clearly defined marker of AnT. Currently, the most widely used non-invasive indicator of AnT is the combined use of ventilatory breakpoint and the ventilatory equivalents for oxygen and carbon dioxide ($\dot{V}_E/\dot{V}O_2$ and $\dot{V}_E/\dot{V}CO_2$ respectively). $\dot{V}_E/\dot{V}O_2$ represents the ratio of air expired to oxygen consumed, while the $\dot{V}_E/\dot{V}CO_2$ is the ratio of expired air to carbon dioxide produced. Figure 9.2 provides an illustration of these AnT

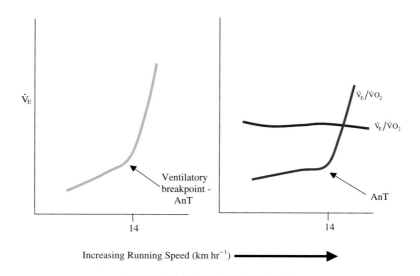

Figure 9.2   Markers of AnT

indicators. As depicted in Figure 9.2, the AnT occurs where there is an increase in $\dot{V}_E/\dot{V}O_2$ without concomitant rise in $\dot{V}_E/\dot{V}CO_2$. In this example, the AnT occurs at a treadmill running speed of 14 km hr^{-1}.

During physiological assessment, for an athlete, AAT is more commonly identified via lactate analysis due to a number of difficulties with the indirect gas exchange assessment of AnT. The ventilatory breakpoint is not always easy to identify making detection of AnT problematic at times. Research has also indicated that for patients with McArdle's syndrome, who produce little or no lactate, there is still a ventilatory breakpoint during incremental exercise, revealing a separation of the LT and AnT. Other studies have confirmed the separation of AnT and LT which places limitations on the validity of non-invasive gas exchange assessment of the AAT. As a consequence, when a higher degree of accuracy is required for the assessment of AAT, lactate concentration analysis is currently the preferred assessment measure.

Key point

Lactic acid was first isolated in sour milk by Carl Scheele. It is for this reason that the 'lact' part of the name for the acid was chosen. Later researchers, Hopkins and Fletcher, along with Hill and Lupton, established that lactic acid was produced during low oxygen conditions during exercise. The terms aerobic (in the presence of oxygen) and anaerobic (without oxygen) arose from this work. In terms of exercise, aerobic refers to an intensity that can be sustained for a longer duration and anaerobic to a higher intensity with shorter sustainability. In aerobic endurance adventure races, performers attempt to race at an intensity level close to the transition from aerobic to anaerobic exercise: the aerobic-anaerobic transition. The closer they can run to this point, the faster they will be able to travel while maintaining performance for the duration of a race. In fitness assessment, a variety of terms have been used to describe this transition point. The most commonly used are the lactate threshold and the anaerobic threshold. The lactate threshold is most commonly assessed by measurement of capillary blood lactate concentration, while the anaerobic threshold is measured using non-invasive gas analysis techniques.

There are a wide variety of lactate measurements for the AAT including mathematical transformations of the data to more clearly identify the transition point. Lactate analysis-based assessments of the AAT include the onset of blood lactate accumulation (OBLA), LT, maximal lactate steady state (MLSS), lactate minimum speed and individual anaerobic threshold. Mathematical data transformations of LT include the log-log transformation and the Dmax method, intended to more clearly identify the LT than visual determination of LT from a LT curve. The most commonly used methods for lactate analysis of AAT are LT (with and without data transformation) and MLSS. The focus of this section will therefore be upon these tests although a brief mention of OBLA is included as this is often linked with, and sometimes seen as synonymous to, LT assessment. A more complete review of the variety of methods for assessment of AAT was completed by Svendahl and MacIntosh (2003).

The *OBLA* test represents an incremental test designed to identify the exercise intensity at which blood lactate accumulation reaches 4 mmol L^{-1}. This assessment point was proposed by Sjödin and Jacobs (1981) as, at a blood lactate concentration of 4 mmol L^{-1}, there appears to be a relationship between lactate concentrations in the muscle and blood, which is not found as clearly as above and below this concentration level. This test has been used widely; however, the assumption that a blood lactate of 4 mmol L^{-1} is synonymous with AAT ignores the wide individual variation in AAT. As a consequence, the use of LT or MLSS tests for AAT identification and monitoring of training-induced changes in AAT would be perhaps more beneficial with athlete populations.

The *LT* represents the point after which lactate begins to accumulate in the blood. The LT test, in common with OBLA and AnT, is conducted as an incremental test. Figure 9.3 provides an illustration of a typical lactate curve, as well as representation of the effects of training on the lactate curve. Threshold training results in a shift of the lactate curve to the right, indicating that an athlete can run, cycle or kayak at a higher intensity before crossing the LT.

The exact protocol for a LT test has varied between researchers. The capillary sampling required for LT assessment has led some researchers to employ discontinuous incremental test protocols. These tests have included a variety of interval length between each increment for lactate sampling. To avoid the effects of employing breaks between increments for lactate sampling, a number of studies have employed continuous incremental tests. Comparison between protocols has indicated differences in the occurrence of LT

according to whether a continuous or discontinuous test was used, although the effect of analysis intervals on the occurrence LT in a discontinuous test appear to be negligible if the break is kept to a maximum of 30 s. The use of portable lactate analysers such as the Lactate Pro illustrated in Chapter 5, provide the opportunity to minimize the length of any break for analysis or even eliminate the need for breaks, enabling a continuous test protocol to be followed. The duration of each increment, along with the exercise intensity increase, will affect the identification of LT.

Ideally, the increments should be as small as practically possible with the duration of each step being around 3 min to enable blood lactate concentrations to best reflect the increment for that step. It is sometimes beneficial during the initial assessment of an athlete's LT that the test is repeated twice, with smaller steps being made during the second test to more closely identify the exercise intensity at which LT occurs for that individual. As for any of the methods described for AAT identification, the mode of exercise should be as specific as possible for the adventure sports performer i.e. a treadmill for a fell runner, kayak ergometer for a marathon paddler and cycle ergometer (ideally a system using their own bike, as with a King cycle) for mountain bikers.

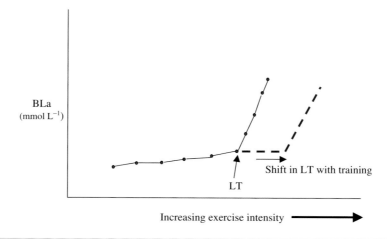

Figure 9.3 Lactate threshold curve and the rightwards shift resulting from training improvements (BLa is blood lactate concentration)

Just as detection of AnT (ventilatory break-point) using the ventilation curve can prove difficult, sometimes the exact exercise intensity at which LT occurs is hard to pin-point by visual inspection. As a consequence, a number of mathematical approaches to LT identification have been developed which are suggested to make the occurrence of LT clearer. Three of these approaches are illustrated in Figure 9.4. Beaver and colleagues (1985) recommended the use of a logarithmic transformation of oxygen consumption and corresponding lactate data (*log-log transformation*) to more clearly identify the LT occurrence.

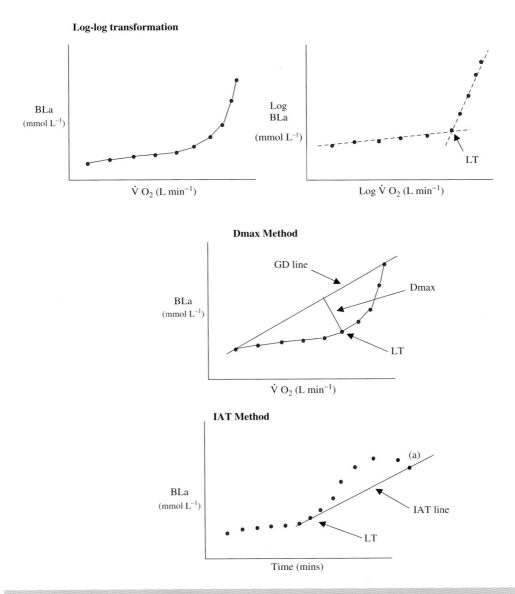

Figure 9.4 Illustration of the log-log transformation, Dmax and IAT methods for LT detection

The *Dmax* method developed by Cheng *et al.* (1992) involves the creation of a plot of oxygen consumption and lactate concentrations during an incremental test. As is illustrated in Figure 9.4, a general direction (GD) line is that drawn between the first and last points on the plot. The LT represents the point at which the greatest distance (distance maximum or Dmax) occurs between the lactate-$\dot{V}O_2$ curve and the GD line. The *individual anaerobic threshold* (IAT) method involves a data collection period beyond the termination of exercise. The IAT test, proposed by Stegman and colleagues (1981), requires lactate samples to be collected post-exercise until blood lactate concentration returns to the level at maximal exercise (as can be seen in Figure 9.4). A plot is then made of the blood lactate concentration for each sample during the duration of the test. The IAT is identified by drawing a tangent from (a) – the point at which lactate returned to the maximal exercise concentration – to the blood lactate curve.

Lactate threshold appears to provide a valid and reliable marker of the AAT. For adventure sports performers the use of a LT test, if carefully conducted, can provide a useful basis for guiding training during the next phase of preparation. Lactate samples are relatively economical to obtain, do not present great practical difficulties and represent a relatively efficient method for AAT identification. The validity and reliability of the threshold point, however, will be determined by the quality of the protocol employed. It may be beneficial, as mentioned above, to consider the use of two LT tests to more precisely identify the LT for a new athlete. Visual inspection of a plot or the use of a log-log transformation appears to provide the most appropriate method for LT identification.

An alternative approach to obtaining the AAT point through the use of lactate sampling is provided by the use of a MLSS test. The tests already described for ATT identification, AnT, OBLA and the various forms of LT are based around the execution of an incremental test. The aerobic-anaerobic transition is identified through MLSS by use of a series of 30 min constant-intensity tests. The MLSS exercise intensity is the maximum that can be maintained for a 30 min period with a less than 1 mmol L^{-1} change in lactate concentration during the last 20 min of the test.

Although developed in the 1980s, one of the most useful methods for the identification of MLSS is that proposed by Dekerle *et al.* (2003). Before the MLSS test can be conducted, an individual must complete a $\dot{V}O_{2\,max}$ test from which the work intensity for the MLSS can be obtained. In the first MLSS assessment (which must be completed on a different day to the $\dot{V}O_{2\,max}$ test) the athlete completes a T_{30} (MLSS 30 min constant exercise intensity) test at 75 per cent of $\dot{V}O_{2\,max}$. Blood lactate samples are collected at the start of testing and every 5 min until the last collection at the end of the 30th minute. On a separate day, the athlete completes a second T_{30}, the intensity of which is determined by the result in the first test. If a MLSS was reached or lactate concentration fell during the final 20 min of the test, the second test is completed at an exercise intensity 5 per cent higher than in the first test. If the lactate concentration rose by more than 1 mmol L^{-1} during the first test the exercise intensity is decreased by 5 per cent. The T_{30} tests would need to be repeated, increasing or decreasing the exercise intensity until a MLSS was identified. The obvious disadvantage with this method relates to the time taken for testing, although this would decrease in future tests and as the experience of the testers increased. Laplaud *et al.* (2006) have developed a single incremental test to estimate MLSS, however, the mean 10 bts min^{-1} HR differences between the MLSS may preclude its use with athletes. A 10 bts min^{-1} difference in HR can represent a large difference when establishing training zones for athletes.

The success behind using LT and MLSS as methods for advising training programme development lies in the close correlation that has been identified between LT or MLSS and performance in 30–60+ min races. Research indicates that both MLSS and LT can be used as better predictors of success in races over these durations than $\dot{V}O_{2\,max}$ test performance. To provide additional information to guide the development of training zones for a phase of training, HR during MLSS and at LT are required as well as exercise intensity. Normally it is not possible or perhaps desirable for lactate samples to be collected during every training session for an adventure sports performer. As a consequence, the HR that corresponds to the MLSS or LT can be used by an athlete to maintain exercise intensity at AAT during training.

Key point

The most commonly used reliable method for identifying the aerobic-anaerobic transition (AAT) appears to be the lactate threshold (LT). A variety of methods have been developed to assess the AAT via lactate analysis, including visual inspection LT, LT log-log transformation, Dmax LT, individual anaerobic threshold (IAT), onset of blood lactate accumulation (OBLA) and maximal lactate steady state (MLSS). The most appropriate to use with adventure sports performers would appear to be the MLSS or LT.

9.3 Physiological response to high-intensity endurance activities

This section includes a discussion of the response to high-intensity endurance activity and the mechanisms behind fatigue for adventure sports that fall into this exercise category. In addition, sections on steady-state exercise, exercise economy, oxygen debt, cardiovascular drift and the 'stitch' are included as they relate to both the response to exercise and fatigue in high-intensity endurance activities.

The general response to exercise will be similar to that explained in Chapter 6 with the CNS and endocrine system stimulating increases in respiratory and cardiovascular functioning as well as increasing the metabolism of fuel substrates (carbohydrate, fat and protein). For high-intensity aerobic endurance activities, the primary macronutrient for energy supply is carbohydrate in the form of glucose. A secondary role in energy metabolism is played by fat and protein due to the intensity of exercise. This has implications for glycogen stores, the enhancement of which is discussed in the following section on nutritional aids to performance.

For longer duration high-intensity endurance activities there will be a gradual shift of blood flow to the extremities to help with internal temperature maintenance. As exercise time progresses, the heat generated through metabolism needs to be dissipated in order to maintain optimal enzyme functioning. Blood is redistributed to the limbs to help dissipate the heat generated during exercise, which leads to a decrease in the volume of blood available for exercise. This is discussed in more detail below in the section on cardiac drift. During longer duration high-intensity aerobic activities, the effects of three additional hormones in response to exercise are important to consider. During high and lower intensity endurance events, the hormones *aldosterone* and *anti-diuretic hormone* are secreted by the adrenal cortex and the pituitary gland, respectively, to stimulate water, Na^+ and K^+ re-absorption from the kidneys to help maintain hydration levels. As the duration of exercise progresses, *endorphins* are released into the bloodstream and these help to block the feelings of pain associated with exercise and promote the feelings of well-being associated with exercise.

The specific response to exercise and the mechanism for fatigue will be determined by the duration of the activity. For those of shorter duration on the left-hand side of the duration continuum in Figure 9.1, a significant proportion of the required energy will be provided by fast rate glycolysis as well as by aerobic mechanisms. As a consequence, limitations to performance will be similar to those described in Chapter 7 and relate to the accumulation of lactate. For the relatively longer duration high-intensity aerobic activities such as kayak or mountain marathon, a major contribution to fatigue will relate to glycogen depletion which was described in Chapter 8. For those interested primarily in the shorter duration high-intensity endurance events, refer to Chapter 7 to identify the specific response to exercise and the mechanisms for fatigue. If you are interested in longer duration events, much of the response to exercise relates to lactate threshold (described earlier in this chapter) and fatigue (Chapter 8).

The upper body plays a significant role in the majority of high-intensity endurance adventure sports and it is therefore important to briefly review the differences in physiological response to exercise of the upper and lower limbs. In untrained individuals, $\dot{V}O_{2\,max}$ for the arms occurs at about 70 per cent of that for the legs. However, for trained upper body athletes such as kayakers, the $\dot{V}O_{2\,max}$ can exceed that for the legs. During sub-maximal exercise at any given workload, arm exercise elicits a higher $\dot{V}O_2$ response than the same workload for the legs. These results appear to be largely a function of the muscle mass involved and differences in mechanical efficiency during exercise for the legs and arms. Heart rate during upper body exercise tends to be around 20 per cent higher than that for the lower body during any given $\dot{V}O_2$.

The respiratory response to arm exercise appears to be greater than that for leg exercise, being expressed through a higher breathing frequency, but often a slightly lower tidal volume. In terms of steady-state exercise, which is discussed in more detail in the following section, the arms take longer to reach a steady state than the legs whether moving from rest to exercise or from exercise level to exercise level. Lactate response for any given exercise level appears to be greater for the upper body than for the legs. This may reflect differences in training frequency or, as was discussed in Chapter 6, variation in muscle fibre type distribution (higher proportions of Type II fibres being found in the arms) establishing a greater capacity for, or reliance upon, fast rate glycolysis. Studies in Denmark and Sweden led the researchers to examine the role of the arms as producers of lactate and the legs as consumers in whole body activities where the upper body provides a significant proportion of the total energy output, such as during rock climbing.

Key point

Many adventure sports involve the upper body in exercise. Research indicates that for untrained individuals, $\dot{V}O_{2\,max}$ for the upper body occurs at around 70 per cent of the values achieved using lower body. In highly trained upper body athletes, $\dot{V}O_{2\,max}$ values for the arms can exceed values achieved by the legs.

Steady-state exercise

The first three chapters in Part II of this textbook involve adventure sport participation where the exercise is either of a short duration and high intensity or intermittent in nature; as a consequence a steady-state exercise level is not normally achieved. Many of the adventure sports that are the focus of this chapter and the next enable a steady state to be achieved due to their sub-maximal exercise levels and longer duration. A *steady-state* exercise level, as illustrated in Figure 9.5, occurs when the energy supply for exercise meets the energy demands. When you start to exercise respiration, heart rate and oxygen consumption rise to meet the body's energy requirements. If an athlete exercises against a constant and sub-maximal workload, heart rate and oxygen consumption will begin to plateau and after 4–5 min a steady state is reached (the nature of the steady-state plateau is an essential part of the Åstrand-Åstrand aerobic capacity test described in Chapter 5). Exercise within the adventure sports described in Chapters 6 and 7 cannot reach steady state because the exercise is maximal and the duration is less than 4 min. The intermittent activities that were the focus of Chapter 8 entail exercise at varying intensities which prohibits the attainment of a steady-state exercise level.

Figure 9.5 shows the plateau associated with steady state for three exercise intensities or workloads. As illustrated, the time to reach steady state increases as workload is increased. As previously

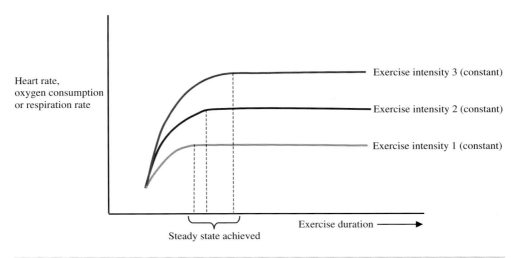

Figure 9.5 Steady-state exercise heart rate, respiration and oxygen consumption plateau

mentioned, it normally takes around 4–5 min for an athlete to reach a steady-state plateau in response to an initial workload. If the workload is subsequently increased, the athlete will only need around 1–2 min to reach a steady state for the second workload. When the workload is increased above a critical level for any adventure sports performer, they will not be able to attain a steady state for heart rate, oxygen consumption or respiration. This point represents the AAT: the focus of the previous section. Research indicates that individuals who train for their adventure sport reach steady state for any given exercise intensity more quickly than sedentary individuals and can exercise against heavier workloads before crossing the threshold above which steady state cannot be achieved. Improved aerobic capacity, demonstrated through increases in the number of mitochondria, capillarization and enzyme functioning, is responsible for improved steady-state attainment.

Exercise economy

Exercise economy relates to the energy demands of a given workload for any individual performer. The relative efficiency of an adventure sports performer will have a significant impact on their performance. Research has shown that more efficient athletes, those with higher exercise economy, finish ahead of their less efficient counterparts. Exercise economy provides the interaction point between physiology and the skilful aspects of performance in adventure sports. An example of efficiency can be found in rock climbing. When comparing climbers of different abilities, the more skilful climber tends to be more efficient and smoother in their movements. Each movement and grasp of a new hold requires less effort and fewer muscle fibres to grip the hold. In a similar way, when learning a new route a climber grips tighter to each hold and requires more effort to make each movement. When the moves have been learnt the movement becomes smoother, more efficient, the effort required to grip each hold decreases and the overall energy demands of the climb are reduced. The more efficient a climber becomes in their movement, the greater exercise economy they possess.

Exercise economy is typically measured by the assessment of oxygen consumption across a range of workloads or exercise intensities. As with other aspects of physiology, exercise economy has been most studied with regard to running and indicates that race performance is improved when the relative oxygen consumption for a given workload is lower than that of other

Biography: Neil Gresham

Adventure sport: Trad rock climbing

Neil Gresham is one of Britain's most well-known all-round climbers. He is equally at home on rock or ice and is one of few climbers to have climbed the coveted grade of E10, with his second ascent of Equilibrium on Peak gritstone. However, Neil still regards his repeat ascent of the notorious Indian Face E9 6c on Cloggy as one of the biggest moments in his climbing. In recent years, he has developed new climbs in countries such as Sweden, Brazil, Mongolia, Cuba and Vietnam using a wide variety of climbing styles. Neil has also been a leading pioneer in Deep Water Soloing and his new route, the Wizard, was the first ground-up 8a in Britain. Since 1995, he has conducted regular 1-to-1 tuition as well as group workshops at climbing walls. Neil also established the first dedicated coaching holidays which he now runs every year, as well as a regular internet training program service. He has also written training articles for the British and American climbing press

Figure 9.6 Neil Gresham (photograph courtesy of Mike Robertson)

since 1993 and currently writes the popular Masterclass column in Climber magazine.

Equilibrium

The beauty of climbing is that it is always up to the individual to decide what they take home on the day. No-one else dictates the pressure and there are times when you will be content just to cruise and soak up the surroundings while sharing a joke with a good pal. On the other hand, there may be days where you use climbing as a tool to search further within yourself than you've ever been before. These situations, far from being fun, are usually utterly harrowing, but the memories will remain etched in your mind forever. My repeat ascent of Equilibrium, Neil Bentley's monolithic E10 at Burbage in the Peak District, is one such a tale.

Equilibrium involves a sequence of intense bouldering moves up a blunt arête, which are partially protected by a cam in a horizontal break. The route starts with easy climbing to a ledge and from there the climbing is intense all the way to the top. A fall from the first few moves above the protection is only an option if the belayer runs back to take in the rope, but a fall from the upper part of the route is almost certainly unsalvageable. After hearing the epic tales of the first ascent, I was intrigued to try the route on a top rope. After a day of futile attempts, I walked away licking my wounded finger tips, repulsed by its sheer difficulty. But the seed had been sown. How could I ever get myself into a condition where I could do this route physically, let alone mentally?

I went away and built replica boulder problems to simulate the moves at my local climbing wall. The move I had struggled with most was a leg press with virtually no assistance from the hands, so I devised a special exercise that involved balancing on the toes of my left foot and standing up in complete control. I was also lacking in reach for the next move so I performed endless yoga exercises to loosen my shoulders and extend my span. I was overwhelmed with apprehension when I returned to the route the following winter. I knew it would still feel desperate but the question was whether or not my preparation would tip the scales. After another day of top roping, I had managed all the moves but the thought of leading it chilled me to the bone. But then again, the conditions had been slightly humid and I rationalized that on a perfect freezing cold and dry day, I might be in with a chance. I would also need to make sure that I was well rested and with good skin on my fingertips, but it would be tricky to do this without losing strength. I would also need to have the right belay partner who would be prepared to take the responsibility. Getting these variables to balance seemed an impossible task and Neil Bentley's name for the route seemed all too poignant.

When the right day came I had everything planned. Warm-up first at the local climbing wall and then drive out quickly to the crag and jog around to stay warm. On my first top rope attempt, I made a very subtle refinement to the ankle position on the leg press move, and this made a lead attempt seem all the more realistic. So I decided to go for it. The look on Charlie's face was one of a belayer who knows the score. He had practised his part too – running down the slope to take the rope in, and he knew all too well that slipping wasn't an option. It's hard to describe what goes through your mind at moments like this. Unless the will to do the route is so strong that it surpasses all then it isn't safe for you to be there. As you go through the moves a final time, the level of focus you put into your visualized map is so strong that you almost feel as if

Figure 9.7 Neil Gresham stretched out on Equilibrium (E10, 7a), Burbage, Peak Gritstone, UK (photograph courtesy of Mike Robertson)

you're already out there involved in the climbing. But even as I completed the final moves in my head, I still couldn't see how so much preparation, or indeed an entire of life of climbing experience, could be compressed into such a tiny moment. But when the time came, my mind was completely empty so that my body could just climb. Guided purely by autopilot, I stepped off the ledge.

The leg press passed in a flash and I watched myself on TV as I made the reach for the sloping break! Got it. I knew this marked the point where I could no longer get away with falling and for a split second my consciousness came back. I contemplated the option of baling, and the next moment I was entirely swept away. When a route like this opens the door you have no choice but to go through. And on reaching the top I realized the answer to my final question. On ascents like this, time becomes almost meaningless. It is truly like entering a dream and waking up on the other side, at last free from the torment; or at least, until the next time.

athletes. In sports such as swimming, which has also been the focus of exercise economy research, similar findings have been achieved. The results from swimming, however, are important for more technical adventure sports such as climbing, kayaking, canoeing and cross-country skiing since they stress the importance of the skilful aspects of the sport. Unsurprisingly, higher level swimmers are more efficient than less skilful swimmers who spend more time and energy maintaining a horizontal body position and pushing up on the water. In kayaking for example, the interaction between the paddler's physiology, their boat awareness, feel for the water and use of the blade are critical to performance. The development of technical aspects alongside the physiological is essential for exercise economy. Research indicates that if you improve exercise economy you delay fatigue and improve performance.

Key point

An athlete reaches steady state when energy supply meets energy demand during exercise. Research indicates that oxygen consumption, heart rate and breathing rate plateau after about 4–5 min of sub-maximal exercise. Exercise economy relates to the energy demands at any given workload. The more efficient an athlete is, the lower the relative energy demand at any given exercise intensity level.

Oxygen debt

When we begin to exercise there is an immediate rise in the body's demand for oxygen, resulting in an increase in heart rate and breathing rate to match oxygen delivery to the exercise requirements. As described in the previous section, it takes time for the supply of oxygen to the active muscles to meet the demands of exercise and reach a steady state during sub-maximal exercise. This delay in the rise of oxygen consumption at the start of exercise, first identified by Archibald Hill and Hartley Lupton in 1922, creates what they termed an *oxygen debt*. The higher the exercise intensity, the longer it takes for the body to reach steady state and the larger the oxygen debt incurred. Figure 9.8 show the oxygen deficit incurred at the start of exercise as the orange shaded area. Oxygen consumption (shown in red) does not rise immediately to meet the demands of the steady-state exercise and, as such, a deficit is incurred. This debt is repaid post-exercise.

When exercise ceases and the additional oxygen for exercise is no longer required, oxygen

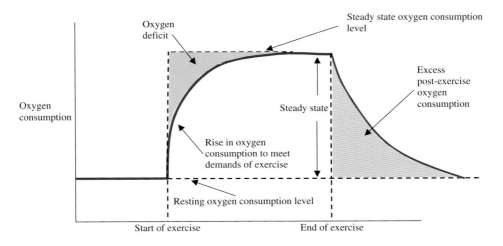

Figure 9.8   Oxygen debt and excessive post-exercise oxygen consumption

consumption should theoretically return to resting levels immediately. As can be seen from Figure 9.8, this is not the case. Oxygen consumption, and indeed heart rate and ventilation rate, remain elevated post-exercise. Brooks *et al.* (1971) termed the post-exercise elevation in oxygen consumption the *excess post-exercise oxygen consumption* (EPOC). The longer the duration and the higher the intensity of the exercise, the longer it takes for oxygen consumption to return to resting levels. The EPOC associated with a short powerful bouldering problem will be of lesser duration that after a 500 m kayak sprint or fell race.

The exact nature of the EPOC mechanisms associated with exercise are not known. However, research has indicated that there are a number of likely reasons behind the slow return of oxygen consumption to resting levels. A major reason behind the elevated levels of oxygen consumption (above resting levels) is to repay the oxygen debt incurred at the start of exercise. Oxygen stores in the blood and myoglobin are thought to be depleted during the start of exercise. Higher than resting level oxygen consumption post-exercise is partly thought to be associated with restoring oxygen stores. The longer the duration of exercise and the higher the intensity, the greater the rise in body

temperature associated with increased metabolism during exercise. High-intensity aerobic exercise, such as that encountered during a kayak marathon, can raise the body temperature by around 3 °C. The rise in body temperature post-exercise causes an increase in metabolic rate which, in turn, requires an increased level of oxygen consumption. As temperature drops over time post-exercise, oxygen consumption drops.

It is believed that a further part of EPOC is associated with the removal of CO_2 and the re-synthesis of lactate to glycogen. As described in Chapter 6, the onset of exercise causes the release of adrenalin and noradrenalin which serve to stimulate an increase in metabolism. These hormones of the sympathetic nervous system are not immediately removed from the blood and will therefore continue to stimulate metabolism, and consequently oxygen consumption, prior to their removal. It is likely that the post-exercise elevation in oxygen consumption is also related to tissue repair and the re-sequestering of Ca^{2+}, K^+ and Na^+ ions to their non-exercising compartments within muscle fibres. Further research is ongoing in this interesting area of physiology. However, it appears there are a number of mechanisms thought to be behind the continued elevation of oxygen consumption post-exercise.

Biography: Tim Brabants

Adventure sport: Kayak sprint racing

Tim, a doctor based in Nottingham, is a K1 kayak sprint racing specialist. He started canoeing at the age of 10 when taken by his mother to a come-and-try session at Elmbridge Canoe Club. The club has a strong tradition in sprint and marathon racing and it wasn't long before Tim was taking part in domestic competitions. His discipline is K1 1000 m and he has won titles and medals at national and international levels. He is the current World Record holder for the 1000 m sprint, a time he set in the heats at the Athens 2004 Olympic Games. He was junior World Champion in 1995, senior European Champion in 2002 and won a bronze medal at the Sydney 2000 Olympic Games. Away from the 1000 m distance, he joined with Conor Holmes in 1998 to take a K2 silver medal at the Marathon World Championships. In 2006, he took Gold again at the European Championships in a time of 3 min 28.59 s.

Figure 9.9   Tim Brabants (photograph courtesy of Trevor Chapman)

Sydney 2000 Olympic Games

Sprint kayak racing is held on a 2000 m regatta lake, racing over a distance of 1000 m or 500 m in single, double or four-man kayaks. Each race has nine lanes with stationary starts and events are organized into heats, semis and finals.

My first shot at qualifying for an Olympic Games was while still a junior in 1995, narrowly missing out on competing in the Atlanta Olympics 1996. Three years later I was fitter, stronger and faster and knew that to qualify for Sydney I would need to finish in the top eight at the World Championships in 1999. I performed well, finishing sixth. From that moment, all my focus was on training and preparing for the Olympic Games in Sydney 2000. I took time out of my medical degree to make more time for training. We work on four year cycles, so this was the big year.

Training two–three times a day, six days a week, is tough. The coldest months are the worst when the water freezes on your clothing, kayak and paddles and hands can be hurting for an hour after as they re-warm. Muscles are aching every day from weight training. I also attended training camps, the World cup and the European Championship races that year. We race and train in a multitude of climates and altitudes so one has to develop effective ways of acclimatizing to each venue quickly and successfully. The physical demands of my sport require a high aerobic capacity combined with good strength endurance as well as an explosive element. Training has to be carefully planned and my coach Eric Farrell was excellent at this. Every day follows the same routine but every day you're closer to your dream of competing in the Olympic Games.

As a nation, we had never medalled in sprint kayak racing at the Olympics. The build-up was different to anything I had experienced. Suddenly there was a large increase in media attention with radio, television and newspaper interviews. I was bombarded with information about the Olympics, travel and boat transport arrangements and being measured for kit. It would be easy to become caught up in all this and get distracted. I knew I wanted to go to Australia and race the best race of my life so far; it wasn't just about competing at the Olympics. To do that would require me to stay focused, train hard and manage my time better with the increased demands. Good quality recovery time enables good quality training.

I'd raced at major events before but I knew this would be different, mainly from a psychological viewpoint. The other main challenges would be the time difference, climate and length of travel. Those we tackled by travelling 4 weeks before competition to fully acclimatize. This allowed a good block of quality training to sharpen up for the start of racing. We stayed away from Sydney for the first 3 weeks to avoid too many distractions. The Olympic village is an amazing place. The best athletes in the world all staying in the same place – everyone having their World Championships at the same time. I found it a very positive and motivating experience. Our competition venue was a 2000 m lake just like everywhere else I race, apart from the 30 000 spectators, high security, TV cameras and media attention.

On race day I was ready. I was acclimatized and used to the different atmosphere at the Olympics. I'd watched other people standing on the Olympic podium and that's where I wanted to be too. The elation and look on their faces was very motivating. When I was in my boat on the water I felt in familiar territory. I was there to do what I'd done many times before and had spent many years training for. The first race went well and then I was drawn in probably the hardest of the three semi-finals, up against the European champion and current Olympic and World Champion. Then it was the final, this was it, my first Olympic final. An Olympic rowing friend Tom Kay had said that if I didn't feel I was going to die when I crossed the finish line then I hadn't done enough. The important thing was to prepare and warm-up like I do for any other race. On the start line when the gun went off my gate didn't drop. For a second I thought that was it, race over. On looking across I was aware that no-one else's had gone down either, a system fault. Time to re-focus, paddle round and get back on the start line. I stuck to my race plan not worrying about where anyone else was. With 50 m to go I was in fifth place, really digging deep, nothing to lose. I gave it everything and in that last 50 m I passed two people to finish third: Britain's first ever Olympic medal in Sprint kayak racing. I'd raced the best race of my life and came home with a medal.

Figure 9.10   Tim Brabants competing for Great Britain at Sydney 2000 Olympic Games (1000 m K1) where he won bronze (photograph courtesy of Trevor Chapman)

Key point

When you start to exercise there is an immediate rise in the demand for oxygen by the working muscles. This results in an increase in heart rate and ventilation to match oxygen demands. The time delay to match oxygen supply to consumption was termed the oxygen debt by Hill and Lupton. The delay in supply is thought to be an important factor in increased oxygen consumption (above normal resting levels) after exercise finishes. This was termed excess post-exercise oxygen consumption (EPOC) by Brooks *et al.* (1971).

Cardiovascular drift

As an individual begins to fatigue during pro-longed exercise, an interesting response is that of

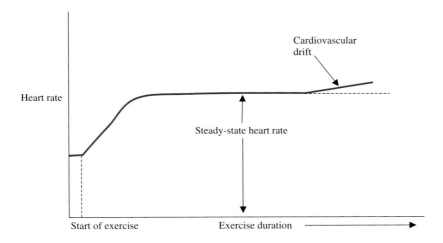

Figure 9.11 Illustration of cardiovascular drift

cardiovascular drift. Cardiovascular drift is associated with long duration exercise and/or exercise in hot environments. Figure 9.11 provides an illustration of cardiovascular drift.

Cardiac output $\dot{Q}$ (Chapter 4) is the product of stroke volume (the blood volume ejected from the heart in one beat) and heart rate (the number of heartbeats per minute). The rise in heart rate associated with *cardiovascular drift* is thought to be caused by a drop in stroke volume, associated with prolonged exercise or exercise in the heat. To maintain $\dot{Q}$ for steady-state exercise intensity, heart rate must rise if the stroke volume drops. The drop in stroke volume is thought to occur for a number of reasons, all or some of which could occur during exercise.

The upright body position during most adventure sports e.g. orienteering, kayaking and mountaineering means that venous return must compete against gravity. It is thought that during exercise at a constant workload over time there is a progressive decrease in venous return, which in turn leads to a decrease in stroke volume and necessitates a rise in heart rate. The rise in body temperature and sweating associated with exercise are also thought to contribute to a decrease in stroke volume during prolonged exercise. As body temperature rises during prolonged exercise or exercise in the heat it is

thought that blood is diverted to the skin for cooling, which results in a decrease in venous return and consequently stroke volume. The incidence of water loss through sweating is believed to lead to a decrease in blood volume and consequently venous return through plasma removal to help maintain tissue hydration levels during exercise.

Key point

Cardiovascular drift refers to a rise in heart rate during exercise that cannot be attributed to an increase in exercise intensity. Cardiovascular drift is associated with exercise in hot environments or over a long duration. During such exercise, at a particular point, venous return is impaired. This results in a decrease in blood flow and oxygen supply to the muscle. To maintain oxygen supply to active muscles heart rate must rise to compensate for the decrease in venous return.

Stitch

A 'stitch' – a sharp pain in the side brought on during running – can become a temporary mechanism of fatigue. Most individuals, however, can quickly overcome a stitch and continue with exercise. A stitch is thought to occur due to the up and down nature of running or jogging, as stitches are not nearly as common in non-running activities such as mountain biking, kayaking or canoeing. The movement of the internal organs with each stride may cause micro-trauma to the digestive system and result in pain. Alternatively, or additionally, the re-distribution of blood away from the digestive tract during exercise may be responsible for the development of a stitch during running. Stitches appear to be more common when exercising shortly after a meal and consequently temporary ischemia (lack of blood flow to the digestive tract) caused by the re-distribution of blood to the exercising muscles may provide the underlying cause of the pain.

Key point

A stitch, occurring most commonly during running-based activities, is thought to be due to movement of the internal organs and possibly through the diversion of blood away from the digestive tract to the working muscles.

9.4 Physiological adaptations to training

The completion of an aerobic training programme will lead to a variety of physiological adaptations in response to the training stress. This section describes the key adaptations that occur in response to aerobic training, providing the basis for improvement in performance for any of the adventure sports in this chapter or the lower intensity activities which are the focus of Chapter 10. The adaptations to aerobic training include alterations in cardiac, respiratory, muscular and metabolic function.

The most obvious response to aerobic training is an improvement in aerobic capacity $\dot{V}O_{2\,max}$

which results in an increased ability to utilise oxygen for metabolism. The intensity, duration, type and specificity of training will all influence the extent of any physiological adaptation. The extent of physiological adaptation to training will also be dependent upon the individual's previous training and their individual response to the training stimulus. For instance, improvements in $\dot{V}O_{2\,max}$ have been shown to be limited to the first 6–12 months; after this time improvements in performance tend to be related to improvements in LT and exercise economy. Consequently, if an individual is new to aerobic training their improvement in $\dot{V}O_{2\,max}$ will be significantly greater than that for an experienced athlete.

Nevertheless, an experienced athlete can still continue to improve their performance through training. Research indicates that individuals respond differently to any training stimulus. Coaches or the athlete themselves need to monitor the response to training and alter a programme if desired results are not being realized. This point should serve to emphasize the importance of including a variety of performance or training assessments at the end of key phases of training.

In Chapter 6, the properties of different muscle fibre types were described in detail. During aerobic exercise, slow twitch Type I fibres are primarily recruited for exercise. In response to aerobic training, Type I fibres tend to decrease slightly in size due to the physiological stresses placed upon them in training. At first this might seem contrary to logic; increasing training should increase muscle size. However, during aerobic training Type I fibres are not required to produce maximal forces, but to complete repeated lower intensity contractions. As described in Chapter 6, strength training is associated with muscle fibre hypertrophy. It is partly to maintain muscle strength and size that aerobic endurance athletes undertake a strength training programme concurrently with aerobic training. By combining strength and aerobic training, an athlete can maintain muscle size and strength as well as make aerobic gains.

The protein degradation within muscle fibres in response to aerobic training is thought to arise from changes in the production of the hormones cortisol and testosterone. In response to aerobic training, levels of cortisol tend to increase. Cortisol has been shown to be associated with protein degradation in Type I muscle fibres. Testosterone is an anabolic hormone (promoting protein synthesis, as opposed to catabolic) and in response to aerobic training, blood levels of testosterone fall. Combined with the increases in cortisol levels, this results in a predominantly catabolic environment within the Type I muscle fibres.

Although aerobic training results in a decrease in the size of Type I muscle fibres, there are a number of structural changes within these fibres that promote increases in aerobic endurance performance. In response to aerobic training there is an increase in the oxidative capabilities of Type I muscle fibres. Endurance training brings about changes in the capillary density, myoglobin concentration and mitochondrial size, number and function, all of which improve the aerobic capability of the trained fibres. Research indicates that the number of capillaries surrounding each muscle fibre can increase by 10–15 per cent through training, promoting oxygen delivery to the active muscle fibres. Linked with this increase in oxygen delivery is increased oxygen storage brought about by an increase in the myoglobin within Type I muscle fibres. Myoglobin concentration in Type I fibres can increase by as much as 75–80 per cent above pre-training levels. Muscle fibre mitochondria, the aerobic powerhouses, have been shown to increase by up to 35 per cent in size and 15 per cent in number in response to aerobic training.

Furthermore, their efficiency has been demonstrated to increase through increased enzyme activity. As an example, the Krebs cycle enzymes – *citrate synthase* which catalyses the binding of acetyl CoA to oxaloacetate to form citrate at the start cycle and *succinate dehydrogenase* (SDH) which catalyses the seventh reaction in the cycle where succinate is degraded to fumarate – increase their activity in response to endurance training. The intensity of training also affects the size of the increase in SDH activity. When an athlete trains at a higher intensity, the increases in SDH activity are more dramatic than for lower intensity aerobic training.

In addition to increases in Krebs cycle enzyme activity, the enzymes involved in β oxidation and fatty acid oxidation also respond to aerobic training. The activity increases in the enzymes involved in fatty acid metabolism have a glycogen sparing effect that promotes endurance performance. The 25 per cent (sometimes higher) improvements in fatty acid oxidation are also supported by increased muscle triglyceride storage. This creates an increased lipid pool immediately available for β oxidation and fatty acid oxidation. Training has been shown to more than double the resting muscle triglyceride stores. These changes result in a sub-maximal decrease in RER values for a given work intensity. The decrease in RER during sub-maximal exercise provides evidence for the shift in metabolism towards fat oxidation as a result of aerobic training. It is thought that the shift to fat oxidation is supported by a decreased sympathetic nervous system (SNS) response to exercise. At the start of exercise the SNS, through the release of the hormones adrenalin and noradrenalin, increases metabolism and the body's reliance on glycogen as a fuel source. As a result of training the SNS response to exercise is blunted, decreasing the emphasis on carbohydrate as the main fuel source for exercise.

As well as improvements in fatty acid storage and oxidation, aerobic training has been shown to result in increases in glycogen synthesis and storage. Training has been shown to improve response to insulin, promoting increased glucose uptake by the muscles. The enhanced glycogen stores have been shown to improve performance and delay fatigue due to glycogen depletion. The improved functioning of the mitochondria in response to training also serves to slow glycogen oxidation and decrease lactate production. In addition to these alterations, primarily in the functioning Type I fibres, research indicates that there are also changes in the properties of Type II fibres which serve to enhance endurance performance. Type $II_{b(x)}$ fibres have been shown to take on the characteristics of Type II_a fibres and a small percentage of Type II_a fibres have been found to become more like Type I fibres in response to aerobic training. These changes in fibre type properties further serve to improve aerobic endurance.

Aerobic training results in a decreased resting heart rate and heart rate response to exercise (maximal HR is normally little altered in response to training). Mean resting heart rate for a sedentary individual is about 72 bts min^{-1}. As a result of aerobic training, resting heart rate can fall below 50 bts min^{-1}. Despite the drop in resting and sub-maximal heart rate response, $\dot{Q}$ (the product of heart rate and stroke volume) has been shown to increase in response to endurance training. The increase in $\dot{Q}$ is a result of improvements in stroke volume. Aerobic training results in an increase in cardiac size and the left ventricular volume, resulting in increases in the volume of blood ejected with each contraction of the heart.

In sustained endurance training, one of the key adaptations is an improvement in LT. In an AAT test, improvements in LT can be found by a shift to the right of the lactate curve as illustrated in Figure 9.3. Aerobic training results in improvements in LT and, research indicates that the most effective way to improve oxidative functioning is to train at or just above the exercise intensity at which LT or AAT occurs. Improvements in LT have been found to be specific to the exercise mode. Running results in a greater shift in the point at which LT occurs in a treadmill test. When measuring improvements in LT the assessment procedure should, as far as possible, mirror the training mode, whether kayak, running or cycle-based.

Key point

Aerobic training results in a variety of cardiovascular, respiratory, structural and metabolic adaptations. In the early stages of training, an individual will see improvements in $\dot{V}O_{2\,max}$. After these initial gains, subsequent improvements in performance will result from improvements in LT and exercise economy. Aerobic training results in improvements in oxygen delivery to the muscles, enzyme activity, mitochondrial function, glycogen sparing and cardiac output, all of which have a positive impact on aerobic endurance performance.

9.5 Nutritional ergogenic aids to aerobic performance

Beyond a balanced healthy diet, the key manipulations that could be beneficial to performance in the adventure sports discussed in this chapter are considered in Chapters 7 and 8. In the very short-term aerobic endurance events such as a 1000 m kayak race, a wave sailing heat or a mountain bike downhill, the use of bicarbonate and β-alanine supplements appear to help with H$^+$ buffering. In the longer duration activities, where glycogen depletion represents a significant possible cause of fatigue, the employment of a glycogen loading strategy and the use of in-race carbohydrate feeding can help to maintain performance. The fatigue delaying effects of buffer supplementation were described in Chapter 7, while the performance benefits of carbohydrate loading and in-race consumption were described in Chapter 8.

In addition to these dietary manipulations, there exists a further supplement which is both legal for athletes to take and has a wealth of research supporting its ergogenic effects: caffeine. One of the most widely consumed drugs, caffeine is found naturally in tea, coffee and chocolate. Tea contains around 65–100 mg per cup, ground coffee 125 mg per cup and milk chocolate in the region of 50 mg per small bar. Caffeine is a purine, similar in structure to adenine which forms the basis for ATP, and

Biography: Jan Sleigh

Adventure sport: Wave sailing – windsurfing

Jan Sleigh has been competing in wave sailing for over 20 years. He began when he was a teenager on a family holiday in France. When he arrived home he sold all his records and anything else he could find to buy his first windsurfer. Jan began a career in marketing while competing at an amateur level in wave sailing. He was placed third at the 1992 Tiree Wave Classic and since then competed all over the world to achieve his aim of taking part in the PWA tour. In 2003, he qualified to take part in all PWA Tour events for that year.

2003 PWA world tour

In 2003 I competed in the wave sailing discipline on the PWA World Tour. This tour would take me to two events on Gran Canaria and Sylt in Northern Germany, but unfortunately the final event on the west coast of Ireland was cancelled.

Figure 9.12 Jan Sleigh (photograph courtesy of Justine Morgan)

I spent ten days on Gran Canaria for the Vargas event, three weeks on Gran Canaria again for the Pzo event and about ten days in Sylt in Germany for the last event. This in itself proved to be a difficult way to compete for several reasons. Firstly, I had no time to acclimatize myself to the climate and the conditions of the locations apart from a week before the Pozo event. Secondly, after each event I returned to my full-time job. This makes keeping in shape and maintaining fitness levels difficult since the only on-water training would be at weekends in the UK if conditions were favourable. Both of these factors create demands on the body, but also the mind because of the fact that I was not training between events like many of my fellow competitors.

Windsurfing competition in waves takes place in strong winds and breaking waves close to the beach. Heats are often as short as 8 min in duration but are intense and are both aerobic and anaerobic in nature. The windsurfer performs jumps and loops in the air on the way out through the waves and then rides the waves back to the beach in much the same way as a surfer.

During competition there is a lot of waiting around. Because the contest is at the whims of the weather, it is often necessary to be at the beach from early in the morning until late afternoon. In the Canary islands it was necessary to contend with strong wind and strong sunshine; in Germany cold winds and rain were the order of the day. The environmental conditions place a further stress on the windsurfer beyond the physical demands.

Figure 9.13 Jan Sleigh catching air during the Professional Windsurfers Association World Tour 2003 (photograph courtesy of Julia Schweiger)

is a central nervous system stimulant. The effects of caffeine are widespread and include increasing mental alertness, concentration, mood state, fatty acid mobilization, catecholamine release (adrenalin and noradrenalin) and muscle fibre recruitment. Caffeine consumption also results in a decrease in perception of effort, time to fatigue and reaction time. As a consequence, caffeine represents the most wide-ranging of the ergogenic aids discussed in each of the nutrition sections in Part II. The increase in mental functioning can impact on all the adventure sports. The reduction in reaction time in conjunction with the increase in muscle fibre recruitment and catecholamine release can result in performance improvements for power and power endurance activities. The lowering of perception of effort and catecholamine release represent potential improvements in performance for anaerobic endurance events. The lowered perception of effort, glycogen sparing effect of increased fatty acid mobilization and usage along with delaying of fatigue can significantly improve performance in intermittent and aerobic activities.

The benefits of caffeine ingestion before exercise have been most widely researched and reported for endurance activities. A classic study by Costill *et al.* (1978) provides an excellent example of the effects of caffeine on endurance performance. In this study, nine cyclists consumed caffeine in one trial and a placebo in another during which they cycled to exhaustion at an exercise intensity of 80 per cent of $\dot{V}O_{2\,max}$. Time to exhaustion was nearly 20 per cent longer during the caffeine trial and the cyclists exhibited a lower RER value indicating a shift towards fat metabolism, thereby sparing glycogen stores. In addition, the athletes reported the caffeine trial as being easier than the placebo trial. Subsequent studies have produced similar findings with caffeine resulting in a 10–20 per cent improvement in time to exhaustion.

At rest, caffeine represents a diuretic and this presented one potential problem with its use as an ergogenic aid. Research, however, indicates that the diuretic effect is negated during exercise as catecholamine release at the start of exercise stimulates the release of anti-diuretic hormone which increases water re-absorption and counteracts the diuretic effects of caffeine. The findings of a variety of studies suggest that an ingestion of 3–5 mg of caffeine per kg bodyweight is sufficient to create an ergogenic effect.

For those who regularly consume caffeine, the ergogenic effect will not be realized unless they avoid caffeine ingestion for a period of time before an event. Research suggests that omission of caffeine from the diet for 5–6 days should be sufficient to establish an ergogenic effect for regular caffeine consumers. For those who do not normally consume caffeine, or when a dose is higher than normally encountered, there are potential side effects. These include restlessness, elevated heart rate, insomnia and headaches. Unless consumed in excess doses, caffeine ingestion does not generally present a health risk and therefore represents an ergogenic aid that can positively affect performance in a wide range of adventure sports.

Caffeine was a restricted substance until 2004 when WADA removed it from the banned list. Although a monitored substance, athletes are now able to use caffeine as a performance supplement. Again, any individual using caffeine as an ergogenic aid has to make a moral decision about its ingestion despite the recent changes in the regulations.

Key point

After the adoption of a balanced healthy diet for short-duration high-intensity aerobic endurance activities, β-alanine and bicarbonate supplementation could be considered. For longer duration events, the use of glycogen loading and glucose in-race feeding should be considered, as both have been found to be beneficial to performance.

Caffeine is a further supplement that could be considered for improvement in high-intensity aerobic endurance activities. Caffeine ingestion, among other benefits, has been found to improve time to exhaustion by 10–20 per cent.

9.6 Physiological challenge of thermal stress

Adventure sports are often performed in environments that place thermal as well as exercise stress upon the athlete. The combined stresses of exercise and heat or cold create multiple threats to homeostasis for adventure sports performers. This section examines the effects of heat and cold on performance during adventure sport participation. As was introduced in the section on water immersion in Chapter 8, the body reacts to its environment to maintain a constant internal temperature. The internal thermal balance is maintained by a variety of mechanisms that are initiated by the hypothalamus in response to sensory input from hot and cold receptors located in the skin, muscles, spinal cord and brain. The thermoregulatory system normally maintains body temperature within one degree of the mean value of 37 °C. To conserve heat, the body can divert blood flow away from the periphery (vasoconstriction) and initiate pilo erection, both of which were described in detail in Chapter 8. In response to cold environmental conditions, the body initially increases muscle tone and, if subsequently required, induces a shivering reaction to increase metabolism and heat production. With regard to adventure sports participation, the exercise necessary to climb an alpine peak or complete a kayak marathon results in an increase in metabolism, therefore presenting another source of heat production. To increase heat loss, the body increases blood flow to the skin (vasodilation) and can increase sweat rate.

Figure 9.14 provides an illustration of the thermoneutral and thermoregulatory zones. For a naked adult standing in air, the thermoneutral zone is 26–30 °C. In water, due to the enhanced conductivity which was described in Chapter 8, the thermoneutral zone is higher than for air and much narrower (35–35.5 °C). Within the thermoneutral zone, body temperature can be maintained by alterations in blood flow alone, keeping skin temperature around 33 °C. Beyond the thermoneutral zone, but within the thermoregulatory zone, core body temperature can be maintained by inducing shivering and sweating. If the naked body is exposed to temperature outside the thermoregulatory zone it will not be able to defend core body temperature. As a consequence, the core temperature will continue to rise (in hot environments) or fall (in cold environments) inducing hyperthermia and hypothermia.

As was explained in Chapter 8, heat represents one form of energy. Heat is exchanged via conduction, convection, evaporation and radiation. The human body temperature can be increased internally through metabolism and externally by conduction, convection and radiation. For example, the sun on a summer's day can provide a major source of radiation heat energy. Body heat can also be lost via conduction, convection, radiation and evaporation.

Conduction involves the transfer of heat from one material to another through direct molecular contact. Heat produced by the muscles during exercise can be conducted to the skin. If you sit on a storage heater to warm up at the end of a day's paddling, heat will be conducted to your body. During cold water immersion, heat is conducted from the body to the boundary layer of water molecules surrounding the body.

Convection is really a specific form of conduction involving the movement of a fluid (gas or liquid) across a warmer surface (solid, liquid or gas). Boundary layer molecules from the fluid are heated by conduction. Normally the warmed boundary layer would then begin

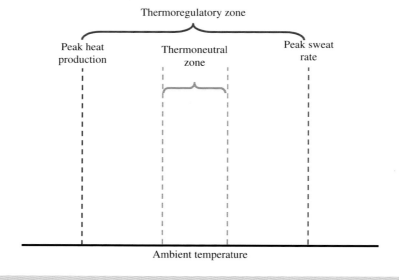

Figure 9.14 Thermoregulation in humans: the thermoneutral and thermoregulatory zones (adapted from Mekjavic and Bligh, 1987)

to act as insulation. However, in convection this heating of the boundary layer reduces its density, creating buoyancy, and the warmed fluid begins to rise. Cool fluid replaces the boundary layer and the process of conduction, movement and replacement continues. Both air and water colder than the skin can transfer heat away from the body by convection, resulting in faster cooling than straightforward conduction.

Water or air molecules heated through contact with the body (conduction) can also be stripped away from the body if there is an air current (wind) or water flow (river current or tidal movement) though the process of *advection*. Advection occurs when the continuous replacement of the boundary layer occurs through an already established flow of the cooling fluid and can result in even faster heat transfer than convection. In a mountaineering setting, this effect is often referred to as 'wind chill' and explains why, on a breezy day, it can be warmer to wear thin insulation under a light windproof than a thicker insulation layer. Heat loss during water immersion is primarily through conduction and convection or advection.

Radiation is emitted from all objects that possess heat energy. Radiation is composed of packets of energy called photons which travel at the speed of light. The sun heats the earth by thermal radiation. Radiation is the only way heat energy can cross a vacuum. However, photons can also travel easily though gases and, to a limited extent, transparent fluids such as water or objects such as glass. The main photons responsible for heat transfer belong in the infrared spectrum, at an energy level just below that of visible light. The energy in infrared radiation will heat solid or liquid objects when it comes into contact with them. When we stand facing the sun we can feel its heat as we are absorbing its radiation. Light-coloured surfaces absorb less thermal radiation than dark surfaces, which is why mountaineers who climb in the Alps in midsummer will often choose white helmets as they are cooler.

As homeotherms (maintaining a constant body temperature), humans also emit radiation in all directions and, at rest, radiation represents the major method through which humans lose heat. During exercise, the relative amount of heat loss by radiation decreases and evaporation becomes the major method for heat dissipation. In water, because we are surrounded by a liquid, heat loss by radiation is greatly decreased (occurring mainly from the head) because infrared radiation through liquid is extremely limited. It is for this reason, along with the thermal conductivity of water, that conduction and convection or advection represent the major methods of heat loss during water immersion.

Evaporation, the use of energy to covert a liquid to a gas, forms the major source of heat loss during exercise. The major source of heat loss for humans at rest, around 60 per cent, is through radiation. During exercise in low humidity conditions, evaporation provides the mechanism for about 80 per cent of our heat loss. In humid conditions, when the water vapour content in air is high, evaporation is reduced but still remains the major source of heat loss. In response to temperature increases due to exercise, hot climatic conditions or a combination of both, the hypothalamus stimulates an increase in the activity of sweat glands. The CNS stimulation of sweat glands results in an increase in sweat production on the skin surface. At the same time, vasodilation increases the conduction of heat to the skin. The thermal energy from the active muscles warms the sweat on the skin to a temperature where it can evaporate. The heat absorbed by sweat for evaporation results in the removal of heat from the body, which assists with core temperature maintenance.

Key point

A number of thermoregulatory mechanisms combine to maintain body temperature close to a normal 37 °C. Heat is transferred by way of conduction, convection, evaporation and radiation. Vasodilation and vasoconstriction of the blood vessels play a major role in the maintenance of core temperature.

The physiological stress of heat

Exercise in the heat provides a dual challenge to the body: thermal stress from the environment and physical stress from the exercise. The primary mechanisms for coping with the thermal stress associated with heat are vasodilation, increased blood flow to the skin and increases in sweat rate. Sensory heat receptors mainly located close to the hypothalamus and sensitive to temperature changes as small as 0.01 °C relay information to the hypothalamus, the body's thermostat. The hypothalamus stimulates blood flow alterations and increases in sweat gland activity. In short-duration adventure sports such as those described in Chapters 6 and 7 (power, power endurance and anaerobic endurance activities) the exposure to a hot environment is minimized and consequently has a minimal impact on performance. For longer duration activities such as the intermittent activities which were the focus of Chapter 8 or for aerobic endurance adventure sports (this and the following chapter), where the exposure to a hot environment is more sustained, heat stress provides a greater threat to homeostasis and a greater challenge to the body's thermoregulatory function.

The longer the duration and the higher the intensity of exercise, the greater will be the challenge to thermal balance. The risks of dehydration and hyperthermia and their deleterious effects on performance are increased as the environmental temperature in which an adventure sport takes place increases. The degree of shade from direct sunlight

and the amount of reflected solar radiation will also affect the thermal stress placed on an adventure sports performer. In addition, the relative humidity and air movement (wind) will affect the degree of heat stress. Research, however, indicates that the better a person's level of fitness, the better they are able to cope with the additional stress of exercise in the heat.

Exercise in the heat creates a dual challenge to physiological function. Exercise creates an increase in demand for oxygen and blood flow to the muscles. To cope with the demands of heat, the body diverts blood flow to the skin for cooling and stimulates increases in sweat rate. The increased activity of the sweat glands increases their demand for oxygen and energy. Exercise in hot environments creates a competitive demand for blood flow and results in an increased rate of glycogen usage. Research shows that exercise in the heat results in higher rates of lactate for a given work load, increases in ventilation and heart rate along with a reduced time to glycogen depletion. The alterations in blood flow in hotter environments result in a decreased venous return and as a consequence, a decrease in stroke volume. To compensate for this and maintain cardiac output, heart rate has to rise during exercise in the heat. This rise in HR, associated with maintenance of cardiac output during exercise, is termed cardiovascular drift (see section on response to exercise, earlier in this chapter).

The key mechanism for heat loss during exercise, and exercise in the heat, is evaporation. When the exercise intensity and environmental temperature are low the sweat rate is low and therefore sodium and chloride ions within sweat are reabsorbed as the sweat makes its journey from the sweat gland to the skin via the sweat duct. As a consequence, the mineral content of sweat during low-intensity exercise in cooler environments is lower than that at higher intensities and temperatures. The re-absorption takes place from the walls of the sweat duct and returns sodium and chloride ions to the bloodstream reducing electrolyte loss.

However, potassium and other minerals such as calcium and magnesium, also lost in sweat, cannot be re-absorbed in this way and must be replaced by dietary intake. At higher intensities of exercise, the production rate of sweat is increased and consequently travels via the sweat ducts to the skin more quickly. As a result, the re-absorption of sodium and chloride is reduced resulting in an increased rate of electrolyte loss.

When exercising at higher intensity levels, fluid loss through sweating can be as high as $2.0 \, L \, hr^{-1}$. For a 70 kg adult this represents a 2.9 per cent drop in body weight through fluid loss. As described in Chapter 8, research indicates a drop of only 1 per cent has been shown to equate to a 2 per cent drop in performance (speed during running). During the course of an endurance race in the heat, such as for a fell race or kayak marathon, performers can lose as much as 6–10 per cent of bodyweight through sweat loss. A 2–3 per cent drop in bodyweight through fluid loss is sufficient to decrease blood plasma volume, contributing to cardiovascular drift, and affecting the sweat response. To offset the negative effects of dehydration during exercise in the heat, adventure sports performers must employ a re-hydration strategy. The use of electrolyte drinks as part of a re-hydration strategy was discussed in Chapter 10. Beyond the negative effects on performance, heat and exercise-induced dehydration contributes to the incidence of heat exhaustion and heat stroke which are potentially life threatening. The use of a fluid re-hydration strategy can help to prevent

Biography: Anna Hemmings

Adventure sport: K1 Kayak Marathon

Anna Hemmings, Britain's leading female marathon canoeist, current World and European Champion, was told by medical experts she might never race again. In 2005 however, she signalled a miraculous return to fitness by regaining her status as the World's leading marathon canoeist at the Marathon Racing World Championships in Perth, Australia. It is her fourth gold medal in the 13 year history of the event, and her ninth World and European Championship medal, confirming her status as Britain's most successful ever female canoeist. Anna's achievements were recognized at the 2005 Sunday Times Sportswoman of the Year Awards, where she won the Champions Award. Anna, 28 years old from Surrey, was diagnosed with Chronic Fatigue Syndrome which had threatened to end her career. However, reverse therapy enabled her to overcome the illness and return to

Figure 9.15 Anna Hemmings (photograph courtesy of Mark Lloyd)

the highest level of marathon performance. After her success at the 2005 World Marathon Championships, Anna has taken the decision to concentrate on sprint racing. While wishing to retain her marathon World title, her goal now is to win Olympic Gold at the Beijing 2008 Olympic Games.

Marathon racing world championships 2005

The sweetest victory

We were about 150 m away from the final portage when I took the lead; I arrived at the portage (where you get out of your kayak and run with it for 200 m before re-entering the water for the next lap), leapt out of my boat, and totally focused I ran, ready to embrace the hardest part of the race.

Imagine having been racing for an hour and three-quarters then running with your boat as fast as you can for 200 m. The first 170 m was on grass around a bend, and then the last 30 m on sand. Just as your legs are getting tired it gets even tougher! You get back in, your legs have ceased up, your arm is tired from carrying the boat, your lungs are bursting and then you have to pull away and paddle as fast as you can. I managed to establish a 100 m lead but I knew that if I relaxed they would catch me. I paddled hard but this was the bit that hurt. The pain was kicking in but I realized I'd been there before in a number of races and in a thousand training sessions. I remembered that this is what I trained for: being able to push through the pain barrier in the last 6 km of the World Championships.

How do you prepare yourself for that? Most people compare marathon canoeing to running a marathon; however, it is more akin to cycling. In the same way that cyclists sit on the slipstream and ride along in the peleton; canoeists sit on the wash (or ride the wave that comes off the side of the boat) and race in a pack. The pack leader changes every so often, thus causing the paddlers to jostle for the best positions in the group. By sitting on the wash you can conserve about 30 per cent more energy than the person leading the pack. So my plan is usually to sit on the wash as much as possible!

The idea of riding waves and racing in groups means that the pace fluctuates; there are numerous sprints, particularly as a group approaches a turn or portage or when the pack leader changes. The start of a marathon race is pretty rapid too. Despite the fact that the race is 18 miles (28.8 km) long, if you want to win, it is essential that you make the front group; otherwise you are playing catch-up for the rest of the race. With between 20–30 canoeists on the start line, the first 1000 m are fairly frantic as everyone sprints to make the front group and fights to gain the best position. Equally, a sprint finish is not uncommon. A successful marathon canoeist, therefore, doesn't only require a high level of endurance; a powerful sprint is also of paramount importance.

In order to prepare my body for this, in a typical training week (during the summer) I would clock up around 50–60 km on the water. This would consist of 10–11 individual training sessions; 8–9 of those would be on the water and 1–2 would be running sessions. I would also normally do weight training twice a week. However, during the summer of 2005 I was challenged with a wrist injury which forced me to omit weight training altogether. Although weight training is important for building power and strength, with a number of years of weight training behind me it wasn't going to be detrimental.

I run a couple of times a week for two reasons. The first being that I believe running is a great way to build general fitness and stamina (crucial for marathon racing). The second reason is because we have to do portages which are usually around 200 m long. Although you don't have to be an elite runner, the more comfortable you are running through the portage, the less it will take out of you and affect you when you get back on the water.

If during one week I did eight sessions on the water, they would consist of:
Five endurance sessions, including:

- 1 × 10 km time trial or 21 km race
- 1 × long intervals at core aerobic pace (CAP)
- 2 × intervals at threshold e.g. 4 × 4 min on 1 min rest, 3 × 5 min on 1 min rest, 2 × 6 min on 1 min rest, with 3–4 min rest between the sets
- 1 × 4 km time trial followed by long intervals at threshold

Three speed sessions, including:

- 1 × speed endurance
- 2 × speed or speed-strength

A speed-strength session is resistance training. In a similar way that a sprinter on the track might run dragging a tyre or some other form of resistance, a canoeist will tie a bungee around the kayak and place 1–3 tennis balls underneath it to create extra drag and resistance.

In my preparation for the world championships, the only time that I paddled the full race distance was at the European Championships. With so many years of endurance training behind me and the solid base that I built at the beginning of my career, it is not necessary for me to regularly race or train over the race distance.

With all this training in place I was finally ready to head to Australia. I flew out to Perth 14 days prior to the race, giving me sufficient time to overcome the jet-lag, acclimatize and complete the finishing touches of the training programme. The race was on a Saturday and I completed my final training session on Tuesday evening, leaving me with three full days to rest and build up my glycogen stores. I consumed a high-carbohydrate diet and plenty of fluids. Race day arrived; I was rested and ready to endure the toughest part of the race.

I raced through the pain barrier and finally I entered the home straight. There was no chance of anyone catching me now, but the challenge was to maintain my focus because I suddenly became distracted by the thought of another world title making me emotional! This was not the time to be getting emotional, I hadn't won yet – I needed to focus.

Figure 9.16   Anna Hemmings, European and World Marathon Champion (K1) in training before regaining her world title in 2005 (photograph courtesy of Billy Black)

After 2 hr 16 min of racing I crossed the line in first place – world title number four!! I threw my arms in the air and gave a yelp of joy and relief. It was the sweetest victory yet. (If you would like to see video footage of this last section of the race please visit www.annahemmings.com.)

the incidence of heat-related illness during adventure sports participation in the heat.

At rest, body temperature is normally maintained within 1 °C of mean body temperature (37 °C). Humans are able to tolerate larger drops below normal temperature than rises above 37 °C. A temperature of 42 °C can be fatal, whereas a human can survive drops in core temperature to 25–27 °C. Heat illnesses, heat exhaustion and heat stroke should therefore be treated seriously. The three causes of hyperthermia (increased body temperature) in combination (high exercise intensity, high environmental temperature and high humidity) increase the likelihood of heat illness. Children, those who are less fit, obese or suffering from illness are at greater risk of heat illness.

Heat exhaustion results from hypovolemia brought about through alterations in blood flow to the skin for cooling and fluid loss through sweating. The competition for blood supply between the thermoregulatory mechanisms and muscles during exercise results in a decrease in blood volume returning to the heart. The decrease in stroke volume impairs cardiac output and results in a decrease in blood pressure leading to circulatory shock. Signs of heat exhaustion include dizziness, fatigue, moist clammy skin, feeling faint, vomiting, low blood pressure and weak rapid pulse. A person displaying symptoms of heat exhaustion should be treated as for shock. If possible, remove the person from the immediate environment to a cooler place, lay the person down with legs raised to increase venous return, provide an electrolyte drink and seek medical help. If the symptoms are untreated then heat exhaustion can progress to heat stroke.

Heat stroke results from a failure of the body's thermoregulatory mechanisms to cope with heat exposure. It is a life-threatening illness and requires immediate medical attention. Heat stroke can result from exposure to hot environmental conditions, but has also occurred for endurance athletes involved in high-intensity exercise in relatively moderate temperatures. The failure of the thermoregulatory system in heat stroke leads to a continuing rise in core temperature which impairs CNS functioning and leads to coma, tissue damage and, if untreated, death. Symptoms of heat stroke include very high body temperature, dry hot skin associated with a cessation of sweating, an increase in blood pressure, confusion and unconsciousness.

Heat acclimatization

Research indicates that the body can acclimatize to exercise in the heat. If an adventure sports performer is going to compete at a venue where the environmental temperature is greater than where they normally train, then an acclimatization strategy will benefit performance. To benefit from a heat acclimatization strategy, the athlete must exercise during the heat of the day. It is not sufficient merely to experience the heat at rest. It appears that an acclimatization period of 5–10 days should be sufficient to alter thermoregulatory response. During this period, the exercise intensity should be reduced to below normal for any given training session and attention should be paid to during and post-exercise re-hydration. This type of acclimatization strategy appears to reduce sodium and chloride loss and increase water retention such that pre-exercise water levels are 10–20 per cent above normal. The hormone aldosterone stimulates increases in mineral and water retention from the sweat glands. Anti-diuretic hormone increases re-absorption of water and minerals from the kidneys. These alterations result in increases in blood volume (plasma), stroke volume and arterial pressure along with decreases in heart rate, commencement of sweat response, core temperature, sodium and chloride content in sweat and feelings of fatigue.

Key point

Heat sensors within the body can detect heat rise as small as 0.01 °C and relay this information to the hypothalamus, the body's thermostat. The body's primary mechanisms for heat loss include vasodilation, moving blood to the surface (towards the skin), increasing sweat rate and as a consequence evaporative cooling. The relative air movement and humidity will affect the rate of evaporative cooling at any given temperature. Increased rates of sweating increase the risks of dehydration. Exercise in the heat can lead to heat exhaustion and heat stroke.

The physiological stress of cold

The physiological stresses presented by exercise in cold environments relate to the air temperature, wind strength, relative humidity and the presence of rain. The lower the temperature, greater the wind strength and lower the relative humidity (or the presence of rain) the greater the physiological threat to homeostasis. When the temperature in any environment falls below that of the skin temperature, the body will lose heat. The presence of air movement (wind) will increase convective and evaporative heat loss. Dry air creates an increased risk of exercise-induced bronchiospasm (asthma) which presents a threat to health and the completion of any exercise task.

The presence of rain, a particular hazard for British adventure sports participants, has been shown to decrease the insulative properties of clothing and consequently decreases the time to fatigue. To meet the challenge of the cold environment, adventure sports performers wear clothing to insulate the body and protect the skin from exposure. The insulative effect of such clothing can result with sweating during exercise in the cold. As the intensity of exercise increases, the sweat rate will increase. Evaporation of sweat produced on the skin or in clothing is enhanced by air movement. Sweating during exercise, particularly in the presence of wind, increases the cooling effect of the environment. This increases the threat to core temperature maintenance. It is for this reason that the adage 'to stay warm in the hills, stay slightly cold' came about. In the section on the physiology of mountaineering in Chapter 10, a number of different clothing strategies are examined to assess their appropriateness for exercise in cold environments.

The body's physiological response to cold air is similar to its response to cold water immersion, described in Chapter 8. In response to sensory stimulation from cold receptors mainly in the skin, the hypothalamus stimulates vasoconstriction and pilo erection to preserve heat. To increase heat production in cold air environments, the hypothalamus stimulates an increase in muscle tone, shivering and an endocrine driven increase in metabolism. During cold water immersion, conduction and convection represent the key mechanisms for heat loss. Specialist insulative clothing such as wetsuits and drysuits (ideally with additional insulation worn underneath) can help to protect against the cooling effects. During exercise in cold air, radiation and evaporation provide the main sources of heat loss. Appropriate use of insulative clothing and changing clothing (to replace wet or damp clothing due to sweat production during exercise) can help to greatly reduce the effects of cooling on core temperature during exercise in cold air environments.

Effects of cold environments on performance

Exercise in the cold affects performance in a variety of ways. Insulative clothing which is often bulky can inhibit movement and decrease efficiency, as well as increasing frictional drag leading to a higher metabolic cost at any given work load. The initiation of increased muscle tone and shivering in cold environments to increase heat production do not come without a metabolic cost. Research indicates that exercise in cold environments can result in a higher metabolic cost for each workload compared to a thermoneutral environment. As with cold water immersion, vasoconstriction and cooling of the skin can lead to a reduction in the functioning of sensory receptors in the hands, resulting in a loss of manual dexterity. As described in Chapter 8, this loss of manual dexterity is sometimes reduced a short time into exercise (hunting reaction). Research indicates that after an initial period of vasoconstriction of the blood vessels in the hand (in response to a cold environment) there is a subsequent vasodilation which counteracts the loss of manual dexterity.

Muscular strength and power have been found to decrease in response to cold stress, due to an increase in time for muscular contraction as well as a decrease in muscle fibre fluid viscosity and enzymatic activity. At any given sub-maximal workload, ventilation, oxygen consumption and lactate production are higher in cold compared to thermoneutral environments. Cardiovascular endurance ability decreases in cold environments due to a number of factors. When the blood temperature falls below 37 °C it has been shown to reduce oxygen dissociation from haemoglobin resulting in decreased oxygen delivery to the muscles. Blood flow to the limbs is reduced due to vasoconstriction in the cold, adding to the decreased oxygen delivered to the muscles. The decreases in oxygen transport result in an inhibition of fat metabolism, complete oxidation of food substrates and increased reliance on glycogen and anaerobic energy production. This results in a rise in blood lactate concentration and an earlier occurrence of fatigue. Through these mechanisms, research has shown that $\dot{V}O_{2\,max}$ is reduced in cold environments.

Exposure to cold environments can lead to hypothermia and frostbite. Hypothermia is most commonly diagnosed at a core temperature of 35 °C or below. Hypothermia can be triggered by a sudden water immersion such as the result of a kayak or canoe capsize or a slower onset such as during a mountain marathon or fell race on a cold day. The symptoms of hypothermia include shivering, greyish-coloured skin, slurred speech, irritability, clumsiness and drowsiness. When the core temperature drops to 31 °C, shivering will cease and this represents a significant point in hypothermia, after which the individual will not re-warm themselves without external heating. In cases of mild hypothermia, removing the person from the wind (a kisu, bothy bag or emergency shelter is excellent for this purpose), replacing wet clothing, increasing insulation and providing food and drink can help the person to recover quite quickly. Decisions should then be made about how best to evacuate. In a case of severe hypothermia, the above measures should be completed where possible but medical assistance should also be sought as quickly as circumstances allow. Frostbite can occur in skin exposed to subzero temperatures. In mild cases, limbs can be re-warmed in body temperature (37 °C) water or, in the case of a hand, by placing it under the armpit. In severe cases of frostbite, especially if there is a danger of the limb refreezing if warmed, medical assistance should be sought for treatment.

Cold acclimatization

Research and experience from adventure sports performers indicates that it may be possible to habituate the body to cold exposure and consequently decrease the effects of exposure on performance. Results of research into cold acclimatization are less conclusive than for hot environments. Research has found a decreased sympathetic response (shivering, vasoconstriction, increases in blood pressure and heart rate) after repeated exposure to the cold. However, other studies have failed to repeat these findings. The incidence of hunter's reaction in fishermen, who

endure frequent periods with their hands in cold water or in contact with cold objects, has been cited as an example of habituation. Cold-water swimmers and marathon runners have been found to have a delayed shivering response and aborigines who sleep in very little clothing during very cold desert nights have been found to tolerate mild hypothermia without shivering response.

Key point

The physiological stress during exercise in a cold environment is moderated by the relative humidity, wind strength and the presence of rainfall. Human responses to cold exposure include vasoconstriction, pilo erection, the addition of extra clothing layers (to preserve heat), increases in muscle tone and shivering (to create heat). Research indicates that $\dot{V}O_{2\,max}$ is reduced in cold environments. Cold exposure can lead to hypothermia and frostbite. Repeat exposure to cold environments appears to lessen the body's response and suggests a habituation that may be useful for adventure sports performers.

9.7 The physiology of land-based adventure sports

Mountain biking

There has been less research on the physiology of mountain biking compared to road cycling. Of the mountain bike research which has been conducted, more has been concentrated on cross-country performance rather than downhill. A study of elite mountain bikers conducted by Nishii *et al.* (2004) found that the use of a full-suspension bike resulted in faster times than through the use of a front-suspension only bike. The front and rear suspension provided with a full-suspension mountain bike has been found to improve the ride comfort, bike control and speed. In cross-country races, many mountain bikers continue to use hard tail bikes, perhaps due to the loss of pedalling power reported with the addition of rear suspension. Seifert *et al.* (1997) found that a front-suspension bike improved time during a 10.44 km cross-country trial. Differences between studies may be reflective of the terrain encountered. The choice of full- or front only suspension for elite cross-country riders should perhaps be another consideration, as should the type of terrain to be encountered.

Research interest in the physiology of mountain bike performance has grown since the late 1990s, perhaps a consequence of the inclusion of mountain bike cross-country in the Olympic Games for the first time at Atlanta (1996). Just as a time and motion analysis might be used in intermittent activities such as canoe polo or surfing, an exercise intensity profile has been conducted with elite cross-country mountain bikers. Impellizeri *et al.* (2002) conducted a study of exercise intensity during four elite mountain bike competitions. The intensity of exercise was measured through heart rate response. From this data, the researchers created three exercise intensity categories: EASY$_{zone}$ (heart rate below LT heart rate), MODERATE$_{zone}$ (heart rate between LT value and OBLA value) and HARD$_{zone}$ (HR above that corresponding to OLBA). For the five cyclists that completed the study, results indicated that the cyclist spent 18 per cent of the time in the EASY$_{zone}$, 51 per cent at moderate intensity and 31 per cent in the HARD$_{zone}$. The average heart rate during the four races was 171 bts min^{-1} which represented 90 per cent of HR$_{max}$ and 84 per cent of $\dot{V}O_{2max}$. These findings confirm that cross-country mountain bike races are conducted at a very high intensity. As for any adventure sport, training for these types of races should reflect the required exercise intensity, which includes a near maximal effort at the beginning of the race during the mass start.

Wilber *et al.* (1997) conducted a study comparing the physiological profiles of elite mountain bike and road cyclists. Performance profiles were established for 20 mountain bikers (10 male, 10 female) and 20 road cyclists (10 male, 10 female). The elite road cyclists had slightly lower body fat percentages: males 4.7 per cent compared to 5.8 per cent for mountain bikers and females 11.9 per cent compared to 13.2 per cent for mountain bikers (all significantly lower than those obtained by sedentary individuals). The road cyclists were slightly heavier than their off-road counterparts which, taking the body fat percentage results into account, is reflective of a slightly higher muscle mass. For males, the power output at $\dot{V}O_{2max}$ was significantly higher for the road cyclists (470 W compared to 420 W), although the $\dot{V}O_{2max}$ figures were very similar. For the females, the $\dot{V}O_{2max}$ was higher for the road cyclists, as was power output at LT (204 W compared to 224 W). The LT occurred at a higher HR for the road cyclists than for the mountain bikers: males 169 bts min^{-1} compared to 166 bts min^{-1}, and females 165 bts min^{-1} compared to 155 bts min^{-1}. The differences between road cyclists and mountain bikers may reflect more the types of tests conducted than meaningful differences in fitness between the two sports. Regardless of the cycling discipline, both a high level of aerobic fitness evident as a high $\dot{V}O_{2max}$ and occurrence of LT at a high percentage of $\dot{V}O_{2max}$ are required for successful performance.

Key point

Mountain bike cross-country races are conducted at a very high intensity due to the relatively short duration compared to some road cycling events. Mountain bike cross-country races typically last between 105–135 min in duration. Research indicates that in a race, mountain bikers attain a mean heart rate of around 171 bts min^{-1} which represents 90 per cent of HR_{max} and 84 per cent of $\dot{V}O_{2\,max}$. The choice of a full suspension or front only suspension bike should be determined according to the type of racing to be undertaken.

The physiology of orienteering, fell running and mountain running

The physiological demands of all-terrain or off-road running have not undergone as much research as track and road running. Information published regarding the physiological demands of fell or mountain running generally appears to be limited to pieces within running magazines, although the studies by Simpson *et al.* (2005) and Kruseman *et al.* (2005) provides some useful findings.

Simpson *et al.* (2005) conducted a study to examine muscle damage and oxidative stress resulting from a 7 km hill race that involved 457 m of ascent and descent. Venous blood samples were taken from 7 club level athletes pre-race, immediately post-race and 48 hours post-race. Results indicated that although reporting incidences of DOMS (delayed on-set muscle soreness) there were no major effects on immune function. The researchers concluded that research regarding longer distance races was required to assess the effects on immune function under such conditions.

The effects of a longer mountain running race, relating to nutritional intake rather than blood markers of immune function damage, was reported by Kruseman *et al.* (2005). In this study, the researchers examined nutritional intake during the Tour des Dents du Midi, a 44 km mountain race in Switzerland. During the race, the 42 participants in the study ascended and descended 2890 m at an altitude of between 860–2494 m with air temperatures between 18–30 °C. The mean time for completion was 7 hr 3 min. Results indicated that the competitors did not intake sufficient carbohydrate or water to match energy expenditure and sweat loss during the race. The competitors were between 1889–2470 kcal deficient during the race and the majority lost 3 per cent of their body mass. Research suggests that athletes taking part in endurance races often find it difficult to replace both fluid and carbohydrate during competition. In these circumstances, the negative effects of dehydration on performance are worthy of consideration by fell and mountain runners and also orienteers.

There is a greater depth of research relating to the physiology of orienteering. However, several of these studies have applications for fell and mountain runners. Bird *et al.* (2003a, b) studied heart rate response during orienteering races in British orienteers. Results revealed differences for 18 female national and club athletes. Mean HR for the club athletes was 158 bts min^{-1} and for the national level athletes was 170 bts min^{-1}. The researchers suggested that these differences most probably occurred due to differences in planning strategies during the race, as well as slowing to navigate and re-locate. For 39 male international, national and club level orienteers no such differences were found. The mean HR for male orienteers was 159 bts min^{-1}. These results suggest that for the British athletes, there was a lesser incidence of slowing down associated with navigation for male orienteers across competition levels.

Creagh and Reilly (1997) concluded that female and male orienteers have a high mean $\dot{V}O_{2\,max}$ of 63 mL kg^{-1} min^{-1} and 76 mL kg^{-1} min^{-1}, respectively, with mean heart rates of 167–172 bts min^{-1}. They found that running on rough terrain increased the energy cost of running, resulting in a 26 per cent increase in oxygen cost. The researchers attributed this decrease in running economy to the altered stride pattern associated with negotiating off-road terrain. Compared to distance runners (track and road) orienteers were found to have a higher incidence of ankle injuries and cuts and bruises.

Jensen *et al.* (1999) found that the terrain encountered during orienteering imposes a greater muscular load than distance running. In a comparative study between distance runners and orienteers using a portable gas analysis system, the researchers found that running economy decreases on heavy terrain. The negative effect on running economy was less pronounced with the orienteers than with the distance runners. They concluded that the difference may relate to some genetic differences between the athletes, but was most likely the effect of the specific heavy terrain training undertaken by orienteers. This result indicates the importance of specificity of training for orienteers. Off-road running on ground similar to that encountered during a race is critical to improved performance.

Key point

Compared to distance running, there appears to be a lack of research relating to the physiological effects of fell running and orienteering. From the studies available, the key findings suggest that off-road running carries a greater oxygen cost than distance running (road or track). The lower than required carbohydrate and water intake during mountain running may have a negative effect on performance. During training, orienteers and fell runners should attempt to match the terrain to that encountered during a race and perhaps develop nutritional strategies to improve carbohydrate and water intake.

9.8 The physiology of water-based adventure sports

Kayaking and canoeing

Paddlesport offers a wide variety of disciplines to the adventure sports performer and the physiological demands can be just as varied. The physiological challenge in freestyle is very different to that of a sea kayaking expedition. Even within a discipline, the nature of the water can change the physical intensity during paddling. For instance, in wildwater racing more demanding stretches of water will increase the skill requirement and can lead to a lowering of exercise intensity, whereas on easier stretches of river the emphasis becomes speed of movement downstream and the intensity increases. In addition, some wildwater races can be as short as 1–2 min in duration while others last for up to 25 min. The majority of research regarding kayaking and canoeing focuses on racing and slalom performance. This is due to the fact that these two disciplines are two of the oldest competition adventure sports and are included within the Olympic programme. Further to this, racing is perhaps the purest form of kayaking (moving from point A to B as quickly as possible) and as such lends itself more readily to physiological analysis. It is not a coincidence that in the BCU's new coaching handbook (2006) the physical aspects of racing paddling are at the top of a list of key aspects for successful performance, above technical and tactical. For slalom, however, the most important aspect is skill after which are listed the key physical demands.

For those involved in disciplines other than racing and slalom, the key features for making decisions about training programme development should be the duration and intensity of the activity. It is essential to identify specific energy demands for successful performance in your discipline. In Chapter 8, the physiology of canoe polo was considered separately from the other paddlesport disciplines as it forms one of the clearest examples of an intermittent adventure sport. In that section, details of a time and motion study of goalkeepers, attackers and defenders was presented. This type of time and motion study, ideally with the inclusion of heart rate monitors to provide details of exercise intensity and variation during performance, could be performed with any canoesport discipline to better inform decisions about training programme development.

This section continues with a review of previous racing and slalom research and is followed by a description of current training methods advocated for racing and slalom development. These ideas could be employed directly or modified to provide useful ideas for any canoesport discipline. For

Biography: Ian Tordoff

Adventure sport: Kayaking

Ian Tordoff began his international career in 1985 coming third in a Spanish Wildwater International race. He has since competed for Great Britain in four different events: wildwater racing, sprint kayaking, marathon paddling and rafting. He has been national wildwater racing champion on ten occasions, achieving third place at the World Championships in 1993 and 1994. He has competed in events all over the world, and in 2005 set a new world record time for crossing the channel in a sea kayak with a time of 3 hr 21 min 54 s.

Figure 9.17   Ian Tordoff (photograph courtesy of Barry Frost)

Cross channel sea kayak world record

I initially became interested in the Channel world record early in 2003 following articles about two brothers who had made an attempt. At the time of the article the record they set seemed to be achievable, but if I wanted to make an attempt I wanted to set a very fast time. The boats I had available in 2003 would be quick enough to beat this time but only just.

In 2004 I was approached with the concept of a new racing sea kayak which was in a prototype form. This was a much faster boat but still needed work. After significant testing and racing in the UK during 2004 and early 2005, Valley Sea Kayaks approached me to finalize the boat and were keen to use the Channel Crossing as a launch for the new racing class boats they had on the drawing board.

The next step was to log the attempt with Guinness World Records. At this point, safety boats and timing people were identified to verify the record. The intention of the attempt at this stage had always been to smash the record by so much that the boat would really make the headlines. Having talked to Guinness World Records, the time set in 2003 was not the actual record. The official record had been set some 30 years earlier and was a full 45 minutes quicker than I had been working on. I was now only 2 months from my attempt date.

In previous events in which I had competed, you are either racing opponents which you can measure your speed against or, in the case of Wild Water Racing, doing a time trial. Although you can't see your competition, due to the event being only 20 min long, your body can operate at the Anaerobic Threshold level for the full 20 min and your feeling of speed comes from the reference points of moving past rocks and the river bank. It's a bit like driving fast down a country lane. It only feels fast because you get so much visual and physical feedback. In relation to crossing the Channel, this information would not be available so a different type of training would be required. With no visual feedback available, the only way to gauge an even pace was to use a Global Positioning System. This gives your actual speed and your average speed at all times and became my new training partner.

Typically for racing, training would be split into components of endurance, speed, time trial/race pace and anaerobic sprint work. This helps to train all the physical elements that enable you to go fast and recover within an event. For the Channel crossing, I needed to maintain a pace that was much quicker than endurance pace, but not so fast that the body went into race mode. It was a case of doing a 3 hr 30 min time trial while stopping the body wanting to go fast.

Although set 30 years previously, the actual Channel crossing time was fast. From the speeds I had produced in training, however, I believed I could break the record and still have 10 minutes in hand for problems. The likely issues on a crossing that would lose time would be fatigue at the end of the crossing, a head wind to slow me down or having to go round any ships that were in the way. In the Channel kayaks have to give way!

Training in the last few weeks was really about testing equipment and getting used to this new pacing which felt a little odd. It was fast and I was able to take advantage of this in several tests of over 2 hr duration, during which time I was able to maintain an average speed of 11.7 km hr^{-1}. This was to be my speed and every session I did in the 6 weeks prior to the date was carried out at this rate.

The day came, and the sea was flat. To reduce the risk of wind and waves, I set off at 5.45 am. At this time of the day, the heat of the sun has not reached the land so thermal wind would be ruled out. As I set off knowing in my heart that I could do it, the event felt more like a routine session rather than a challenge.

After 10 min the first real feedback hit me: I was going too fast at 12.9 km hr^{-1}. I felt fine, so it could have been adrenalin, but after 10 min I would have expected it to have settled. Then it occurred to me: the one thing I had not considered was salt water. Although I had paddled the boat on the sea on a number of occasions, all the speed tests had been on fresh water. The effect of salt water is that due to its increased water density caused by the salt, the boat floats higher in the water and thus reduces the drag of the boat in the water. Although in theory friction increases as well, the floatation gives far greater benefit.

I monitored speed for the next 40 min and accepted that the new average speed was going to be 12 km hr^{-1}. The rest of the trip was fairly uneventful. In terms of hydration and energy, I planned to take on board carbohydrate every 40 min and set a watch to tell me. This was in the form of a carbo gel which is easy to carry on the deck lines and meant that I could carry squash in my liquid drink system to maintain hydration and to keep my mouth fresh after salt water spray. From my marathon racing, I had found hydration was far more important than carbohydrate. If you eat well in the lead up to a 3 hr race, your body will typically have enough glycogen stores.

Towards the end, the pace was still the same like a metronome, and I could see from the GPS the kilometres counting down. At the point when I had 10 km to go, things became tough. I had been able to see France from the very start of the 36 km crossing, but with 10 km to go it looked very close. It looked like only 10 min away, but my head knew it was closer to 50 min. With 2 km to go, I was keen to speed up but I could feel that my balance in the boat was not what I would have hoped. The constant movement of the sea combined with fatigue was enough to throw my balance completely. Only my years of paddling kept me up. This was confirmed as I launched up the beach to find upon exiting the boat that I was unable to stand up. After several failed attempts, I decided to get back in the boat (it was safer there) and paddled back out to the safety craft.

The time I set was a new World Record on 3 hr 21 min and 54 s, taking 12 min off the old time. Although the shortest distance is 36 km, due to the tidal drift of the incoming and outgoing tide I paddled 40 km.

Was it hard? Sort of! I had in my mind what I had to do, I had all the equipment to break the record and I had completed several training sessions close to the actual distance in the weeks leading up to the day. The biggest gamble was always going to be the weather. Without the best conditions I couldn't have beaten the record, but on the day I was lucky.

Figure 9.18 Ian Tordoff setting off on his record-breaking crossing of the English Channel, 3 hr 21 min 54 s (photograph courtesy of Barry Frost)

instance, the *20–100 test* developed for slalom could readily be adapted for use with canoe polo or freestyle paddlers.

One of the earliest papers published on the subject examined the physiological characteristics of whitewater (slalom) kayakers. Sidney and Sheppard (1973) examined the physiology of twelve members of the Canadian slalom team including three who went on to represent Canada at the 1972 Munich Olympic Games (the first time slalom was included in the Olympics). In this study, the mean treadmill (running) $\dot{V}O_{2\,max}$ for the paddlers was 57 mL kg^{-1} min^{-1} for the males and 50 mL kg^{-1} min^{-1} for the females. Wakeling and Saddler (1978) published details of one of the earliest field tests of $\dot{V}O_{2\,max}$ using an on-the-water kayak protocol with seven British slalom paddlers. They found mean kayak-based $\dot{V}O_{2\,max}$ values of 51 mL kg^{-1} min^{-1} for the males and 41 mL kg^{-1} min^{-1} for the females. The lower values attained by the paddlers in this study can most likely be attributed to smaller muscle mass involved in the kayak tests as opposed to the running assessments carried out for the Canadian kayakers.

Comparing the two studies, the upper body $\dot{V}O_{2\,max}$ were between 85–92 per cent for males and 82 per cent for females of the running-based $\dot{V}O_{2\,max}$. These percentage differences between upper body and whole body $\dot{V}O_{2\,max}$ are similar to results published by Tesch (1983) for six elite male (members of the Swedish Olympic Team) racing paddlers. Tesch found that kayak $\dot{V}O_{2\,max}$ represented 86–89 per cent of treadmill tests. In his study, however, the mean $\dot{V}O_{2\,max}$ values reported for the sprint paddlers were higher than those for slalom paddlers, with mean treadmill $\dot{V}O_{2\,max}$ being 67 mL kg^{-1} min^{-1} and kayak based $\dot{V}O_{2\,max}$ 58 mL kg^{-1} min^{-1}. These values most likely reflect differences in the requirements for each discipline and consequential differences in training. Tesch's study also included a 1000 m kayak race assessment, which resulted in high post-race blood lactate concentrations of 11.0–17.5 mmol L^{-1}. Taken together, these results reflect the high demands placed upon aerobic and anaerobic energy production during sprint kayak and canoe performance. Tesch concluded that success in sprint racing requires a high aerobic capacity, anaerobic endurance and upper-body strength.

Research findings for elite racing paddlers categorize them as being taller than average with longer limb length, increased upper body muscular strength and with low body fat percentages. In one of the early muscle biopsy studies, Tesch found that 500 m sprint paddlers had higher percentages of Type II (fast twitch) fibres (50–59 per cent) than 10 000 m paddlers (26–52 per cent). These results are consistent with findings for biopsy studies conducted with elite athletes (sprint and distance runners). Research regarding training volumes suggests that elite racing paddlers complete 3000–5000 km each year in training, along with supplementary land-based conditioning and strength training.

The British Canoe Union's (BCU) recently published Coaching Handbook (2006) suggests that the key physical components for racing performance success are aerobic endurance, strength, anaerobic power and functional stability. These aspects reflect those originally identified by Tesch (1983), with the exception of functional stability. In the past, stability work was subsumed within the strength and technique development exercises undertaken by sprint paddlers. Injury profiles for elite kayak and canoe paddlers have led to a clearer understanding of the importance of maintaining functional stability as a specific part of training designed to optimize efficiency and reduce injury risk. The large volumes of training required to maximize racing performance necessitate attention to stabilizing muscles to maximize their ability to perform their role during exercise. With improved functional stability, a paddler will become more biomechanically efficient resulting in an improvement in paddling speed and therefore performance.

As described by Anna Hemmings in her piece about the 2005 Kayak Marathon World Championships, wash riding can reduce the physiological stress during a race. In longer distance kayak and canoe races where there are no lanes it is possible for trailing paddlers to conserve energy by riding the wash of the leader. Research by Gray et al. (1995) and Pérez-Landaluce et al. (1998) supports the energy-saving possibilities associated with wash riding. Gray et al. (1995) found wash riding led to a 10 per cent decrease in the oxygen demands of the activity and a 5 per cent reduction in heart rate. Pérez-Landaluce et al. (1998) suggest even greater benefits of wash riding, with riding as the trail boat in a diamond of kayaks being of greatest benefit to the paddler. Whether paddling on the wing or as the trail boat in a diamond formation, wash riding can result in a 18–30 per cent energy saving, 20 bts min^{-1} fall in heart rate and reduction in required power from 190 W as the head boat to 129 W as the trail boat. For tactical marathon paddlers, wash riding represents a significant energy saver. The research carried out by Pérez-Landaluce et al. (1998) is featured in more detail as the research summary for this chapter.

Chapter 5 presented a generalized model for differentiating adventure sport conditioning training. The general training zones for adventure sports are shown in Table 5.8. As with any adventure sport, the development of a training programme should be specific to the particular demands of the activity. Table 9.2 provides an illustration of the training zones that can be identified for performance development in race training.

The training zones identified in Table 9.2 have been identified through work with elite athletes and are specific to kayak and canoe race improvement. However, this developed model provides a useful training guide for any of the high-intensity aerobic endurance activities that are the focus of this chapter. The specific adaptations brought about through training at each intensity level are shown in the aims of training column. A comparative summary of terminology used in the model presented in Table 5.8 and that used by race paddlers is presented in Table 9.3. The specific language used to describe these zones in slalom paddling is slightly different and therefore slalom-appropriate terminology is also included in Table 9.3.

In this textbook, the term long slow distance is used to describe training at 50–70 per cent of maximal heart rate. The equivalent for racing is core aerobic pace (CAP) i.e. training at 60 per cent of 1000 m maximum speed and base aerobic work in slalom paddling where lactate levels post-exercise are no higher than

2–3 mmol L^{-1}. Regardless of the terminology used, training at this intensity develops efficiency of the cardiovascular system, improves fat metabolism and enhances recovery from training and competition.

For kayakers, the work of van Someren et al. (2000) provides support for the use of kayak ergometers for training and assessment purposes. Although training and testing should be carried out in-boat where possible due to the highly specific nature of adaptations to training, in circumstances where land-based performance is necessary van Someren's research indicates that an appropriately calibrated kayak ergometer can provide results similar to those that would be achieved in-boat. In a study using K1 ERGO kayak ergometer, van Someren et al. (2000) found no significant differences between $\dot{V}O_{2peak}$ on open water and on the ergometer. In subsequent research, van Someren and Oliver (2002) found similar blood lactate concentrations for in-boat and kayak ergometer exercise at a heart rate which equated with lactate threshold.

With regard to physiological assessment, a performance measure developed for British slalom kayakers by the World Class Slalom Coaches (in conjunction with Gordon Burton and Barney Wainwright, the World Class physiologists for canoesport) may provide a useful test that could be used with a variety of kayak and canoe disciplines such as canoe polo, surf and freestyle. The 20–100 test was developed as a flat water method through which to mimic the exercise demands in a slalom race. The test is a repeat sprint protocol with varied rest intervals, requiring very little equipment for performance. Three posts set with one at the start of the course, a second 20 m away and the third a further 30 m away mark out the course. The kayaker completes six 20 m sprints followed by a 100 m sprint and a further six 20 m sprints. The exact protocol is shown in Table 9.4 and GB Squad times are listed in Table 9.5 to provide a comparison for test results obtained yourself or by the paddlers you coach. The 100 m sprint is completed as an out-and-back from the 0 m

Table 9.2 Training zones for racing paddling development. A technique emphasis or goal should be attached to every session. More recently, the Core Aerobic Pace (CAP) category has been split into a CAP I with an intensity of 60–70 per cent and CAP II at 70–80 per cent of 1000 m boat speed

Pace name	Abbreviation	Specific physiological and technical aims of training	Speed (% of 1000 m race pace)	Stroke rate (str min^{-1})	Test distance	Lactate (mmol L^{-1})	Perceived exertion
Core aerobic pace	CAP	General endurance: general improvements to aerobic system; improving fat metabolism; technique foundation	60–79	60	10 km	1–2	Moderately comfortable; intermittent conversation
Threshold pace	THR	Specific aerobic conditioning: higher intensity aerobic improvement; start of some anaerobic work and lactate tolerance; acceleration of the blade	80	75	5 km	3–4	Slightly uncomfortable; breathing heavily; more concentration
Sub-race pace	SRP	Aerobic power: improving maximum ability to consume oxygen; increasing lactate tolerance; anaerobic metabolism increases; consolidation of near race pace technique	90	95	2000 m	6–8	Hard; breathing very heavily but under control
Race pace	RP	Race endurance: race practice, increase ability to sustain high work rate; maximum oxygen consumption, high lactate levels; improving race technique	100	110–115	750 m	10+	Painful; breathing extremely hard
Peak race pace	PRP	Speed endurance: improving ability to sustain maximum speed; improving anaerobic system and lactate tolerance; keeping strong technique	110	130	200 m	8+	Very hard; local muscular pain
Maximum pace	MP	Maximum speed: maximum speed and power development; improving anaerobic abilities - lactic and ATP-PCr; increasing maximum stroke rate	112	130–140	100 m	6	Fast and powerful but physiologically comfortable

Table 9.3 Terminology to describe the conditioning zones in kayak and canoe racing and slalom

Terminology used in Table 5.8	Racing training zone equivalent*	Slalom training zone equivalent*
Long slow distance	Core aerobic pace (CAP)	Base aerobic
Fartlek	Speedplay or Fartlek	Tech repeats/Progressives and speed tech
Medium-paced continuous training	Threshold pace (THR)	Threshold
Fast-paced continuous training	Sub-race pace (SRP)	$\dot{V}O_{2\,max}$ peak
Fast-paced continuous training/Interval training	Race pace (RP)	Full runs, race practice
Interval training	Peak race pace (PRP)	Lactate tolerance
Sprint training	Maximum pace (MP)	Power ups

*For further information about the types of training for each of these levels, see BCU Coaching Handbook (2006)

Table 9.4 Protocol for the 20–100 test

Paddle distance (m)	Recovery interval
20	2 min
20	5 min
20	10 s
20	10 s
20	10 s
20	10 s
20	10 s
20	5 min
100	10 s
20	10 s
20	10 s
20	10 s
20	10 s
20	10 s

line to the 50 m line. By setting the recovery intervals as they are, it is possible for the coach or performer to obtain an idea of maximum short sprint time (a measure of the ATP-PCr system performance) before obtaining information about anaerobic endurance performance during the remainder of the test.

Key point

The majority of published paddlesport research concerns racing and slalom, because these are the oldest and Olympic disciplines and, certainly in the case of racing, it represents the purest forms of canoesport. There are, however, important findings that can be considered for any paddlesport. The key aspects when considering the physiology of any paddlesport is to concentrate on the duration and intensity of the activity, along with the relative importance of skill-related

Table 9.5   Target times: GB Squad results for the 20–100 test

Age group	Class	Fastest 20 m (s)	Fastest 100 m (s)	Fastest total time (s)
Junior	MK1	6.3	34.5	119.8
	WK1	7.4	42.1	141.5
	C1	7.5	44	142
Senior	MK1	5.8	33.3	114.9
	WK1	6.5	37.6	129
	C1	7	40.2	135
	C2	6.1	36	119.9

factors. Tests such as the *20–100 test* used in slalom could easily be adapted for paddlers from a variety of disciplines. When considering the physical demands of any paddlesport, the importance of environmental stresses should be taken into account. For instance, much of the whitewater paddling that takes place in Britain occurs during the winter months when water levels are normally at their best. This has obvious implications for a cold water immersion associated with a capsize or for spending a full day paddling in low ambient temperatures. The environmental as well as the physiological demands of any paddlesport should be considered together when planning a development programme.

Getting into research: wash riding by Pérez-Landaluce *et al.* (1998)

When kayaking or canoeing, a significant proportion of the effort expended is used to overcome the resistance provided by the water. During kayak and canoe training or races beyond sprint distance, competitors are not restricted to lanes and as a consequence, the movement of other paddlers through the water is more closely felt. Wash riding describes the purposeful use of a leader's water to assist in forward movement. Research indicates that as a canoe or kayak moves through the water, the movement of the boat decreases the pressure and resistance in the water behind them. The resistance encountered by a following paddler will be decreased due to the already moving water that the trailing paddler encounters. Providing a paddler stays close enough to a lead boat, he or she will be able to ride the wash of the leader. Wash riding results in a reduction in the required effort to maintain the same speed as the boat in front. The performer contribution written by Anna Hemmings, regarding marathon paddling, describes this advantage in a race context.

Previous studies have examined riding on the wing of another boat; however, the study by Pérez-Landaluce *et al.* went on to examine the effects on each paddler moving in a diamond formation as shown in Figure 9.19. The purpose of their study was to examine the energy demands of each position within a racing paddling diamond. Eight male Spanish international kayakers volunteered to take part in the study. The participants had a mean age of 21.5 years, height 1.82 m and mass 81 kg. The study involved both on-the-water and laboratory-based testing using a kayak ergometer. During the field test, the paddlers completed four 2000 m repetitions at a speed of 13.58 km hr^{-1}. A 30 min rest was given between each repetition and the speed was set at a common training speed for the kayakers. During each of the on-the-water tests, each kayaker's heart rate (HR), blood lactate concentration (BLa), rating of perceived exertion (RPE) and stroke rate (SR) were recorded.

As shown in Figure 9.19, when riding on the wing the paddler had to remain within 1–2 m (left or right) of the lead paddler with their bow in-line with the cockpit rear. In the trailing position, the kayaker paddled with his bow in-line with the rear of the cockpits of the wing boats and behind the lead boat. Three days later in the researchers' laboratory, the HR, BLa and RPE for each of the positions were used to establish a matched exercise intensity on a kayak ergometer. During exercise at each intensity, the oxygen consumption ($\dot{V}O_2$ in mL kg^{-1} min^{-1}) and power (W) output were recorded for each paddler.

The key results obtained by Pérez-Landaluce *et al.* are shown in Table 9.6. As can be seen from the results, the data for BLa, HR and RPE indicate a good level of agreement between the field and laboratory exercise intensities. To support this finding, statistical analysis revealed non-significant differences between conditions for each of the parameters.

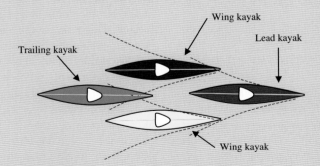

Figure 9.19 Diamond formation illustrating the position for each boat to take advantage of wash-riding behind a lead boat (adapted from Pérez-Landaluce *et al.*, 1998)

Table 9.6 Mean blood lactate concentrations, HR, RPE, $\dot{V}O_2$ and power outputs for field and laboratory kayaking in lead, wing or trailing positions

	Lead kayak	Wing position	Trailing kayak
Field lactate (mmol L^{-1})	4.2	2.1	1.5
Laboratory lactate (mmol L^{-1})	4	2.3	1.77
Field HR (bts min^{-1})	172	159	152
Laboratory HR (bts min^{-1})	173	161	149
Field RPE	15	12.6	9.7
Laboratory RPE	15.1	13.5	10.9
$\dot{V}O_2$ (mL kg^{-1} min^{-1})	46.61	38.87	32.47
Power (W)	190	156	129

As far as was possible, the exercise intensities were matched between the on-the-water and laboratory-based protocols. The results for oxygen consumption and power output reveal significant energy savings during wash riding. Paddling in the wing and trailing positions resulted in a 15.8 per cent and 30.3 per cent reduction in oxygen consumption, respectively. In agreement with this result, the power output required to maintain the 13.58 km hr^{-1} speed was significantly reduced when paddling on the wing (18.26 per cent) and trailing position (31.9 per cent). The results of this study provide importance evidence for paddlers and their coaches involved in endurance racing. The position during paddling can significantly affect the energy required for any given speed and wash riding enables an athlete to save energy during training or racing.

9.9 Summary and study questions

Aerobic-anaerobic transition

While it is traditional to describe the metabolic processes by starting with short-term high-intensity exercise and then moving towards endurance activities, high-intensity aerobic endurance activities actually work the other way round. These activities have an aerobic base but then utilize anaerobic metabolism during short bursts of activity or to provide a finishing 'kick'. In the earlier stages of a race, participants try to maximize the level of performance that can be achieved through aerobic metabolism with a minimal reliance upon anaerobic metabolism as this would lead to early fatigue. This point in the exercise continuum is the Aerobic-Anaerobic Threshold (AAT), below which the by-products of any anaerobic metabolism can be dealt with without lactate accumulation. Above the AAT, increasing reliance on anaerobic metabolism means that lactate will be produced at a faster rate than it can be metabolically re-synthesized. The term Lactate Threshold is often used to describe this concept, but in fact lactate threshold has a very specific meaning and shouldn't be used as an umbrella term in this way.

Gas analysis has been used to determine the anaerobic threshold (AnT) or ventilatory threshold.

During exercise, $\dot{V}_E$ rises linearly with increasing exercise until a breakaway occurs; this was believed to occur due to lactate and H^+ buffering and indicated the point at which lactate began to accumulate. Later research identified problems with this theory, and more clearly identified indicators of AnT were sought. The most commonly used indicator is to look at ventilatory breakpoint and the ventilatory equivalents for oxygen and carbon dioxide ($\dot{V}_E / \dot{V}O_2$ and $\dot{V}_E / \dot{V}CO_2$).

During the assessment of athletes, it is more common to use lactate analysis to identify AAT. There are a variety of ways to determine AAT from lactate measurements, ranging from visual identification of the deflection point of the plotted curve to mathematical data transformations intended to make this process more accurate. Maximal lactate steady state (MLSS) is also a popular analysis used to identify the AAT.

Lactate threshold testing protocols vary between researchers although all are incremental. Some protocols are discontinuous (with breaks for blood sampling); however, it seems a better solution to use a continuous protocol now that technology such as portable lactate analysers are available. The smaller the intervals between exercise intensities, the more accurate a test should be; however, the duration of each interval needs to be around 3 min in order to allow lactate concentrations to accurately reflect the effects of each step. Athletes should be tested using a form of exercise specific to their adventure sport.

The point at which LT occurs is a particularly useful piece of information when it comes to setting an athlete a training programme. Research has found a much closer relationship between MLSS or LT and performance in $30-60^+$ min races than if $\dot{V}O_{2\,max}$ is used as a predictor. Heart rate corresponding to LT or MLSS can be used when setting training zones for performers.

Responses to high-intensity endurance exercise

The primary macronutrient for high-intensity aerobic endurance exercise is carbohydrate in the form of glucose, with fat and protein providing a lesser contribution. In many high-intensity endurance adventure sports the upper body plays a significant role. The response of the upper and lower limbs to exercise can be quite different especially in untrained individuals. $\dot{V}O_{2\,max}$ scores for the upper body are generally around 70 per cent of the score for the legs. However, some trained athletes such as kayakers can record higher scores using the upper body. Any given sub-maximal work load will elicit a higher $\dot{V}O_2$ response from the arms than the legs and heart rates tend to be around 20 per cent higher at the same $\dot{V}O_2$.

Steady state is the state a performer achieves when performing sub-maximal exercise in which energy production can keep pace with the demands. This state is normally achieved after working for $4-5$ min initially or $1-2$ min for a change of intensity. If the workload is increased beyond a critical level (the lactate threshold), then it is no longer possible to achieve steady state.

When exercise begins, the time taken for steady state to be achieved means that initial oxygen delivery does not match the exercise requirements. This delay creates an oxygen debt which is repaid when the exercise stops. When a performer stops exercise, oxygen consumption rate and ventilation remain elevated during what is known as Excess Post-exercise Oxygen Consumption (EPOC) which is still not fully understood. It is believed that this is due to the need to replenish oxygen stores, remove CO_2 and allow the re-synthesis of lactate to glycogen. In addition, the raised body temperature of the performer will cause increased metabolic rate which will also create an increased demand for oxygen.

Fatigue

The mechanisms underlying fatigue in high-intensity endurance activity are related to the specific intensity and duration of the event. For shorter duration activities with a correspondingly higher intensity, fatigue will occur through the accumulation of lactate. As the intensity eases and the duration lengthens, glycogen depletion will be the predominant mechanism behind fatigue.

As a performer begins to fatigue during prolonged exercise, they are likely to experience cardiovascular drift. This is especially true if the environment is hot. Cardiovascular drift is thought to be caused by a drop in stroke volume which means that heart rate must rise to compensate. Reduced venous return as fatigue sets in, the diversion of blood to the skin for cooling and a loss of blood plasma volume through sweating are all believed to contribute to the reduction in stroke volume.

Exercise economy is the point at which skill and physiology meet. A more efficient performer will be able to complete more work for the same physiological cost.

Adaptations to training

An effective aerobic training programme will result in a number of physiological changes. The most obvious improvement will be in aerobic capacity $\dot{V}O_2$; however, research indicates that improvements in $\dot{V}O_2$ occur mainly in the first 6–12 months of training. Improvements in experienced athletes are more likely to be due to enhanced LT and increased exercise economy.

Aerobic training tends to reduce the size of Type I fibres since they are required to contract repeatedly rather than maximally. Endurance athletes are often advised to perform some strength training in order to maintain muscle size and strength. Capillary density, myoglobin concentration and mitochondrial size and number all increase as a result of endurance training. Mitochondrial functioning can also increase due to improved enzyme activity. Muscle fibres become better at fat oxidization and lipid storage is increased. Training has also been shown to increase glucose uptake by the muscles which can delay fatigue induced through glycogen depletion. Muscle fibres can adapt to training activities and Type $II_{x(b)}$ fibres can take on the properties of Type II_a fibres. A small percentage of Type II_a fibres can become more like Type I fibres. All these changes improve aerobic endurance.

Nutritional and ergogenic aids

In very short-term aerobic endurance events, supplements to enhance buffering of H^+ can help. In longer duration events, glycogen loading and carbohydrate feeding can be effective. Caffeine is another supplement that is both legal and has a large base of scientific support. Caffeine has widespread effects that are beneficial to high-intensity aerobic endurance and results show a 10–20 per cent improvement in time to exhaustion.

Thermal stress

Metabolic functioning increases during aerobic endurance events and this means that the body converts a considerable amount of its chemical energy stores into heat. Since efficient metabolic functioning can only occur within a narrow band, this additional heat can provide a threat to the performer especially if the environmental conditions already place the body under thermal stress. In addition to the heat generated through metabolic processes, heat can be gained or lost through conduction, convection and radiation. Evaporation during sweating and breathing also provide a source of heat loss.

Exercise in a hot environment creates competitive blood flow demands as blood is diverted to the muscles to maintain performance and to the skin in order to aid cooling. Sweat gland activity also increases: an activity which requires glycogen. Exercise in the heat generates higher rates of lactate production, increases in heart rate and respiration and a reduced time to glycogen depletion for the same work load. The body's ability to deal with higher than normal body temperatures is quite limited and heat exhaustion and heatstroke are real risks if high-intensity aerobic activity is carried out in hot environments without adequate preparation and vigilance. It is possible to improve exercise performance in heat if an acclimatization strategy is adopted over 5–10 days before the event.

Adventure activities in cold environments generally require specialist clothing to provide insulation and limit heat loss. One of the key issues for performers is to balance clothing so that heat build up doesn't trigger cooling responses while maintaining core body temperature. Additionally the bulk of clothing can inhibit performance and reduce efficiency. Muscular strength and power decrease when cold, as does cardiovascular endurance. Evidence for acclimatization to cold environments is less strong than for hot environments.

Mountain biking

Most of the research in mountain biking has been in the cross-country arena. Time and motion analysis has shown that cross-country races are conducted at very high intensity with almost one-third of the time spent above the AAT. Comparisons with road cyclists have shown elite mountain bikers have lower overall mass, slightly higher body fat levels, similar $\dot{V}O_{2\,max}$ scores but lower power output scores at LT and $\dot{V}O_{2\,max}$.

Kayaking and canoeing

The physiological requirements of paddlesport are very varied which reflects the diversity of activities themselves. One of the key concepts that a coach or performer needs to bear in mind during the planning of any development programme is specificity. In addition, techniques such as wash riding during racing can reduce the physiological load on paddlers, resulting in up to 30 per cent less energy expenditure.

Orienteering, fell running and mountain running

The present research on mountain running is sparse; however, it seems that insufficient fluid and carbohydrate intake may be a common problem in these kinds of events. Research on orienteering suggests that the physiological intensity of the event may be higher for national level athletes. This seems to be due to the reduced time lost in navigation and route planning which provides resting opportunities for less skilful performers. Running on rough terrain appears to increase the oxygen cost of the activity. Muscular loads are also higher; however, it seems that practice can improve running efficiency on rough terrain.

Study questions

1. Define the Aerobic-Anaerobic Transition. Why is knowledge of the intensity at which this occurs important when training a high-intensity endurance athlete?

2. Why do the increases in exercise intensity need to be relatively small when conducting an incremental lactate threshold test? Why will some physiologists have a performer complete a second LT test with smaller intervals?

3. What is meant by cardiovascular drift? Explain why it may occur during high-intensity aerobic endurance activity?

4. Why is it important to consider the activity of an athlete before conducting aerobic capacity testing? Why is it difficult to meaningfully compare the $\dot{V}O_{2\,max}$ scores of sea kayakers and fell runners?

5. What is an oxygen debt? Why might the continued elevation of oxygen consumption continue after the oxygen debt appears to be repaid?

6. What happens to levels of cortisol and testosterone as a result of aerobic training? How does this impact on Type I muscle fibre?

7. How does sweating increase the rate of heat loss during exercise? Why does sweat contain higher concentrations of minerals during vigorous exercise in hot environments?

8. Why is oxygen transport to the muscles inhibited in cold environments? What impact does this have on energy production?

10

Aerobic endurance part 2: lower-intensity activities

10.1 Introduction

The adventure sports discussed in this chapter on lower intensity (aerobic) endurance activities are listed in Table 10.1. Despite being referred to as lower intensity endurance activities, there are differences in both the intensities and durations of the adventure sports covered in this chapter. Figure 10.1 provides an illustration of the relative intensity and duration of each of the adventure sports in this chapter on an intensity-duration continuum. It is important to recognize that decisions regarding the placement of each adventure sport on the continuum have been made based upon an average for each activity. The level of the performer and their goals, along with intentions of the session, trip or expedition, can greatly alter the intensity and duration of any activity. It is important that each performer, or their coach, make decisions about the relative intensity and duration and therefore the physiological demands of the adventure sport for themselves.

The predominance of aerobic metabolism for many adventure sports, a key consideration in this chapter, makes an understanding of the mechanisms underlying aerobic endurance vital to adventure sports performers. This chapter begins with a

more detailed description of the processes involved in aerobic metabolism which were introduced in Chapter 4. The nature of response to exercise and fatigue during aerobic endurance adventure sports is covered with a brief description of testing and training adaptations associated with such activities. This is followed by a section discussing nutritional and ergogenic aids to improve performance.

A review of the effects of hyperbaric and hypobaric conditions on physiological functioning is contained within this chapter, due to the relevance to lower intensity endurance in e.g. diving and mountaineering. A summary of the physiological demands and findings of research into mountaineering is presented in the final section. In addition, performer contributions by Dave MacLeod (ice climbing), Bob Sharp (hill walking/Munro climbing), Des Marshall (caving) and Matt Berry (inland kayaking) are included throughout the text to illustrate first hand the physiological demands of a variety of aerobic endurance adventure sports. A summary of a classic study into the effects of living high and training low by Levine and Stray-Gundersen is also included in this chapter, which begins with a discussion of the aerobic system.

Adventure Sport Physiology Nick Draper and Chris Hodgson
© 2008 John Wiley & Sons, Ltd

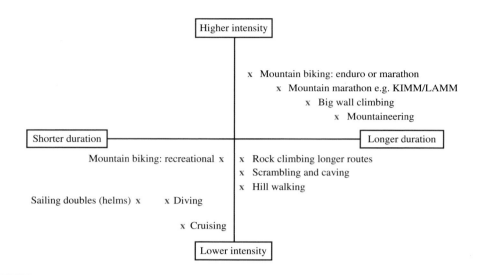

Figure 10.1 Duration and intensity continuum for lower intensity endurance adventure sports

Table 10.1 Lower intensity endurance adventure sports

Water-based activities	Land-based activities
Inland kayaking	Mountaineering
Open canoeing	Big wall climbing
Sea kayaking	Climbing (trad or sport, longer routes, lower relative grade)
Dinghy sailing (double helms)	Scrambling
Cruising or recreational sailing	Hill walking
Diving	Mountain marathon (KIMM or LAMM*)
	Mountain biking recreational, marathon or Enduro**

*KIMM: Karrimor International Mountain Marathon 2 day event; LAMM: Lowe Alpine Mountain Marathon 2 day event. **Mountain Biking Enduro: 12–24 hr endurance event raced solo, paired or as a team.

10.2 Aerobic system

The aerobic system, also known as the oxidative system or referred to as aerobic metabolism, represents the predominant energy system for the high-intensity endurance activities described in Chapter 9 and for the lower intensity endurance activities that are the focus of this chapter. In addition, aerobic metabolism plays a vital role in energy production and recovery for the intermittent activities covered in Chapter 8. Regardless of the name used, the key component for aerobic endurance activities is the presence of oxygen. The phosphagen system described in Chapter 5 and glycolysis described in Chapter 7 provide the anaerobic mechanisms for ATP production for very short and short-duration activities. The aerobic system, using oxygen delivered to the cells of the body via the respiratory and cardiovascular systems, provides the third gear or overdrive for energy production. The mechanisms of aerobic metabolism provide the energy necessary for an ascent of mountain peaks such as Kilimanjaro or Aconcagua, competing in a kayak or mountain marathon, a day scrambling or climbing a Munro, or for sea kayaking and canoeing expeditions. Due to the often sustained nature of adventure activities, the aerobic system is probably the most important energy system for adventure sports performers.

The mechanisms that play a role in aerobic metabolism are shown in Figure 10.2. Power and power-endurance (anaerobic) activities fuelled by the phosphagen system rely heavily upon PCr to sustain ATP production, and anaerobic endurance adventure sports can only utilize glucose/glycogen as the source for ATP re-synthesis during fast rate glycolysis. In contrast, the aerobic system can directly utilize all three dietary

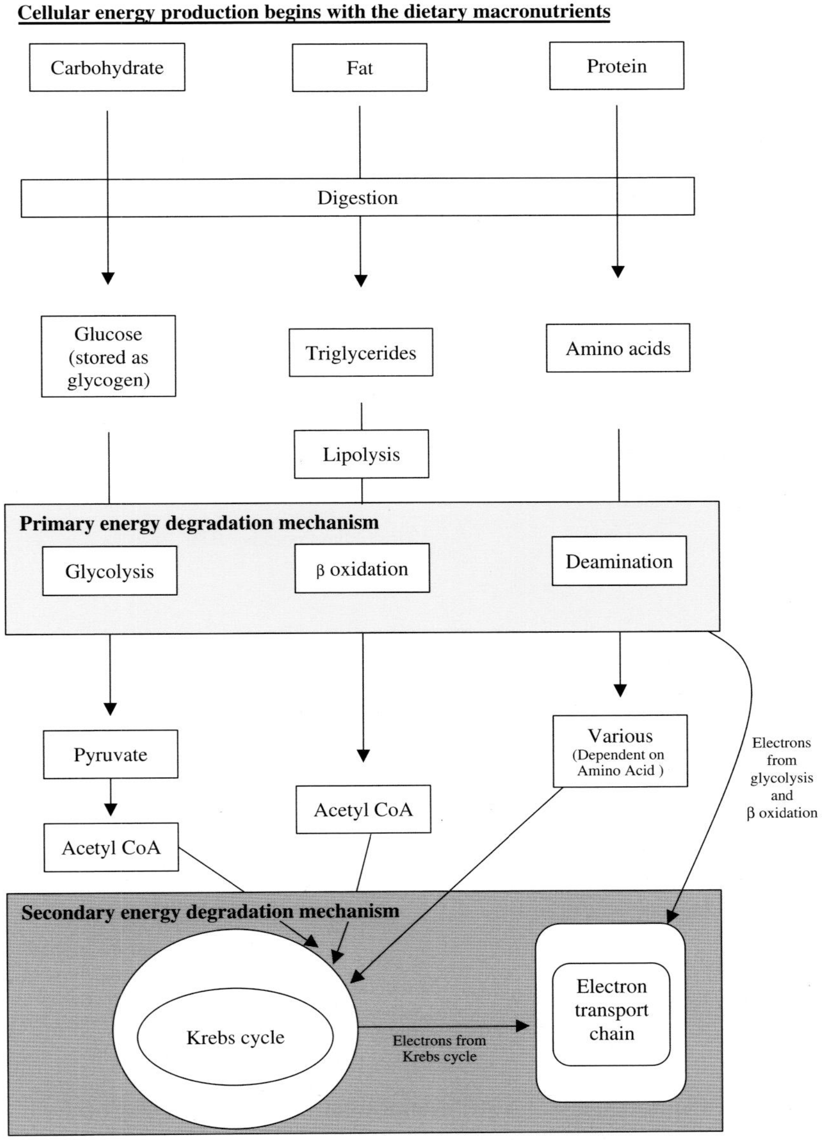

Cellular energy production begins with the dietary macronutrients

Figure 10.2 Metabolic mechanisms within the aerobic system

macronutrients (carbohydrate, fat and protein) to synthesize ATP. As described in the section on gluconeogenesis in Chapter 7 fat and protein can be converted to glucose for use within metabolism. As can be seen in Figure 10.2, the use of carbohydrate, fat and protein as sources for ATP production relies on a primary breakdown mechanism (beyond digestion which was described in Chapter 2).

After digestion, glucose and its storage form glycogen, triglycerides and amino acids can be used for ATP production, but have to be further degraded before ATP can be realized. The primary degradation mechanism for glucose and glycogen, *glycolysis*, was described in Chapter 7; the mechanism for the catabolism of triglycerides, β *oxidation*, is described in this chapter. To enable the use of amino acids for ATP production, the nitrogen-containing amine group must be removed from the carbon skeleton. The processes of *deamination* and transamination have been partially covered in Chapters 2 and 7, in reference to the alanine-glucose cycle and gluconeogenesis. After preparation through the primary breakdown mechanisms (glycolysis, β oxidation and deamination), the products of these processes are further degraded in the mitochondria (the powerhouses or powerplants for each cell).

An illustration of an individual mitochondrion is shown in Figure 10.3. The secondary mechanisms for aerobic generation of ATP involve the parallel processes of the Krebs cycle and the electron transport chain. The secondary mechanisms for aerobic energy production, along with β oxidation, take place within the mitochondria. Krebs cycle and β oxidation take place within the inner matrix of the mitochondria and all the enzymes required for these processes are located within this compartment. The reactions of the electron transport chain occur across the inner membrane of the mitochondria. Although aerobic metabolism takes longer to provide energy for exercise, it provides (relatively) inexhaustible supplies of ATP to sustain performance.

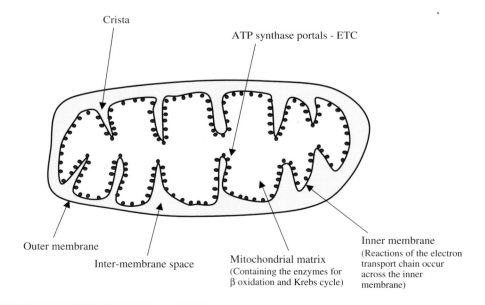

Crista

ATP synthase portals - ETC

Outer membrane

Inter-membrane space

Mitochondrial matrix
(Containing the enzymes for
β oxidation and Krebs cycle)

Inner membrane
(Reactions of the electron
transport chain occur
across the inner
membrane)

Figure 10.3　Structure of mitochondria. The ATP synthase portals provide the channel for H^+ ions to return to the matrix. They comprise an F_0 stalk which is integral to the inner membrane and an F_1 nodule which protrudes into the matrix and has ATP synthase attached. ATP synthase is the enzyme which catalyses the reaction to form ATP during oxidative phosphorylation

Key point

Carbohydrate, fat and protein in the diet are broken down to their energy-rich component parts (glucose, fatty acids and amino acids) during digestion. The primary aerobic catabolism mechanisms for the substrates are glycolysis, β oxidation and deamination. Of these processes, only glycolysis results in the production ATP, and does so anaerobically i.e. without the need for oxygen. Glucose and fatty acids are broken down by acetyl CoA and amino acids to a variety of substrate intermediates for entry into Krebs cycle. The body's major aerobic ATP producer, the electron transport chain, runs parallel to and utilizes the hydrogen ions and electron by-products of these processes.

Primary mechanisms for aerobic substrate catabolism

As a function of digestion, as described in Chapter 2, dietary carbohydrate, fat and protein are catabolized to glucose, triglycerides and amino acids. The primary aerobic mechanisms for the further degradation of these substrates to release ATP to sustain exercise in adventure sports are glycolysis, β oxidation and deamination.

Glycolysis

The process of glycolysis, the second of the body's anaerobic energy systems, was described in Chapter 7. During fast rate glycolysis, required for high-intensity short-duration activities, glucose and the body's carbohydrate storage form glycogen are broken down to pyruvate and ultimately to lactic acid in the cytosol of the muscle fibre. For events lasting between 10–90 s, fast rate glycolysis represents the predominant energy system for ATP

production. For events lasting beyond 90 s and up to 10 min, and for the finishing 'kick' in events lasting longer than this, fast rate glycolysis provides a significant proportion of the required energy for ATP synthesis.

During lower intensity aerobic exercise slower rate glycolysis, as well as providing ATP to meet the energy demands of the activity, also results in the degradation of glucose to pyruvate for entry into the Krebs cycle, one of the secondary aerobic metabolism mechanisms. The mitochondrial outer membrane is easily permeable to pyruvate, but to enter the inner matrix of the mitochondria a protein carrier is required to cross the less permeable inner membrane. Once inside the matrix, pyruvate is further degraded to a two-carbon acetyl CoA molecule. The CoA to which the two remaining pyruvate carbons are bonded is a derivative of pantothenic acid, one of the B complex vitamins. The formation of acetyl CoA is achieved through a series of reactions which includes the formation of the intermediary substrate acetoacetyl CoA, important in protein deamination and the entry point of some proteins to the Krebs cycle; see Table 10.2. This formation is catalysed by the enzyme *pyruvate dehydrogenase* (PDH). During this process, a carbon atom is removed from pyruvate with two oxygen atoms forming carbon dioxide which is removed from the body through respiration. In addition, through the PDH catalysed reactions two further hydrogen atoms are removed from the structure and transported to the ETC by NAD^+. The removed hydrogen atoms result in the indirect synthesis of three further ATP from each pyruvate molecule. Once pyruvate (formed during slower rate glycolysis) has been converted to acetyl CoA, it can enter the Krebs cycle within the mitochondrial matrix.

Key point

Glycolysis takes place within the cytosol. Slower rate glycolysis in the presence of sufficient oxygen results in the formation of pyruvate. Two three-chain pyruvate molecules are formed from each glucose molecule during glycolysis. To enter the mitochondria for further degradation and energy release in the Krebs cycle, pyruvate must be further broken down to the two-carbon based acetyl CoA. Once acetyl CoA is formed, it can cross the inner membrane of the matrix for entry into Krebs cycle.

β *oxidation*

The storage of fat within the specialist adipocytes (fat cells) represents the body's long-term energy store. Stores of glucose and glycogen are limited within the body and are relatively short-term stores requiring frequent replenishment. Adipocytes, however, possess a plentiful supply of energy-rich triglycerides. The properties of fat and its use within the body were discussed in more detail in Chapter 2. Triglycerides, the major storage form for fat, provide over 90 per cent of the lipids for exercise and are stored as three fatty acids bonded at three separate points to a glycerol molecule. The primary mechanism for the aerobic catabolism of a fatty acid is through β oxidation. However, before this process can take place, triglyceride molecules require catabolism to the component parts through a process known as lipolysis. Once a fatty acid is freed from the glycerol backbone, it can be transferred into the matrix of the mitochondria where the enzymes for β oxidation are sequestered. Through the process of β oxidation, the long carbon chain fatty acids are broken down to many two-carbon chain acetyl CoA molecules for entry into the Krebs cycle.

Lipolysis The process of *lipoylsis* takes place within the adipocyte cytosol and requires a series of three reactions catalysed by the enzyme hormone-sensitive lipase (HSL). Through the hydrolysis of a triglyceride, water added to the reactants enables one of the fatty acids (or acyl molecules) to be removed from the structure. Each

Table 10.2 Entry point of amino acids to aerobic metabolism following transamination or deamination

Aerobic energy system entry point	Amino acids
Pyruvate	Alanine (n)
	Cysteine (n)
	Glycine (n)
	Serine (n)
	Tryptophan (n)
Acetoacetyl CoA (An intermediary substrate synthesised before the formation of Acetyl CoA)	Leucine (e)
	Lysine (e)
	Phenylalanine (e)
	Tryptophan (e)
	Tyrosine (n)
Acetyl CoA	Leucine (e)
	Isoleucine (e)
	Tryptophan (e)
α ketoglutarate	Arginine (n)
	Glutamine (n)
	Glutamic acid (n)
	Histidine (e/n)
	Proline (n)
Succinyl CoA	Isoleucine (e)
	Methionine (e)
	Threonine (e)
	Valine (e)
Fumarate	Phenylalanine (e)
	Tyrosine (n)
Oxaloacetate	Asparagine (n)
	Aspartic acid (n)

Notes:

1. The larger amino acids (isoleucine, leucine, phenylalanine, tryptophan and tyrosine) can be catabolized to a variety of aerobic metabolism intermediaries and therefore appear more than once in the table.

2. (e) essential amino acid, (n) non-essential amino acid. Histidine is essential for children, but can be synthesized in adults.

of the three reactions removes one fatty acid from the glycerol backbone. In the final reaction, HSL is assisted in its catalyst role by a further enzyme monoglyceride lipase. The hormones that can influence the rate of lipolysis through action on HSL are adrenalin, noradrenalin, glucagon, human growth hormone and cortisol, which increase the rate of hydrolysis whereas insulin inhibits the action of HSL. These lipid-hydrolysis promoting hormones have the effect of favouring lipolysis, thereby sparing the limited glycogen stores.

The reactions in lipolysis are shown in Figure 10.4. The first reaction normally cleaves the first fatty acid from the third carbon molecule of glycerol while the remaining fatty acids are attached to carbons 1 and 2 of glycerol, hence 1,2-Diacylglycerol. The second reaction removes the fatty acid from the first glycerol carbon leaving 2-monoacylglycerol. In the final reaction, the remaining fatty acid is split from the glycerol molecule.

Transport to the matrix As water-insoluble structures, the fatty acids combine with the blood protein albumin for transport to the skeletal muscles for degradation through β oxidation and Krebs cycle. The remaining glycerol molecule is not reused within the adipocyte but is transported via the blood, primarily to the liver, for use within glycolysis. On reaching the skeletal muscle fibres (cells), the fatty acids pass through the sarcolemma (muscle cell membrane) via active transportation but must also pass through the mitochondrial outer and inner membranes before their energy-rich stores can be realized through β oxidation and Krebs cycle. The fatty acids can easily pass through the outer membrane but each has to be bonded with a CoA molecule and then attached to the small peptide carrier carnitine for transport across the inner membrane. The enzyme acyl CoA synthetase catalyses the reaction to bond a fatty acid to a CoA molecule, with a cost of two ATP. After the formation of acyl CoA, acyltransferase catalyses the temporary bond of the fatty acyl CoA to the carnitine carrier to form acylcarnitine. Carnitine acyltransferase is found both within the inter-membrane space (see Figure 10.2) where it catalyses the bonding of carnitine to acyl CoA and in the mitochondrial matrix were it catalyses the detachment of carnitine in the reverse reaction. As soon as the acyl CoA molecule reaches the matrix, the fatty acid molecule is detached from its carnitine transport molecule and begins degradation through the process of β oxidation.

A fatty acid, detached from a triglyceride in an adipocyte and transported via the blood to a muscle fibre, is then attached to a CoA molecule. On

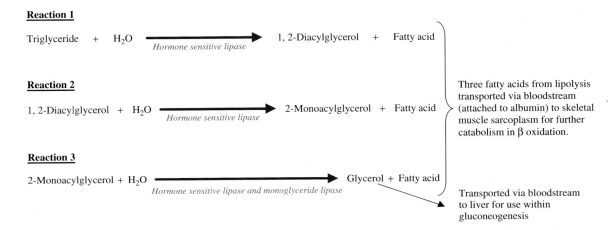

Figure 10.4 The three reactions of lipolysis and the fate of the three fatty acids and glycerol

crossing the mitochondrial inner membrane (temporarily attached to carnitine) it is immediately released and undergoes β oxidation in the matrix of the mitochondria. All the necessary enzymes for β oxidation are held within the matrix (see Figure 10.3). β *oxidation*, as already mentioned, is a process by which pairs of carbon molecules are detached from the main fatty acid (acyl CoA) until all the energy-rich pairs are released (fatty acids always contain even numbers of carbon atoms e.g. the 16 in palmitic acid). The process is called β oxidation because the break in the acyl CoA molecule to form the two-carbon acetyl CoA molecule occurs at the β carbon – the second carbon in the chain. The process of fatty acid oxidation involves the four reactions shown in Figure 10.5.

The cycle of reactions in β oxidation continue until the entire acyl CoA molecule has been oxidized to acetyl CoA pairs of carbon molecules. In the case of palmitic acid, seven cycles of β oxidation would be required for the complete catabolism of the palmitic acyl CoA to create eight acetyl CoA molecules. The four reactions of β oxidation are as follows.

1. In the first reaction, catalysed by acyl CoA dehydrogenase, two hydrogen atoms are removed and transferred to the electron transport chain by the co-enzyme carrier flavin adenine dinucleotide (FAD). In this reaction FAD, a derivative of the B vitamin riboflavin and closely related to NAD^+, is reduced to $FADH_2$.

2. In the second reaction, the addition of water hydrates the transenoyl CoA molecule to form 3-hyroxyacyl CoA, catalysed by the enzyme enoyl CoA hydratase.

3. The third reaction is similar to the first in that the enzyme, in this case 3-hydroxyacyl CoA dehydrogenase, catalyses the removal of two hydrogen atoms but the carrier in this instance is NAD^+.

4. In the final reaction, the addition of a CoA to attach to the end two carbons during their split from the acyl CoA molecule, catalysed by acyl CoA thiolase, creates the acetyl CoA molecule. The acetyl CoA molecule enters Krebs cycle and the β oxidation cycle begins again acting on the next β carbon bond.

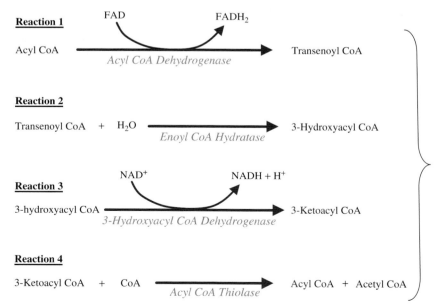

The four reactions of β oxidation, leading to the release of a pair of carbon atoms in the formation of acetyl CoA. The reactions take place within the matrix of the mitochondria. The acetyl CoA molecule produced immediately enters Krebs cycle for further degradation. Krebs cycle also occurs within the matrix. The remaining acyl CoA molecule then undergoes the same four reactions to release the next pair of carbon atoms as acetyl CoA. The process of β oxidation continues until all the energy-rich carbon pairs from a fatty acid are released.

Figure 10.5 The four reactions of β oxidation and the fate of acetyl CoA

For the process of β oxidation to continue, oxygen must be present in sufficient quantities to oxidize the hydrogen ions released to FAD and NAD$^+$ and transferred to the electron transport chain. β oxidation, in common with Krebs cycle and the electron transport chain, is an oxygen-dependent process.

Key point

β oxidation can only occur during conditions of sufficient oxygen. It is a process of aerobic metabolism. For β oxidation to occur, fatty acids must undergo considerable processing before they can enter the mitochondrial matrix, the location for this primary aerobic catabolism mechanism. A triglyceride molecule stored in a fat cell must be broken down to its component parts and the freed fatty acids combined with albumin for transport via the bloodstream to the skeletal muscles. Once within a muscle fibre, a fatty acid must be combined with a CoA molecule to form acyl CoA and then temporarily combine with the small carrier protein carnitine to cross the mitochondrial inner membrane. Once across this membrane, the acyl CoA molecule is released and β oxidation can begin immediately. During β oxidation, two carbon molecules are released from the main fatty acid body, until all the carbons are released. Each of these two carbon chains is combined with CoA molecule to from acetyl CoA for entry into Krebs cycle.

Deamination

The removal of the nitrogen-containing amine group from an amino acid, described in Chapter 2, enables the carbon-based skeleton to be utilized as an energy source for aerobic metabolism. The removal of the amine group (NH_2) can occur through an isolated deamination or through the deamination that enables a transamination (the removal of an amine group from one amino acid to attach to the carbon-based skeleton of another) which occurs during the alanine-glucose cycle.

Twenty amino acids are necessary for human growth and health, as listed in Table 2.6. The amino acids contained within the liver, bloodstream and skeletal muscle serve as the body's amino acid store. As a result of the differences in chemical composition of the body's twenty amino acids, the removal of the amine group reveals a variety of carbon-based skeletons. The Krebs cycle, illustrated in Figure 10.6, describes a cycle of nine reactions resulting in the degradation of acetyl CoA to create CO_2 and release electrons for use in the electron transport chain.

The composition of each carbon-based skeleton determines its entry to aerobic metabolism as pyruvate, acetoacetyl CoA, acetyl CoA or as an intermediary substrate in Krebs cycle. Table 10.2 provides the aerobic metabolism entry point for the carbon skeletons of the twenty amino acids found in the body. Once deaminated, the carbon-based skeleton remaining from each amino acid can contribute to aerobic metabolism. Amino acid deamination takes place mainly within the liver, but some amino acids can be deaminated within skeletal muscle fibres. The carbon-based skeletons remaining after deamination are transported to the mitochondria and into the matrix via the same carrier protein processes as described for the movement of pyruvate to the matrix. Once within the matrix, the carbon-based skeletons can enter the Krebs cycle according to the particular substrate entry point as shown in Table 10.2.

Key point

Deamination occurs as a process of changing the structure of amino acids as required by the body and to provide a carbon-based

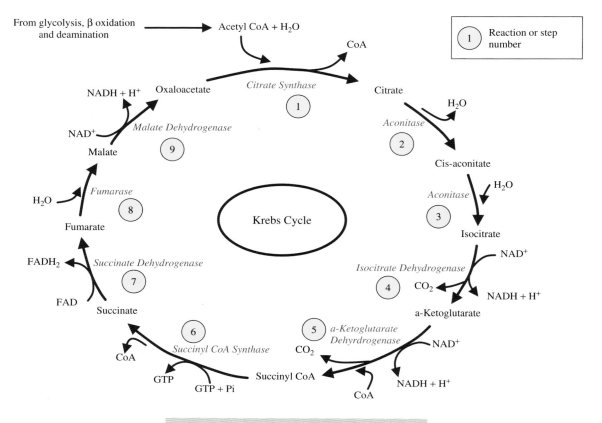

Figure 10.6 The reactions in the Krebs cycle

skeleton for use as an energy source. Deamination involves the removal of the nitrogen-containing amine from the amino acid. The differential structure of the twenty amino acids essential for health and growth results in the formation of a variety of carbon-based skeletons. These carbon-based skeletons can create products to be converted to acetyl CoA for entry to Krebs cycle at the same point as carbohydrate and fat or as Krebs cycle intermediaries for later entry to the same process.

Krebs cycle

The nine reactions within the Krebs cycle, also known as tricarboxylic acid cycle or citric acid cycle, occur within the matrix of the mitochondria. The location for Krebs cycle is shown in Figure 10.3. The product of primary carbohydrate and fat aerobic catabolism, acetyl CoA, is the entry point to Krebs cycle for these energy-rich macronutrients. Protein, catabolized for energy production, enters the Krebs cycle at a variety of points according to the structure of its carbon-based skeleton. The entry points to the Krebs cycle for protein are shown in Table 10.2. The nine reactions in the Krebs cycle (shown in Figure 10.6

and named after Hans Krebs who first identified the pathway) result in the breakdown of acetyl CoA to CO_2. This process, as with glycolysis and β oxidation, releases hydrogen ions and electrons for use within the electron transport chain. Krebs cycle employs both FAD and NAD^+ as co-enzyme carriers for the hydrogen ions produced. The chemical composition of each of the intermediary substrates within the Krebs cycle is illustrated in Figure 10.7.

One complete cycle of Krebs results in the complete breakdown of the acetyl CoA molecule, with the release of the CoA in the first reaction, the two carbon atoms as CO_2 in reactions four and five and the hydrogen atoms and their associated electrons to the co-enzyme carriers FAD and NAD^+ for entry to the electron transport chain. Acetyl CoA

enters the Krebs cycle when it is bonded with a four-carbon chain oxaloacetate molecule to form a six-carbon citrate molecule. As a result of this reaction, catalysed by citrate synthase, the CoA component is removed from the acetyl structure and is released to be able to combine with α ketoglutarate to form succinyl CoA in reaction five of Krebs cycle. Reactions two–eight create a variety of Krebs cycle intermediates, before the recreation of a four-carbon oxaloacetate molecule in reaction nine and the completion of one rotation of the Krebs cycle.

The enzymes that catalyse each of the reactions in the Krebs cycle are shown in red in Figures 10.6 and 10.7. One molecule of water is consumed in the first reaction, while one is released in reaction two for a net consumption of zero, but a further two

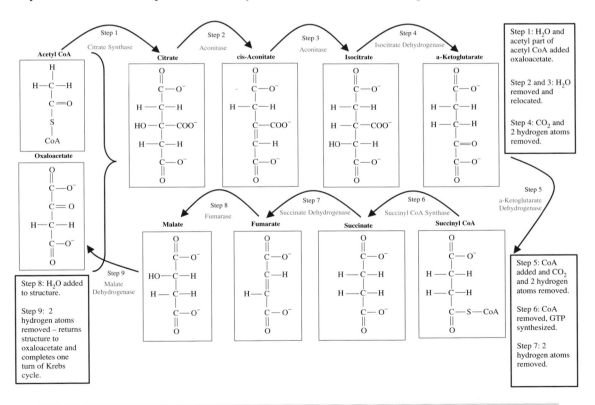

Figure 10.7   The chemical composition and structure of the components within the Krebs cycle

are consumed in reactions three and eight resulting in a net water consumption of two molecules to facilitate the complete oxidation of one acetyl CoA molecule. It is interesting to note that the intermediary substrates are not changed by the Krebs cycle; they merely host the acetyl group through the rotation of reactions to harvest the electrons and produce one molecule of GTP.

As with ATP, GTP is another of the nucleotides found in the body, the composition and functioning of which is described in more detail in Chapter 6. The GTP molecule produced through one turn of the Krebs cycle can be used to form ATP or to provide the energy source for protein synthesis. The dehydrogenase enzymes catalyse reactions that remove hydrogen atoms and their associated electrons from the intermediary substrate. The enzymes responsible for these hydrogen atom removal reactions are isocitrate dehydrogenase (reaction four), α-ketoglutarate dehydrogenase (reaction five), succinate dehydrogenase (reaction seven) and malate dehydrogenase (reaction nine). In reactions four, five and nine, the hydrogen atoms and their electrons reduce NAD^+ to $NADH + H^+$; in reaction seven FAD is reduced to $FADH_2$. The removed hydrogen atoms, the most important product of the Krebs cycle, are released for use within the electron transport chain.

Key point

The Krebs cycle is closely linked to the electron transport chain (ETC). Its major product is further hydrogen atoms and their associated electrons for use within the ETC. In addition, the Krebs cycle results in the production of one substrate phosphorylated ATP molecule (after conversion from GTP) to add to those created through glycolysis. Further ATP for exercise, and indeed the majority from aerobic energy sources, are created through the ETC. The Krebs cycle provides an important supply of electrons for this purpose.

Electron transport chain and oxidative phosphorylation

The electron transport chain (ETC) is also referred to as cellular respiration, oxidative phosphorylation, oxidative metabolism or as the terminal respiratory chain. It functions in parallel to the processes of glycolysis, β oxidation and the Krebs cycle, utilizing the hydrogen atom by-products of the reactions within these processes. Figure 10.8 provides an overview of the energy production in the electron transport chain, a mitochondrial process. The complete oxidation of carbohydrate, fat or protein actually involves two coupled processes (the electron transport chain and oxidative phosphorylation), hence the variety of names that are used for this aspect of aerobic metabolism. The electron transport chain comprises five complex reactions which pass electrons down a chain. The final two complexes (complexes IV and V), sometimes referred to as oxidative phosphorylation, result in a coupled reaction and form the point when ATP and water are created: the end-products of the ETC.

Hydrogen and the co-enzyme carriers:

Hydrogen atoms, as described in Chapter 2, are the smallest atoms and comprise a proton and an electron. If the electron is removed from a hydrogen atom it becomes a positively charged hydrogen ion (H^+), sometimes referred to by what is left: a proton. If a hydrogen atom receives an additional electron it becomes a negatively charged hydride ion (H^-). The pairs of hydrogen atoms released through glycolysis, β oxidation and the Krebs cycle are transported to the ETC by the carrier co-enzymes NAD^+ and FAD. The nature of these two co-enzymes creates a difference in how hydrogen atoms are bound to the carrier

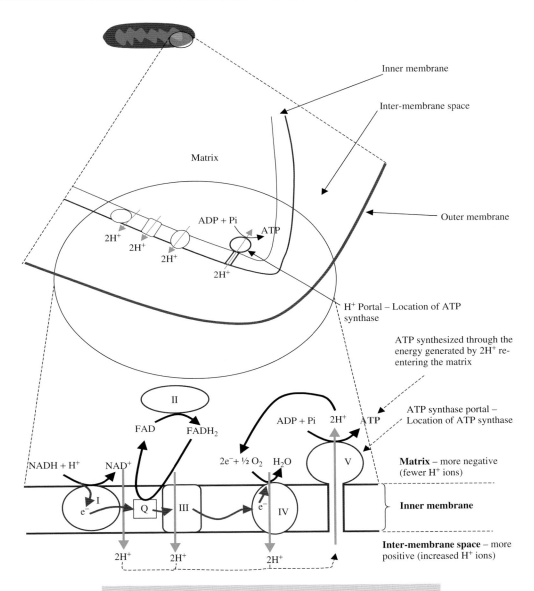

Figure 10.8 Overview of ETC and oxidative phosphorylation

during transport to the ETC. As a positively charged ion, NAD^+ binds with two hydrogen ions such that a negatively charged hydride ion is formed and bonds with NAD^+. The remaining hydrogen ion (that has donated its electron to create the hydride ion) is carried with NADH to form $NADH + H^+$. FAD, on the other hand, carries the hydrogen atoms with their electrons intact and forms $FADH_2$. This difference in co-enzyme storage creates a difference in the potential energy within the stored hydrogen. Stored as hydride and hydrogen ions bound to NAD^+, the hydrogen atoms have a higher potential energy to drive the ETC than for the hydrogen binding with FAD. It is for this reason that FAD enters electrons one step down in the ETC and results in the synthesis of one fewer ATP molecules.

The *electron transport chain* describes the movement of electrons passed from $NADH + H^+$ and $FADH_2$ to a series of electron carriers. As the electrons are passed down the ETC, energy

is released to pump pairs of H^+ ions out of the mitochondrial matrix and create a potential electrical difference between the inside and outside of the matrix (the ETC pumps result in the movement of H^+ from inside the matrix to outside). As electrons move down the ETC the amount of potential energy they contain decreases, such that the energy stored in step or complex I is higher than that retained in complex II, and so on. The energy cascade during the ATP production process of the ETC is illustrated in Figure 10.9. The energy released during the passage of electrons down the ETC is harnessed to pump six hydrogen ions from the matrix of the mitochondria (four hydrogen ions for FAD).

Oxidative phosphorylation refers to the actual process of ATP creation. At the end of the ETC the

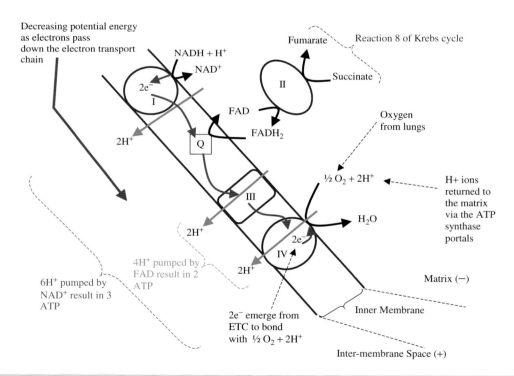

Figure 10.9 The energy cascade as electrons flow down the electron transport chain. Electrons are passed from NAD^+ to proteins in complex I and from FAD to Coenzyme Q. Note that the different start point for electron transfer by NAD^+ and FAD results in NAD^+ enabling the pumping of two additional H^+ ions from the matrix. It is the differential starting points that result in electron transfer from NAD^+ leading to the creation of 3 ATP and FAD leading to the creation of 2 ATP

ATP synthase portals, illustrated in Figures 10.3, 10.8 and 10.10, represent the essential structures for oxidative phosphorylation. As explained in Figure 10.3, the ATP synthase portals comprise an F_0 stalk which is integrated within the inner membrane and an F_1 nodule which is located in the inside wall of the inner membrane (see Figure 10.8 or 10.10). While the ETC pumps protons (H^+) out from the matrix to create a electrical difference between the inside and the outside of the cell, ATP

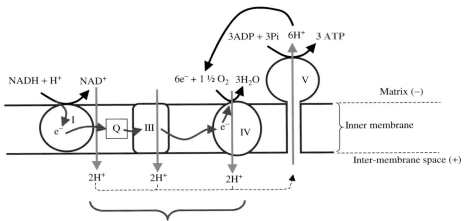

NAD$^+$ enables the pumping of six hydrogen ions from the matrix to the inter-membrane space. This leads to the synthesis of three ATP.

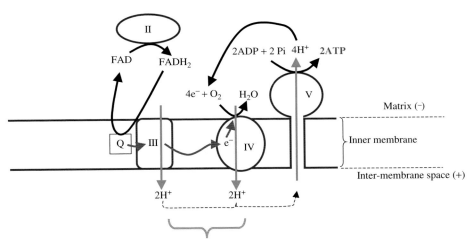

FAD enables the pumping of four hydrogen ions from the matrix to the inter-membrane space. This leads to the synthesis of two ATP.

Figure 10.10 Oxidative phosphorylation of electrons from NAD$^+$ and FAD resulting in the creation of 3 and 2 (respectively) molecules of ATP

synthase portals enable H^+ to flow back into the matrix. The movement of hydrogen ions back into the matrix, established as a result of the charge difference on either side of the inner membrane, drives the phosphorylation of ADP and Pi to ATP. In order for this reaction to proceed, oxygen is required to accept the electrons released at the end of the ETC. The released electrons are re-united with the hydrogen ions re-entering the matrix while being oxidized via the ATP synthase portals. The end product of this reaction between oxygen, the ETC electrons and hydrogen ions is water. In this way oxygen, direct from respiration, facilitates the phosphorylation (adding of phosphate to) ADP to ATP; the term oxidative phosphorylation is very appropriate.

It is important to understand that glycolysis, β oxidation and the Krebs cycle do not require oxygen to be taken into the body through respiration for the series of reactions involved in each to occur. For each of these processes to continue over a long duration, such as that during a higher or lower intensity aerobic endurance activity, oxygen is required to oxidize the electrons emerging from the ETC. The co-enzyme carriers NAD^+ and FAD transport hydrogen atoms produced during glycolysis, β oxidation and Krebs cycle to the ETC where the oxidation process begins. In this way, the ETC is a truly aerobic system or mechanism because it directly consumes molecular oxygen (from respiration) for ATP creation. Glycolysis, β oxidation and the Krebs cycle rely on the ETC for removal of hydrogen atoms and, as a consequence, are dependent on the oxygen consumed during the ETC for their continuation. Without the presence of oxygen in sufficient quantities for ETC functioning, the Krebs cycle and β oxidation cannot operate. Glycolysis can continue (Chapter 7) for a short duration without the oxygen required by the ETC because the hydrogen atoms produced during the process can be passed to pyruvate creating lactic acid.

The electron transport chain is the key mechanism for ATP production and is responsible for the creation of over 90 per cent of the body's ATP. It is able to create so much ATP because the electrons it accepts at the start of the chain from $NADH + H^+$

and after complex II from $FADH_2$ have a high-energy transfer potential. The ETC utilizes this energy to pump hydrogen ions outside the matrix, creating an electrical difference across the inner membrane. It is the difference created by the ETC that enables the phosphorylation of ADP to ATP. The oxygen required by the ETC is supplied to the muscles via the blood. Type I fibres are especially adept at this process because they are densely covered with capillaries, have the highest numbers of mitochondria and contain high levels of myoglobin for oxygen storage within the muscle.

The coenzyme carriers NAD^+ and FAD are reduced to $NADH + H^+$ and $FADH_2$ when they accept hydrogen ions from the reactions in glycolysis, β oxidation and the Krebs cycle. The processes of β oxidation and the Krebs cycle take place within the matrix of the mitochondria and, consequently, the hydrogen ions they produce can easily be transferred from the carrier co-enzymes to the ETC. The hydrogen ions produced during glycolysis, which occurs in the cytosol, have to be shuttled to the mitochondrial matrix before they can be transferred to the ETC. The glycerol-phosphate shuttle is the mechanism through which $NADH + H^+$ and $FADH_2$ enter the mitochondria. In this process, the hydrogen atoms and their electrons are transferred to glycerol-phosphate in the inter-membrane space to cross the inner membrane. Once inside the matrix, hydrogen ions from glycolysis can enter the ETC.

The ETC, illustrated in Figure 10.9, shows the cascade in potential energy as electrons are transferred down the chain. The steps in the ETC involve a series of five complexes (or reactions) that are also illustrated in Figures 10.8 and 10.10. The five complexes are as follows.

- Complex I: The electrons from $NADH + H^+$ are transferred to flavin mononucleotide. The energy produced during this reaction is sufficient to pump out 2 H^+ from the matrix.

- Complex II: The electrons are passed from flavin mononucleotide to co-enzyme Q (labelled Q in Figure 10.9). It is at this point that FAD can transfer its hydrogen atoms, which have a lower potential energy than those held by NAD^+, to the ETC (see the box on the nature of hydrogen atoms for an explanation of this difference). As illustrated in Figures 10.9 and 10.10, the later entry by FAD to the ETC results in the re-synthesis of one less ATP molecule than is enabled by NAD^+. It is for this reason that FAD results in the production of two ATP and NAD^+ in the creation of three ATP. From co-enzyme Q, the electrons are transferred to a series of protein-based cytochromes which form the basis for complex III.

- Complex III: The transfer of electrons to the first of the cytochrome groups (cytochrome $b-c_1$) enables the removal of two further free hydrogen ions from the matrix.

- Complex IV: The electrons are passed within a final cytochrome group (cytochrome $a-a_3$). As result of the transfers within complex IV, sufficient energy is released to pump two further protons (hydrogen ions) from the matrix. The pumping of hydrogen ions across the inner membrane (into the inter-membrane space), achieved as a result of the transfer of electrons down the ETC, creates an electrical potential either side of the membrane. The inter-membrane space (outside the matrix) becomes more positively charged than the inside of the matrix. The electrical difference creates the mechanism to drive oxidative phosphorylation of ADP. The events at the end of complex IV and those of complex V should be seen as coupled. The final reaction of complex IV involves the oxidation (joining with molecular oxygen from respiration) of the electrons that emerge from the ETC.

- Complex V: Protons (H^+) re-enter the matrix via complex V – the ATP synthase portals (illustrated in Figures 10.8 and 10.10). As the hydrogen ions re-enter the matrix via the ATP synthase portals they immediately bond with the oxidized electrons to form water ($4e^- + O_2 + 4H^+ = 2H_2O$). The action of the hydrogen ions 'rushing' back to the matrix (moving from an area of higher electrical charge to lower electrical charge) to bond with the oxidized electrons liberates sufficient energy to enable

the phosphorylation of ADP and Pi to ATP. The enzyme that catalyses the reaction is, unsurprisingly, ATP synthase.

Thus the end-products of the ETC are water and ATP, three molecules of ATP for each $NADH + H^+$ and two molecules for each $FADH_2$ (Figure 10.10 provides an illustration of this difference). Figure 10.8 illustrates this oxidative phosphorylation at the end of the ETC, where the presence of ATP synthase portals provides the H^+ ions for the creation of water and the location for ATP synthase which catalyses the reaction to phosphorylate ADP to ATP.

ATP created through the ETC is transferred from the matrix to the inter-membrane space across the inner membrane via an antiport (a protein-based membrane channel). The antiport enables an exchange between ATP and ADP, such that as an ATP molecule departs the matrix an ADP molecule is brought into the matrix for phosphorylation. A similar type of channel enables phosphate (Pi), essential to phosphorylation, to enter the matrix for use within the ETC. From the inter-membrane space, ATP can easily pass through the outer membrane to provide energy for muscular contraction during exercise. As mentioned previously, the series of reactions in the ETC provide the body's main source for ATP. Table 10.3 provides examples of the amount of ATP produced from a molecule of glucose and of a typical molecule of fat, palmitic acid. The importance of the ETC for ATP production can be clearly seen from the table, where substrate synthesis of ATP (direct from glycolysis and the Krebs cycle) results in only around 10 per cent for the total energy possible through aerobic metabolism. Figure 10.11 provides an illustration of the figures in Table 10.3 for glucose, glycogen and palmitic acid to show the production of ATP through the electron transport chain.

Although ATP production through aerobic metabolism is relatively efficient, only about 40 per cent of the total energy released through the complete oxidation of carbohydrate or fat is harnessed to create ATP. The remaining energy

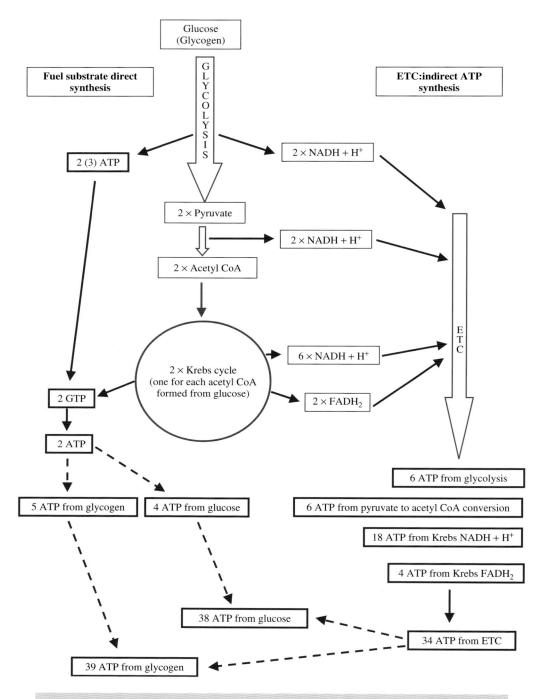

Figure 10.11   ATP formation through (a) glycolysis and (b) oxidation of palmitic acid

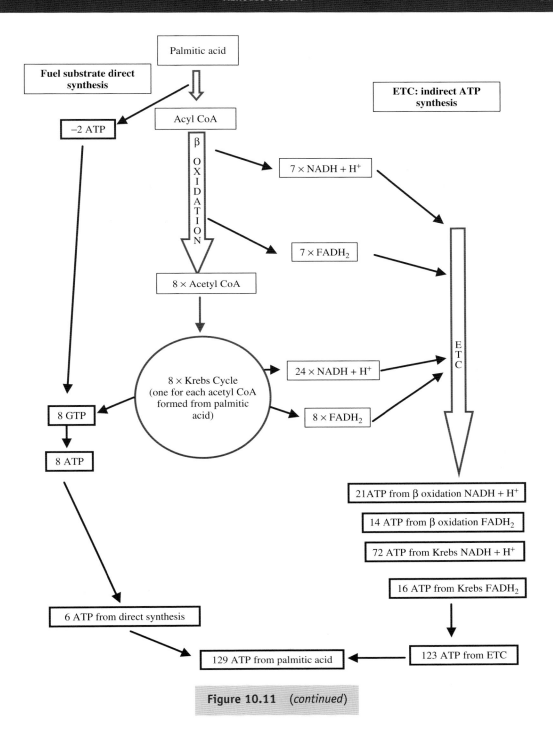

Figure 10.11 (*continued*)

Table 10.3 ATP production from glucose, glycogen and palmitic acid via direct substrate creation and the ETC

Pathway	Glucose $(C_6H_{12}O_6)$	Glycogen	Palmitic Acid $(C_{16}H_{32}O_2)$
Glycolysis – direct synthesis	2	3^1	–
Fatty acid conversion	–	–	-2^2
β oxidation – direct synthesis	–	–	0
Krebs cycle	2^3	2^3	8^4
ETC	34^5	34^5	123^6
Total	38	39	129

Notes:

1. When glycogen is the substrate for glycolysis, one fewer ATP molecules is consumed during the degradation process and so three rather than two molecules of ATP are produced directly.

2. To convert a fatty acid to acyl CoA (to enter the mitochondria prior to β oxidation) costs two ATP molecules.

3. Glycolysis results in the formation of two acetyl CoA molecules to drive two rotations of the Krebs cycle, leading to the production of two ATP molecules (or strictly speaking GTP molecules which can be converted to ATP).

4. Through β oxidation, eight acetyl CoA molecules are formed resulting in the production of eight molecules of ATP (from GTP) through the Krebs cycle.

5. Glycolysis results in the formation of two NADH + H$^+$ resulting in 6 ATP molecules (three from each NADH + H$^+$); the conversion from pyruvate to acetyl CoA results in the formation of 6 ATP (resulting from the formation of two NADH + H$^+$, one for each pyruvate), Krebs cycle through two rotations results in the formation of 18 ATP from NADH + H$^+$ and 4 ATP from FADH$_2$, producing a total of 34 ATP.

6. During each of the seven cycles of β oxidation for palmitic acid, five ATP are produced in the ETC from the FADH$_2$ and NADH + H$^+$ produced in each cycle making a total of 35 ATP from β oxidation, along with 88 from the Krebs cycle making a grand total through ETC of 123 molecules of ATP.

created through the reactions within glycolysis, β oxidation, Krebs cycle and the ETC escapes as heat energy, which is why temperature rises during exercise.

Key point

The electron transport chain is the body's main mechanism for ATP production. Through a series of reactions (complexes I–V) within the mitochondrial matrix, electrons from NADH + H$^+$ and FADH$_2$ are transferred down a chain of reactions to create ATP. Each NADH + H$^+$ results in the creation of three ATP and each FADH$_2$, which enter the ETC further down the chain, results in the synthesis of two ATP molecules. Molecular oxygen, not directly required for the reactions in glycolysis, β oxidation or the Krebs cycle, is essential in the final step in the ETC. Through the process of oxidative phosphorylation, coupled reactions involving complexes IV and V, electrons from the ETC are transferred to oxygen from respiration (molecular oxygen) and combine with H$^+$ (re-entering the matrix through the ATP synthase portals) to form water. At the same time, the energy potential created by the electrical difference between the inside and outside of the inner membrane is harnessed to re-synthesize ATP. The end products of the ETC are ATP and water.

Metabolic rate

Heat can be measured using a calorimeter. Recall the use of a bomb calorimeter to determine the calorific content of macronutrients in Chapter 2. Our metabolism represents the sum of the reactions that occur in our bodies. The catabolic reactions that take place to provide the energy for work are generally exergonic, meaning they liberate energy in the form of heat. For instance, the reactions within aerobic metabolism, as described in the last section, are about 40 per cent efficient, the rest of the energy being liberated as heat. A calorimeter could be used to directly measure the heat produced through the reactions taking place in our bodies: our *metabolic rate*. The process of direct measurement of metabolism through the use of a calorimeter is difficult because humans do not liberate all the heat they produce. As a consequence, researchers have developed methods for indirect determination of metabolism. One of the most commonly used methods for indirect determination of an individual's metabolic rate is through gas analysis. The measurement of oxygen consumption $\dot{V}O_2$ can be utilized, not only to gain information about an athlete's fitness ($\dot{V}O_{2\,max}$) and the predominant macronutrient for performance (RER), but it can also serve to inform the physiologist about an individual's metabolic rate.

In the original studies of metabolism, researchers brought participants into a laboratory overnight to examine their basal metabolic rate (BMR). The participants in such study had to refrain from eating for 12 hr, be in a completely rested state and be assessed in a supine (on their back) position for a 10 min period during which the gas collection is made. Analysis of the expired air sample was then used to determine the BMR. Subsequent research has shown there is little difference between a resting metabolic rate (RMR) and BMR. As a consequence, due to the cost involved in the determination of BMR, most researchers identify an individual's RMR before looking at the effect of exercise on the rate of metabolism.

Research has shown that when 1 L of oxygen is used to oxidize a mixture of carbohydrate, fat and protein, about 4.8 kcal of energy is liberated. At rest, a litre of oxygen is consumed every 3 min resulting in a consumption of around 20 L of oxygen an hour, or around $450–480\ \text{L}\,\text{day}^{-1}$. An average daily energy expenditure at RMR equates to around 2100–2300 kcal. The Department of Health UK estimated average calorie intake statistics (see Chapter 2) for recommended nutritional intakes, developed from similar estimates of average RMR. To maintain your weight a balance must be maintained between calorific intake and energy expenditure.

A variety of factors can influence resting metabolic rate. Higher levels of fat-free mass, body surface area, body temperature and stress all serve to increase RMR. Increasing age, on the other hand, has been shown to result in a linear decrease in RMR. It is thought, however, that much of the identified decrease in RMR with age is related to a decrease in fat-free mass and an increase in fat mass associated with a sedentary lifestyle. Research has indicated that for individuals with an active lifestyle – those who exercise or train regularly – the age-related decrease in RMR is halted or at least slowed. The presence of thyroxine and adrenalin, associated with the onset of exercise, also increase metabolism. As well as the stimulation of hormonal increases in metabolic rate, exercise stimulates a general rise in metabolism, resulting in an increase in energy expenditure. Table 10.4 provides an overview of estimated energy expenditures for performers of 65 and 75 kg taking part in a variety of adventure sports.

A further term often encountered with regard to energy expenditure is that of the MET. A *MET*, or metabolic equivalent, represents a further system for classifying the intensity of exercise based on a resting metabolic rate. One MET is equal to a resting oxygen consumption of $3.5\ \text{mL}\,\text{kg}^{-1}\,\text{min}^{-1}$. Exercise intensities can then be graded according to the number of MET units required for participation. For example, participation in orienteering represents a high energy expenditure of 9 METs, recreational canoeing or kayaking around 4–5 METs and sailing (helm for doubles or match racing) entails an energy expenditure of 3 METs.

The linear relationship between oxygen consumption and heart rate at sub-maximal exercise intensities can be employed as a guide to energy expenditure. In a variety of adventure sports, where it may not be possible to collect oxygen consumption data in the field, this relationship could offer a method for estimation of energy expenditure. If a physiologist knows from laboratory testing an individual's oxygen consumption and heart rate across a range of exercise intensities,

Table 10.4   Estimates of energy expenditure for a variety of adventure sports (note that exercise intensity and terrain can dramatically alter the following figures)

Adventure sport	Estimated energy expenditure for a 65 kg performer (kcal hr^{-1})	Estimated energy expenditure for a 75 kg performer (kcal hr^{-1})
Canoe or kayak racing	780	940
Diving	700	800
Orienteering	600	720
Mountaineering	600	720
Rock climbing	580	690
Hill walking (15 kg rucksack)	530	660
Skiing or snowboarding (recreational)	400	450
Inland kayaking	330	400
Surfing	300	360
Canoeing or kayaking (recreational)	270	330
Windsurfing (recreational)	270	300
Sailing (match racing or doubles helm)	160	195

it is possible to infer the energy expenditure for exercise in the field, eliciting an equivalent heart rate. The limitations with this method occur due to the specificity of exercise. The exercises in the laboratory must match those in the field. For instance, research has indicated that for rock climbing – where there is a relatively high loading on the upper body – the use of treadmill running tests in the laboratory over-predict the oxygen consumption. Mermier *et al.* (1997) described this as an apparent breakdown in the $\dot{V}O_2$ heart rate relationship. For the relationship between heart rate and $\dot{V}O_2$ to be beneficial to the physiologist, the exercise in the laboratory must be specific to the field-based activity. The HR-$\dot{V}O_2$ relationship may have limited usefulness for researchers, but can provide a guide to the energy demands of adventure sports for practitioners, especially during the development of a body of research knowledge for a particular activity.

Key point

Metabolism represents all the reactions that take place within the body. The reactions that take place liberate heat. A calorimeter could be used to measure heat and can therefore be used to measure our metabolism. Due to the problems associated with heat liberation in humans, metabolism is usually estimated using indirect calorimetry via gas analysis. The minimum level of physiological functioning to maintain life is referred to as our basal metabolic rate (BMR). Some researchers use the MET (metabolic equivalent) to provide a measure of the metabolic cost of a variety of different activities, including adventure sports performance.

10.3 Physiological response to lower intensity endurance activities

The physiological response to exercise in lower intensity endurance activities follows the same general pattern as that described in Chapter 6. The onset of exercise results in an increase in metabolism, cardiovascular function and respiratory rate to meet the demands of exercise. The CNS is responsible for initiating the response to exercise, with the endocrine system reinforcing the required cardiovascular and metabolic responses. It is possible, and indeed often necessary, for all adventure sports performers to establish a steady state to match the duration of the activity with a specific intensity response. In other words, an athlete normally adopts an intensity of exercise that

enables them to complete the activity. The duration of the activity e.g. the ascent and descent of an alpine peak, determines the intensity set. In adventure sports, however, environmental factors such as an imminent electrical storm can alter decisions made regarding the intensity and, if escape routes are possible, the duration of an activity.

The predominant energy system for lower intensity endurance activity performance is the aerobic system. The relative intensity of the adventure sports that fall into this category would determine the predominant energy substrate: carbohydrate or fat. The higher the intensity of the adventure sport, the greater is the reliance of the body on glycogen stores. At an exercise intensity above 50–60 per cent of maximum, fat stores alone cannot provide the energy for exercise. Activities above the duration line in the duration-intensity continuum for lower intensity aerobic endurance adventure activities (Figure 10.1) would normally be expected to comprise an intensity of exercise (or periods of higher intensity exercise) such that glycogen supplies would be required in addition

Biography: Dave MacLeod

Adventure sport: Climbing

Dave MacLeod is 27 and has been a climber for 12 years. Over the past few years he has been putting the Scottish mountains back on the map in terms of world rock and ice climbing standards. Dave has used his educational background in sport science to manage the demands of pushing several climbing disciplines at once, leading to a current position as the best all-round climber in the UK. His ascents include:

Figure 10.12   Dave MacLeod (photograph courtesy of Steve Gordon)

- The Hurting XI, 11, Cairngorm, first ascent. Hardest traditional mixed route in the world

- Rhapsody E11 7a, Dumbarton Rock, first ascent. First E11 grade

- Holdfast E9 7a, Glen Nevis, first ascent. Second hardest traditional route in Scotland

- Pressure Fb8b, Dumbarton Rock, first ascent. Hardest boulder problem in Scotland

- Devastation F8c, Dumbuck, first ascent. Hardest sport climb in Scotland

- Hurlyburly F8b, Birnam Quarry. Hardest free-solo in the UK

The Hurting

Scottish winter climbing is considered a separate discipline in climbing's broad spectrum of challenges for good reason. The cliffs of Scotland's mountains are short enough that high-standard technical moves on steep, ice-plastered faces create the challenge. The fierce, wet and unpredictable highland winter is responsible for the reputation of Scottish winter climbing as one of the most challenging disciplines in world mountaineering.

Frequent freeze/thaw cycles plaster the cliffs in frost and wet snow, covering every nook and cranny in hard ice. This makes progress on a route extremely arduous. For every move upwards, ice must be laboriously scraped off with an ice axe to locate each axe hook in the rock and placements for protection (such as pitons and camming devices). All this effort must be made while hanging on by the other ice axe, demanding an exceptional level of upper body endurance to sustain this level of effort for several hours at a time. The second and more significant challenge arises from the ephemeral nature of good conditions. Where most other sports take place on set dates or periods during the year, allowing athletes to peak for competition, the winter climber's objective may only come into condition one or two days per winter. More often than not, our sport takes place during the foulest winter storms with severe gales, blizzards and avalanche danger thrown into the list of dangers.

In the late 1990s, the top winter climbers around began to experiment with more imaginative training techniques, incorporating the basic principles of training as well as what had been learnt from rock climbing disciplines. Training for mixed climbing (and indeed many adventure sports) carries an inherent specificity limitation. How can you expose yourself to a sport environment that only occurs a few days a year? Climbers realized that the different strands of ability required could be broken down and trained systematically at different times of the year to achieve the specificity needed yet still achieve a training overload. Techniques for climbing with ice axes were trained in summer by 'dry tooling' on bare rock or specially adapted climbing walls. Early ice climbing world cup events demonstrated clearly that upper body and core strength from steep rock climbing were of surprising importance, with podium places being filled by competition rock climbers who were new to ice climbing. This was much to the embarrassment of the 'old guard' of top ice climbers, but was also a welcome wake up call. Mental control aspects could be practiced on hard summer rock climbs. For those who recognized this way forward, all the ingredients were there for a step forward.

At that time I was climbing with the best Scottish winter climber around, Alan Mullin. Alan had used dry tooling on disused railway bridges during summer to hone his technique and confidence with ice tools. I was immediately converted, but I could see that where Alan failed on his projects in the mountains, it was down to lack of routine experience on dangerous routes all year round and lack of physical strength. If I applied myself I could gain the skills Alan had and fill in the gaps he was missing. This was an opportunity not to be missed.

A hard summer rock climb on Cairngorm called 'The Hurting' was an obvious challenge: steep, sustained, technical and downright dangerous. Perfect. On my first attempt I reached to within a few feet of the top, falling due to a combination of the wild blizzard, a slight route-finding

Figure 10.13 Dave MacLeod on a Cairngorm climb, The Hurting, XI, 11 (photograph courtesy of Steve Gordon)

error and loss of focus due to the shock of the new. A few days later the blizzard was still howling but I was rested, focused and ready to bring my preparation together. I pushed on forcefully beyond my highpoint, but not without a dry mouth and a little 'elvis leg' due to fear both of the fall but also the anticipation of failing a second time just inches from the top. I lunged for the sanctuary of an icy ledge and the finishing slopes.

The route had brought a small step forward in the level of the sport. Successful mountaineers have always been those who have an inherent ability to bring together and use the many diverse skills involved at the crucial moment. I realized that if those skills were more ruthlessly and thoroughly explored by applying sport science, there was room to gain an advantage. I suspect the situation is the same for many adventure sports.

to fat metabolism. This model, however, is only a guide and activities below the line could easily be completed at an intensity level that necessitates a metabolic contribution from glycolysis.

Fatigue in lower intensity activities occurs primarily due to glycogen depletion or as a result of the duration of the activity (which can also include glycogen depletion). The mechanism behind glycogen depletion and its effect on performance was described in Chapter 8. In some activities, such as mountaineering where night time starts may be required to take advantage of the best ice conditions, the added factor of sleep deprivation can impact on performance. This introduces a further aspect to fatigue that has not as yet been covered.

In Chapters 6–9 the mechanisms behind fatigue have been described within a physiological context as they pertain to substrate depletion and impaired functioning at a muscular level. These aspects of fatigue constitute *peripheral fatigue*. Since the start of the 20th century, researchers have become increasingly interested in the effects of *central fatigue* on performance. Research indicates that sleep deprivation does not have an impact on exercise metabolism (as long as substrate supplies are maintained), cardiovascular performance or muscular strength but it does significantly increase feelings of fatigue. The feelings of fatigue at the end of a race, expedition or day's activity, which can include pain, nausea and tiredness, are centrally located expressions of fatigue. The complex nature of the physiology and biochemistry of the brain mean that the mechanisms behind central fatigue are not fully understood. Any adventure sports performer, however, who has made an alpine start to ascend a peak or has felt the pain in their muscles at the end of a kayak marathon has felt the symptoms of central fatigue.

Key point

Fatigue in lower intensity aerobic endurance activities is likely to be the result of glycogen depletion or central fatigue. An athlete perceiving fatigue normally precedes physiological fatigue. Central fatigue can be expressed through feelings of tiredness, pain and nausea.

10.4 Physiological adaptations to training

The physiological adaptations to aerobic training were described in Chapter 9. These adaptations would apply equally to the lower intensity aerobic endurance activities that are the focus of this chapter. However, the extent of any adaptation will be dependent upon the duration, intensity and type of training undertaken. Research indicates that most improvements in $\dot{V}O_{2\,max}$ will have been realized after a relatively short period of time (i.e. 6–12 months after training commences). After this period, an individual can continue to improve their AAT through training at or just above the exercise intensity at which LT occurs. After testing to identify an individual's LT, the heart rate at which LT occurs can be used to guide the training intensity for future exercise. To continue improvements with LT and performance, interval and continuous training provide the most useful methods to sustain training adaptations.

This section continues with a brief review of overreaching and overtraining – when training goes too far. A balanced training programme must incorporate adequate rest periods and be flexible enough to respond to the individual needs of an athlete. Research clearly indicates that when a coach or an athlete places the training requirements over health maintenance, it is detrimental to performance and wellbeing.

Overreaching and overtraining

If the intensity and volume of training are too great an athlete can experience overreaching and overtraining, both of which have a negative impact on performance. Overreaching and overtraining can be factors that impact on any adventure sports performer, regardless of the duration of their activity. It is the intensity and duration of training that can lead to incidences of overreaching and overtraining. The performer or their coach should attempt to ensure that the periodized training load for any adventure sports performer is balanced such that they do not break down in training. This is easy to understand in theory, however, at the elite level athletes have to push hard and sometimes exist on a knife edge between maximal performance gains and overtraining. This section is included here because much of the early work carried out in the area of overtraining was conducted with and as a result of symptoms reported by endurance athletes. It is important to recognize that, as mentioned above, overtraining syndrome can affect any athlete regardless of the duration of their activity.

Overreaching describes a short-term expression of overtraining. When the stresses of training and everyday life result in a short-term decrement in performance the athlete is said to have overreached. If training continues, regardless of the symptoms of overreaching, longer term negative effects on performance will be realized. It is essential that an athlete's training load incorporates periods of rest to allow the body to recover from the stresses encountered. After an instance of overreaching, an athlete normally recovers in a period of 3–14 days; the recovery after overtraining often takes several months or longer. The symptoms of overreaching and the longer term overtraining syndrome are decrements in performance, persistent feelings of general fatigue, sustained muscle soreness, elevated resting heart rate, mood state issues, loss of competitive drive, insomnia, loss of appetite and weight loss. As well as training overloads, psychological stresses, poor diet and insufficient recovery periods can contribute to the occurrence of overreaching or overtraining. A well-planned periodized training programme with clear goals for the year ahead as well as an on-going dialogue between the athlete and their coach are the best measures to avoid an incidence of either overreaching or overtraining. A periodized programme should include variety for the adventure sports performer (to avoid staleness), pre-planned recovery periods and regular nutritional reviews to ensure a healthy diet for the athlete.

Key point

After initial gains in aerobic capacity, the major gains in performance for an endurance athlete are seen through improvements in LT. Overreaching and overtraining generally result from too intense training, insufficient rest and poor diet, or a combination of all three. Athletes need balance in their lives as well as their training and diet to maintain and improve performance. A periodized training programme with specific goals as well as regular communication and trust between an athlete and their coach can help to avoid the problems associated with overtraining.

10.5 Nutritional ergogenic aids to aerobic performance

Employment of the carbohydrate loading strategies described in Chapter 8 provides a starting point, beyond a healthy balanced diet, for dietary manipulation by adventure sports performers involved in lower intensity endurance activities. As well as increasing carbohydrate intake prior to exercise, the use of carbohydrate drinks to maintain performance during an event, day trip or expedition could be considered in some adventure sports. The long duration of many of the adventure sports that are the focus of this chapter perhaps necessitates the consumption of food during activity to maintain performance. In this context, the ingestion of a range of carbohydrate sources appears to provide the best mechanism for maintaining longer term aerobic performance. The consumption of food items comprising complex as well as simple carbohydrates appears to help maintain blood glucose supply over time. A combination of whole wheat breads, sandwiches, biscuits, fruit, snack bars and sweets can help to maintain carbohydrate levels during a sea paddle or an alpine ascent.

In addition the use of carbohydrate drinks in longer duration adventure sports, particularly in hot environments where water loss through sweating can be substantial, the consumption of a combined electrolyte and carbohydrate replacement drink could be considered. Sodium, potassium and chlorine (in the form of chloride ions)

Biography: Matt Berry

Adventure sport: Kayaking

Matt began kayaking when he was 14 with his local club, Barking and Dagenham. Although being based in east London was not ideal for whitewater paddling, there were a number of keen paddlers at the club who took Matt to learn the skills of whitewater paddling in north Wales and elsewhere around the country. Matt's passion for whitewater paddling grew from there and by the time he was 18 he had paddled in the Alps and Himalayas. Since then, he has paddled throughout Europe, in the US and Canada. Matt is a BCU level 5 coach and a Senior Lecturer in Adventure Education.

Figure 10.14   Matt Berry

Kayaking the Grand Canyon of the Colorado River

The Grand Canyon of the Colorado River is a must on any whitewater kayaker's list. Running 280 miles from Lees Ferry to Lake Mead, the Colorado takes 2 weeks of continuous kayaking amidst breathtaking scenery, enormous waves and some very challenging grade 5 rapids. (Rivers are graded 1 to 6, where grade 1 is easy to paddle with no obstacles and grade 6 requires exceptional ability coupled with huge amounts of luck!) The Canyon is a beautiful but committing place for there are very few ways out. If you get into trouble with a snake bite or scorpion sting or even a bad case of toothache, most often the only way out is to carry on down river.)

With these thoughts in mind I checked the inside of my boat for scorpions and set off down river with a few friends and a small team of rafts that accompanied us to carry the equipment. The kayaking in the canyon has long sections of flat water, so there are sustained periods of forward paddling, generating a lot of heat. With the air being very arid, dehydration sets in fairly quickly. Some afternoons on the early stages of the river where there are many flatter stretches, I found I was getting headaches through dehydration, being a bit too lazy to keep taking my spraydeck off to get the water bottles from inside the boat.

One surprise on the river was the water temperature. Being released from the base of the dam, the water was extremely cold and a stark contrast from the baking heat of the desert sun being reflected off the rich red rocks of the canyon walls. The rapids on the early sections were particularly cold. With huge waves crashing over the boat it took some time after each rapid to reach a comfortable temperature again, particularly as I had only brought a short sleeve paddle top with me! With the water being so cold it was a real shock to the system after getting so hot on the flat sections and, I forgot how hot or thirsty I was as soon as the first waves hit me. Over successive rapids, the accumulative effect was to become more and more dehydrated as the day wore on without really realizing it until it was too late and the headache set in. (Kayakers are also notoriously bad at taking on fluids because they hate having to get out to have a wee.)

Many of the 14 days of kayaking and the several hundred rapids tend to blur in my memory but one particular day sticks in my mind. It was the 13th day: Lava day. Lava is a particular grade five rapid that has an enormous 'V' shaped wave and a huge re-circulating wave called a stopper running almost the width of the river. This stopper would hold a kayak under indefinitely which must be avoided for obvious reasons! We approached the beginning of the rapid marked by a huge block of volcanic lava rock and paddled to the side to get out and inspect the rapid and try to look for the best line down. From our vantage point high up on the canyon side, the stoppers and waves still looked enormous and I stood mesmerized by the river-wide stopper that would surely consume a 20 ft raft with ease, let alone an 8 ft kayak!

We chose our route which was to run directly through the guts of the huge 'V' wave, skirting the deadly stopper on our left. It seemed the best line as long as I could still see the route clearly down at water level. I was nervous but confident I could run it as long as I came out of the 'V' wave the right way up. I walked back to my boat and got in. It took a second attempt to get my deck on and the first few strokes away from the bank were not as clean and powerful as I would have liked given what was to follow. My heart was pounding and I could hear the enormous roar of the rapid increase in volume as I drew closer with the swiftness of the current at this point. I saw the deep hole at the centre of the 'V' wave and knew I was in the right place. I paddled hard with deep, powerful stokes hoping I would punch through the other side the right way up but was held by the water coming back on itself. I braced hard with my paddle, placing the blade deep into the wave, searching for the blade to purchase the green water rather than foam. After a few seconds, still unable to see, I sensed I was through but definitely not the right way up. I knew I needed to roll up quickly because the flow after the 'V' wave was awkward, taking you between two boulders that could pin the boat under. Not as swiftly as I would have liked, I did roll up and get a breath. I then tried to clear water from my nose and eyes so I could quickly scan for the line away from the boulders.

I saw the line of standing waves I was looking for and sprinted across the river. This section was straightforward enough as I powered downstream 300 m or so towards the large calm pool down on the left of the river where I could catch my breath. Here I sat, heart still pounding but ecstatic while I waited for the others. They all made it and we excitedly exchanged stories of our route

Figure 10.15   Paddling through the big volume water of Crystal Rapid during descent of Colorado River (photograph courtesy of Jerry Baird)

through the 'V' wave or, in Bill's case, his upside down route through the boulders!

With a jubilant grin and aching shoulders we all pushed on towards Diamond Creek and the last remaining rapids of the final day of white water, looking forward to the end of the canyon, a hot shower and a night in a bed.

comprise the body's electrolytes. Electrolytes serve to maintain the fluid balance between the body's various intracellular and extracellular compartments. Longer duration adventure sports, especially those where either the clothing necessary or the environmental temperature encountered result in prolonged periods of sweating, can lead to a marked loss of sodium. In activities such as a long day's paddle or an alpine ascent, where the duration of activity can be as long as 10–15 hr, sodium loss through sweating can present a significant problems for adventure sports performers.

Hyponatremia, resulting from the ingestion of too much plain water (in a given time) and sodium loss through sweating represents a potentially life-threatening condition. There have been incidences of hyponatremia recorded after ironman and ultramarathon races. When extreme levels of sodium are lost through sweating, there is a reduction in the total concentrations available within the body to maintain intracellular and extracellular electrolyte balance. The ingestion of a high volume of plain water (with a low sodium content) can lead to a dilution of extracellular sodium levels within the body. The dual effect of decreased sodium levels and intake of plain water can lead to an upset of sodium regulation in the body. The consumption of plain water can lead to a further reduction of extracellular sodium as sodium leaks into the intestine. The incidence of hyponatremia can be prevented through the use of an electrolyte or combined electrolyte-carbohydrate drink during prolonged aerobic exercise.

In multi-day events such as mountaineering, research indicates that sodium levels can be maintained through normal dietary intake.

Hyponatremia is an extreme and relatively uncommon condition, but a possibility that should be considered by adventure sports performers involved in long duration activities particularly in the heat.

This information should not prevent adventure performers re-hydrating, for the effects and incidence of dehydration are far more significant to performance during prolonged aerobic activities. Each individual, however, should make informed decisions about the nature of the fluid they take on board during a day's exercise.

In drawing a conclusion to the sections on nutritional and ergogenic aids to performance that can be found in each chapter, it is worth considering one further factor that can result in a performance improvement: the placebo effect. A *placebo effect* occurs when the body's response to a substance is determined by the individual's expectation of effect rather than by any physiologic effect from the substance. A placebo is often given during blind drug trials and represents an inert substance that appears identical to the substance under test. This type of control, which should ideally be double-blind to ensure that the investigator who administers the substances does not know which is the placebo and therefore does not present any subliminal signals to the participants, is essential when investigating the ergogenic effects of a new supplement. It is perhaps because of a lack of this type of control that some products released as ergogenic aids later, in controlled double-blind studies, cannot reproduce the same performance benefits.

A classic study which revealed the effect a placebo can have on performance was conducted by Ariel and Saville (1972). In this clever study the researchers recruited experienced weightlifters to take part in an anabolic steroid study. From a pool of 15 volunteers, who they had told would be selected to be administered steroids based on their performance during a 4 month pre-study period, 8 participants were randomly selected (improvements during the 4 month pre-study period were ignored) to enter the study. To further emphasize the reality of the study to the participants, 2 of the 8 volunteers were then rejected during a sham medical screening. The remaining 6 participants were then given a placebo, but were told it was in fact the anabolic steroid dianabol. The participants took the dianabol placebo for a period of 4 weeks. The volunteers continued their normal weight-training schedules during the duration of the 4 months of the pre-test and the 4 weeks of the study. Strength assessments were conducted during the final 7 weeks of the pre-treatment and during the study. The competitive nature of the pre-treatment period meant, even though the volunteers were experienced weightlifters, they made improvements during this phase of the experiment. The group improved their lifts by 11 kg during the final 7 weeks of this phase. However, during the treatment phase when they thought they were taking an anabolic steroid, they improved their lifts by a further 45 kg. This represented a great improvement for experienced weightlifters and clearly demonstrated the placebo effect.

Key point

For lower intensity aerobic endurance activities, the duration of events makes replacement of carbohydrate stores an essential part of any nutritional strategy. Carbohydrate intake should be from a variety of complex and simple sources to maintain delivery over a longer duration. In addition, the use of electrolytes drinks should be considered, particularly when exercising in the heat for a long duration, to avoid the risk of hyponatremia. Generally, whatever the nutritional supplements an athlete uses, if they believe it works it will have a positive effect on their performance. The placebo effect has been shown to have a considerable influence on performance.

10.6 The effects of hyperbaric and hypobaric condition on performance

The physiological challenges associated with the ascent of mountain peaks above 3000 m or underwater exploration involve entry into hypobaric and hyperbaric environments. Hypobaric environments, encountered at high altitude, have a lower environmental pressure than encountered at sea level. Freediving and SCUBA (self-contained underwater breathing apparatus) diving, whether for pleasure as sport diving or for commercial purposes, present a number of physiological challenges to the diver, a major one being increased atmospheric pressure. Operating at atmospheric or environmental pressures that are higher or lower than sea level presents additional stress to the adventure sports performer. This section begins with a review of the physiological challenge of exercise at altitude and then examines the effects of the underwater environment on physiological functioning during diving.

Exercise in a hypobaric environment

The sport of mountaineering involves exercise in an environment that presents a thermal and hypobaric challenge to the adventure sports performer. As described in Chapter 1, the physiological challenge of the mountain environment has been of great interest to physiologists for many centuries. As a comparative environment, altitude study has provided researchers with an improved understanding of physiological functioning at sea level. Subsequently, research has focused on improving knowledge of the challenge of the hypobaric environment and the physiological response to prolonged stays at altitude. Perhaps in no other area of physiology has the development of physiological understanding been so closely linked with the development of an adventure sport. There are many examples of physiologists who have been integral members of mountaineering parties.

Ascent to altitude provides at least three physiological stressors: exercise, altitude and cold. The general physiological response to exercise has been covered widely throughout this textbook and the thermal challenge of cold environments was described in Chapter 9. This section examines the physiological challenge of exercise at altitude. The challenge of the mountain environment is primarily concerned with the decreasing atmospheric pressure associated with increasing altitude. *Altitude* is commonly defined as commencing at a height 1500 m. Below 1500 m, increases in height have a minimal effect on performance. Above this level, rises in altitude have an increasing affect on physiological function. *High altitude* refers to heights between 1500–3000 m above sea level, *very high altitude* between 3000–5500 m and *extreme altitude* from 5500 m and above. Consequently, Mitre Peak in New Zealand represents a high altitude peak, the summit of Mont Blanc (France) is at very high altitude and an ascent of Aconcangua (Chile) would represent an extreme altitude expedition. Around 15 million people worldwide are thought to live at an altitude above 3000 m, with the limit of permanent human habitat being at around 5300–5500 m in the Andes Mountains. Humans visit extreme altitude for relatively short periods of time. The main stress associated with exercise at altitude occurs due to the sub-normal (sea level) oxygen pressure or hypoxia.

The effect of altitude on the oxygen cascade

As altitude increases the atmospheric pressure falls. Atmospheric pressure represents the weight or pressure of air pressing down on us. If you ascend a mountain the amount of air pressing down you decreases, so atmospheric pressure falls. Atmospheric pressure is measured with a barometer which, as was described in Chapter 1, was developed by Evangelista Torricelli in 1644. The earliest experiment to investigate the effects of altitude on atmospheric pressure was carried out by Blaise Pascal in 1648. In his experiment, Pascal showed that atmospheric pressure was lower at the top of the 1465 m Puy de Dôme (the venue for many Tour de France stage finishes) than in the valley below. Subsequent research has shown that atmospheric pressure continues to fall with increasing altitude. At sea level, average atmospheric pressure is 760 mm Hg. Table 10.5 provides an illustration of the fall in atmospheric pressure with increasing altitude. For every 1000 m gain in altitude there is a 86–37 mm Hg fall in atmospheric pressure. If the atmospheric pressure had been 760 mm Hg at sea level on the day of Blaise Pascal's experiment, then the pressure on the summit of the Puy de Dôme would have been around 635 mm Hg, more than sufficient to reveal a clear drop in pressure to Pascal.

It is not, however, the fall in atmospheric pressure that creates the physiological stress of altitude. The oxygen cascade was described in Chapter 4. As air moves from the atmosphere to the blood (trachea, alveoli, blood) the partial pressure of oxygen (PO_2) falls at each stage. The PO_2 is critical for diffusion of oxygen into the bloodstream. The PO_2 in the alveoli must be higher than that of the blood for diffusion to take place (oxygen diffuses from high to lower concentration). The atmospheric PO_2 and tracheal

Table 10.5 Changes in atmospheric pressure and partial pressure of oxygen with altitude ascent

Altitude (m)	Atmospheric pressure	PO_2 (mm Hg)	Tracheal PO_2 (mm Hg)	Mountain with similar altitude	1st ascent
Sea level	760	159	149		
1000	674	141	131	Yr Wyddfa, Snowdonia, Wales (1085 m)	Not known
2000	596	125	115	Mitre Peak, South Island, New Zealand (2000 m)	1910
3000	526	110	100	Mount Dixon, South Island, New Zealand (3019 m)	1931
4000	462	97	87	Piz Bernina, Alps, Switzerland (4049 m)	1850
5000	405	85	75	Mont Blanc, Alps, France (4808 m)	1786
6000	354	74	64	Kilimanjaro, Tanzania (5895 m)	1889
7000	308	64	55	Aconcagua, Andes, Chile (6959 m)	1897
8000	267	56	46	Shishapangma, Himalaya (8027 m)	1964
9000	231	48	38	Chomolangma (aka Sagarmatha or Everest) (8848 m)	1953

PO_2 are shown in columns 3 and 4 of Table 10.5. A secondary environmental oxygen cascade is clearly evident with increasing altitude. Whereas the PO_2 at sea level is 159 mm Hg, at the summit of Chomolangma (Mount Everest) the PO_2 would be around 48 mm Hg.

At sea level, the oxygen cascade results in a fall in PO_2 from 159 mm Hg to 103 mm Hg as air moves from the atmosphere to the alveoli. This decrease in PO_2 is shown in more detail in Table 4.2. The partial pressure of oxygen in blood arriving at the capillaries surrounding the alveoli is around 40 mm Hg which creates a pressure gradient of 63 mm Hg (alveolar PO_2 is 103 mm Hg – capillary PO_2 of 40 mm Hg = 63 mm Hg). Oxygen dissolves and diffuses across the gaseous exchange membrane (between the alveoli and the capillaries) as a direct consequence of this pressure or concentration gradient. This results in an almost complete oxygen saturation of haemoglobin. At sea level oxygen saturation of blood, the extent to which haemoglobin combines with oxygen, is around 97–98 per cent.

Christian Bohr (Chapter 1) found that the level of haemoglobin oxygen saturation was determined by the PO_2. He found that graphically, the oxygen saturation of haemoglobin occurs in a sigmoid (S–shaped) curve. An illustration of the oxyhaemoglobin dissociation curve is shown in Figure 10.16. The Sigmoid curve reveals that up to around 80 per cent haemoglobin saturation rises rapidly to relatively small increases in PO_2. Beyond 80 per cent saturation the curve flattens as the haemoglobin oxygen saturation moves closer to 100 per cent saturation. At a PO_2 of 103, the oxyhaemoglobin saturation is nearly 100 per cent. An increase in body temperature above 37 °C or a decrease in pH (a rise in acidity), such as occurs during exercise, cause the curve to shift to the right. This results in a decrease in oxyhaemoglobin saturation for any given level of PO_2 and is known as the *Bohr effect*.

At altitude, as illustrated in Table 10.5, the atmospheric PO_2 falls with ascent. The PO_2 at an altitude of 1000 m is 141 mm Hg, whereas at 6000 m the PO_2 is 74 mm Hg. The fall in atmospheric pressure directly affects PO_2 throughout the oxygen cascade and results in decreasing oxyhaemoglobin saturation with increasing altitude. For instance, on the summit of Mount Dixon at an altitude of 3019 m, the atmospheric PO_2 is around 110 mm Hg leading to an alveolar PO_2 of 71 mm Hg. This results in a decrease in the pressure gradient for oxygen transfer from 63 mm Hg at sea level to 31 mm Hg on Mount Dixon (alveolar PO_2 at an altitude of 3000 m is 71 mm Hg – capillary PO_2 40 mm Hg = 31 mm Hg). It is the combined effect

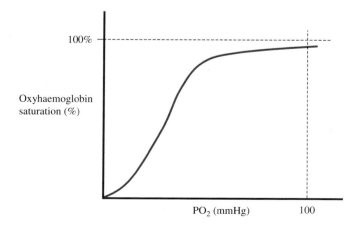

Figure 10.16   The oxyhaemoglobin dissociation curve at a normal pH 7.4 and body temperature (37 °C)

of decreased PO_2 and oxyhaemoglobin saturation that presents the physiological challenge and the hypoxia associated with exercise at altitude.

Key point

The PO_2 decreases as you ascend to higher altitudes. The thinner air at altitude presents the greatest challenge to an adventure sports performer. The relationship between the PO_2 and oxygen saturation of the blood is described by the oxyhaemoglobin dissociation curve.

Physiological response to altitude

The body's response to the hypoxia experienced at altitude can be divided into immediate responses and later adaptations that occur in the course of a stay at any particular altitude. This section examines the body's immediate response to exposure to high altitude and in the following section on acclimatization we examine the adaptations that occur.

The body alters respiratory and cardiovascular functioning in response to the lower PO_2 experienced at altitude. On arrival at altitude, the body's chemoreceptors detect the lower PO_2 in the thinner air as lower atmospheric pressure air contains fewer oxygen molecules. The chemoreceptors, located in the aorta and common carotid arteries, stimulate the alterations in cardiovascular and respiratory function.

The respiratory driver, located in the medulla oblongata and the pons (part of the brain stem), interpret the sensory information from the chemoreceptors and increase pulmonary ventilation (the rate of breathing) in response to the lower PO_2 encountered at altitude. The hyperventilation associated with arrival at altitude increases the alveolar PO_2 towards the levels in ambient air and facilitates oxygen transfer to the bloodstream. In this way, hyperventilation serves to improve oxygen supply to the muscles to counter the hypoxia experienced at altitude. The increase in pulmonary ventilation, while beneficial to oxygen supply, upsets the acid balance in the blood. Hyperventilation at altitude increases the PO_2 but at the same time decreases alveolar PCO_2. At high altitude, the PCO_2 in the lungs can fall to 20 mm Hg compared to a sea level mean value of 39 mm Hg (see Table 4.2). With pulmonary PCO_2 remaining at 46 mm Hg, this increases the pressure gradient from 7 mm Hg to 26 mm Hg. The rise in partial pressure differential results in an increased unloading of carbon dioxide at the lungs. As was described in the section on buffering in Chapter 7, carbon dioxide in the blood can be carried as carbonic acid which provides a pH lowering effect to maintain blood pH. With increased pulmonary ventilation and unloading of carbon dioxide, the level of carbonic acid in the blood falls (converted to carbon dioxide and water) resulting in a rise in alkalinity of the blood and upsetting the acid-base balance. This is known as respiratory alkalosis.

In response to the decreased PO_2, the cardiovascular system makes a number of alterations in functioning to improve oxygen supply to the muscles. After a few hours of altitude exposure, a decrease in blood plasma volume is initiated. Plasma volume can be decreased by over 20 per cent compared to sea level and this condition is maintained for several weeks after arrival at altitude. By decreasing the blood plasma volume, the remaining blood has an increased concentration of erythrocytes (red blood cells) and an increased relative oxygen-carrying capacity. This brings about an enhanced oxygen delivery to the tissues.

Within a short time of arrival at altitude, to compensate for the lowered PO_2, cardiac output ($\dot{Q}$) is increased. Cardiac output, as described in Chapter 4, is the product of stroke volume and heart rate i.e. the amount of blood ejected with each heart beat multiplied by the heart rate. By increasing $\dot{Q}$, the body can enhance oxygen transport around the body. The decrease in blood plasma volume initiated after arrival at altitude results in increased blood viscosity. The rise in blood 'thickness' – ratio of cellular components (red blood cells, etc.) to plasma – leads to a decrease in stroke volume so any rise in $\dot{Q}$ has to be brought about through an increase in heart rate. On arrival at altitude, HR rises at rest and in response to any given sub-maximal exercise intensity compared to sea level.

The increased load on the heart is only sustained for a short period of time (a few days) after which there is an increase in oxygen extraction by the muscles to compensate for the lowered blood PO_2. This enables the heart rate response to fall towards normal. The increased oxygen release to the muscles is brought about through increases in erythrocyte production of 2,3-diphosphoglycerate (2,3-DPG). Produced during erythrocyte glycolysis, 2,3-DPG has been shown to couple with haemoglobin decreasing its affinity for oxygen. This action results in a greater oxygen release to the tissues of the body and reduces the demands on the heart to increase the circulation of blood.

An additional mechanism that is thought to bring about compensation for reduced PO_2 occurs through a rise in pulmonary artery blood pressure. The pulmonary arteries are responsible for transport of the blood from the heart's right ventricle to the lungs for gaseous exchange. It appears that on arrival at altitude, pulmonary blood pressure rises through vasoconstriction. The rise in blood pressure results in an increase in blood flow to the upper sections of the lungs, enhancing the volume of blood available for gaseous exchange and oxygen loading.

Despite these compensatory mechanisms, exposure to altitudes above 2500 m can lead to a number of altitude-related illnesses. The most commonly occurring altitude illness is *acute mountain sickness* (AMS). Many people experience the symptoms of AMS when they arrive at altitude or move to a higher altitude. Symptoms normally occur within 24 hours of arrival, but subside within a week as the individual acclimatizes. The symptoms of AMS include headache, mood change, nausea, fatigue, sleep disturbance, loss of appetite and dizziness. It is thought that AMS is triggered by short-term decreases in cerebral oxygen saturation. If symptoms are intolerable, descent to a lower altitude has been shown to induce a fast recovery from AMS. Slow rates of ascent, rest days at specific altitudes and climb high/sleep low

strategies have been found to reduce the incidence of AMS.

High altitude pulmonary edema (HAPE) occurs in only a minority of individuals who travel to altitude. It is thought that HAPE occurs more commonly in response to rapid mountain ascents without sufficient time for acclimatization and as a result of hard physical exertion at extreme altitude. Symptoms of HAPE include stronger headaches (than experienced with AMS), excessive hyperventilation, disruption of bowel and bladder function and decreases in trunk control. It is thought that pulmonary edema (an excess build-up of fluid in cells or tissues of the lungs) causes damage to the gaseous exchange barrier. Descent to lower altitude also initiates recovery for an individual suffering from HAPE, although this process can take several days.

High altitude cerebral edema (HACE), although rare, is the most serious of altitude-related illnesses and requires immediate medical treatment. It develops from the symptoms of AMS which can make early diagnosis difficult. Symptoms include coordination and walking problems, severe fatigue, chest pain and impaired mental functioning lapsing to unconsciousness. It appears that HACE, which leads to increased fluid pressure on the brain, results from cerebral vasodilation leading to increased fluid and protein movement across the blood–brain barrier. To recover from this potentially life-threatening illness, descent from altitude and medical attention are essential.

Exercise performance is generally impaired at altitude. Not surprisingly, in response to the decreased PO_2, aerobic capacity as measured through $\dot{V}O_{2\,max}$ is impaired at altitude. Research indicates that $\dot{V}O_{2\,max}$ falls by 5–10 per cent for each 1000 m increase in altitude, although the percentage decline increases more rapidly at higher altitudes. At the summit of Aconcagua, at an altitude of around 7000 m, $\dot{V}O_{2\,max}$ is approximately half that at sea level. On the summit of Chomolangma, it is reduced to between 10–20 per cent of sea level values. In the same way endurance performance declines with increasing altitude resulting in longer times for race completion compared to sea level. In contrast, performance in anaerobic events (particularly power and power endurance events) is either unaffected by high altitude exposure or is improved as a result of the thinner air.

Key point

The body detects the reduced PO_2 at altitude through chemoreceptors that relay the information to the brain. In response to altitude, the body's respiratory driver stimulates hyperventilation which leads to an increase in oxygen levels but a reduction in CO_2 levels, upsetting the body's acid-base balance. To further improve oxygen availability, the body reduces the plasma volume to increase the relative red blood cell content per litre of blood. To maintain cardiac output, the body has to increase heart rate to counter the reduced stroke volume which results from the increased viscosity of the blood. Medical problems associated with altitude include AMS, HAPE and HACE, the symptoms of which can be alleviated by descent to a lower altitude.

Acclimatization

Adaptation to altitude can be simulated in a laboratory, known as *acclimation* or by a period spent living at altitude (*acclimatization*). Full acclimatization to high altitude appears to take around 2 weeks, with another week required for every additional 600 m ascended. Acclimatization to altitude occurs by a range of respiratory, cardiovascular and metabolic adaptations to the reduced PO_2. Alkalosis by hyperventilation (which remains elevated even during prolonged stays at altitude) is compensated for by increases in the excretion of bicarbonate (HCO_3^-). The decrease in blood bicarbonate associated with reducing respiratory alkalosis is thought to be partly responsible for the

decreased anaerobic performance and reduced lactate concentration experienced during prolonged stays at altitude. The decrease in blood bicarbonate concentration, however, restores the acid-base balance in the blood.

Perhaps the most significant adaptation to altitude comes through the stimulation of red blood cell production, due to the release of the hormone erythropoietin from the kidneys. Erythropoietin secretion is stimulated in response to the decreased PO_2 experienced at altitude and results in an increase in red blood cell production. Altitude-induced polycythemia (an increase in the number of red blood cells above normal) can realize increases in haematocrit from sea level values of 40–45 per cent to 60–65 per cent at very high altitude (see Chapter 4 for more information on sea level haematocrit). The rise in erythrocyte number, and as a consequence the haemoglobin concentration, enhances the oxygen transport capacity of blood. Athletes who perform at sea level have become interested in altitude training primarily as a result of this response to lowered PO_2. Erythrocyte numbers do not immediately return to sea level concentrations on return to sea level after a sojourn at altitude, providing an athlete with an increased oxygen-carrying capacity at sea level relative to pre-altitude training levels. It is consequently thought by many athletes that altitude training, which is the focus of the next section, has an ergogenic effect on performance.

One impact of acclimatization to altitude that is commonly reported is that of weight loss. As a result of a stay at altitude both fat mass and fat-free mass have been found to decline. The weight loss associated with prolonged exposure to altitude is related to loss of appetite, increased energy expenditure, impaired nutrient absorption from digestion, increased water loss due to higher rates of evaporation in the cool dry air at altitude, increased urination to decrease plasma volume and inadequate drinking to replace fluid loss. A significant factor for athletes who train at altitude, to avoid impairment of performance, is to maintain fluid intake. Dehydration is a common problem associated with a period spent at altitude.

Research into acclimatization has identified a number of additional effects associated with prolonged exposure. However, some of these findings are most likely due to the different altitudes and methods employed between studies. It is generally agreed that muscle fibres decrease in size at altitude, which is attributed to decreased training and exercise intensities associated with reduced oxygen availability. Some researchers have reported increased capillarization in response to chronic (continued) altitude exposure, while others have suggested that the increase in capillary density is the result of decreases in muscle fibre size and capillaries consequently being closer together. With regard to improvements in aerobic metabolism, researchers have found increases in mitochondrial size and number in response to a sojourn at altitude, as well as reported incidences of some level of increased aerobic enzyme activity (succinate dehydrogenase and citrate synthase). These adaptations, along with increases in myoglobin concentration, appear to contribute to improving oxidative mechanisms in response to the decreased PO_2.

Catecholamine (adrenalin and noradrenalin) secretion increases in response to altitude exposure. As you adapt to altitude, catecholamine secretion is reduced and this has an important link with an apparent *lactate paradox*. On arrival at altitude, blood lactate response is found to increase for a given form of exercise, attributed to the decreased PO_2 and increased reliance on fast rate glycolysis. One might expect lactate response to sub-maximal exercise at altitude to remain higher than at sea level, despite acclimatization, as a result of the decreased buffering capacity (associated with bicarbonate buffer excretion) and the fact that $\dot{V}O_{2\,max}$ does not increase with prolonged altitude exposure. The lactate paradox occurs because blood lactate levels at altitude decrease to a level similar to equivalent exercise intensity at sea level. This paradox is thought to exist due to the decreased adrenalin response associated with acclimatization. Adrenalin secretion with acute exposure to altitude stimulates increased glycogen catabolism (glycogenolysis) which, when combined with the lower PO_2, results in an increase in fast rate glucose metabolism and the formation of lactate. It appears that the improved oxidative functioning associated with acclimatization (increased red blood cell production, etc.) is sufficient to blunt the stimulus for adrenalin secretion which results in a decrease in lactate production. In addition, it has been speculated that after acclimatization to altitude, muscles oxidize more lactate directly resulting in a decreased release into the blood.

Key point

Acclimatization to a particular altitude takes around 2 weeks, with every additional 600 m ascent requiring a further week. The process of acclimatization leads to a reduction in bicarbonate levels in the blood to redress the acid-base balance, increased red blood cell numbers, weight loss and reduction in catecholamine response, leading to a decreased lactate response at altitude.

Altitude training

The rise in the number of red blood cells, and the consequential improvement in oxygen transport capacity associated with prolonged exposure to altitude, have led athletes to experiment with altitude training as a method for improving endurance performance. The theory is that if an athlete trains at altitude for a period of time before returning to compete at sea level, the improved oxygen transport capacity should improve their endurance performance. In other words, training at altitude results in an increase in the number of red blood cells. This leads to increased oxygen delivery to the working muscles and an improvement in sea level race performance.

The results of studies, however, have been equivocal as to the success of altitude training. Some studies have reported significant improvements in sea level endurance performance after altitude training. Others researchers, however, have not been able to produce the same results with similar groups of athletes. One major reason for the conflicting results is due to the decreased training intensity that can be achieved. Several studies have clearly shown that the exercise intensity that athletes can maintain at altitude is reduced compared to sea level. The lower PO_2 results in a decreased ability to train at the intensities that would normally be performed at sea level. The reduced training load at altitude has been found to offset the benefits of improved oxygen transport capacity. In addition, the recipe for successful altitude training is further complicated by the need to identify:

1. how long an athlete should stay at altitude to produce the best results;

2. the most appropriate altitude at which to stay;

3. how to offset the upset in an athletes normal routine enforced by a stay at altitude; and

4. the intensities of training that bring the best results.

The equivocal results achieved with pure altitude training have led researchers to investigate the benefits of live-high train-low (hi-lo) strategies. In this type of training, athletes spend most of their time living at altitude to stimulate the physiological benefits associated with prolonged altitude exposure, but return to sea level for training, thereby maintaining the intensity of training. In different studies, participants have lived at altitude for acclimatization or alternatively resided within a de-pressurized house, while some have slept in a hypobaric chamber or tent. These simulated altitude alternatives provide an opportunity for an athlete to experience the benefits of living at altitude without having to move away from their normal training environment. A study by Levine and Stray-Gundersen (1997) on the effects of hi-lo training has been summarized to provide an insight into the performance benefits of such training.

Key point

Altitude training has been used increasingly by athletes as a method to improve performance. By residing at altitude, an athlete can improve their oxygen-carrying capacity which on return to sea level has been shown to improve performance. Some research has indicated that the reduction in training intensity encountered when training at altitude offsets the improved oxygen capacity. As a consequence, the idea of living high and training low (hi-lo) has emerged as an advanced form of altitude training that maximizes the benefits of living at altitude while maintaining sea level training intensity.

Getting into research: Live high, train low (Levine and Stray-Gundersen, 1997)

Results regarding the effects of altitude training on subsequent sea level performance have been equivocal. It has been suggested that the decreased training intensity associated with exercise at altitude lessens the effect of increased oxygen-carrying capacity. An ideal solution to this problem, to obtain the benefits of altitude living while maintaining the exercise intensity associated with sea level training, would be to adopt a live high train low strategy. Levine and Stray-Gundersen conducted a classic study regarding the effects of a live high train low (hi-lo) regime compared to live high, train high (hi-hi) and live low train low (lo-lo) and their relative effects on performance.

The study, based in Dallas, Texas (150 m), but also using centres at Deer Valley Utah (high altitude, 2500 m) and Chula Vista, San Diego (sea level), was conducted with 39 male and female distance runners (27 and 12 respectively). The exact protocol for the study is show in Figure 10.17 and included a 2 week familiarization (lead-in) phase where the athletes were brought together and introduced to the types of training to be used in the test. At the same time, athletes were familiarized with all the tests to be conducted during the study. After a 4 week base line training phase in Dallas, the athletes were randomly assigned to one of the three treatments. Twenty-six of the athletes travelled to Utah for hi-hi or hi-lo training, while the remaining 13 athletes travelled to Chula Vista, designed to match the climatic conditions in Utah, but without the altitude. Tests conducted through the study included 5000 m time trials conducted on a 400 m running track, various blood measures (plasma volume, blood volume, red cell mass and haemoglobin content) as well as $\dot{V}O_{2\,max}$.

The results indicated that the hi-hi and hi-lo protocols resulted in increases in red cell mass (from 27.9–31.7 mL kg^{-1} for hi-hi and 26.2–29.6 mL kg^{-1} for hi-lo compared to a slight decline for the SL athletes after the treatment phase) and haemoglobin content (hi-hi 13.8–15.0 mg dL^{-1}, hi-lo 13.3–14.8 mg dL^{-1}, lo-lo 13.6–14.1 mg dL^{-1}). The alterations in oxygen-carrying capacity led to increases in $\dot{V}O_{2\,max}$ for the hi-hi (64.2–67 mL kg^{-1} min^{-1}) and hi-lo (62.4–66.3 mL kg^{-1} min^{-1}) athletes, whereas the lo-lo athletes actually decreased slightly (64.4–63.7 mL kg^{-1} min^{-1}).

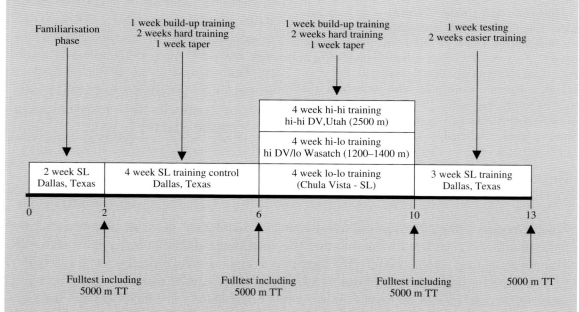

Figure 10.17 The study phases during the 13 week testing period. SL: sea level; DV: Deer Valley, Utah; hi-hi: live high train high; hi-lo: live high train low; lo-lo: live low train low; TT: time trial (adapted from Levine and Stray-Gundersen, 1997)

The physiological changes upon return to SL, however, led to improved performances (5000 m race time) for only the hi-lo athletes. The performance of the hi-hi and lo-lo athletes decreased with slightly slower mean times recorded than were achieved prior to the treatment phase. The 13.4 s improvement for the hi-lo athletes represented a statistically significant difference than for the hi-hi and lo-lo athletes times. The results obtained in this study indicate that hi-lo training has a performance advantage for endurance athletes. Living at altitude (2500 m), but training at a lower altitude (1200–1400 m) results in improved endurance performance compared to hi-hi and lo-lo training strategies. This study provides important findings for consideration by coaches and performers taking part in competitive endurance adventure sports.

Exercise in a hyperbaric environment

Breath-hold diving represents the earliest form of exploration of the environment below the open water surface. The earth's hyperbaric environment presents a unique physiological challenge to humans. As was described in the section on hypobaria, the body is under an average pressure of air while standing at sea level that is sufficient to raise a column of mercury 760 mm. The pressure is created by the air in the atmosphere 'pressing' down on objects at sea level. In diving terms, this is known as one atmosphere or 1 bar. The density of water is 775 times greater that of air. As a consequence, you need only descend a short distance to reach a point at which the pressure exerted on the body is double that encountered at sea level. Figure 10.18 provides an illustration of the impact of underwater depth on pressure with its effect on the volume of a gas. As can be seen from Figure 10.18, the environmental pressure doubles at a water depth of 10 m. The pressure at this point is referred to as 2 bar and represents that of two atmospheres (one atmosphere for air and one atmosphere for the 10 m of water).

The pressure associated with increasing depth of dive presents one of the main challenges to the diver. An increase in water depth results in a linear decrease in the volume of any gas, as demonstrated in Figure 10.18. The law that describes the relationship between gas volume and pressure was discovered by Robert Boyle (Chapter 1). When a gas held at a constant temperature is placed under increasing pressure its volume decreases (the gas molecules are forced closer together into a smaller area). The effects of increases in pressure on the volume of a gas are shown in Figure 10.18, such that at 2 atmospheres the gas volume halves and at 4 atmospheres the gas volume is a quarter of its sea level volume. As well as the pressure associated with underwater exploration, and its effect on the body's gases, the key challenge is lack of oxygen available underwater. Freedivers overcome this challenge by breath-holding during submersion, whereas SCUBA divers and submariners carry an oxygen supply with them. A third challenge associated with underwater exploration is that of the cold temperatures encountered during water immersion; this aspect of diving was covered in Chapter 8. This section examines the physiological challenge of breath-hold diving and SCUBA diving.

Freediving (breath-hold diving)

Breath-hold diving has a recorded history dating back around 5000 years. Carried out for pleasure, food, salvage and military purposes, perhaps the most famous breath-hold divers are the Japanese women Ama divers. Although there are less than 1000 Ama divers today, in the 1920s there were as many as 13 000 Ama divers who worked collecting pearls. Ama divers descend to a depth of around 20 m, carrying a lead weight on a cord attached to their waists which they use to assist their

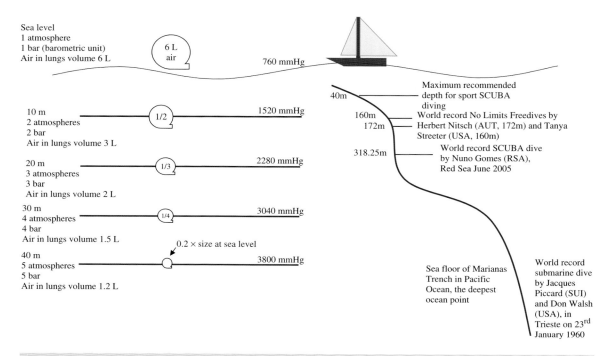

Figure 10.18   Increasing atmospheric pressure with increasing depth and its relationship with gas volume

descent. On return to the surface the Ama divers off-load collected shells and ready themselves for another dive, often making 50–100 dives per day. Unsurprisingly, a relatively high incidence of ear drum ruptures and ear infections have been recorded for Ama divers.

From these origins, breath-hold diving has grown into the competitive sport of freediving which is governed by the International Association for the Development of Apnea (absence of breathing) now often referred to as the International Association for the Development of Freediving (AIDA). There are eight different categories of freediving world record and a world championship held every year. The fifth AIDA World Championships were held in December 2006 in Eygpt. The world records for men and women in each of the eight AIDA disciplines are shown in Table 10.6.

The *static apnea* record involves breath-hold without moving in a swimming pool or open water with the respiratory tract (mouth and nose) covered while submerged. This is the only record where

the measurement is of time rather than distance. *Dynamic apnea* (with and without fins) can also be conducted in a swimming pool, although the minimum pool length must be 25 m and involves swimming underwater for as far as possible. When using fins, the record can be attempted using long fins or a monofin. *Constant weight* freediving (with and without fins) is the most basic of the depth descents where the competitor uses their body and any ballast they wish to carry to assist the descent. With a rope to guide the dive, they must return to the surface carrying the same weight with which they descended. *Free immersion* involves the use of a rope, without swimming propulsion, to pull hand-over-hand as far down as possible before using the rope to ascend. In *variable weight* freediving, a weight is attached to the diver or a 'sled' is used to enhance their descent. In *no limits* diving a sled is used for descent and ascent is completed using a compressed air-filled balloon or compartment in the diver's wetsuit, released on reaching the maximal descent point.

Table 10.6 Freediving world records

Discipline	Men's record	Year	Record holder	Women's record	Year	Record holder
Static apnea	08:58:00 min	2004	Tom Sietas (GER)	7:30 min	2006	Natalia Molchanova (RUS)
Dynamic apnea (no fins)	180 m	2006	Tom Sietas (GER)	131 m	2005	Natalia Molchanova (RUS)
Dynamic apnea (fins)	212 m	2005	Tom Sietas (GER)	200 m	2006	Natalia Molchanova (RUS)
Constant weight (no fins)	80 m	2005	Martin Stepanek (CZE)	55 m	2005	Natalia Molchanova (RUS)
Constant weight (fins)	108 m	2005	Martin Stepanek (CZE)	86 m	2005	Natalia Molchanova (RUS)
Free immersion	106 m	2006	Martin Stepanek (CZE)	80 m	2006	Natalia Molchanova (RUS)
Variable weight	140 m	2006	Carlos Coste (VEN)	122 m	2003	Tanya Streeter (USA)
No limit	172 m	2005	Herbert Nitsch (AUT)	160 m	2002	Tanya Streeter (USA)

These records are exceptional, when it is considered that average breath-hold in a normal untrained adult is around 1–2 min. One of the key aspects that enables such endurance records to be set, as is used by Ama divers, is hyperventilation. Hyperventilation, as was introduced in the section on the hypobaric environment, results in a 'blowing-off' of carbon dioxide and a reduction in the alveolar and blood PCO_2. In breath-hold diving, hyperventilation is used to decrease the carbon dioxide levels within the body just prior to submerging. As described in Chapter 4, carbon dioxide levels detected by chemoreceptors in the common carotid arteries and close to the medulla oblongata and pons (the location of the respiratory driver) provide the primary stimulus for increasing breathing rate during exercise. By hyperventilating prior to submersion, breath-hold divers are able to reduce carbon dioxide levels in their blood which helps to decrease the respiratory drive and extend dive time.

This technique is not without risks, however, and has led to the death of those involved in breath-hold swimming. The decreased carbon dioxide levels brought about through hyperventilation, while delaying the desire to breathe, can lead to fainting before the PCO_2 has increased to a level sufficient to stimulate the desire to take a breath. These black-outs appear to be linked to a fall in PO_2 below a critical level in the alveoli and blood. It appears that hyperventilation increases the risk of oxygen depletion during a dive which can lead to fainting. A further risk, linked particularly to diving in colder waters, is the slowing in heart rate associated with the dive reflex. The dive reflex is described in more detail in Chapter 8 in the section on water immersion. In breath-hold diving, it is believed that performers can face problems associated with cardiac arrhythmias due to the diving reflex. Success in freediving has been linked to total lung volume, with those having larger lung volume showing significantly improved breath-hold times.

Key point

Breath-hold diving is the oldest form of exploration of the hyperbaric environment. In the early days of breath-hold diving, descents were mainly made for commercial or military purposes. The sport of freediving emerged during the 1970s as an increasingly popular adventure sport for which there are eight world record categories maintained by AIDA, the international governing body for the sport.

SCUBA diving

In the early days of diving, the air for breathing was supplied from the surface. It was not until 1943 that two French divers, Jacques Cousteau and Emile Gagnan, developed a system whereby the air supply was carried with the diver. The development of the SCUBA system enabled divers to explore the underwater environment without the need for a supply line from the surface. Originally developed for commercial purposes, one of the earliest uses of SCUBA involved de-mining at the end of the second world war. After the excellent underwater films produced by Jacques Cousteau, SCUBA diving became a popular recreational adventure sport.

The SCUBA system comprises four components that enable breathing underwater: an air tank, two regulators and a breathing valve. The compressed air for breathing is contained in one or more tanks which are carried on the back. The air tanks are normally compressed to around 200 bar (3000 psi). This pressure is far too great for breathing purposes and so two regulators are fitted to the tank to down regulate the pressure. The first regulator decreases the pressure of the air to about 9.5 bar (140 psi) and the second, which supplies air on demand, matches the air pressure to the pressure of the water around the diver. Finally, a one-way mouthpiece enables air to be inhaled from the regulator and then exhaled into the water, rising to the water surface as bubbles. This type of SCUBA is called open-circuit, because the exhaled air is released into the water. For military stealth purposes a second expired air collection tank, normally worn on the chest, has been used to avoid the release of air bubbles into the water. This type is known as a closed-circuit system.

A variety of air mixtures are used in diving, including: pure oxygen, nitrox (oxygen and nitrogen) and heliox (helium and oxygen) to extend the duration and depth of diving possible. Using a compressed air system, the length of a dive is normally limited by any previous dives that day, the number of air tanks carried, depth of dive and efficiency of movement and air usage. 'Over-breathing', or inefficient use of air during a dive, can more than half the possible submersion time. The pressure encountered in SCUBA diving provides the main potential hazard associated with this type of adventure sport. The complications that can be associated with SCUBA diving include the ear, sinus and mask squeezes, pneumothorax, nitrogen narcosis and decompression sickness.

A *squeeze* can occur when you descend and pressure increases upon air trapped within the body or between your face and the face mask. The pain is caused by the air trying to equalize pressure on either side of a block. Normally as you descend you need to clear your ears at regular intervals by blowing gently with your nose blocked, stretching your jaw muscles or yawning to allow the pressure to equalize between your middle ear and the external pressure. If you have a blockage of the eustachian tube it may not be possible to equalize pressure, causing pain. If you cannot equalize the pressure you should not continue to dive; it is possible to rupture the ear drum if the air pressure is not equalized. Normally with eye and nose masks you do not get a squeeze with SCUBA diving; however, if you cannot clear your sinuses, not only will it lead to pain but can also create the potential for a face mask squeeze. To avoid the pain associated with a squeeze you should not dive when you have a cold or other upper respiratory tract infection.

When ascending from a dive, it is important to continue to breathe normally as you rise. While exploring the underwater environment during a dive you are breathing air under pressure equal to that at the depth you are moving. Divers at times of stress have sometimes held their breath on ascent, leading to serious lung problems. The volume of air in your lungs increases during ascent but if you continue to breathe normally and concentrate on exhaling, the air will leave your body without any problems. If, however, you breath-hold on ascent the volume of air inside your lungs will increase which can a cause the alveoli to burst. When the alveoli burst it can lead to air entering the pleural space (see section on the structure of the lungs in Chapter 4). As an ascent continues, the air in the pleural space will continue to expand creating pressure on the lung tissue and causing a sudden collapse of the lung known as a *pneumothorax*. By simply maintaining exhalation on ascent, a diver can avoid this problem.

In deeper and longer duration dives, nitrogen (an inert gas during respiration) can begin to have an intoxicating and anaesthetic effect on the body. The increase in pressure associated with deeper dives establishes a higher partial pressure of nitrogen (PN_2) leading to more entering the body. Known as *nitrogen narcosis* or *rapture of the deep*, the euphoria associated with this condition can impair judgement and functioning. It has led divers to continue to descend when they should ascend, believing they have sufficient air for a deeper dive, and on occasions to remove their SCUBA mouthpiece believing they no longer need it. As such, it is a potentially life-threatening risk associated with SCUBA. However, the symptoms quickly disappear if the diver ascends.

Another potential nitrogen-associated problem during deep and prolonged dives occurs when the diver ascends too quickly. During a longer dive, nitrogen slowly enters the tissues of the body due to the increased PN_2. When ascending slowly, the nitrogen can pass harmlessly from the tissues as your rise. If, however, you rise rapidly by mistake or in an emergency, the nitrogen dissolved in your body does not have time to leave your body harmlessly. In a rapid ascent, nitrogen that was in solution forms bubbles in the tissues, fluids and joints of your body. This condition is known as *decompression sickness* or *the bends* due to the effect it has on the body. A diver on the surface with the bends is often seen to bend over to try to relieve the pain from nitrogen in the joints. Treatment for decompression sickness is essential and requires immediate use of a recompression chamber. In a recompression chamber, the pressure can be returned to a level such that the nitrogen bubbles re-dissolve into the tissues and can then leave the body slowly as the pressure in the chamber is gradually reduced. Although more common now, recompression chambers are not normally available to sport divers and therefore great attention should be paid to the rate of ascent to avoid an incidence of the bends.

Key point

One of the main physiological challenge in self-contained underwater breathing apparatus (SCUBA) diving relates to the pressure encountered in the earth's hyperbaric environment. SCUBA, developed by Jacques Cousteau and Emile Gagnan, enables humans to move underwater without the restrictions of a connection tube with the surface. SCUBA includes two regulators that decrease the pressure in the compressed air tank to that of the water at any particular depth. The chief problems associated with diving relate to the effects of pressure on gases and close monitoring of dive time along with a slow ascent can avoid these concerns.

10.7 The physiology of mountaineering

Mountaineering potentially presents the greatest number of challenges to homeostasis of any adventure sport. The stresses associated with mountaineering include the physical stress of exercise, altitude, cold and upset of diurnal rhythm due to the need for alpine starts as well as difficulties encountered with maintaining hydration status due to increased evaporation at altitude and decreased water intake. On some routes, the alpine ascent style of climbing (fast up and fast back) has led to climbers limiting the amount of food they carry with them which can clearly lead to problems with glycogen depletion during the 10–15 hr or more required for a typical ascent. So why do

Biography: Des Marshall

Adventure sport: Caving

Des, who began his fascination with the outdoor environment when he started mountain walking at the age of 7, went on to take up rock climbing and caving in his teenage years. He has climbed and caved throughout Great Britain and around the world including expeditions to Kenya, Hong Kong, Nicaragua, Morocco and the Himalayas. He holds the Caving Instructor Certificate and Mountain Instructor Award. He has been involved in instructing outdoor education for over 30 years and remains a very keen caver.

Untamed river caving expedition

Some summers ago I was invited to take part in a caving expedition to Papua New Guinea to explore the huge underground river of Nare on the Island of New Britain. This came as a complete surprise to me and the horror of how was I going to afford this quickly set in. I was invited on the trip as much for

Figure 10.19   Des Marshall

my climbing expertise as for my caving. The area was not a new one for caving; French cavers had been there 4 years before us but were repulsed about a kilometre from the entrance by the huge underground river which, at the time, was the largest such known river in the world. The trip presented a logistical nightmare to get people and equipment to the cave site in the middle of the rain forest. I knew that illness and food had been a problem with the French cavers, which probably had an effect on their failure to reach the end of the cave which they called 'La Rivière Indomptée'. The challenge had been set as there was much friendly rivalry between the French and British cavers. As a result 'The Untamed River Expedition' was born.

The sheer logistics of getting 5 ton of equipment to the cave half-way around the world presented a major problem. Fundraising and obtaining sponsorship became everyday occurrences which took a toll on day-to-day living. Not only had one to earn money to live, money had to be found to fund the trip. Relief about funding the expedition dissipated once we had gained approval from the Royal Geographical Society, as many sponsors came forth to offer us all manner of things including air fares and a container from a shipping company. As a result, the expedition ended up with a remarkable set of kit and some superb food, which was augmented by food obtained from the forest. This was found by the local indigenous Ira population who were instrumental in keeping us healthy by their deliveries of fresh forest foods. Many of these leaves and plants were totally unknown to us: aibika, for example, a fern-type leaf that when cooked left a juice which was a most refreshing drink as well as tasting like spinach. Food was of paramount importance. Without good wholesome meals, illness would have been a major possibility. Our support team even baked fresh bread in an oven made in the side of a dried-up stream bed. This was a real morale booster.

New caving techniques had to be evolved prior to our departure to enable the team to cross the Nare from time to time. Many training exercises were undertaken at various large rivers in North Wales to perfect these skills. Even during the training sessions, there were several narrow escapes and the thought of something going wrong on the expedition was somewhat worrying. After 6 days of travel, including a 2 day hike through rain forest, the cave was located. We set our base camp conveniently 10 min away from the cave and our daily initial 'commute' down the 305 m entrance shaft to the thunderous Nare River. Although taking only 25 min to abseil down this awesome shaft, it was an absolute minimum of an hour to prussik out! This added greatly to the effort of exploring the cave with some days in excess of 10 hr of continual physical effort to be completed.

After several days of this, each of us was exhausted. As such it was obvious that an underground camp was necessary and the first was set up at the base of the shaft to enable us to push harder and further for longer without the need to return to the surface. Freshly baked bread from the camp oven and bananas were lowered to us each day and eagerly eaten. The team at the surface base camp supplied our every need by lowering all food supplies and caving kit down the huge shaft. Spirits also rose when mail arrived from England – it took only 8 days from Manchester to the bottom of our shaft in the middle of this 99 per cent rain forest covered island! Due to the continual pressure of pushing into the cave and ensuring that everything was securely fixed, progress was only slowly made. Each time we arrived back in camp at night we were wet, very wet. Although the temperature inside the cave was constant at around 15 °C, the atmosphere was always damp. We were always putting on wet clothes for our pushing trips but, fortunately, due to the effort involved in the exploration and the concentration, we very quickly warmed up and forgot about the misery. Nothing could dry in that environment. It was possible, but only just, to keep spare clothing dry to change into upon our return to camp.

We were spending up to 5 days underground at any one time but sometimes one or another of us had enough as it was all extremely nerve-wracking: the noise, the doubt of where we were going, the worry of flooding and whether we were all going to go back home. When someone felt like this we simply headed out for a breather, contemplation, relaxation and there were always unlimited food supplies and huge meals back on the surface. Underground evening meals were great as were breakfasts but lunch was more often than not a chocolate bar, a few nuts and a swig of water. Even though water was everywhere, underground dehydration was a problem. After 2–3 days on top we descended once more into the exploration phase for another 5 day spell, refreshed, calmed and eager.

Figure 10.20   Des Marshall caving in Papua New Guinea

Even with constant care and concentrating to ensure safe attachments, there were a number of incidents which halted us from time to time. Ropes had to be strung across the river to enable us to keep above the crashing torrent. Failure of any fixing would mean death to anyone on the 'tyrolean'. This type of action focuses the mind and is always mentally stressful as well as physically demanding.

A couple of the team had near-drowning experiences and falling rocks nearly wiped me out at the bottom of the entrance shaft. In these instances, the team above back at base camp became as close as family. After the falling rock incident, camp was sensibly moved and established under the 150 m high entrance portal of the cave. There were 12 of us on the trip with a split between a pushing team of 6 and a support base camp team of 6. Being an integral member of the pushing team, I was always extremely grateful of the surface team and their efforts in every respect. It was largely due to good logistical support from them that we were eventually to reach the end of the cave.

After a month of hard caving we soon found out why the French had stopped. I had never seen so much water anywhere and never have since: a roaring, swirling maelstrom crashing down a 10 m tube at over 22 cumecs. (Once past the huge entrance, the passage until then had been up to 40 m high and 20 m wide). We found a sneaky way to cross over the river before the tube. This then allowed us to traverse a wall to bypass this seeming impasse, to be met by an even greater swirling torrent. This was the end as we know it – there was *no* way past this one. We had gone further than the French team, but only by 200 m. It was a very important 200 m, however, and one that we were absolutely overjoyed with.

mountaineers undertake the challenge? There is still perhaps no better reason than that given by George Mallory when asked about why he wanted to climb Everest: 'Because it's there'.

Hohlrieder *et al.* (2005) recently investigated the benefits of avalanche transceivers in avalanche accidents in Austria. The researchers conducted a retrospective study examining the use of transceivers in 194 accidents that involved snow burial of 278 casualties (skiers, snowboarders and mountaineers). The investigation revealed that over 50 per cent of those buried during an avalanche had used a location transceiver (156 out of 278). The use of a transceiver was linked with a significant decrease in burial time and mortality. The use of transceivers did save lives when carried during an avalanche. The study also indicated a significant rise in mortality rate with depth of burial over 1.5 m, so the researchers stressed the importance of avoidance measures over reliance on an avalanche transceiver.

Several studies have examined the effects of different supplements or prescription medicines in the prevention or reduction of symptoms of acute mountain sickness (AMS). Free radicals (Chapter 2) have been linked to damage to the blood-brain barrier which in turn has been implicated in the incidence of AMS. As a consequence, there have been many investigations to examine the potential of antioxidants to combat the action of free radicals on the blood-brain barrier and examine if, by this action, they decrease the incidence of AMS. A study by Bailey and Davies examined the prophylactic effects of the antioxidants *L-ascorbic acid* (Vitamin C), *dl-α-tocopherol acetate* (the synthetic form of Vitamin E found in many supplement tablets) and *α-lipoic acid* (a common dietary supplement associated with vitamin C and E that has an antioxidant effect) on the incidence of

AMS. In a double-blind study, 18 participants were divided into two groups and provided with either antioxidant tablets or a placebo for 3 weeks at sea level and over a 10 day ascent to Chomolangma basecamp at 5180 m. Their results indicated that those using the antioxidant supplement scored lower on the Lake Louise scale for symptoms of AMS (http://www.medex.org.uk/new_page_5.htm) and had higher oxygen saturation levels. From this study it would appear that antioxidant supplements provide a potentially effective intervention to counteract AMS.

In another supplementation study, Bartsch *et al.* (2004) investigated the use of ginkgo biloba and theophylline in the prevention of AMS. Ginkgo biloba, derived from trees, is reported to improve blood circulation and decrease-free radical damage. Theophylline, found naturally in green and black tea and structurally similar to caffeine, is used to treat respiratory diseases such as asthma. The researchers found that neither ginkgo biloba nor theophylline prevented the symptoms of AMS, although theophylline did have an effect on reducing the periodic breathing associated with sleep at altitude.

Bartsch *et al.* (2004) concluded that, at the time of the study, the best substance to treat the symptoms of AMS was acetazolamide (sold commercially under the name Diamox®). This finding was supported by research conducted by Birmingham Medical Research Expeditionary Society (BMRES) who conducted a study on the effects of acetazolamide on a variety of measures during a trek to 4846 m. For the 21 participants divided into two groups (acetazolamide and placebo) in a double-blind study, beneficial effects on exercise performance and preservation of muscle mass during ascent were reported for the acetazolamide group.

In other studies of AMS, researchers have examined the use of a pulse oximeter as an alternative indicator of altitude sickness. During a research expedition to Falcon Kangri (8047 m) in the Karakorum conducted by Tannheimer *et al.* (2002), pulse oximetry measures were recorded for the 13 participants. Falcon Kangri, first climbed in 1957, was originally named K3 because it was the

third mountain measured in the Karakoram range. It is now more commonly referred to as Broad Peak as its summit is over 1 mile long. A pulse oximeter, which is used to measure arterial blood oxygen saturation, is depicted in Figure 10.21.

The researchers reported a dramatic reduction in oxygen saturation compared to sea level measures (97–80 per cent saturation) when the participants reached an altitude of 7100 m. Oxygen saturation decreased to 59–65 per cent at this altitude. In addition, they reported a significant correlation between pulse oximetry saturation levels and the incidence of altitude sickness. They concluded that oxygen saturation levels could be used as an alternative AMS diagnosis tool. However, O'Connor *et al.* (2004) found that only heart rate was correlated with reported AMS incidence (using Lake Louise score), concluding that pulse oximetry does not assist in the diagnosis of AMS.

This type of disagreement between studies can also be found in research relating to the lactate paradox (introduced earlier in this chapter). Although the lactate response to exercise at altitude increases initially compared to sea level equivalent exercise, after acclimatization lactate response at altitude falls close to sea level response. It gets its name because the athlete is still exercising in a reduced oxygen environment. Van Hall *et al.* (2001) reported a difference in the lactate paradox after 8 weeks at altitude. This research group found that while lactate response pattern matched that established during early acclimatization, after 8–9 weeks at altitude the lactate response increased again.

Pronk *et al.* (2003) conducted a follow-up study to further investigate this finding and reported that, for their participants, no such secondary lactate rise occurred. The disagreement between research groups over any particular finding, whether it be related to the lactate paradox, use of pulse oximetry, use of antioxidants to prevent AMS or any other aspect of research, may relate to differences in the participants (between studies), the study design, data collection methods or any other variable. The emphasis for us as readers of published research is not to accept the results of one study at face value. We must be critical readers: how do the

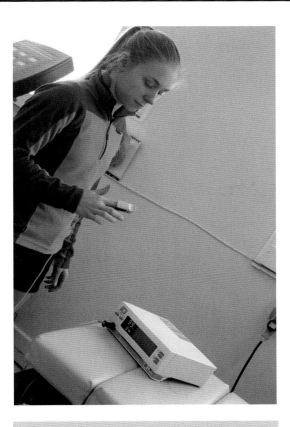

Figure 10.21 Illustration of a pulse oximeter

findings fit in with the current body of knowledge, what are the implications of using the findings for altering our performance strategy in our adventure sport and do other studies report conflicting findings? Because something is printed in a peer-reviewed paper or journal does not mean that it is either true or, if conducting the research yourself, that you would find similar results. It is often useful to find out whether the research has been sponsored by a particular institution or company as this can influence the nature of a study design and the reporting of results.

The discontinuous nature of mountaineering, resulting from exercise and rest intervals or pitched climbing, presents a challenge to clothing systems employed during a mountain ascent. A growing body of research has examined the most appropriate clothing system to cope with the dynamic thermal challenge associated with mountaineering. A balance must be maintained between the heat production by the mountaineer in response to the exercise demands and the heat dissipation from the clothing worn. During exercise the primary concern remains dissipation of heat, whereas during a stationary period, e.g. belaying, the emphasis becomes heat conservation. The challenge of any clothing system for mountaineering is to cope with these potentially contradictory demands.

Adventure sport clothing manufacturers have generally proposed a layering system to maintain temperature during work and rest intervals. The basic layering system consists of three layers: a wicking base-layer, an insulative mid-layer and a protective outer-shell. Wicking materials are designed to move moisture away from the skin to avoid moisture build-up, wetting out of clothing and the deleterious effects of moisture evaporation from the skin. The mid-layer, normally composed of synthetic fleece, has relatively long fibres which are designed to trap air and improve the clothing system's insulative properties. The most technically complex layer is perhaps the outer-shell which must combine protection against rain and wind with release of sweat produced during exercise (breathability). Heat regulation is maintained by adding or removing layers during exercise. Research has highlighted the benefits of this system for wicking moisture away from the body, thereby protecting core temperature.

This system has recently been criticized due to the inconvenience of adding and removing mid-layers during exercise. The discontinuous nature of mountaineering can necessitate frequent removal of outer and mid layers to maintain an appropriate level of heat dissipation or conservation. In a mountaineering context, this system can become impractical due to the time taken and the heat loss during clothing changes. The layering system can be further complicated by the need to wear a harness during an ascent. Each clothing change could require releasing the harness to remove or add a layer. This can lead to a mountaineer missing out layering changes which most problematically

can result in excess sweat loss with implications for wetting out insulation leading to problems with body temperature maintenance, particularly when stationary. In response to the problems associated with a classical layering system, climbers such as Mark Twight (1999) have suggested an alternative that perhaps provides a more practical approach for mountaineering.

The Twight layering system involves using a base-layer and an outer-layer but no mid-layer. In this way the mountaineer remains cooler during exercise, thereby reducing sweat rate and build-up within clothing. During a rest period, or when belaying, the climber would add an outer synthetic or down layer to maintain core temperature. Another advantage of this system is its practicality, in that a down or synthetic layer can be added over a harness, reducing the need for removal of the climber's harness.

Preliminary research into this method suggests that such a system has a major potential advantage for the mountaineer. In this study, nine participants completed a 1 hr intermittent exercise protocol (20 min walking, 10 min stationary, 20 min walking, 10 min stationary). Core temperature, thermal comfort and heart rate were record throughout the study. Figure 10.22 provides a clear illustration

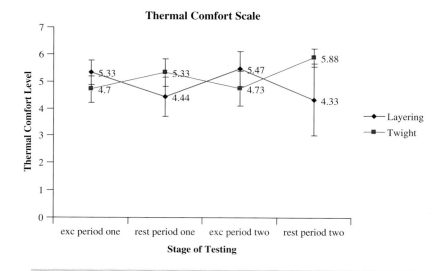

Figure 10.22 Mean standard deviation for the thermal comfort scale

of the advantages in thermal comfort achieved using the Twight layering system over classical layering. Thermal comfort was reduced during exercise but increased relative to a traditional layering system during the rest periods. The use of the Twight layering system may provide a beneficial alternative to traditional layering for many forms of mountaineering. The adage of 'you need to stay slightly cold to stay warm in the mountains' works very well for maintenance of core temperature and fits with the use of the Twight layering system.

Key point

Research regarding mountaineering is growing, especially since the establishment of specialist journals such as High Altitude and Medicine Biology. Acetazolamide (Diamox®) has been shown to reduce the symptoms of AMS. Antioxidant supplementation, such as Vitamins C and E, taken prior to and during high-altitude ascent, may provide an alternative to acetazolamide for the prevention of AMS. Avalanche transceivers do appear to result in decreased mortality rate after avalanche burial, although prevention measures are stressed by researchers. The Twight layering system may provide a useful alternative strategy for maintaining core temperature compared to classical layering.

Biography: Bob Sharp

Adventure sport: Hill walking/Munro bagging

Bob Sharp has a doctorate in skill psychology and was a student of John Whiting at Leeds University in the early 1970s. He has been a Lecturer, Course Director and Head of Department at Jordanhill (Strathclyde University) for 32 years and is presently employed as a Reader in the Department of Sport, Culture and the Arts. A long time lover of the outdoors, he developed one of the UK's first degrees in Outdoor Education. He has served as Vice-

Figure 10.23 Bob Sharp

President of the Mountaineering Council of Scotland, Chair of the Scottish Mountain Safety Forum, Director of the Scottish Mountain Leader Training Board and Secretary to the Mountain Rescue Committee of Scotland. He leads one of Scotland's mountain rescue teams and is Munroist number 981!

Confessions of a Munro Bagger

If you're a Munro Bagger then you're one of few hardy outdoor types (some would say, sad) whose mission in life is to climb every Scottish mountain over 3000 ft (of which there are 277). The Munros were first climbed in 1901 and in the following 80 years, some 240 had climbed every mountain. At the time of writing, over 4000 people have achieved this remarkable feat. Munro bagging is a very popular activity! My memories of the first round (I'm half-way through the second) are mixed with emotions and experiences in abundance: pain, pleasure, rescue, hunger, competition, apprehension, humour, thirst, heat, cold, fatigue, inebriation, friendship, tiredness, cooperation, risk, anxiety, worry, history, satisfaction, absurdity and so on. If you want to explore the rich tapestry of life and open every sense to the full, then try Munro bagging!

A lot happened on my first excursion to the Scottish Highlands. Following an off-the-cuff comment over coffee about being a 'bit of a mountaineer', my work colleagues talked me into joining them on a trip to Glencoe. I suddenly found myself having to prove my worth. January 1979 proved to be exceptionally cold, but a blocking high-pressure system made for clear skies, brilliant views and reliable snow and ice. An early start from home saw the three of us tumble out of the car at the bottom of Buachaille Etive Mhor around 7.30 am. The first lesson was an easy one: a hard day on the hill should be preceded by a good night's sleep and strict attention to nutritional needs. We had failed on both accounts.

The snow-covered heather made movement difficult from the start. After only a few hundred metres of stumbling, I was just about all in. But my friends didn't seem to bother and chatted happily as I toiled and perspired furiously. At least the views kept my attention focused every time I stopped to recover. The mountain never seemed to get closer and it wasn't until the ground began to steepen that I began to sense an increasing closeness. With height, the snow hardened and progress became more comfortable. I also remember there came a point when things seem to get much easier. My body was beginning to adapt to the underfoot conditions, temperature and pace.

The chosen route (Curved Ridge) is a popular scramble in summer but a serious undertaking in winter conditions and my 'friends' had failed to alert me to the risks about to unfold. But at least they moved in close as we tiptoed across the ice sheets and skirted crags with menacing drops. Eventually we came to an impasse with steep cliffs all around. At this point we all stopped to don helmets and harnesses. Why did I brag that I was a climber? One of my colleagues took off up the nearest crag trailing a rope. After a short time he called me to tie on and follow. I found the going surprisingly easy. The short lesson in using crampons paid immediate dividends and I 'romped' up the snow and ice-encrusted rock with surprising ease and confidence. But the pleasure was short lived. At the top, I untied and suddenly went into serious oxygen debt. It became obvious that the effort of climbing the near-vertical 100 feet had been carried out with scant attention to breathing. I suspect the fear and apprehension of climbing in such an unusual manner had masked my normal breathing pattern. It took a while to recover. The remaining ascent involved spells of strenuous muscular activity followed by regular walking sections that permitted a degree of recuperation. But I do remember that as the summit appeared and, with only a few minutes to go, I felt absolutely jiggered.

This had not only been an unusual experience in wild surroundings but a physical strain of monstrous proportions. We had rarely stopped to eat or drink and rests were infrequent and brief. In later years I learned of the need for speed when travelling in winter, but also the need to replenish fluid and carbohydrate loss on a regular basis. Further, I was not fit for climbing, especially in such arduous, physical conditions. I considered myself to be a runner/athlete which clearly demanded a different form of fitness. Indeed, I recall some years later taking a group of students on a summer walk to bag a few Munros. Two in the party were professional footballers at the time – and did they struggle!

The views from the summit were simply stunning, but the biting northerly wind hastened our departure. Exposure to the cold and wind chill are easy killers in the winter mountains. The brief stop had eased my aching muscles but the lactic acid had remained. The descent was tortuous in the extreme – broken ground, steep crags, ice sheets, loose snow. I stumbled and slid and constantly lost my footing. However, my colleagues kept a watchful eye on me and the banter kept spirits high.

We had been on the hill for a good seven hours and daylight was fast disappearing as we began the final haul uphill to the road. To climb up again was utter purgatory. My legs would hardly move and every step was painful. In contrast, my colleagues chatted away with no apparent discomfort. It took several days to recover, but during this time, the initial feeling that I would never repeat such an ordeal was replaced with a burning desire to do it again!

In the process of becoming a Munro bagger, I had experienced a comprehensive lesson in exercise physiology. I had learned about the relationship between exercise and fluid balance and the need to restore blood sugar levels when working over a long period. I had learned about the specificity

Figure 10.24 Bob Sharp, Munro enthusiast, bagging a winter peak

of training both as an individual and in regard to type of activity. As new Munros were climbed over the following months and years, I began to appreciate the long-term benefits of exercise and the need to pace. I could now climb and chat to my pals at the same time and rarely suffered fatigue even on multi-day expeditions. I also learned to cope with extreme weather conditions by choosing suitable routes and using appropriate equipment and clothing. But one lesson remained until the final Munro – how to descend a mountain safely with half a bottle of whiskey sloshing around in your belly!

10.8 Summary and study questions

Aerobic energy production

The aerobic energy system is the dominant energy system for many adventure activities. Although it is a slower way to provide energy for exercise, it provides a relatively inexhaustible energy supply. It also enables the body to realise ATP directly from all three macronutrients (carbohydrates, fats and protein). During digestion, carbohydrates, fats and protein are catabolized into glucose, triglycerides and amino acids.

The primary mechanism for aerobic catabolism of fatty acids is through β oxidization. This process is dependent on the availability of oxygen. In order to derive energy from protein, the amine group needs to be removed from the molecule. This deamination leaves a carbon-based skeleton which can be used to generate energy. There are 20 amino acids and so the carbon-based skeleton will enter the Krebs cycle at an appropriate point for its chemical structure.

The Krebs cycle consists of nine chemical reactions which take place within the matrix of the mitochondria. Carbohydrates and fats are broken down and enter the Krebs cycle as acetyl CoA whereas each catabolized protein will enter at a point corresponding to its carbon-based skeleton. One turn of the Krebs cycle produces one molecule of ATP along with hydrogen ions and their electrons that are used within the electron transport chain. Since most of the energy produced aerobically is created through the electron transport chain, the hydrogen ions are the most important product of the Krebs cycle.

The electron transport chain (ETC) consists of five chemical reactions that end with the production of ATP and water as a result of complexes IV and V. As electrons pass down the electron transport chain, they release chemical energy which is used to pump H^+ out of the mitochondrial matrix. The subsequent electrical difference means that these hydrogen ions flow back into the matrix and this action drives the phosphorylation of ADP. The electrons that are released at the end of the ETC are combined with H^+ and oxygen to produce water, which is why oxygen is vital to the functioning of the electron transport chain. As one ATP molecule leaves the mitochondrial matrix through an antiport, an ADP molecule is allowed to enter the matrix and is then available for phosphorylation. Phosphate (Pi) essential for phosphorylation is also allowed to enter the matrix though a similar channel. Even though aerobic production of ATP is relatively efficient, 60 per cent of the energy realized from carbohydrate or fat is converted to heat which is why body temperature increases during aerobic activities.

Metabolic rate

Gas analysis is one of the most common methods used to determine metabolic rate. At rest, typical oxygen consumption is $20 \, L \, hr^{-1}$ which equates to a daily energy expenditure of 2100–2200 kcal. This is the resting metabolic rate (RMR) and exercise, in any form, will increase energy expenditure. Higher levels of fat-free mass, body surface area, body temperature or stress all result in increases in RMR. Energy expenditure during exercise is sometimes expressed as a MET value. One MET (metabolic equivalent) is equal to $3.5 \, mL \, kg^{-1} \, min^{-1}$, the assumed RMR. An activity that entails 3 METs energy expenditure has approximately tripled the rate of metabolism.

It is possible to estimate metabolic rate from heart rate. In order to do this, a physiologist must determine oxygen consumption and heart rate across a range of exercise intensities. It is important that the activity during testing equates to the activity in the field, for instance performance on a kayak ergometer could be used to estimate metabolic rate during sea kayaking.

Response to lower intensity aerobic endurance activity

During lower intensity aerobic endurance exercise, a performer will generally establish an exercise level that allows them to reach a steady state and complete the activity. The onset of exercise will increase metabolism, cardiovascular function and respiratory rate. These responses are initiated by the CNS and reinforced by endocrine responses. The intensity of the exercise will impact on the proportions of carbohydrate and fat metabolized to support the activity. Higher intensity exercise will mean greater reliance on glycogen stores.

Fatigue

Fatigue in lower intensity activity will generally occur as a result of glycogen depletion or as a result of the extended duration of activity. Substrate depletion occurs at a muscular level and will lead to peripheral fatigue. Factors such as sleep deprivation can also lead to the onset of central fatigue, which involves increased feelings of fatigue without the impaired metabolic functioning of peripheral fatigue. The mechanisms behind central fatigue are not yet fully understood.

Adaptations to training

The extent and exact nature of adaptations to lower intensity aerobic endurance training will depend on the duration, intensity and type of training undertaken. Most of the potential improvement in $\dot{V}O_{2 \, max}$ should have been realized during the first 6–12 months of training. After this, it will still be possible to improve aerobic performance and LT through appropriate training. It is important that training loads are appropriate and that adequate rest is built into a programme. If an athlete is continuously pushed too hard in training then overreaching and ultimately overtraining can occur. Although this balance seems easy to achieve in principle, in practice an elite athlete will need to work very close to the edge of their capability to maximize their performance. Endurance athletes can be particularly prone to overtraining but it is important to realize that any form of training can result in overtraining if levels are excessive.

Overreaching is a short-term effect while overtraining results from a more long-term problem. The symptoms of both overreaching and overtraining include a decline in performance, persistent feelings of general fatigue or muscle soreness, elevated resting heart rate and mood state issues. Psychological stress and diet issues can contribute to both problems.

Ergogenic aids

Carbohydrate loading strategies provide a good start for performance enhancement in lower intensity aerobic endurance activity. Carbohydrate drinks and some solid food during the activity should also be considered. A combination of simple and complex carbohydrate seems to be the best way of maintaining performance over time. Electrolyte drinks can also be consumed as a way of maintaining hydration, particularly in warmer environments. Sodium loss can present a real problem to adventure sports performers during prolonged activities, and ingesting only plain water can lead to hyponatremia as concentration levels in the body decrease.

It is important to realize that any expectation of a benefit from a supplement can lead to a placebo effect where performance does improve.

This placebo effect is responsible for the popular use of many substances and supplements that actually have no scientific backing.

Hyperbaric and hypobaric effects

Mountain-based activities involve exercise at low atmospheric pressures (hypobaric conditions) that present an additional stress to the performer. Once height exceeds 1500 m, rises in altitude increase their effect on physiological functioning. The main stress associated with altitude is the decreased oxygen pressure. For oxygen to diffuse into the blood, the partial pressure of oxygen must be higher in the air of the alveoli than that of the blood. The PO_2 at the summit of Mount Everest (Chomolangma) is less than one-third than at sea level and the tracheal PO_2 is equivalent to the normal blood PO_2 at sea level. Since there must be a pressure gradient for diffusion to occur, although there is oxygen transfer into the blood oxyhaemoglobin saturation is severely compromised at this kind of altitude.

Responses to altitude

The body's responses to hypoxia and altitude can be grouped into immediate responses and longer-term adaptations. The immediate response to the lower PO_2 of altitude is an increase in pulmonary ventilation. This hyperventilation increases pulmonary PO_2 which aids oxygen transfer to the blood and therefore the supply to muscles and other organs. Pulmonary blood pressure increases, which results in an increase in blood flow to the upper sections of the lungs. The increase in pulmonary ventilation also leads to an unloading of CO_2 at the lungs which can upset the pH of the blood. After a few hours, blood plasma volume is decreased which effectively concentrates the blood and enhances oxygen delivery to the muscles. Cardiac output is also increased in response to the lowered PO_2, which enhances oxygen transport. Since the reduction in blood plasma increases the viscosity of the blood, stroke volume is actually decreased and the increase

in cardiac output is due to an increase in heart rate. After a few days, the ability of the muscles to extract oxygen from the blood increases and the heart rate can return to more normal levels. Despite these compensatory mechanisms, altitude-related illnesses include Acute Mountain Sickness (AMS), High Altitude Cerebral Edema (HACE) and High Altitude Pulmonary Edema (HAPE). Exercise performance is generally reduced at altitude and aerobic capacity is approximately halved at 7000 m.

Acclimatization or adaptation to altitude appears to take about 2 weeks and occurs through changes in respiratory, cardiovascular and metabolic functioning. These adaptations include the restoration of blood pH levels through the excretion of bicarbonate and an increase in the production of red blood cells. Additional effects have sometimes included enhanced aerobic enzyme activity and increased mitochondrial size and number. The potential positive adaptations to altitude have lead to the idea of altitude training where athletes train at high altitude before returning to sea level to compete.

One of the greatest challenges in diving activities is the increase in pressure of the environment (hyperbaric condition). At 10 m, the environmental pressure doubles. Other challenges include the necessity of breath-holding, or a safe supplementary oxygen supply, and the effects of cold water. Freediving, or breath-hold diving, is the most basic and oldest response to the oxygen problem in diving. Considering that the average maximal breath-hold of an adult is between 1–2 min, the performance of trained breath-hold divers is astounding. One of the key techniques for this kind of diving is to 'blow off' CO_2 by hyperventilating before breath-holding. Reduced CO_2 levels inhibit the respiratory drive and extend dive time. This technique is not without risks and has lead to the deaths of breath-hold divers.

SCUBA equipment enables divers to carry an oxygen supply with them. The gas mixtures that can be used for diving include pure oxygen, nitrox (oxygen and nitrogen) and heliox (oxygen and helium). The main hazard for SCUBA divers arises from the increased pressure during dives. Complications can include ear, sinus and mask squeezes, pneumothorax, nitrogen narcosis and decompression sickness. A squeeze is a problem of pressure equalization when air is trapped inside the body; if it is not possible to equalize the pressure then the dive should be aborted. Other problems can occur if SCUBA divers breath-hold during ascent in which case the lungs can be damaged and a collapse of the lung (pneumothorax) can occur.

At high pressures, nitrogen can begin to have intoxicating effects known as nitrogen narcosis. Increased pressure also leads to higher partial pressures of nitrogen resulting in more nitrogen entering the blood and body tissues. Nitrogen narcosis has lead to divers making gross errors of judgement, sometimes resulting in their deaths. The increased nitrogen levels pose another threat to the diver: a rapid ascent will mean that nitrogen in suspension will become depressurized and form bubbles in the tissues. This condition is known as the bends or decompression sickness, and treatment requires the diver to be quickly re-pressurized in a decompression chamber and then decompressed slowly. Decompression sickness can be avoided altogether is the diver ascends slowly enough for the nitrogen in the body's tissues to be slowly eliminated.

Research into supplements and clothing systems for mountaineering

To date the most consistent aid to altitude acclimatization is the drug Acetazolamide. A number of studies have supported its use. Other studies have examined the prophylactic effects of vitamin C, vitamin E and α-lipoic acid with some success. Gingo biloba has also been trailed although without much success; however, theophylline did alleviate periodic breathing during sleep at altitude. Blood saturation levels have been used as a predictor for AMS, but results are equivocal.

The best clothing system for mountaineering is under debate with the traditional layering system, in which the outer layer acts as a windproof and waterproof shell, being trialled against systems where the outer layer is an easily adjusted insulation layer. It seems that adopting an easily adjusted system with a waterproof and windproof mid-layer could have advantages over an outer shell system.

Study questions

1. What are the degradation mechanisms for glycogen, triglycerides and amino acids that allow their use for ATP production in the Krebs cycle?

2. What are the two products of the Krebs cycle that can be used to support lower intensity aerobic endurance activity? How do these two products contribute to energy production?

3. How is the electron cascade related to the movement of hydrogen ions during the process of the electron transport chain? Why is the movement of H^+ out of the mitochondrial matrix important for ATP production?

4. Why is oxygen essential to the functioning of the electron transport chain and the process of oxidative phosphorylation?

5. What is meant by Resting Metabolic Rate? What is the daily expected energy expenditure at RMR?

6. Why is a high level $\dot{V}O_{2\,max}$ advantageous for a high-altitude mountaineer, even though activities may take place at a slow walking pace during an ascent? Why is their blood volume likely to decrease at high altitude?

7. Explain what causes decompression sickness in a SCUBA diver who returns to the surface too quickly? Why might a diver feel particularly eurphoric during a deep dive?

References

Chapter 1

Ashcroft, F. (2001). *Life at the extremes: The science of survival*. Flamingo, London.

Boas Hall, M. (1965). *Robert Boyle on natural philosophy: An essay with selected writings from Robert Boyle*. Bloomington, Indiana.

Cook, P. and Webb, B. (1977). *The Complete Book of Sailing*. Ward Lock, London.

English, B. (1984). *Total telemarking*. East River Publishing, Colorado.

Ferrero, F. (2002). *British Canoe Union canoe and kayak handbook*. Presda Press, Wales.

Foster, M. (1970). *Lectures on the history of physiology during the 16th, 17th & 18th centuries*. General Publishing Company Ltd, Toronto.

Golden, F. and Tipton, M. (2002). *Essentials of sea survival*. Human Kinetics, Champaign, IL.

Hale, T. (2003). *Exercise physiology: A thematic approach*. John Wiley & Sons Ltd., Chichester.

Hall, T.S. (1975). *History of general physiology 600 B.C. to A.D. 1900*. University of Chicago Press, Chicago.

Hart, P. (1988). *Improve your windsurfing*. Willow Books, London.

Harvey, W. (1628). Exercitatio anatomica de motu cordis et sanguinis. In R. Willis, (1952). *Great books of the western world*. Encyclopaedia Britannica, Inc., Chicago.

Hutchinson, D. (1994). *The complete book of sea kayaking*. A & C Black, London.

Jones, A.M. and Poole, D.C. (2005). (Eds.) *Oxygen uptake kinetics in sport, exercise and medicine*. Routledge, London.

Lopes, B. and McCormak, L. (2005). *Mastering mountain bike skills*. Human Kinetics, Champaign, IL.

Lowe, J. (1996). *Ice world*. Cordee, Leicester.

McArdle, W.D., Katch, F.I. and Katch, V.L. (2001). *Exercise physiology: Energy, nutrition and human performance*. Lippincott, Williams & Wilkins, Baltimore.

Rothschuh, K.E. (1973). *History of physiology*. Robert E. Krieger, Huntington, New York.

Shackleton, R. (1985). *All about wave skis*. Surfside Press, NSW, Australia.

Sleight, S. (1999). *Complete sailing manual*. Dorling Kindersley, London.

Sparrow, A. (1997). *The complete caving manual*. Crowood, Ramsbury.

Starling, E.H. and Visscher, M.B. (1926). The regulation of the energy output of the heart. *Journal of Physiology*, **62**, 243–261.

Strauss, M.B. and Aksenov, I.V. (2004). *Diving science*. Human Kinetics, Champaign, IL.

Twight, M.F. and Martin, J. (1999). *Extreme alpinism*. The Mountaineers, Seattle, WA.

Wilson, K. (1978). *The games climbers play*. Baton Wicks, London.

http://www.en.wikipedia.org

http://scienceworld.wolfram.com

http://www.chemheritage.org

http://nobelprize.org/medicine/

Chapter 2

Armstrong, L.E., Costill, D.L. and Fink, W.J. (1985). Influence of diuretic-induced dehydration on competitive running performance. *Medicine and Science in Sports and Exercise*, **17**, 456–461.

Baechle, T.R. and Earle, R.W. (2000). *Essentials of strength training and conditioning*. Human Kinetics, Champaign, IL.

Bloomfield, J., Fricker, P.A. and Fitch, K.D. (Eds.) (1995). *Science and medicine in sport*. Blackwell Science, Carlton, Australia.

Clayton, P. (2004). *Health defence*. Accelerated Learning Systems, Aylesbury.

Department of Health (1991). *Dietary reference values for food energy and nutrients in the UK*. HMSO, London.

Department of Health (2004). *Choosing Health? Choosing a better diet*. HMSO, London.

Draper, H.H. (1977). The Aboriginal Eskimo Diet in Modern Perspective. *American Anthropologist*, **79**(2), 309–316.

Garrett, W.E. and Kirkendall, D.T. (Eds.) (2000). *Exercise and sports science*. Lippincott, Williams & Wilkins, Baltimore, MD.

Hale, T. (2003). *Exercise physiology: A thematic approach*. John Wiley & Sons Ltd., Chichester.

Holum, J.R. (1998). *Fundamentals of general, organic and biological chemistry*. John Wiley & Sons, Chichester.

Houston, M.E. (1995). *Biochemistry primer for exercise science*. Human Kinetics, Champaign, IL.

Katch, F.I. and McArdle, W.D. (1988). *Introduction to nutrition, exercise and health*. Williams and Wilkins, Baltimore, MD.

Maughan, R. and Gleeson, M. (2004). *The biochemical basis of sports performance*. OU Press, Oxford.

McArdle, W.D., Katch, F.I. and Katch, V.L. (2001). *Exercise physiology: Energy, nutrition and human performance*. Lippincott, Williams & Wilkins, Baltimore, MD.

Seeley, R.R., Stephens, T.D. and Tate, P. (2000). *Anatomy and physiology*. McGraw-Hill Higher Education, San Francisco, CA.

Sherwood, L. (2001). *Human physiology: From cells to systems*. Brooks/Cole, Pacific Grove, CA.

Tortora, G.J. and Derrickson, B. (2006). *Principles of anatomy and physiology*. John Wiley & Sons, Chichester.

Williams, M.H. (1995). *Nutrition for fitness and sport*. Brown and Benchmark, Dubuque, IA.

Wilmore, J.H. and Costill, D.L. (2004). *Physiology of sport and exercise*. Human Kinetics, Champaign, IL.

http://www.bhf.org.uk/
http://www.dh.gov.uk/
http://www.nhsdirect.nhs.uk
http://www.nutrition.org.uk
http://www.vegsoc.org
http://www.en.wikipedia.org

Chapter 3

Åstrand, P.E. and Rodahl, K. (1986). *Textbook of work physiology: Physiological bases of exercise*. McGraw-Hill, New York, NY.

Åstrand, P.E., Rodahl, K., Dahl, H.A. and Strømme, S.B. (2003). *Textbook of work physiology: Physiological bases of exercise*. Human Kinetics, Champaign, IL.

Bloomfield, J., Fricker, P.A. and Fitch, K.D. (Eds.) (1995). *Science and medicine in sport*. Blackwell Science, Carlton, Australia.

Bray, J.J., Cragg, P.A., Macknight, A.D.C., Mills, R.G. and Taylor, D.W. (1994). *Lecture notes on human physiology*. Blackwell Science, Oxford.

Garrett, W.E. and Kirkendall, D.T. (Eds.) (2000). *Exercise and sports science*. Lippincott, Williams & Wilkins, Baltimore, MD.

Green, J.H. (1978). *An introduction to human physiology*. OU Press, Oxford.

Hale, T. (2003). *Exercise physiology: A thematic approach*. John Wiley & Sons Ltd., Chichester.

Holum, J.R. (1998). *Fundamentals of general, organic and biological chemistry*. John Wiley & Sons, Chichester.

Jones, A.M. and Poole, D.C. (2005). (Eds.) *Oxygen uptake kinetics in sport, exercise and medicine*. Routledge, London.

Maughan, R. and Gleeson, M. (2004). *The biochemical basis of sports performance*. OU Press, Oxford.

McArdle, W.D., Katch, F.I. and Katch, V.L. (2001). *Exercise physiology: Energy, nutrition and human performance*. Lippincott, Williams & Wilkins, Baltimore.

Rowett, H.G.O. (1988). *Basic anatomy and physiology*. John Murray Ltd., London.

Seeley, R.R., Stephens, T.D. and Tate, P. (2000). *Anatomy and physiology*. McGraw-Hill Higher Education, San Francisco, CA.

Sherwood, L. (2001). *Human physiology: From cells to systems*. Brooks/Cole, Pacific Grove, CA.

Tortora, G.J. and Derrickson, B. (2006). *Principles of anatomy and physiology*. John Wiley & Sons, Chichester.

Wilmore, J.H. and Costill, D.L. (2004). *Physiology of sport and exercise*. Human Kinetics, Champaign, IL.

Chapter 4

Åstrand, P.E. and Rodahl, K. (1986). *Textbook of work physiology: Physiological bases of exercise*. McGraw-Hill, New York, NY.

Åstrand, P.E., Rodahl, K., Dahl, H.A. and Strømme, S.B. (2003). *Textbook of work physiology: Physiological bases of exercise.* Human Kinetics, Champaign, IL.

Bloomfield, J., Fricker, P.A. and Fitch, K.D. (Eds.) (1995). *Science and medicine in sport.* Blackwell Science, Carlton, Australia.

Bray, J.J., Cragg, P.A., Macknight, A.D.C., Mills, R.G. and Taylor, D.W. (1994). *Lecture notes on human physiology.* Blackwell Science, Oxford.

Garrett, W.E. and Kirkendall, D.T. (Eds.) (2000). *Exercise and sports science.* Lippincott, Williams & Wilkins, Baltimore, MD.

Green, J.H. (1978). *An introduction to human physiology.* OU Press, Oxford.

Hale, T. (2003). *Exercise physiology: A thematic approach.* John Wiley & Sons Ltd., Chichester.

Holum, J.R. (1998). *Fundamentals of general, organic and biological chemistry.* John Wiley & Sons, Chichester.

Jones, A.M. and Poole, D.C. (Eds.) (2005). *Oxygen uptake kinetics in sport, exercise and medicine.* Routledge, London.

Maughan, R. and Gleeson, M. (2004). *The biochemical basis of sports performance.* OU Press, Oxford.

McArdle, W.D., Katch, F.I. and Katch, V.L. (2001). *Exercise physiology: Energy, nutrition and human performance.* Lippincott, Williams & Wilkins, Baltimore.

Rowett, H.G.O. (1988). *Basic anatomy and physiology.* John Murray Ltd., London.

Seeley, R.R., Stephens, T.D. and Tate, P. (2000). *Anatomy and physiology.* McGraw-Hill Higher Education, San Francisco, CA.

Sherwood, L. (2001). *Human physiology: From cells to systems.* Brooks/Cole, Pacific Grove, CA.

Starling, E.H. and Visscher, M.B. (1926). The regulation of the energy output of the heart. *Journal of Physiology*, **62**, 243–261.

Tortora, G.J. and Derrickson, B. (2006). *Principles of anatomy and physiology.* John Wiley & Sons, Chichester.

Wilmore, J.H. and Costill, D.L. (2004). *Physiology of sport and exercise.* Human Kinetics, Champaign, IL.

Chapter 5

Andersen, R. and Montgomery, D. (1988). Physiology of alpine skiing. *Sports Medicine*, **6**, 210–221.

Ayalon, A., Inbar, O. and Bar-Or, O. (1974). Relationships among measurements of explosive strength and anaerobic power. In R.C. Nelson and C. A. Morehouse (Eds.), *International series on sports science: Volume 1. Biomechanics IV: Preceedings of the fourth International Seminar on Biomechanics.* University Park Press, Baltimore, MD.

Baechle, T.R. and Groves, B.R. (1998). *Weight training: Steps to success.* Human Kinetics, Champaign, IL.

Baechle, T.R. and Earle, R.W. (2000). *Essentials of strength training and conditioning.* Human Kinetics, Champaign, IL.

Balyi, I. and Way, R. (1995). Long-term planning of athlete development. *British Columbia Coach*, **Summer**, 2–10.

Balyi, I. and Hamilton, A. (1998). Long-term planning can make a difference. *Soccer Journal*, **July/August**, 18–23.

Billat, V., Palleja, P., Charlaix, T., Rizzardo, P. and Janel, N. (1995). Energy specificity of rock climbing and aerobic capacity in competitive sport rock climbers. *Journal of Sports Medicine and Physical Fitness*, **35**, 20–24.

Bishop, P.A., Smith, J.F., Kime, J.C., Mayo, J.M. and Tin, Y.H. (1992). Comparison of a manual and an automated enzymatic technique for determining blood lactate concentrations. *International Journal of Sports Medicine*, **13**, 36–39.

Booth, J., Marino, F., Hill, C. and Gwinn, T. (1999). Energy cost of sport rock climbing in elite performers. *British Journal of Sports Medicine*, **33**, 14–18.

Dassonville, J., Beillot, J., Lessard, Y., Jan, J., Andre, A.M., LePourcelet, C., Rochcongar, P. and Carre, F. (1998). Blood lactate concentrations during exercise: effect of sampling site and exercise mode. *Journal of Sports Medicine and Physical Fitness*, **38**, 39–46.

Elliot, B. (Ed.) (1998). *Training in Sport.* John Wiley & Sons Ltd., Chichester, UK.

Feliu, J., Ventura, J.L., Segura, R., Rodas, G., Riera, J., Estruch, A., Zamora, A. and Capdevila, L. (1999). Differences between lactate concentration of samples from ear lobe and the fingertip. *Journal of Physiology and Biochemistry*, **55**, 333–340.

Forsyth, J.J. and Farrally, M.R. (2000). A comparison of lactate concentration in plasma collected from toe, ear, and fingertip after simulated rowing exercise. *British Journal of Sports Medicine*, **34**, 35–38.

Fox III, S.M., Naughton, J.P. and Haskell, W.L. (1971). Physical activity and the prevention of coronary heart disease. *Annals of Clinical Research*, **3**, 404–432.

Gerbert, W. and Werner, I. (1999). Blood lactate response to competitive climbing. *Communication to the 1st International Conference on Science and*

Technology in Climbing and Mountaineering, Leeds, UK, April.

Grant, S., Hynes, V., Whittaker, A. and Aitchison, T. (1996). Anthropometric, strength, endurance and flexibility characteristics of elite and recreational climbers. *Journal of Sports Sciences*, **14**, 301–309.

Grant, S., Hasler, T., Davies, C., Aitchison, T.C., Wilson, J. and Whittaker, A. (2001). A comparison of the anthropometric, strength, endurance and flexibility characteristics of female elite and recreational climbers and non-climbers. *Journal of Sports Sciences*, **19**, 499–505.

Hale, T. (2003). *Exercise physiology: A thematic approach*. John Wiley & Sons Ltd., Chichester.

Heller, J., Bunc, V. and Novotny, V. (1991). Effect of site of capillary blood sampling on lactate concentrations during exercise and recovery. *Acta Universitatis Carolinae Gymnica*, **27**, 29–38.

Heller, J., Bíl, M., Pultera, J. and Sadilová, M. (1994). Functional and energy demands of elite female kayak slalom: A comparison of training and competition performances. *Acta Universitatis Carolinae Gymnica*, **30**, 59–74.

Heyward, V.H. (1998). *Advanced fitness assessment and exercise prescription*. Human Kinetics, Champaign, IL.

Hoffman, J. (2002). *Physiological aspects of sports training and performance*. Human Kinetics, Champaign, IL.

Leger, L.A. and Lambert, J. (1982). A maximal multi-stage 20m shuttle run test to predict $\dot{V}O_{2\,max}$. *European Journal of Applied Physiology*, **49**, 1–5.

Magel, J.R., Foglia, G.F., McArdle, W.D., Gutin, B., Pechar, G.S. and Katch, F.I., (1975). Specificity of swim training on maximal oxygen uptake. *Journal of Applied Physiology*, **38**, 151–155.

Margaria, R.M., Aghemo, P., and Rovelli, E. (1966). Measurement of muscular power (anaerobic) in man. *Journal of Applied Physiology*, **21**, 1662–1664.

Matveyev, L.P. (1966). *Periodisation of sports training*. Fiscultura i sport, Moscow.

McArdle, W.D., Katch, F.I. and Katch, V.L. (2001). *Exercise physiology: Energy, nutrition and human performance*. Lippincott, Williams & Wilkins, Baltimore, MD.

Medbø, J.I., Mamen, A., Holt-Olsen, O. and Evertsen, F. (2000). Examination of four different instruments for measuring blood lactate concentration. *Scandinavian Journal of Clinical Laboratory Investigation*, **60**, 367–380.

Mermier, C.M., Robergs, R.A. McMinn, S.M. and Heyward, V.H. (1997). Energy expenditure and physiological responses during indoor rock climbing. *British Journal of Sports Medicine*, **31**, 224–228.

Mermier, C.M., Janot, J.M. Parker, D.L. and Swan, J.G. (2000). Physiological and anthropometric determinants of sport climbing performance. *British Journal of Sports Medicine*, **34**, 359–366.

Minkler, S. and Patterson, P. (1994). The validity of the modified sit-and-reach test in college age students. *Research Quarterly for Exercise and Sport*, **65**, 189–192.

Orvanova, E. (1987). Physical structure of winter sports athletes. *Journal of Sports Sciences*, **5**, 197–248.

Pyne, D.B., Boston, T., Martin, D.T. and Logan, A. (2000). Evaluation of the Lactate Pro blood analyser. *European Journal of Applied Physiology*, **82**, 112–116.

Richardson, R.S., White, A.T., Seifert, J.D., Porretta, J.M. and Johnson, S.C. (1993). Blood lactate concentrations in elite skiers during a series of on-snow downhill ski runs. *Journal of Strength and Conditioning*, **7**(3), 168–171.

Robergs, R.A. and Landwehr R. (2002). The surprising history of the HRmax = 220 age equation. *Journal of Exercise Physiology*, **5**(2), 1–10.

Rockport Walking Institute. (1986). *Rockport fitness walking test*. Rockport Walking Institute, Marlboro, MA.

Rundell, K.W. (1995). Treadmill roller ski test predicts biathlon roller ski race results of elite US biathlon women. *Medicine and Science in Sport and Exercise*, **27**, 1677–1685.

Rundell, K.W. and Bacharach, D.W. (1995). Physiological characteristics and performance of top US biathletes. *Medicine and Science in Sport and Exercise*, **27**, 1302–1310.

Semenick, D. (1990). Tests and measurements: The line drill test. *National Strength and Conditioning Association Journal*, **12**, 47–49.

Seyle, H. (1956) *The stress of life*. McGraw-Hill, New York, NY.

Sidney, K. and Shephard, R.J. (1973). Physiological characteristics and performance of the white-water paddler. *European Journal of Applied Physiology*, **32**, 55–70.

Sklad, M., Krawczyk, B. and Majle, B. (1994). Body build profiles of male and female rowers and kayakers. *Biological Sport*, **11**, 249–256.

Tanaka, H., Monahan, K.G. and Seals, D.S. (2001). Age-predicted maximal heart rate revisited. *Journal of the American College of Cardiology*, **37**, 153–156.

Thoreau, H.D. (1854). *Walden; or life in the woods.* Richard Lenat, Pennsylvania, USA.

Vaccaro, P., Gray, P.R., Clarke, D.H. and Morris, A.F. (1984). Physiological characteristics of world class white-water slalom paddlers. *Research Quarterly for Exercise and Sport*, **55**, 206–210.

Wallin, D., Ekblom, B., Grahn, R. and Nordenberg, T. (1985). Improvement of muscle flexibility. *American Journal of Sports Medicine*, **13**, 263–268.

Watts, P.B. (2004). Physiology of difficult rock climbing. *European Journal of Applied Physiology*, **91**, 361–372.

Watts, P.B., Martin, D.T. and Durtschi, S. (1993). Anthropometric profiles of elite male and female competitive sport rock climbers. *Journal of Sports Sciences*, **11**, 113–117.

Williams, J.R., Armstrong, N. and Kirby, B.J. (1992). The influence of the site of sampling and assay medium upon the measurement and interpretation of blood lactate responses to exercise. *Journal of Sports Sciences*, **10**, 95–107.

Chapter 6

Åstrand, P.E. and Rodahl, K. (1986). *Textbook of work physiology: Physiological bases of exercise.* McGraw-Hill, New York, NY.

Åstrand, P.E., Rodahl, K., Dahl, H.A. and Strømme, S.B. (2003). *Textbook of work physiology: Physiological bases of exercise.* Human Kinetics, Champaign, IL.

Balsom, P.D., Ekblom, B., Soderlund, K., Sjodin, B. and Hultman, E. (1993). Creatine supplementation and dynamic high-intensity intermittent exercise. *Scandanavian Journal of Medicine and Science in Sports*, **3**, 143–149.

Bloomfield, J., Fricker, P.A. and Fitch, K.D. (Eds.) (1995). *Science and medicine in sport.* Blackwell Science, Carlton, Australia.

Brooks, G.A., Fahey, T.D. and Baldwin, K.M. (2005). *Exercise physiology: Human bioenergetics and its applications.* McGraw-Hill, New York, NY.

Casey, A., Constantin-Teodosiu, D., Howell, S., Hultman, E. and Greenhaff, P.L. (1996). Creatine ingestion favourably affects performance and muscle metabolism during maximal exercise in humans. *American Journal of Physiology*, **271**, E31–37.

Connolly, D.A.J., Sayers, S.P. and McHugh, M.P. (2003). Treatment and prevention of delayed onset of muscle soreness. *Journal of Strength Training and Conditioning*, **17**(1), 197–208.

Evans, W.J. and Cannon, J.G. (1991). The metabolic effects of exercise-induced muscle damage. *Exercise and Sport Sciences Reviews*, **19**, 99–125.

Garrett, W.E. and Kirkendall, D.T. (Eds.) (2000). *Exercise and sports science.* Lippincott, Williams & Wilkins, Baltimore, MD.

Girard–Eberle, S. (2000). *Endurance sports nutrition: eating plans for optimal training, racing and recovery.* Human Kinetics, Champaign, IL.

Gonyea, W.J. (1980). Role of exercise inducing increases in skeletal muscle fibre number. *Journal of Applied Physiology*, **48**, 421–426.

Gonyea, W.J., Sale, D.G., Gonyea, F.B. and Mikesky, A. (1986). Exercise induced increases in muscle fibre number. *European Journal of Applied Physiology*, **55**, 137–141.

Greenhaff, P.L., Casey, A., Short, A.H., Harris, R., Soderlund, K. and Hultman, E. (1993). Influence of oral creatine supplementation on muscle torque during repeated bouts of maximal voluntary exercise in man. *Clinical Science*, **84**, 565–571.

Harris, R.C., Soderlund, K. and Hultman, E. (1992). Elevation of creatine in resting and exercised muscle of normal subjects by creatine supplementation. *Clinical Science*, **83**, 367–374.

Holum, J.R. (1998). *Fundamentals of general, organic and biological chemistry.* John Wiley & Sons, Chichester.

Howarth, H. (2006). *Effects of wall angle on the physiological demands of bouldering.* Unpublished dissertation, University of Chichester.

Hultman, E., Soderlund, K., Timmons, J.A., Cederbald, G. and Greenhaff, P.L. (1996). Muscle creatine loading in men. *Journal of Applied Physiology*, **81**(1), 232–237.

Jeukendrup, A. and Gleeson, M. (2004). *Sport nutrition: An introduction to energy production and performance.* Human Kinetics, Champaign, IL.

Leenders, N.M., Lamb, D.R. and Nelson, T.E. (1999). Creatine supplementation and swimming performance. *International Journal of Sports Nutrition*, **9**(3), 251–262.

Maughan, R. and Gleeson, M. (2004). *The biochemical basis of sports performance.* OU Press, Oxford.

McArdle, W.D., Katch, F.I. and Katch, V.L. (2001). *Exercise physiology: Energy, nutrition and human performance.* Lippincott, Williams & Wilkins, Baltimore.

Mujika, I., Chatard, J.C., Lacoste, L., Barale, F. and Geyssant, A. (1996). Creatine supplementation does not improve sprint performance in competitive swimmers. *Medicine and Science in Sports and Exercise,* **28**(11), 1435–1441.

Ranvier, L.A. (1873). Structure and function differences in the red and white muscles of rabbits and ray fish. *Les Comptes Rendus de l'Académie des Sciences,* **77**, 1030–1034.

Schwane, J.A., Johnson, S.R. Vandenakker, C.B. and Armstrong, R.B. (1983). Delayed onset muscular soreness and plasma CPK and LDH activities after downhill running. *Medicine and Science in Sports and Exercise,* **15**, 51–56.

Scott, K.E., Rozenek, R., Russo, A.C. Crussemeyer, J.A. and Lacourse, M.G. (2003). Effects of delayed onset of muscle soreness on selected physiological responses to submaximal running. *Journal of Strength Training and Conditioning,* **17**(4), 652–658.

Seeley, R.R., Stephens, T.D. and Tate, P. (2000). *Anatomy and physiology.* McGraw-Hill Higher Education, San Francisco, CA.

Sherwood, L. (2001). *Human physiology: From cells to systems.* Brooks/Cole, Pacific Grove, CA.

Talag, T.S. (1973). Residual muscular soreness as influenced by concentric, eccentric and static contractions. *Research Quarterly,* **44**, 458–469.

Tortora, G.J. and Derrickson, B. (2006). *Principles of anatomy and physiology.* John Wiley & Sons, Chichester.

Vandenberghe, K., Goris, M., Van Hecke, P., Van Leemputte, M., Vangerven, L. and Hespel, P. (1997). Long-term creatine intake is beneficial to muscle performance during resistance training. *Journal of Applied Physiology,* **83**(6), 2055–2063.

Voet, D. and Voet, J. (2004). *Biochemistry.* John Wiley & Sons, Chichester.

Williams, M.H. (1998). *The ergogenics edge: Pushing the limits of sports performance.* Human Kinetics, Champaign, IL.

Wilmore, J.H. and Costill, D.L. (2004). *Physiology of sport and exercise.* Human Kinetics, Champaign, IL.

Climb Magazine

Climber Magazine

Chapter 7

Åstrand, P.E. and Rodahl, K. (1986). *Textbook of work physiology: Physiological bases of exercise.* McGraw-Hill, New York, NY.

Åstrand, P.E., Rodahl, K., Dahl, H.A. and Strømme, S.B. (2003). *Textbook of work physiology: Physiological bases of exercise.* Human Kinetics, Champaign, IL.

Bergh, U. (1982). *Physiology of cross-country ski racing.* Human Kinetics, Champaign, IL.

Bergh, U. (1987). The influence of body mass in cross-country skiing. *Medicine Science and Sports Exercise,* **19**, 324–331.

Bergman, B.C., Wolfel, E.E., Butterfield, G.E., Lopaschuk, G., Casazza, G.A., Horning, M.A. and Brooks, G.A. (1999). Active muscle and whole body lactate kinetics after endurance training in men. *Journal of Applied Physiology,* **87**, 1684–1696.

Billat, V., Palleja, P., Charlaix, T., Rizzardo, P. and Janel, N. (1995). Energy specificity of rock climbing and aerobic capacity in competitive sport rock climbers. *Journal of Sports Medicine and Physical Fitness,* **35**, 20–24.

Bloomfield, J., Fricker, P.A. and Fitch, K.D. (Eds.) (1995). *Science and medicine in sport.* Blackwell Science, Carlton, Australia.

Booth, J., Marino, F., Hill, C. and Gwinn, T. (1999). Energy cost of sport rock climbing in elite performers. *British Journal of Sport Medicine,* **33**, 14–18.

Brooks, G.A., Dubouchaud, H., Brown, M., Sicurello, J.P. and Butz, C.E. (1999). Role of mitochondrial lactate dehydrogenase and lactate oxidation in the "intra-cellular lactate shuttle". *Proceedings of the National Academy of Sciences,* **96**, 1129–1134.

Brooks, G.A., Fahey, T.D. and Baldwin, K.M. (2005). *Exercise physiology: Human bioenergetics and its applications.* McGraw-Hill, New York, NY.

Connolly, D.A.J. (2002). The energy expenditure of snowshoeing in packed vs. unpacked snow at low-level walking speeds. *Journal of Strength and Conditioning Research,* **16**(4), 606–610.

Costill, D.L., Verstappen, F., Kuipers, H., Janssen, E. and Fink, W. (1984). Acid-base balance during repeated bouts of exercise: Influence of HCO_3^-. *International Journal of Sports Medicine,* **5**, 228–231.

Cutts, A. and Bollen, S.R. (1993). Grip strength and endurance in rock climbers. *Journal of Engineering in Medicine,* **207**, 87–92.

Di Prampero, P.E. (1986). The energy cost of human locomotion on land and in water. *International Journal of Sports Medicine,* **7**, 55–72.

Doyon, K.H., Perrey, S., Abe, D. and Hughson, R.L. (2001). Field testing of VO^2 peak in cross-country skiers with portable breath-by-breath system. *Canadian Journal of Applied Physiology,* **26**(1), 1–11.

Eriksson A., Forsberg, A., Nillson, J. and Karlsson, J. (1978). Muscle strength, EMG activity, and oxygen uptake during downhill skiing. *Biomechanics VI–A, Baltimore Md., University Park Press,* 54–61.

Garrett, W.E. and Kirkendall, D.T. (Eds.) (2000). *Exercise and sports science*. Lippincott, Williams & Wilkins, Baltimore, MD.

Gerbert, W. and Werner, I. (1999). Blood lactate response to competitive climbing. *Communication to the 1st International Conference on Science and Technology in Climbing and Mountaineering*, Leeds, UK, April.

Girard-Eberle, S. (2000). *Endurance sports nutrition: eating plans for optimal training, racing and recovery*. Human Kinetics, Champaign, IL.

Goddard, D. and Newmann, U. (1993). *Performance rock climbing*, Stackpole Books, Mechanicsburg, PA.

Grant, S., Hynes, V., Whittaker, A. and Aitchison, T. (1996). Anthropometric, strength, endurance and flexibility characteristics of elite and recreational climbers. *Journal of Sports Sciences*, **14**, 301–309.

Grant, S., Hasler, T., Davies, C., Aitchison, T.C., Wilson, J. and Whittaker, A. (2001). A comparison of the anthropometric, strength, endurance and flexibility characteristics of female elite and recreational climbers and non-climbers. *Journal of Sports Sciences*, **19**, 499–505.

Grant, S., Shields, C., Fitzpatrick, V., Ming Loh, W., Whitaker, A., Watt, I. and Kay, J.W. (2003). Climbing-specific finger endurance: a comparative study of intermediate rock climbers, rowers and aerobically trained individuals. *Journal of Sports Sciences*, **21**, 621–630.

Hague, D. and Hunter, D. (2006). *The self coached climber*, Stackpole Books, Mechanicsburg, PA.

Harris, R.C., Tallon, M.J., Dunnett, M., Boobis, L., Coakley, J., Kim, H.J., Fallowfield, J.L., Hill, C.A., Sale, C. and Wise, J.A. (2006). The absorption of orally supplied β alanine and its effect on muscle carnosine synthesis in human vastus lateralis. *Amino Acids*, **30**(3), 279–289.

Hedman, R. (1957). The available glycogen in man and the connection between rate of oxygen intake and carbohydrate usage. *Acata Physiologica Scandinavia*, **40**, 305–321.

Holum, J.R. (1998). *Fundamentals of general, organic and biological chemistry*. John Wiley & Sons, Chichester.

Hörst, E. (1994). *Flash training*. Chockstone Press, Inc., Colorado.

Hörst, E.J. (1997). *How to climb 5.12*. Falcon Publishing Co., Inc. Montana.

Hörst, E.J. (2003). *Training for climbing: the definitive guide to improving your climbing performance*. The Globe Pequot Press, Connecticut.

Hurni, M. (2003). *Coaching climbing: a complete program for coaching youth climbing for high performance and safety*. The Globe Pequot Press, Connecticut.

Hurni, M. and Ingle, P. (1988). *Climbing fit*, The Crowood Press, Wiltshire.

Janot, J.M., Steffen J.P., Porcari, J.P. and Maher, M.A. (2000). Heart rate responses and perceived exertion for beginner and recreational sport climbers during indoor climbing. *Journal of Exercise Physiology online*, **3**(1).

Jette, M, Thoden, J.S. and Spence, J. (1976). The energy expenditure of a 5 km cross-country ski run. *Journal of Applied Physiology*, **20**, 425–431.

Jeukendrup, A. and Gleeson, M. (2004). *Sport nutrition: An introduction to energy production and performance*. Human Kinetics, Champaign, IL.

Karlsson, J. (1984). Profiles of cross-country and alpine skiers. *Clinics in Sports Medicine*, **3**(1), 245–271.

Karlsson, J., Eriksson, A., Forsberg, A., Kallberg, L. and Tesch, P. (1977). *Physiology of alpine skiing*. Park City, UT, The U.S. Ski Coaches Association.

Kascenska, J.R., Dewitt, J. and Roberts, T. (1992). Fitness guidelines for rock climbing students. *JOPERD*, **March**, 73–79.

Kornexl, E. (1975). Anthropometrische Untersuchungen in alpinen schirennlauf. *Leibesübungen-Leibeserziehung*, **29**, 196–201.

Koukoubis, T.D., Cooper, L.W., Glisson, R.R., Seaber, A.V. and Feagin Jr, J.A. (1995). An electromyographic study of arm muscles during climbing. *Knee Surgery, Sports Traumatology, Arthroscopy*, **3**, 121–124.

Kreider, M.E., Stumvoll, M., Meyer, C., Overkamp, D., Welle, S. and Gerich, J. (1997). Steady-state and non-steady-state measurements of plasma glutamine turnover in humans. *American Journal of Physiology*, **272**, E621–E627.

Maughan, R. and Gleeson, M. (2004). *The biochemical basis of sports performance*. OU Press, Oxford.

McArdle, W.D., Katch, F.I. and Katch, V.L. (2001). *Exercise physiology: Energy, nutrition and human performance*. Lippincott, Williams & Wilkins, Baltimore.

Mermier, C.M., Robergs, R.A., McMinn, S.M. and Heyward. V.H. (1997). Energy expenditure and physiological responses during indoor rock climbing. *British Journal of Sports Medicine*, **31**, 224–228.

Mermier, C.M., Janot, J. M., Parker, D.J. and Swan, J.G. (2000). Physiological and anthropometric determinants of sport climbing performance. *British Journal of Sports Medicine*, **34**, 359–366.

Mygind, E., Andersen, L.B. and Rasmussen, B. (1994). Blood lactate and respiratory variables in elite cross–country skiing at racing speeds. *Scandinavian Journal of Medicine Science and Sports*, **4**, 243–251.

Noe, F. (2006). Modifications of anticipatory postural adjustments in a rock climbing task: the effect of supporting wall inclination. *Journal of Electromyography and Kinesiology*, **16**(4), 336–341.

Perriello, G., Nurjhan, N., Stumvoll, M., Bucci, A., Welle, S. Dailey, G., Bier, D.M., Toft, I., Jenssen, T.G. and Gerich, J.E. (1997). Regulation of gluconeogenesis by glutamine in normal postabsorptive humans. *American Journal of Physiology*, **272**, E437–E445.

Quaine, F., Vigouroux, L. and Martin, L. (2003). Finger flexors fatigue in trained rock climbers and untrained sedentary subjects. *International Journal of Sports Medicine*, **24**(6), 424–427.

Rohrbough, J.T., Mudge, M.K. and Schilling, R.C. (2000). Overuse injuries in the elite rock climber. *Medicine Science and Sports Exercise*, **32**(8), 1369–1372.

Roth, D.A. and Brooks, G.A. (1990). Lacate transport is mediated by a membrane-bound carrier in rat skeletal muscle sarcolemmal vesicles. *Archives of Biochemistry and Biophysics*, **279**, 377–385.

Rougier, P. and Blanchi, J.P. (1991). Évaluation du niveau d'expertise en escalade par l'analyse de la relation posturo-cinétique. *Science et motricité*, **14**, 3–12.

Sagar, H.R. (2001). *Climbing your best: training to maximize your performance*. Stackpole Books, Mechanicsburg, PA.

Schöeffl, V., Klee, S. and Strecker, W. (2004). Evaluation of physiological standard pressures of the forearm flexor muscles during sport specific ergometry in sport climbers. *British Journal of Sports Medicine*, **38**, 422–425.

Schöeffl, V.R., Möckel, F., Köstermeyer, G., Roloff, I. and Küpper, T. (2006). Development of a performance diagnosis of the anaerobic strength endurance of the forearm flexor muscles in sport climbing. *International Journal of Sports Medicine*, **27**, 205–211.

Seeley, R.R., Stephens, T.D. and Tate, P. (2000). *Anatomy and physiology*. McGraw-Hill Higher Education, San Francisco, CA.

Sheel, A.W. (2004). Physiology of sport rock climbing. *British Journal of Sports Medicine*, **38**, 355–359.

Sheel, W., Seddon, N., Knight, A., McKenzie, D.C., Warburton, R. and Darren, E. (2003). Physiological responses to indoor rock-climbing and their relationship to maximal cycle ergometry. *Medicine and Science in Sports and Exercise*, **35**(7), 1225–1231.

Sherwood, L. (2001). *Human physiology: From cells to systems*. Brooks/Cole, Pacific Grove, CA.

Stanley, W.C. Gertz, E.W., Wisneski, J.A., Morris, D.L., Neese, R. and Brooks, G.A. (1986). Lactate metabolism in exercising human skeletal muscle: Evidence for lactate extraction during net lactate release. *Journal of Applied Physiology*, **60**, 1116–1120.

Stumvoll, M., Meyer, C., Perriello, M. Welle, S. and Gerich, J. (1998). Human kidney and liver gluconeogenesis: Evidence for organ substrate selectivity. *American Journal of Physiology*, **274**, E817–E826.

Subudhi, A.W., Davis, S.L., Kipp, R.W. and Askew, E.W. (2001). Antioxidant status and oxidative stress in elite alpine ski racers. *International Journal of Sport Nutrition and Exercise Metabolism*, **11**, 32–41.

Tortora, G.J. and Derrickson, B. (2006). *Principles of anatomy and physiology*. John Wiley & Sons, Chichester.

Vigouroux, L. and Quaine, F. (2006). Fingertip force and electromyography of finger flexor muscles during a prolonged intermittent exercise in elite climbers and sedentary individuals. *Journal of Sports Sciences*, **24**(2), 181–186.

Voet, D. and Voet, J. (2004). *Biochemistry*. John Wiley & Sons, Chichester.

Wall, C.B., Starek, J.E., Fleck, S.J. and Byrnes, W.C. (2004). Prediction of indoor climbing performance in women rock climbers. *Journal of Strength and Conditioning Research*, **18**(1), 77–83.

Watts, P.B. (2004). Physiology of difficult rock climbing. *European Journal of Applied Physiology*, **91**, 361–372.

Watts, P.B. and Drobish, K.M. (1998). Physiological responses to simulated rock climbing at different angles. *Physical Fitness and Performance*, **30**(7), 1118–1122.

Watts, P.B., Martin, D.T. and Durtschi, S. (1993). Anthropometric profiles of elite male and female competitive sport rock climbers. *Journal of Sports Sciences*, **11**, 113–117.

Watts, P., Newbury, V. and Sulentic, J. (1996). Acute changes in handgrip strength, endurance, and blood lactate with sustained sport rock climbing. *Journal of Sports Medicine and Physical Fitness*, **36**, 255–260.

Watts, P.B., Daggett, M., Gallagher, P. and Wilkins, B. (1999). Metabolic responses during sport rock climbing and the effects of active versus passive recovery. *International Journal of Sports Medicine*, **21**, 185–190.

Watts, P.B., Joubert, L.M., Lish, A.K., Mast, J.D. and Wilkins, B. (2003). Anthropometry of young competitive sport rock climbers. *British Journal of Sports Medicine*, 37, 420–424.

Williams, E.S., Taggart, P. and Carruthers, M. (1978). Rock climbing: observations on heart rate and plasma catecholamine concentrations and the influence of oxprenolol. *British Journal of Sports Medicine*, 12(3), 125–128.

Williams, M.H. (1998). *The ergogenics edge: Pushing the limits of sports performance*. Human Kinetics, Champaign, IL.

Wilmore, J.H. and Costill, D.L. (2004). *Physiology of sport and exercise*. Human Kinetics, Champaign, IL.

Climb Magazine

Climber Magazine

Chapter 8

Ahmaidi, S. Granier, P. Taoutaou, Z. Mercier, J. Dubouchaud, H. and Prefaut, C. (1996). Effects of active recovery on plasma lactate and anaerobic power following repeated intensive exercise. *Medicine and Science in Sports and Exercise*, 28, 450–456.

Ainsworth, B.E., Serfass, R.C. and Leon, A.S. (1993). Effects of recovery duration and blood lactate level on power output during cycling. *Canadian Journal of Applied Physiology*, 18, 19–30.

Andersson, J., Schagatay, E., Gislén, A. and Holm, B. (2000). Cardiovascular responses to cold–water immersions of the forearm and face, and their relationship to apnoea. *European Journal of Applied Physiology*, 83, 566–572.

Armstrong, L.E. (2000). *Performing in extreme environments*. Human Kinetics, Champaign, IL.

Åstrand, P.E. and Rodahl, K. (1986). *Textbook of work physiology: Physiological bases of exercise*. McGraw-Hill, New York, NY.

Åstrand, P.E., Rodahl, K., Dahl, H.A. and Strømme, S.B. (2003). *Textbook of work physiology: Physiological bases of exercise*. Human Kinetics, Champaign, IL.

Baker, S and Atha, J. (1981). Canoists' disorientation following cold immersion. *British Journal of Medicine*, 15, 111–115.

Bangsbo, J., Graham, T., Johansen, L. and Saltin, B. (1994). Muscle lactate metabolism in recovery from intense exhaustive exercise: Impact of light exercise. *Journal of Applied Physiology*, 77, 1890–1895.

Belcastro, A.N. and Bonen, A. (1975). Lactic acid removal rates during controlled and uncontrolled recovery exercise. *Journal of Applied Physiology*, 39, 932–936.

Bergström, J. and Hultman, E. (1967). Muscle glycogen synthesis after exercise: An enhancing factor localised to the muscle cells in man. *Nature*, 210, 309–310.

Bergström, J., Hermansen, L., Hultman, E. and Saltin, B. (1966). Diet, muscle glycogen and physical performance. *Acta Physiologica Scandinavia*, 71, 140–150.

Bloomfield, J., Fricker, P.A. and Fitch, K.D. (Eds.) (1995). *Science and medicine in sport*. Blackwell Science, Carlton, Australia.

Bogdanis, G., Nevill, M.E., Lakomy, H.K.A., Graham, C.M. and Louis, G. (1996). Effects of active recovery on power output during repeated maximal sprint cycling. *European Journal of Applied Physiology*, 74, 461–469.

Briner, W.W. (1996). Tympanic membrane vs rectal temperature measurement in marathon runners. *Journal American Medical Association*, 276(3), 194.

Brooks, G.A., Fahey, T.D. and Baldwin, K.M. (2005). *Exercise physiology: Human bioenergetics and its applications*. McGraw-Hill, New York, NY.

Burke, D.G., MacNeil, S.A., Holt, L.E., MacKinnon, N.C. and Rasmussen, R.L. (2000). The effect of hot or cold water immersion on isometric strength training. *Journal of Strength and Conditioning Research*, 14(1), 21–25.

Bussell, C. (1995). *Motion analysis into the work rates of canoe polo players*. Unpublished M.Sc. Dissertation, Liverpool John Moores University.

Castagna, O. and Brisswalter, J. (2007). Assessment of energy demand in Laser sailing: influences of exercise duration and performance. *European Journal of Applied Physiology*, 99, 95–101.

Christensen E.H., Hedman R., Saltin B. (1960). Intermittent and continuous running. A further contribution to the physiology of intermittent work. *Acta Physiol Scand*, 50, 269–286.

Connolly, D.A.J., Brennan, K.M. and Lauzon, C.D. (2003). Effects of active versus passive recovery on power output during repeated bouts of short term, high intensity exercise. *Journal of Sports Science and Medicine*, 2, 47–51.

Corder, K.P., Potteiger, J.A., Nau, K.L., Figoni, S.F. and Hershberger, S.L. (2000). Effects of active and passive recovery conditions on blood lactate, rating of perceived exertion and performance during resistance exercise. *Journal of Strength and Conditioning Research*, 14, 151–156.

Cunningham, P. (2004). The physiological demands of elite single-handed dinghy sailing. Unpublished Ph.D. Thesis, University of Chichester.

Cunningham, P. and Hale, T. (2007). Physiological responses of elite Laser sailors to 30 minutes of simulated upwind sailing. *Journal of Sports Sciences*, **25**, 1109–1116.

Darling, J.L., Linderman, J.K. and Laubach, L.L. (2005). Energy expenditure of continuous and intermittent exercise in college–aged males. *Journal of Exercise Physiology*, **8**(4), 1–8.

De Vito, G., Di Filippo, L., Rodio, A., Felici, F. and Madaffari, A. (1997). Is the Olympic boardsailor an endurance athlete? *International Journal of Sports Medicine*, **18**(4), 281–284.

Devienne, M.F. and Guezennec, C.Y. (2000). Energy expenditure of sailing. *European Journal of Applied Physiology*, **82**, 499–503.

Douris, P., McKenna, R., Madigan, K., Cesarski, B., Costiera, R. and Lu, M. (2003). Recovery of maximal isometric grip strength following cold immersion. *Journal of strength and conditioning research*, **17**(3), 509–513.

Draper, N., Bird, E.L., Coleman, I. and Hodgson, C. (2006). Effects of active recovery on lactate concentration, heart rate and RPE in climbing. *Journal of Sports Science and Medicine*, **5**(1), 97–105.

Dupont, G. and Berthoin, S. (2004). Time spent at a high percentage of VO$_2$ max for short intermittent runs: active versus passive recovery. *Canadian Society for Exercise Physiology*, **29**, S3–S16.

Edwards, R.H.T., Ekelund, L.G., Harris, R.C., Hesser, C.M., Hultman, E., Melcher, A. and Wigertz, O. (1973). Cardiorespiratory and metabolic costs of continuous and intermittent exercise in man. *Journal of Physiology*, **234**, 481–497.

Eglin, C.M. and Tipton, M.J. (2005). Repeated cold showers as a method of habituating humans to the initial responses to cold water immersion. *Journal of Applied Physiology*, **93**, 624–629.

Essén, B., Hagenfeldt, L. and Kaijser, L. (1977). Utilization of blood-borne and intramuscular substrates during continuous and intermittent exercise in man. *Journal of Physiology*, **265**, 489–506.

Fulbrook, P. (1993). Core temperature measurement in adults: a literature review. *Journal of Advanced Nursing*, **18**, 1451–1460.

Gaitanos, G.C., Williams, C., Boobis, L.H. and Brooks, S. (1993). Human muscle metabolism during intermittent maximal exercise. *Journal Applied Physiology*, **75**(2), 712–719.

Garrett, W.E. and Kirkendall, D.T. (Eds.) (2000). *Exercise and sports science*. Lippincott, Williams & Wilkins, Baltimore, MD.

Girard-Eberle, S. (2000). *Endurance sports nutrition: eating plans for optimal training, racing and recovery*. Human Kinetics, Champaign, IL.

Golden, F. and Hervey, G.R. (1981). The afterdrop and death after rescue from immersion in cold water. In *Hypothermia ashore and afloat*, J.A. Adam (Ed). Aberdeen University Press, Aberdeen, UK.

Golden, F. and Tipton, M. (2002). *Essentials of sea survival*. Human Kinetics, Champaign, IL.

Graham, J.E., Boatwright, J.D., Hunskor, M.J. and Howell, D.C. (2003). Effect of active vs. passive recovery on repeat suicide run time. *Journal of Strength and Conditioning Research*, **17**(2), 338–341.

Gray, S. and Nimmo, M. (2001). Effects of active, passive or no warm-up on metabolism and performance during high-intensity exercise. *Journal of Sports Sciences*, **19**, 693–700.

Hargreaves, M., McKenna, M.J., Jenkins, D.G., Warmington, S.A., Li, J.L., Snow, R.J. and Febbraio, M.A. (1998). Muscle metabolites and performance during high-intensity, intermittent exercise. *Journal of Applied Physiology*, **84**(5), 1687–1691.

Harrison, J., Burstyn, P., Coleman, S. and Hale, T. (1988). A comparison of heart rate/oxygen uptake relationship for cycle and dinghy ergometry. *Journal of Sports Science*, **6**, 160–164.

Hayes, P.A. and Cohen, J.B. (1987). *Further development of a mathematical model for the specification of immersion clothing insulation*. RAF, IAM Report 653.

Hayward, J.S., Lisson, P.A., Collis, M.L. and Eckerson, J.D. (1978). *Survival suits for accidental immersion in cold water: Design–concepts and their thermal protection performance*. University of Victoria Report, Victoria, B.C.

Hermansen, L., and Stensvold, I. (1972). Production and removal of lactate during exercise in man. *Acta Physiologica Scandinavica*, **86**, 191–201.

Holum, J.R. (1998). *Fundamentals of general, organic and biological chemistry*. John Wiley & Sons, Chichester.

Jemni, M., Sands, W.A., Friemel, F. and Delamarche, P. (2003). Effect of active and passive recovery on blood lactate and performance during simulated competition in high level gymnasts. *Canadian Journal of Applied Physiology*. **28**(2), 240–256.

Jeukendrup, A. and Gleeson, M. (2004). *Sport nutrition: An introduction to energy production and performance*. Human Kinetics, Champaign, IL.

Karlsson J., Hermansen L., Agnevite G., Saltin B. (1967). *Energikraven vid Lopning/Idrottsfysiologi*. Stockholm, Sweden: Framtiden, Rapport 4.

Lee, S.M.C., Williams, W.J. and Fortney Schneider, S.M. (2000). Core temperature measurement during supine exercise: esophageal, rectal, and intestinal temperatures. *Aviation, Space, and Environmental Medicine*, **71**(9), 939–945.

Lewis, T. (1930). Observations upon the reactions of the vessels of the human skin to cold. *Heart*, **15**, 177–208.

Livingstone, S.D., Grayson, J., Frim, J., Allen, C.L. and Limmer, R.E. (1983). Effect of cold exposure on various sites of core temperature measurements. *Journal of Applied Physiology: Respiration, Environment and Exercise Physiology*. **54**(4), 1025–1031.

Lowdon, B.J. (1980). The somatotype of international surfboard riders. *Australian Journal of Science Medicine Sports*, **12**, 34–39.

Lowdon, B.J. and Pateman, N.A. (1980). Physiological parameters of international surfers. *Australian Journal of Sports Medicine*, **12**, 30–33.

Lowdon, B.J., Pateman, N.A. and Pitman, A.J. (1983). Surfboard-riding injuries. *Medical Journal of Australia*, **2**, 613–616.

Lowdon, B.J., Pitman, A.J., Pateman, N.A. and Ross, K. (1987). Injuries to international competitive surfboard riders. *Journal Sports of Medicine and Physical Fitness*, **27**, 57–63.

Lowdon, B.J., Bedi, J.F. and Horvath, S.M. (1989). Specificity of aerobic fitness testing of surfers. *Australian Journal of Science Medicine Sports*, **21**, 7–10.

Margaria, R.M., Aghemo, P., and Rovelli, E. (1966). Measurement of muscular power (anaerobic) in man. *Journal of Applied Physiology*, **21**, 1662–1664.

Margaria R., Oliva R.D., Di Prampero P.E., Ceretelli P. (1969). Energy utilization in intermittent exercise of supramaximal intensity. *Journal of Applied Physiology*, **26**, 752–756.

Maughan, R. and Gleeson, M. (2004). *The biochemical basis of sports performance*. OU Press, Oxford.

McArdle, W.D., Katch, F.I. and Katch, V.L. (2001). *Exercise physiology: Energy, nutrition and human performance*. Lippincott, Williams & Wilkins, Baltimore.

Meir, R.A., Lowdon, B.J. and Davie, A.J. (1991). Heart rates and estimated energy expenditure during recreational surfing. *Australian Journal of Science Medicine Sport*, **23**, 70–74.

Mendez-Villanueva, A. and Bishop, D. (2005). Physiological aspects of surfboard riding performance. *Sports Medicine*, **35**(1), 55–70.

Mendez-Villanueva, J.A., Bishop, D. and Hamer, P. (2003). Activity patterns of elite surfing competition. *Journal Science of Medicine Sports*, **6**, 11–15.

Mohr, M., Krushtrup, P. and Bangsbo, J. (2005). Fatigue in soccer: a brief review. *Journal of Sports Sciences*, **23**(6), 593–599.

Moran, D.S. and Mendal, L. (2002). Core temperature measurement methods and current insights. *Sports and Medicine*, **32**(14), 879–885.

Morton, R.H. and Billat, L.V. (2004). The critical power model for intermittent exercise. *European Journal of Applied Physiology*, **91**, 303–307.

Muir, I.H., Bishop, P.A., Lomax, R.G. and Green, J.M. (2001). Prediction of rectal temperature from ear canal temperature. *Ergonomics*, **44**(11), 962–972.

O'Brien, C., Hoyt, R.W., Buller, M.J., Castellani, J.W. and Young, A.J. (1998). Telemetry pill measurement of core temperature in humans during active heating and cooling. *Medicine & Science in Sports and Exercise*, **30**(3), 468–472.

Pitcher, J.B. and Miles, T.S. (1997). Influence of muscle blood flow on fatigue during intermittent human handgrip exercise and recovery. *Clinical and Experimental Pharmacology and Physiology*, **24**, 471–476.

Saltin B., Essen B. (1971). In: Pernow, B., and Saltin, B., (Eds.). Muscle glycogen, lactate, ATP and CP in intermittent exercise. *Muscle metabolism during exercise: Advances in experimental medicine and biology, Vol 2*. New York: Plenum Press, 1971: 419–424.

Seeley, R.R., Stephens, T.D. and Tate, P. (2000). *Anatomy and physiology*. McGraw-Hill Higher Education, San Francisco, CA.

Sherman, W.M., Costill, D.L., Fink, W.J. and Miller, J.M. (1981). Effect of exercise-diet manipulation on muscle glycogen and its subsequent utilisation during performance. *International Journal of Sports Medicine*, **2**, 114–118.

Sherman, W.M., Plyley, M.J. Sharp, R.L., van Handle, P.J, McAllister, R.M., Fink, W.J. and Costill, D.L. (1982). Muscle glycogen storage and its relationship to water. *International Journal of Sports Medicine*, **3**, 22–24.

Sherwood, L. (2001). *Human physiology: From cells to systems*. Brooks/Cole, Pacific Grove, CA.

Sparling, P.B., Snow, T.K. and Millard-Stafford, M.L. (1993). Monitoring core temperature during exercise: ingestible sensor vs. rectal thermistor. *Aviation, Space, and Environmental Medicine*, **64**, 760–763.

Spurway, N.C. and Burns, R. (1993). Comparison of dynamic and static fitness training programs. *Medical Science Research*, **21**(14), 865–867.

Šrámek, P., Šimečková, M., Janský, L., Šavlíková, J. and Vybíral, S. (2000). Human physiological responses to immersion into water of different temperatures. *European Journal of Applied Physiology*, **81**, 436–442.

Stamford, B.A., Weltman, A., Moffat, R. and Sady, S. (1981). Exercise recovery above and below anaerobic threshold following maximal work. *Journal of Applied Physiology*, **51**, 840–844.

Tikuisis, P., Eyolfson, D.A., Xu, X. and Giesbrecht, G.G. (2002). Shivering endurance and fatigue during cold water immersion in humans. *European Journal of Applied Physiology*, **87**, 50–58.

Tipton, M.J. (1989). The initial responses to cold-water immersion in man. *Clinical Science*, **77**, 581–588.

Tomlin, D.L. and Wenger, H.A. (2001). The relationship between aerobic fitness and recovery from high intensity intermittent exercise. *Sports Medicine*, **31**(1), 1–11.

Tortora, G.J. and Derrickson, B. (2006). *Principles of anatomy and physiology*. John Wiley & Sons, Chichester.

Toubekis, A.G., Douda, H.T. and Tokmakidis, S.P. (2005). Influence of different rest intervals during active or passive recovery on repeated sprint swimming performance. *European Journal of Applied Physiology*, **93**, 694–700.

Voet, D. and Voet, J. (2004). *Biochemistry*. John Wiley & Sons, Chichester.

Vogiatzis, I., Roach, N.K. and Trowbridge E.A. (1993). Cardiovascular, muscular and blood lactate responses during dinghy hiking. *Medical Science Research*, **21**, 861–863.

Vogiatzis, I., Spurway, N.C. and Wilson, J. (1994). On-water oxygen uptake measurements during dinghy sailing. *Journal of Sports Science*, **12**, 153–157.

Vogiatzis, I., De Vito, G., Rodio, A., Madaffari, A. and Marchetti, M. (2002). The physiological demands of sail pumping in Olympic level windsurfers. *European Journal of Applied Physiology*, **86**, 450–454.

Watts, P.B., Daggett, M., Gallagher, P. and Wilkins, B. (2000). Metabolic response during sport rock climbing and the effects of active versus passive recovery. *International Journal of Sports Medicine*, **21**, 185–190.

Weltman, A., Stamford, B.A., Moffatt, R.J. and Katch, V.L. (1977). Exercise recovery, lactate removal and subsequent high intensity exercise performance. *Research Quarterly*, **48**, 786–796.

Williams, M.H. (1998). *The ergogenics edge: Pushing the limits of sports performance*. Human Kinetics, Champaign, IL.

Wilmore, J.H. and Costill, D.L. (2004). *Physiology of sport and exercise*. Human Kinetics, Champaign, IL.

http://www.canoepolo.org.uk/

http://nsf-pad.bme.uconn.edu

http://www.sailing.org

http://www.surfwells.com

http://www.aspworldtour/profiles.com

Climb Magazine

Climber Magazine

Chapter 9

Almarwaey, O.A., Jones, A.M. and Tolfrey, K. (2004). Maximal lactate steady state in trained adolescent runners. *Journal of Sports Sciences*, **22**, 215–225.

Armstrong, L.E. (2000). *Performing in extreme environments*. Human Kinetics, Champaign, IL.

Åstrand, P.E. and Rodahl, K. (1986). *Textbook of work physiology: Physiological bases of exercise*. McGraw-Hill, New York, NY.

Åstrand, P.E., Rodahl, K., Dahl, H.A. and Strømme, S.B. (2003). *Textbook of work physiology: Physiological bases of exercise*. Human Kinetics, Champaign, IL.

Baker, S. and Atha, J. (1981). Canoeists' disorientation following cold immersion. *British Journal of Sports and Medicine*, **15**, 111–115.

Baldari, C., Videira, M., Madeira, F., Sergio, J. and Guidetti, L. (2005). Blood lactate removal during recovery at various intensities below the individual anaerobic threshold in triathletes. *The Journal of Sports Medicine and Physical Fitness*, **45**, 460–466.

Baron, R. (2001). Aerobic and anaerobic power characteristics of off-road cyclists. *Medical Science of Sports Exercise*, **33**(8), 1387–1393.

Beaver, W.L., Wasserman, K. and Whipp, B.J. (1985). Improved detection of lactate threshold during exercise using a log–log transformation. *Journal Applied Physiology*, **59**(6), 1936–1940.

Beaver, W.L., Wasserman, K. and Whipp, B.J. (1986). A new method for detecting anaerobic threshold by gas exchange. *Journal Applied Physiology*, **60**(6), 2020–2027.

Beneke, R. (1995). Anaerobic threshold, individual anaerobic threshold, and maximal lactate steady state in rowing. *Medicine and Science in Sports and Exercise*, **27**(6), 863–867.

Bird, S., George, M., Balmer, J. and Davison, R.C.R. (2003a). Heart rate responses of women aged

23–67 years during competitive orienteering. *British Journal of Sports and Medicine*, **37**(3), 254–257.

Bird, S., George, M., Theakston, J., Balmer, J. and Davison, R.C.R. (2003b). Heart rate responses of male orienteers aged 21–67 years during competition. *Journal of Sports Sciences*, **21**, 221–228.

Bishop, D. (2000). Physiological predictors of flat-water kayak performance in women. *European Journal of Applied Physiology*, **82**, 91–97.

Bishop, D. (2004). The validity of physiological variables to assess training intensity in kayak athletes. *International Journal of Sports Medicine*, **25**(1), 68–72.

Bloomfield, J., Fricker, P.A. and Fitch, K.D. (Eds.) (1995). *Science and medicine in sport*. Blackwell Science, Carlton, Australia.

British Canoe Union. (2006). *Coaching Handbook*. Pesda Press, Caernarfon, Wales.

Brooks, G.A. (1971). Temperature, skeletal muscle mitochondrial functions and oxygen debt. *American Journal of Physiology*, **220**, 1053–1059.

Brooks, G.A. (1985). Anaerobic threshold: review of the concept and directions for future research. *Medicine and Science in Sports and Exercise*, **17**(1), 22–31.

Brooks, G.A., Fahey, T.D. and Baldwin, K.M. (2005). *Exercise physiology: Human bioenergetics and its applications*. McGraw-Hill, New York, NY.

Brynes, W.C. and Kearney, J.T. (1997). Aerobic and anaerobic contributions during simulated canoe/kayak events. *Medicine and Science in Sports and Exercise*, **29**, S220.

Bunc, V. and Heller, J. (1993). Ventilatory threshold and work efficiency during exercise on cycle and paddling ergometers in young female kayakists. *European Journal of Applied Physiology*, **68**, 25–29.

Cheng, B., Kuipers, H., Snyder, A.C., Keizer, H.A., Jeukendrup, A. and Hesselink, M. (1992). A new approach for the determination of ventilatory and lactate thresholds. *International Journal of Sports and Medicine*, **13**(7), 518–522.

Costill, D.L., Dalsky, G.P. and Fink, W.J. (1978). Effects of caffeine ingestion on metabolism and exercise performance. *Medicine and Science in Sports and Exercise*, **10**(3), 155–158.

Creagh, U. and Reilly, T. (1997). Physiological and biomechanical aspects of orienteering. *Sports Medicine*, **24**(6), 409–418.

Daanen, H.A.M. (2003). Finger cold-induced vasodilation: a review. *European Journal of Applied Physiology*, **89**, 411–426.

Davis, J.A. (1985). Anaerobic threshold: review of the concept and directions for future research. *Medicine and Science in Sports and Exercise*, **17**(1), 6–18.

Dawson, E.A., Shave, R., George, K., Whyte, G., Ball, D., Gaze, D. and Collinson, P. (2005). Cardiac drift during prolonged exercise with echocardiographic evidence of reduced diastolic function of the heart. *European Journal of Applied Physiology*, **94**, 305–309.

Dekerle, J., Baron, B., Dupont, L., Vanvelcenaher, J. and Pelayo, P. (2003). Maximal lactate steady state, respiratory compensation threshold and critical power. *European Journal of Applied Physiology*, **89**, 281–288.

Fiore, D.C. and Houston, J.D. (2001). Injuries in whitewater kayaking. *British Journal of Sports and Medicine*, **35**, 235–241.

Flodgren, G., Hedelin, R. and Henriksson Larsen, K. (1999). Bone mineral density in flatwater sprint kayakers. *Calcified Tissue International*, **64**(5), 374–379.

Fry, R.W. and Morton, A.R. (1991). Physiological and kinanthropometric attributes of elite flatwater kayakists. *Medicine and Science in Sports and Exercise*, **23**(11), 1297–13-1.

Garrett, W.E. and Kirkendall, D.T. (Eds.) (2000). *Exercise and sports science*. Lippincott, Williams & Wilkins, Baltimore, MD.

Gerbert, W. and Werner, I. (1999). Blood lactate response to competitive climbing. *Communication to the 1st International Conference on Science and Technology in Climbing and Mountaineering*, Leeds, UK, April.

Girard-Eberle, S. (2000). *Endurance sports nutrition: eating plans for optimal training, racing and recovery*. Human Kinetics, Champaign, IL.

Gray, G.L., Matheson, G.O. and McKenzie, D.C. (1995). The metabolic cost of two kayaking techniques. *International Journal of Sports medicine*, **16**, 250–254.

Hahn, A.G., Pang, P.M., Tumilty, D.McA. and Telford, R.D. (1988). General and specific aerobic power of elite marathon kayakers and canoeists. *EXCEL*, **5**(2), 14–19.

Heck, H., Mader, A., Hess, G., Mücke, S., Müller, R. and Hollmann, W. (1985) Justification of the 4-mmol/l lactate threshold. *International Journal of Sports and Medicine*, **6**, 117–130.

Hespel, P. Maughan, R.J. and Greenhaff, P.L. (2006). Dietary supplements for football. *Journal of Sports Sciences*, **24**(7), 749–761.

Hill, A.V., and Lupton, H. (1923). Muscular exercise, lactic acid, and the supply and utilization of oxygen. *Quarterly Journal of Medicine*, **16**, 135.

Hishii, T., Umemura, Y. and Kitagawa, K. (2004). Full suspension mountain bike improves off-road cycling performance. *Journal Sports Medicine and Physical Fitness*, **44**, 356–360.

Holum, J.R. (1998). *Fundamentals of general, organic and biological chemistry*. John Wiley & Sons, Chichester.

Impellizeri, F., Sassi, A., Rodriguez-Alonso, M., Mognoni, P. and Marcora, S. (2002). Exercise intensity during off-road cycling competitions. *Medical Science of Sports Exercise*, **34**(11), 1808–1813.

Jensen, K., Johansen, L. and Kärkkäinen, O.P. (1999). Economy in track runners and orienteers during path and terrain running. *Journal of Sports Sciences*, **17**, 945–950.

Jeukendrup, A. and Gleeson, M. (2004). *Sport nutrition: An introduction to energy production and performance*. Human Kinetics, Champaign, IL.

Jones, A.M. and Doust, J.H. (1998). The validity of the lactate minimum test for determination of the maximal lactate steady state. *Medicine and Science in Sports and Exercise*, **30**(8), 1304–1313.

Klassen, G.A., Andrew, G.M. and Becklake, M.R. (1970). Effect of training on total and regional blood flow and metabolism in paddlers. *Journal of Applied Physiology*, **28**, 397–406.

Kruseman, M., Bucher, S., Bovard, M., Kayser, B. and Bovier, P.A. (2005). Nutrient intake and performance during a mountain marathon: an observational study. *European Journal of Applied Physiology*, **94**, 151–157.

Kuphal, K.E., Potteiger, J.A., Frey, B.B. and Hise, M.P. (2004). Validation of a single-day maximal lactate steady state assessment protocol. *Journal of Sports Medicine and Physical Fitness*, **44**, 132–140.

Laplaud, D., Guinot, M., Favre-Juvin, A. and Flore, P. (2006). Maximal lactate steady state determination with a single incremental test exercise. *European Journal of Applied Physiology*, **96**, 446–452.

Larsson, P., Burlin, L., Jakobsson, E. and Henriksson-Larsén, K. (2002). Analysis of performance in orienteering with treadmill tests and physiological field tests using a differential global positioning system. *Journal of Sports Sciences*, **20**, 529–535.

MacRae, H.S., Hise, K.J. and Allen, P.J. (2000). Effects of front and dual suspension mountain bike systems on uphill cycling performance. *Medical Science of Sports Exercise*, **32**, 1276–1280.

Maughan, R. and Gleeson, M. (2004). *The biochemical basis of sports performance*. OU Press, Oxford.

McArdle, W.D., Katch, F.I. and Katch, V.L. (2001). *Exercise physiology: Energy, nutrition and human performance*. Lippincott, Williams & Wilkins, Baltimore.

Mekjavic, I.B. and Bligh, J. (1987). The pathophysiology of hypothermia. *International Reviews of Ergonomics*, **1**, 201–218.

Nielens, H. and Lejeune, T.M. (2001). Energy cost of riding bicycles with shock absorption systems on a flat surface. *Medical Science of Sports Exercise*, **22**, 400–404.

Nimmo, M. (2004). Exercise in the cold. *Journal of Sports Sciences*, **22**, 898–916.

Nishii, T., Umemura, Y. and Kitagawa, K., (2004). Full suspension mountain bike improves off-road cycling performance. *Journal of Sports Medicine and Physical Fitness*, **17**, 356–360.

Olsen, J. (1996). Bicycle suspension systems. In: Burke, E.R. editor. *High-tech cycling*. Champaign: Human Kinetics, 45–64.

Orr, G.W., Green, H.J., Hugson, R.L. and Bennett, G.W. (1982). A computer linear regression model to determine ventilatory anaerobic threshold. *Journal Applied Physiology: Respiration Environment Exercise Physiology*, **52**(5), 1349–1352.

Pendergast, D.R. (1989). Cardio-vascular, respiratory and metabolic responses to upper body exercise. *Medicine and Science in Sports and Exercise*, **21**(5), S121–124.

Pérez-Landaluce, J., Rodríguez-Alonso, M., Fernandez-Garcia, B., Bustillo-Fernandez, E. and Terrados, N. (1998). Importance of wash riding in kayaking training and competition. *Medicine and Science Sports Exercise*, **30**(12), 1721–1724.

Peters, E.M., Robson, P.J., Kleinveldt, N.C., Naicker, V.L. and Jogessar, V.D. (2004). Hematological recovery in male ultramarathon runners: the effect of variations in training load and running time. *Journal Sports Medicine and Physical Fitness*, **44**, 315–321.

Piehl-Aluin, K., Gimby, G. and Karlsson, J. (1977). Sailing. *Idrottsfyiologi rapport*, **16**, 28–57.

Pringle, J.S.M. and Jones, A.M. (2002). Maximal lactate steady state, critical power and EMG during cycling. *European Journal of Applied Physiology*, **88**, 214–226.

Sawka, M.N. (1989). Introduction: Upper body exercise: Physiology and practical considerations. *Medicine and Science in Sports and Exercise*, **21**(5), S119–120.

Seeley, R.R., Stephens, T.D. and Tate, P. (2000). *Anatomy and physiology.* McGraw-Hill Higher Education, San Francisco, CA.

Seifert, J.G., Luetkemeier, M.J., Miller, D. and Burke, E.R. (1997). The effect of mountain bike suspension systems on energy expenditure, physical exertion, and tie trail performance during mountain bicycling. *Medical Science of Sports Exercise,* **18**, 197–200.

Shephard, R.J. (1987). Science and medicine in canoeing and kayaking. *Sports Medicine,* **4**, 19–33.

Sherman, W.M., Costill, D.L., Fink, W.J. and Miller, J.M. (1981). Effect of exercise-diet manipulation on muscle glycogen and its subsequent utilisation during performance. *International Journal of Sports Medicine,* **2**, 114–118.

Sherman, W.M., Plyley, M.J. Sharp, R.L., van Handle, P.J, McAllister, R.M., Fink, W.J. and Costill, D.L. (1982). Muscle glycogen storage and its relationship to water. *International Journal of Sports Medicine,* **3**, 22–24.

Sherwood, L. (2001). *Human physiology: From cells to systems.* Brooks/Cole, Pacific Grove, CA.

Sidney, K. and Shephard, R.J. (1973). Physiological characteristics and performance of the white-water paddler. *European Journal of Applied Physiology,* **32**, 55–70.

Simpson, R.J., Wilson, M.R., Black, J.R., Ross, J.A., Whyte, G.P., Guy, K. and Florida-James, G.D. (2005). Immune alterations, lipid peroxidation, and muscle damage following a hill race. *Canadian Journal of Applied Physiology,* **30**(2), 196–211.

Sjödin, B. and Jacobs, I., (1981). Onset of blood lactate acculation and marathon running performance. *International Journal of Sports Medicine,* **2**, 23–26.

Sklad, M., Krawczyk, B. and Majle, B. (1994). Body build profiles of male and female rowers and kayakers. *Biology of Sport,* **11**(4), 249–256.

Stegman H., Kindermann W., and Schabel A. (1981) Lactate kinetics and individual anaerobic threshold. *International Journal of Sports Medicine,* **2**, 160–165.

Svendahl, K. and MacIntosh, B.R. (2003). Anaerobic threshold: the concept and methods of measurement. *Canadian Journal of Applied Physiology,* **28**(2), 299–323.

Tegtbur, U., Busse, M.W. and Braumann, K.M. (1993). Estimation of an individual equilibrium between lactate production and catabolism during exercise. *Medicine and Science in Sports and Exercise,* **25**(5), 620–627.

Tesch, P.A. (1983). Physiological characteristics of elite kayak paddlers. *Canadian Journal of Applied Sports Science,* **8**(2), 87–91.

Tesch, P.A. and Kindeberg, S. (1984). Blood lactate accumulation during arm exercise in world class kayak paddlers and strength trained athletes. *European Journal of Applied Physiology,* **52**, 441–445.

Tesch, P., Piehl, K., Wilson, G. and Karlsson, J. (1976). Physiological investigation of Swedish elite canoe competitors. *Medicine and Science in Sports and Exercise,* **8**, 214–218.

Tortora, G.J. and Derrickson, B. (2006). *Principles of anatomy and physiology.* John Wiley & Sons, Chichester.

van Hall, G., Jensen-Urstad, M., Rosdahl, H., Holmberg, H.C., Saltin, B. and Calbet, J.A.L. (2002). Leg and arm lactate and substrate kinetics during exercise. *American Journal of Physiology, Endocrinology and metabolism,* **284**, E193–205.

van Someren, K.A., Phillips, G.R.W. and Palmer, G.S. (2000). Comparison of physiological responses to open water kayaking and kayak ergometry. *International Journal of Sports Medicine,* **21**, 200–204.

van Someren, K.A. and Oliver, J.E. (2002). The efficacy of ergometry determined heart rates for flatwater kayak training. *International Journal of Sports Medicine,* **23**, 28–32.

Vrijens, J., Hoekstra, P., Boukaert, J and van Trank, P. (1975). Effects of training on maximal working capacity and hemodynamic response during arm and leg exercise in a group of paddlers. *European Journal of Applied Physiology,* **34**, 113–119.

Voet, D. and Voet, J. (2004). *Biochemistry.* John Wiley & Sons, Chichester.

Wakeling, P. and Saddler, S. (1978). Aerobic capacities of some British slalom and wild-water racing kayak competitors of international status. *Research Papers in Physical Education,* **3**(4), 16–18.

Wassermann, K., and McIlroy, M.B. (1964). Detecting the threshold of anaerobic metabolism in cardiac patients during exercise. *American Journal of Cardiology,* **14**, 844–852.

Wilber, R.L., Zawadzki, K.M., Kearney, J.T., Shannon, M.P. and Disalvo, D. (1997). Physiological profiles of elite off-road and road cyclists. *Medical Science of Sports Exercise,* **29**, 1090–1094.

Williams, M.H. (1998). *The ergogenics edge: Pushing the limits of sports performance.* Human Kinetics, Champaign, IL.

Wilmore, J.H. and Costill, D.L. (2004). *Physiology of sport and exercise.* Human Kinetics, Champaign, IL.

Yeh, M.P., Gardner, R.M., Adams, T.D., Yanowitz, F.G. and Crapo, R.O. (1983). 'Anaerobic threshold': problems of determination and validation. *Journal Applied Physiology,* **55**(4), 1178–1186.

http://www.chemheritage.org
http://encyclopedia.jrank.org
http://www.pubmedcentral.nih.gov/
http://www.britishcycling.org.uk/web/site/BC/gbr/
gbteam_home.asp
http://www.fellrunner.org.uk/international05.htm
http://en.wikipedia.org
http://www.britishorienteering.org.uk/asp/homepage.asp
Climb Magazine
Climber Magazine

Chapter 10

Araneda, O.F., García, C., Lagos, N., Quiroga, G., Cajigal, J., Salazar, M.P. and Behn, C. (2005). Lung oxidative stress as related to exercise and altitude. Lipid peroxidation evidence in exhaled breath condensate: a possible predictor of acute mountain sickness. *European Journal of Applied Physiology*, **95**, 383–390.

Ariel, G. and Saville, W. (1972). Anabolic steroids: the physiological effects of placebos. *Medicine and Science in Sports and Exercise*, **4**, 124–126.

Armstrong, L.E. (2000). *Performing in extreme environments*. Human Kinetics, Champaign, IL.

Åstrand, P.E. and Rodahl, K. (1986). *Textbook of work physiology: Physiological bases of exercise*. McGraw-Hill, New York, NY.

Åstrand, P.E., Rodahl, K., Dahl, H.A. and Strømme, S.B. (2003). *Textbook of work physiology: Physiological bases of exercise*. Human Kinetics, Champaign, IL.

Bailey, D.M. and Davies, B. (2001). Acute mountain sickness; prophylactic benefits of antioxidant vitamin supplementation at high altitude. *High Altitude Medicine and Biology*, **2**(1), 21–29.

Bartsch, P., Bailey, D.M., Berger, M.M., Knauth, M. and Baumgartner, R.W. (2004). Acute mountain sickness: controversies and advances. *High Altitude Medicine and Biology*, **5**(2), 110–124.

Bloomfield, J., Fricker, P.A. and Fitch, K.D. (Eds.) (1995). *Science and medicine in sport*. Blackwell Science, Carlton, Australia.

Bradwell, A.R. and Coote, J.H. (1987). The BMRES 1984 medical research expedition to the Himalayas. *Postgraduate Medical Journal*, **63**(737), 165–167.

Brooks, G.A., Fahey, T.D. and Baldwin, K.M. (2005). *Exercise physiology: Human bioenergetics and its applications*. McGraw-Hill, New York, NY.

Bunting, C.J. and Tolson, H. (2000). Physiological stress response of the neuroendocrine system during

outdoor adventure tasks. *Journal of Leisure Research*, **32**(2), 191–207.

Corry, C. (2006). *The physiological effects of cold on the body, while wearing different clothing systems in a simulated alpine environment*. Unpublished dissertation, University of Chichester.

Dehnert, C., Hütler, M., Liu, Y., Menold, E., Netzer, C., Schick, R., Kubanek, B., Lehmann, M., Böning, D. and Steinacker, J.M. (2002). Erythropoiesis and performance after two weeks of living high and training low in well trained triathletes. *International Journal of Sports Medicine*, **23**, 561–566.

Felici, F., Rosponi, A., Sbriccoli, P., Scarcia, M., Bazzucchi, I. and Iannattone, M. (2001). Effect of human exposure to altitude on muscle endurance during isometric contractions. *European Journal of Applied Physiology*, **85**, 507–512.

Garrett, W.E. and Kirkendall, D.T. (Eds.) (2000). *Exercise and sports science*. Lippincott, Williams & Wilkins, Baltimore, MD.

Girard-Eberle, S. (2000). *Endurance sports nutrition: eating plans for optimal training, racing and recovery*. Human Kinetics, Champaign, IL.

Green, H.J., Roy, B., Grant, S., Hughson, R., Burnett, M., Otto, C., Pipe, A., McKenzie, D. and Johnson, M. (2000). Increases in submaximal cycling efficiency mediated by altitude acclimatization. *Journal of Applied Physiology*, **89**, 1189–1197.

Hill, D.W., Borden, D.O., Darnaby, K.M. and Hendricks, D.N. (1994). Aerobic and anaerobic contributions to exhaustive high-intensity exercise after sleep deprivation. *Journal of Sports Sciences*, **12**, 455–461.

Hohlrieder, M., Mair, P., Wuertl, W. and Brugger, H. (2005). The impact of avalanche transceivers on mortality from avalanche accidents. *High Altitude Medicine and Biology*, **6**(1), 72–77.

Holum, J.R. (1998). *Fundamentals of general, organic and biological chemistry*. John Wiley & Sons, Chichester.

Jeukendrup, A. and Gleeson, M. (2004). *Sport nutrition: An introduction to energy production and performance*. Human Kinetics, Champaign, IL.

Joanny, P., Steinberg, J., Robach, P., Richalet, J.P., Gortan, C., Gardette, B. and Jammes, Y. (2001). Operation everest III (comex '97): The effect of simulated severe hypobaric hypoxia on lipid peroxidation and antioxidant defence systems in human blood at rest and after maximal exercise. *Resuscitation*, **49**, 307–314.

Kreider, R.B., Fry, A.C. and O'Toole, M.L. (1998). *Overtraining in sport*. Human Kinetics, Champaign, IL.

Levine, B.D. and Stray-Gundersen, J. (1997). 'Living high–training low': effect of moderate-altitude acclimatization with low-altitude training on performance. *Journal of Applied Physiology*, **83**(1), 102–112.

MacDonald, M.J., Green, H.J., Naylor, H.L., Otto, C. and Hughson, R.L. (2001). Reduced oxygen uptake during steady state exercise after 21-day mountain climbing expedition to 6,194 m. *Canadian Journal of Applied Physiology*, **26**(2), 143–156.

Maughan, R. and Gleeson, M. (2004). *The biochemical basis of sports performance*. OU Press, Oxford.

Mazzeo, R.S., Wlofel, E.E., Butterfield, G.E. and Reeves, J.T. (1994). Sympathetic response during 21 days at high altitude (4,300 m) as determined by urinary and arterial catecholamines. *Metabolism*, **43**(10), 1226–1232.

McArdle, W.D., Katch, F.I. and Katch, V.L. (2001). *Exercise physiology: Energy, nutrition and human performance*. Lippincott, Williams & Wilkins, Baltimore.

O'Connor, T., Dubowitz, G. and Bickler, P.E. (2004). Pulse Oximetry in the diagnosis of acute mountain sickness. *High Altitude Medicine and Biology*, **5**(3), 341–348.

Peltonen, J.E., Tikkanen, H.O. and Rusko, H.K. (2001). Cardiorespiratory responses to exercise in acute hypoxia, hyperoxia and normoxia. *European Journal of Applied Physiology*, **85**, 82–88.

Pronk, M., Tiemessen, I., Hupperets, M.D.W., Kennedy, B.P., Powell, F.L., Hopkins, S.R. and Wagner, P.D. (2003). Persistence of the lactate paradox over 8 weeks at 3800 m. *High Altitude Medicine and Biology*, **4**(4), 431–443.

Reilly, T. and Piercy, M. (1994). The effect of partial sleep deprivation on weight-lifting performance. *Ergonomics*, **37**, 107–115.

Rusko, H.K., Tikkanen, H.O. and Peltonen, J.E. (2004). Altitude and endurance training. *Journal of Sports Sciences*, **22**, 928–945.

Savourey, G., Garcia, N., Caravel, J.P., Gharib, C., Pouzeratte, N., Martin, S. and Bittel, J. (1998). Pre-adaptation, adaptation and de-adaptation to high altitude in humans: hormonal and biochemical changes at sea level. *European Journal of Applied Physiology*, **77**, 37–43.

Savourey, G., Launay, J.C., Besnard, Y., Guinet, A., Bourrilhon, C., Cabane, D., Martin, S., Caravel, J.P., Péquignot, J.M. and Cottet-Emard, J.M. (2004). Control of erythropoiesis after high altitude acclimatization. *European Journal of Applied Physiology*, **93**, 47–56.

Seeley, R.R., Stephens, T.D. and Tate, P. (2000). *Anatomy and physiology*. McGraw-Hill Higher Education, San Francisco, CA.

Sherwood, L. (2001). *Human physiology: From cells to systems*. Brooks/Cole, Pacific Grove, CA.

Sutton, J.R., Coates, G. and Houston, C.S. (Eds.) (1992). *Hypoxia and mountain medicine*. Queen City Printers, Burlington, Vermont.

Symons, J.D., Bell, D.G., Pope, J. van Helder, T. and Myles, W.S. (1988). Electro-mechanical response times and muscle strength after sleep deprivation. *Canadian Journal of Sports Sciences*, **13**, 225–230.

Tannheimer, M., Thomas, A. and Gerngroß, H. (2002). Oxygen saturation course and altitude symptomatology during an expedition to broad peak (8047 m). *International Journal of Sports Medicine*, **23**, 329–335.

Tortora, G.J. and Derrickson, B. (2006). *Principles of anatomy and physiology*. John Wiley & Sons, Chichester.

Twight, M.F. and Martin, J. (1999). *Extreme alpinism climbing light, fast and high*. The Mountaineers, Seattle, WA.

van Hall, G., Calbet, J.A.L., Sondergaard, H. and Saltin, B. (2001). The re-establishment of the normal blood lactate response to exercise after prolonged acclimatization to altitude. *Journal of Physiology*, **536**(3), 963–975.

van Helder, T. and Radomski, M.W. (1989). Sleep deprivation and the effect on exercise performance. *Sports Medicine*, **7**, 235–247.

Vasankari, T.J., Kujala, U.M., Rusko, H., Sarna, S. and Ahotupa, M. (1997). The effect of endurance exercise at moderate altitude on serum lipid peroxidation and antioxidative functions in humans. *European Journal of Applied Physiology*, **75**, 396–399.

Voet, D. and Voet, J. (2004). *Biochemistry*. John Wiley & Sons, Chichester.

Williams, M.H. (1998). *The ergogenics edge: Pushing the limits of sports performance*. Human Kinetics, Champaign, IL.

Wilmore, J.H. and Costill, D.L. (2004). *Physiology of sport and exercise*. Human Kinetics, Champaign, IL.

http://www.aida-international.org/

http://www.medex.org.uk/new_page_5.htm

http://www.high-altitude-medicine.com/AMS-LakeLouise.html

Climb Magazine

Climber Magazine

INDEX